Fourth Edition

LAND RESOURCE ECONOMICS

THE ECONOMICS OF REAL ESTATE

Raleigh Barlowe

Michigan State University

D0209774

Prentice-Hall, *Englewood Cliffs, N.J. 07632*

Library of Congress Cataloging-in-Publication Data

Barlowe, Raleigh.
  Land resource economics.

  Includes bibliographies and index.
  1. Land use. 2. Land use—United States. I. Title.
HD111.B25 1985   333   85-16786
ISBN 0-13-522541-8

Editorial/production supervision and
  interior design: Kathryn Pavelec and Mary Bardoni
Cover design: Joe Curcio
Manufacturing buyer: John Hall

Prentice-Hall International (UK) Limited, *London*
Prentice-Hall of Australia Pty. Limited, *Sydney*
Prentice-Hall Canada Inc., *Toronto*
Prentice-Hall Hispanoamericana, S.A., *Mexico*
Prentice-Hall of India Private Limited, *New Delhi*
Prentice-Hall of Japan, Inc., *Tokyo*
Prentice-Hall of Southeast Asia Pte. Ltd., *Singapore*
Editora Prentice-Hall do Brasil, Ltda., *Rio de Janeiro*
Whitehall Books Limited, *Wellington, New Zealand*

To
George S. Wehrwein and V. Webster Johnson
scholars, public servants, champions
of good land use

CONTENTS

PREFACE

My objective with this edition of *Land Resource Economics*, as earlier, has been to present a systematic description of the basic economic and public policy concepts that influence people in their decisions concerning possession and use of land and real estate resources. Since meaningful and realistic analysis of land resource issues and problems calls for recognition of the cross-workings of many types of factors, I have emphasized a holistic political economic approach. Frequent questions are raised concerning what is physically and biologically possible, economically and technically feasible, and institutionally acceptable.

Following the introductory chapter, the book is divided into four groups of chapters. Chapters 2 through 4 deal with the adequacy of our land resource base to provide for our emerging demands and needs. Chapters 5 through 10 focus on the operation of the economic principles that affect human decisions regarding the acquisition, development, management, and use of real estate resources. Chapters 11 through 16 emphasize the role institutional considerations, such as concepts of property rights, have on decisions concerning land use. The final two chapters highlight the principal public policy issues associated with land use planning and public efforts to direct the uses made of land resources.

My level of presentation is designed mostly for the general reader. Though the book is intended primarily as an introduction to land economics, I naturally hope that advanced students will find it helpful and will be motivated by it to further study issues not dealt with in detail.

This book could not have been written without the helpful assistance of others. My thinking as a land economist is rooted in a rich heritage of ideas. Much of it stems from the extensive literature on land and real estate issues and from contacts with Lewis C. Gray, George S. Wehrwein, and Leonard A. Salter, Jr. In addition to these three great teachers of land economics, my association with V. Webster Johnson and Kenneth H. Parsons, with other teachers, with colleagues and fellow land economists, and with graduate and undergraduate students both in and out of the classroom have added to my perception and understanding of the issues discussed here.

Special acknowledgements are due to several people who have assisted with the preparation of this edition, Among my colleagues at Michigan State University, J. Paul Schneider and Patricia Nishan have been helpful in assisting with the preparation of the graphs and diagrams. Daniel E. Chappelle, Leighton L. Leighty, Lawrence W. Libby, A. Allan Schmid, and Milton H. Steinmueller have provided ideas and suggestions that are incorporated in this edition. Valuable assistance and suggestions also have been received from Richard Barrows of the University of Wisconsin, H. Thomas Frey, William H. Heneberry and Roger W. Hexem of the U.S. Department of Agriculture, Philip D. Gardner of the University of California at Riverside, Robert G. Healy of the Conservation Foundation, Frank J. Popper of Rutgers University, Giles T. Rafsnider of Colorado State University, Philip M. Raup of the University of Minnesota, and Clifford E. Tiedemann of the University of Illinois at Chicago. Special thanks are due to my dear wife, Jean, for her forbearance and enthusiastic support in pushing this project to completion.

Raleigh Barlowe
Michigan State University

I

LAND ECONOMICS: A STUDY OF LAND AND PEOPLE

Land and people are our two most basic resources. As human beings, we are ever inclined to stress the importance of human resources. But we must also be mindful of the essential role land resources play in supporting our existence and our day-to-day activities. Viewed broadly as the ground, water, air, and other natural resources tied to the earth's surface, land provides us with living space, with the primary products that support our material needs, and with opportunities and satisfactions dear to our ways of life. As Henry George observed: "... land is the habitation of man, the store-house upon which he must draw for all his needs, the material to which his labor must be applied for the supply of all his desires. ... On the land we are born, from it we live, to it we return again—children of the soil as truly as is the blade of grass or the flower of the field. Take away from man all that belongs to land, and he is but a disembodied spirit."[1]

History speaks eloquently of the high regard our ancestors felt for land. The ancient Minoans and Greeks prayed to an earth goddess, a reverence that has come down to us in the respect we hold for Mother Nature. For long centuries, frequent wars were fought for possession of land; and average men and women everywhere lived in close association with the fields, forests, and fishing grounds that supplied their sustenance. Rights held in land often determined one's economic, social, and political status. Hunger for land and its ownership brought thousands of immigrants to the Americas and still colors the aspirations and thinking of millions of people.

Concern over the close relationship between people and land has influenced the writings and teachings of some of the world's greatest thinkers, including the early

[1] Henry George, *Progress and Poverty*, 1879 (New York: Robert Schalkenbach Foundation, 1958), pp. 295–96.

economists who saw land as a key factor of production. Much of this sense of dependency on land has disappeared in recent decades as technological advances have tempered the threat of famine and as widespread urbanization has separated most people from the intimate contacts they once had with land as producers of food and raw materials. Home ownership and possession of real estate still rate high as individual goals; but most Americans and Canadians have lost much of the land hunger felt by their ancestors. Except for expressions of concern about the quality of their environment, most people now view land resources as a readily available input in their scheme of values.

But while most Americans are no longer as land-conscious as they once were, it should not be forgotten that land resources still make important contributions to the nation's economy. Investments in land improvements account for two-thirds of the wealth of the United States. They are the source of around a sixth of the nation's personal income. New structures on land accounted for annual expenditures of between 9.6 and 11.3 percent of the nation's gross national product during the 1960–1980 period. Expenditures on housing are the most important single item in most family budgets; real estate loans account for more than half of our private noncorporate credit; and taxes on real estate provide most of the revenue for local governments.

Our rich land resource base deserves much of the credit for the high standards of life enjoyed in the United States and Canada. Unfortunately, many nations enjoy smaller shares of this bounty. Two-thirds of the world's people live in the shadow of want and hunger. With the continuing upward surge in world population numbers, the problem of providing adequate supplies of needed food and raw materials for them can easily worsen. This situation has serious implications for attainment of our worldwide goals of maintaining peaceful relationships, facilitating economic growth and trade, and expanding opportunities for individual development and the enjoyment of human rights.

Throughout much of the Western world, our major land problem is more one of securing effective and efficient utilization of the land resources needed to provide operators and others with high levels of living than it is one of simply providing people with adequate sustenance. Attainment of better operating and living conditions within these areas calls for actions and policies that deal with the different sets of land-use problems found at the urban, suburban, and rural levels. Urban communities often suffer from congestion, pollution, declining attractiveness, inadequate and insufficient housing, environmental degradation, outward migration of their economic and tax bases, and poor growth management.

With the outward sprawl of most urban centers, the problems of cities have spilled over into suburban and urban-fringe communities. Many areas regarded as rural a few years ago now face problems comparable to those of nearby cities. In addition, they must provide schools, water and sewerage systems, paved streets, and other public services. Action is needed to prevent undesired developments and to integrate local programs with those of neighboring communities.

Rural residents who own and operate farms, forests, and other rural lands also face numerous problems in the acquisition, management, improvement, conservation, and eventual transfer of their properties. Some of these center on the individual operator's choice of enterprises, managerial decisions, and willingness to bring new areas into use. Others involve the acceptance of new technologies, competition with other producers, use of credit and leasing arrangements, maintenance of family-sized operations, and protection of rural land uses from invading urban-oriented uses.

Land resource problems are also an important province of government. Units of government in the United States now control and administer 42 percent of the nation's land area. They depend on property taxes for much of their revenue. They carry on active research and educational programs dealing with land and real estate problems. They have sponsored reclamation, urban development, and highway construction programs and have concerned themselves with reforestation, flood control, public housing, rent control, area development, and many other activities involving land use. They have taken important steps to plan for the more orderly and effective use of land resources and to use zoning ordinances and other public measures to control and direct land-use practices in the public interest.

Discussion in the chapters that follow centers on the principles and problems that affect people in their use of land and real estate resources. In the balance of this chapter, emphasis is focused on the scope and content of land economics and on some basic land economic concepts.

SCOPE AND CONTENT OF LAND ECONOMICS

Land economics deals with the economic relationships people have with others respecting land.[2] It is concerned with our economic use of the surface resources of the earth and the physical and biological, technological and economic, and institutional factors that condition and control our use of these resources. As L. A. Salter once observed, "Land economics is a social science that deals with those problems in which social conduct is strategically affected by the physical, locational or property attributes of whole surface units."[3]

[2] Land economics was first recognized as a course for collegiate study in 1892 when Richard T. Ely offered a seminar on Landed Property at the University of Wisconsin. Formal recognition as a separate field came in 1919 when a Division of Land Economics was established in the U.S. Department of Agriculture. Foundations were established in the 1920s for much of the work in urban and rural land economics and in real estate economics that has followed. The first course materials dealing specifically with this field were published in 1922.

For more-detailed accounts of the history of land economics, see Leonard A. Salter, Jr., *A Critical Review of Research in Land Economics* (Minneapolis: University of Minnesota Press, 1948), chap. 2; Coleman Woodbury, "Richard T. Ely and the Beginnings of Research in Urban Land and Housing Economics," *Land Economics*, 25 (February 1949), 55–66; V. Webster Johnson, "Twenty-five Years of Progress: Division of Land Economics," *Journal of Land and Public Utility Economics*, 31 (February 1945), 54–64; and Henry C. and Anne D. Taylor, *The Story of Agricultural Economics in the United States* (Ames: Iowa State College Press, 1952), Part VI.

[3] Leonard A. Salter, Jr., "The Content of Land Economics and Research Methods Adapted to Its Needs," *Journal of Farm Economics*, 24 (February 1942), 235.

Like general economics, land economics is concerned with the allocation and use of scarce resources. Its chief focus is on one particular type of resource—land or real estate. But land economists do not give exclusive attention to the land factor for the simple reason that land by itself has little economic value until it is used in conjunction with inputs of capital, labor, and management. Land economics involves a wide variety of economic relationships; but it is always concerned with problems and situations in which land, its use, or its control are regarded as factors of strategic or limiting importance. This factoral approach can be compared with the attention given to the factors of capital, labor, and management in the fields of money and banking, labor economics, and business management, respectively.

Land Economics: A Branch of Political Economy

Land economics is an applied branch of economics. Economists who work in this area are naturally interested in applications of economic theory to land problems. They are concerned with the impacts economic concepts such as costs, returns, prices, profits, and value have on decisions concerning land use. At the same time, the emphasis they give to finding solutions for land and real estate problems causes them to take a very practical position regarding applications of economic concepts under real-life conditions. Like the nineteenth-century economist Richard Jones, they must "look and see." They recognize the importance of economic theory, but they also remember that economic activity does not take place in a vacuum and that it frequently involves conditions that do not match the tight assumptions often assumed in economic analysis.

Land economics is often characterized by its practical, institutional, and problem-solving approach. In their attempt to explain human behavior with respect to land, land economists frequently find it expedient to use working tools from history, law, political science, psychology, and sociology as well as economics. Similarly, when they consider land resource problems, they employ concepts used by business operators, geographers, soil scientists, planners, architects, engineers, foresters, and geologists. In their use and integration of these various lines of thought, land economists often step beyond the bounds of economics to operate as social science land specialists in the broader field of political economy.

Threefold framework affecting land use. Our use of land and real estate resources takes place within a threefold framework. This framework involves the impacts that (1) physical and biological factors, (2) technological and economic considerations, and (3) institutional arrangements have on private and public decisions relative to land use. Together, these three sets of factors (see Figure 1-1) set the limits concerning what individuals, groups, and governments can accomplish in their development, utilization, and conservation of land resources.

Briefly stated, the *physical and biological framework* is concerned with the natural environment in which operators find themselves and with the quantity, nature, and characteristics of the resources with which they work. Physical and biological

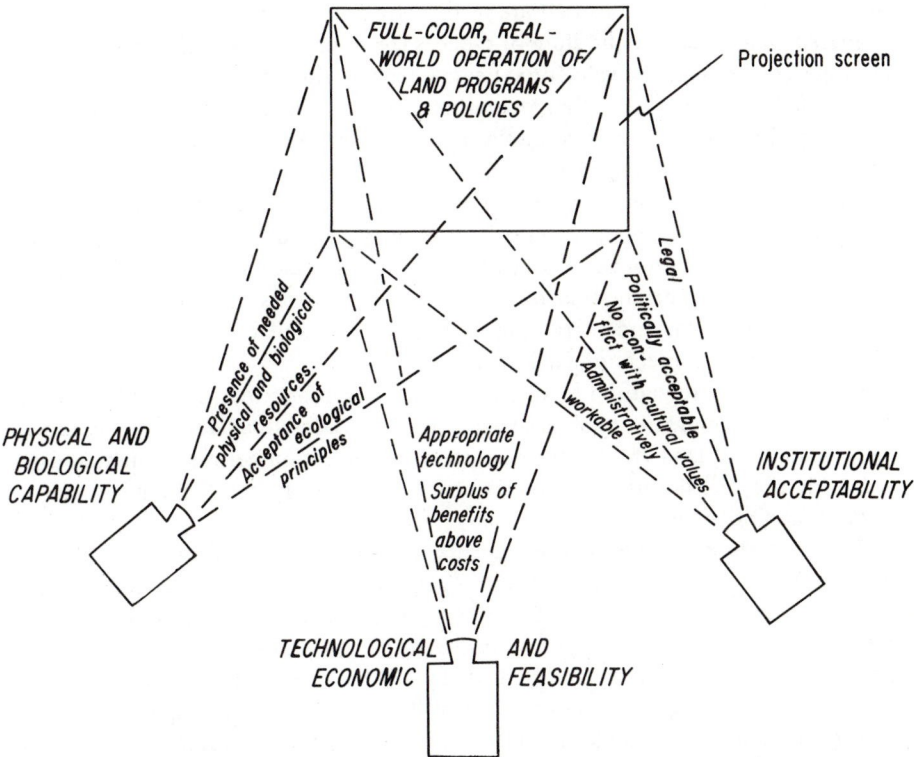

FIGURE 1-1.   Illustration of the simultaneous operation of the threefold framework that affects decisions concerning use of land resources.

factors provide the physical sites, support, and raw materials for our activities. At the same time they provide not only the inanimate resources of the earth but also the vegetative, bacterial, insect, fish, animal, and human resources that both help and hinder us in our use of land. This framework has important effects on the total supply of land resources and the demands we make of them.

The physical and biological framework sets definite limits on what operators can and cannot do in exploiting and utilizing land resources. Successful resource-use policies and programs must respect the physical and biological limits of resource capability both in the short run and over time. Unlike the situation in times past, mankind now has the technical ability to deface, pollute, poison, and destroy significant portions of the natural environment. Tempting as the prospect of short-run benefits from some types of resource exploitation may appear, society must be wary of actions that can destroy fragile and nonreplaceable resources or seriously disrupt normal ecological processes. Those who violate the ecological laws of nature must expect a lessening of the productive capacity of the earth resources on which the human race depends.

With the *technological and economic framework*, the important test is feasibility not capability. Many actions that are technically possible are not feasible in a realistic market sense because the state of the arts (technology) has not been sufficiently developed or because the value of expected benefits does not warrant the necessary cost outlays. It is essential that operators have sufficient technical knowledge to make beneficial uses of their resources and that they also be able to work out productive input-output arrangements that enable them to realize a surplus of economic returns above their production costs. New scientific and technological developments can have major impacts on economic feasibility and make projects and policies practicable that were not so under earlier circumstances. Changes in product prices; consumer demands; production responses to combinations of inputs; and raw material, transportation, and marketing costs can have similar effects in determining whether production is economically feasible.

In addition to its natural emphasis on production and marketing relationships, the economic framework involves social welfare considerations. It is concerned with the availability of land and other resources; their allocation between potential users; and with the distribution of production benefits and incomes among operators, workers, and others in society.

The *institutional framework* is concerned with the role cultural environments and forces of social and collective action play in influencing the behavior of people as individuals and as members of families, groups, and communities. To be workable, land-use programs and policies must pass the test of institutional acceptability. They must be constitutional and legal. They must recognize the significance of property rights. Unless their proponents are willing to work against uphill odds, they must be politically acceptable and not conflict with accepted cultural attitudes, customs and traditions, or widely held beliefs. Their ultimate success also requires that they be administratively workable.

These three frameworks set the stage on which man's use of land takes place at any given time. Each framework has its own special impacts on land-use programs and policies, and each can be examined separately. In the real world, however, the three frameworks are interlinked and work together. They do not operate independently of one another; nor are they applied in a one-two-three sequence. In many respects, their operations resemble those of the three projectors used to transmit the red, blue, and green segments of the color spectrum needed to provide full-color television pictures. Programs and policies must be physically and biologically possible, technically and economically feasible, and institutionally acceptable if they are to work out in practice. Resource managers and policymakers must respect the constraints posed by each of the three frameworks if their programs are to prove successful.

The subject matter of land economics overlaps and includes that of real estate economics. At times, it is divided into two subfields: rural land economics and real estate or urban land economics. Where this division exists, rural land economics is often considered as a phase of agricultural economics, while real estate or urban land economics is regarded as a branch of general economics or business. From a research standpoint, real estate and urban land economists have given most of their attention to problems of housing; urbanization; urban land development and redevelopment; industrial and commercial location; and urban real estate appraisal, finance, and marketing. Rural land economists, in turn, have concentrated on issues such as land and water utilization, land settlement and development, reclamation, land classification, property valuation, land tenure, resource conservation, taxation problems, land-use planning, zoning and land-use controls, and public land management.

Organization of book. This volume deals with the political economy of both urban and rural land resource use. From the standpoint of overall organization, the chapters that follow are divided into four parts. Chapters 2 through 4 deal with adequacy of our land resource base and how it affects the economic supply of land resources, overall demand for land and its products, and future land requirements.

Chapters 5 through 10 are concerned with the economic framework within which land resources are used. Consideration is given in these chapters to the economic roles that input-output relationships, land rent, resource development, management and conservation decisions, and location factors play in influencing decisions concerning land use and investments in real estate.

The institutional framework is examined in Chapters 11 through 16. Emphasis is given in these chapters to the general impact of institutional factors on land use; the nature of the rights people hold in property; and the problems that arise with the acquisition, transfer, leasing, mortgaging, and taxation of land holdings.

Chapters 17 and 18 deal with the public direction of land use. Attention is given in these chapters to the land-resource planning process and to the approaches governments can employ in influencing and guiding individual and group decisions respecting land use.

SOME BASIC LAND ECONOMIC CONCEPTS

Like most fields, land economics has several specialized concepts and terms that must be understood if they are to be useful as tools of analysis. These terms are introduced and explained in the chapters that follow. At this point, emphasis is given to four basic ideas: (1) the economic concept of land and land resources, (2) a classification of land uses by type, (3) the concept of land use-capacity, and (4) the concept of highest and best use.

Economic Concept of Land

The term *land* suggests different things to different people, depending on their outlook and their interests at the moment. In its most widely accepted use, this term refers to the solid portion of the earth's surface. But it may also apply to a nation, a people, or a political division of the earth's surface. People often refer to ground, soil, or earth as land and speak of land as something on which they can walk, build a house, plant a garden, or grow a crop. These commonly accepted definitions of land should not be confused with the more technical concepts used by lawyers and economists.

From a legal standpoint, land or real estate may be considered as any portion of the earth's surface over which ownership rights might be exercised. These rights relate not just to surface area but also to things such as trees, which are attached to the surface by nature; to buildings and other improvements attached by man; and to those objects of value that lie either above or below the surface.

Because of their concern over distinctions between land and capital, economists often differ in their opinions regarding the nature of land. Many economists accept broad definitions similar to those used by lawyers; others treat certain aspects of this broad concept as capital. For our purposes, the *economic concept of land* can be viewed as being synonymous with the legal concept of real estate. It involves the natural and man-made resources that individuals, groups, or communities control through possession of portions of the earth's surface.

This broad concept of land includes all of the earth's surface—water and ice as well as ground. In addition to building sites, farm soil, growing forests, mineral deposits, and water resources, it also involves such natural phenomena as access to sunlight, rain, wind, and changing temperatures and location with respect to markets and other areas. Moreover, it includes all those man-made improvements that are attached to the surface of the earth and cannot be easily separated from it.

Concepts of land. The term *land* often means different things depending upon the context in which it is used and the circumstances under which it is considered. Like a cut diamond, it has many facets. Most important among these are the views of land as (1) space, (2) nature, (3) a factor of production, (4) a consumption good, (5) situation, (6) property, and (7) capital. Other facets of land may also be noted. Some cultural groups view land as a deity that possesses itself and that can exercise certain controls over the people who use it. Investors sometimes see it as a store of wealth that possesses unique advantages over alternative areas of investment. Others have viewed it as a gene bank—a potential source of new species and products—and as a source of pleasure and recreation.

Land may be thought of as *space*—as room and surface within which and on which life takes place. In this sense, land is fixed in quantity and is indestructible because space cannot be destroyed or increased. Land as space includes not only the surface of the earth with the oceans, mountains, valleys, and plains, which provide physical support for people and their activities, but also cubic space. It involves the space beneath the surface within which minerals are found and from which they might be removed, the space people occupy in their daily living, and the space above and about them.

When land is considered as *nature*, it may be closely associated with the natural environment. As such, it is conditioned by access to sunlight; rainfall; wind; changing climatic conditions; and different evaporation, soil, and topographic conditions.

Because of the past and present workings of nature, some areas are rich in soil, forest, fish, and other resources, whereas others appear bleak and barren. Operators can change or modify many of the characteristics of land as nature. Its more basic features however, still lie beyond the tampering hand of man.

Economists frequently refer to land along with labor, capital, and management as basic factors of production. When land is considered as a *factor of production*, it is usually thought of as the nature-given source of the food, fibers, building materials, minerals, energy resources, and other raw materials used in modern society. This concept is closely allied to that of land as a *consumption good*. Land is often held and coveted not only because it adds directly to production but because it has value as a consumer good. Building lots, parks, recreation and residential properties are frequently treated as consumption goods even though they may also be regarded as factors of production.

Considerable importance is attached in the modern world to the concept of land as *situation*. This concept involves location with respect to markets, geographic features, other resources, and other countries. It is significant because the value and use of land is determined largely by its location and accessibility and also because of the economic and political significance that is often attributed to the control of strategic sites.

The concept of land as *property* involves real estate and has legal connotations. It is concerned with the areas over which individuals, groups, or sovereign powers exercise rights of possession and use and with the nature of the rights and responsibilities they hold. Property institutions change with time, but the interpretations accepted at any given moment always wield a powerful influence in shaping attitudes and actions concerning land and how it can or should be used.

While land may be treated as a unique and separate factor of production, it is often realistic to view it as a type of *capital*. The early classical economists argued that land was a free gift of nature, whereas capital was man-made. Capital represented past savings and stored-up production and was expendable, whereas land was seen as durable. These distinctions have some validity but tend to break down and blend together when they are applied in twilight cases. One might ask, for instance, if one field is land because its owner settled on productive soils while a second is capital because the owner started with poor soil and built up a productive farm unit. Similarly, one may question how free land actually is and how durable soil, forest, and mineral resources are in comparison with capital goods. These problems have led many economists to regard land as a species of capital, while others assert that space and situation are the only singular characteristics of land.

Whether land should be clearly separated from capital is an academic issue and has little bearing on the issues considered here. Suffice it to say that the characteristics of land can be similar to those of capital. Land may be fixed in quantity, durable in nature, and a nature-supplied good from the standpoint of society; but the average investor sees it as a resource that must be purchased or leased like other capital goods. In this sense, land is capital to the individual even though it may be viewed differently from the standpoint of society.

Land resources. Because of occasional confusion and lack of agreement regarding the precise meaning of the term *land* when applied in economics, it is often desirable to speak of land resources or real estate rather than land. With this substitution of terms, it is possible to clarify the general meaning of this central concept and at the same time avoid quibbling over details. As used here, the concepts of *land resources* and *real estate* are treated as roughly comparable to the economic concept of *land*. These terms are used interchangeably throughout with no distinction in meaning. The first two terms are concepts representing a merging of the economic and legal concepts of land and definitely include buildings and other capital improvements attached to the land as well as the natural characteristics of land.

As a descriptive term in economics, *land resources* is both broader and narrower than the term *natural resources*. It is broader because it includes the man-made improvements that are attached to land. Natural resources can involve a broader concept in that it includes all nature-given resources from the center of the earth to the highest heavens, whereas the term *land resources*, when treated as an economic concept, is limited to surface resources together with the thin layer of subsurface and suprasurface resources that people use in their daily lives.

Principal Types of Land Use

Various classification schemes can be used to describe the principal types of land use found throughout the world. One of the more workable and more inclusive of these—and the system followed throughout this book—calls for the following ten-fold classification of land uses:[4]

Residential lands	Mineral land
Commercial and industrial sites	Recreation land
Cropland	Transportation lands
Pasture and grazing land	Service areas
Forestland	Barren and waste

Residential lands and *commercial and industrial sites* account for most of the land area of cities but cover only a small proportion of the earth's surface. These uses and their various subclasses are particularly important in the modern world because they represent the areas where three-fourths of the residents of the developed nations and almost a third of those of the developing countries live and where most

[4] For other discussions of land-use classifications, cf. Marion Clawson and Charles L. Stewart, *Land Use Information* (Baltimore, Md.: Johns Hopkins University Press and Resources for the Future, 1965), chaps. 7 and 8; Urban Renewal Administration and Bureau of Public Roads, *Standard Land Use Coding Manual* (Washington, D.C.: Government Printing Office, 1965); *A Review of the New York State Land Use and Natural Resources Inventory* (Ithaca, N.Y.: Cornell University Center for Aerial Photographic Studies, 1970); James R. Anderson, "Land-Use Classification Schemes," *Photogrammetric Engineering*, 37 (April 1971), 379-87; and A. P. A. Vink, *Land Use in Advancing Agriculture* (New York: Springer-Verlag, 1975), chaps. 1-5.

economic activity takes place. They involve the areas most subject to intensive human use and the sites of highest market value.

Because of their contributions to agricultural production, the next three classifications are often grouped together as agricultural uses. These uses account for most of the world's surface land area that has economic value. *Cropland* includes the cultivated areas used in the production of food, feed, fibers, and other crops. As a land-use concept, it includes not only cropland harvested but also planted areas that suffer from crop failure and cropland areas that are temporarily idle or fallow.

The concept of *pasture and grazing land* is somewhat complicated because it really involves two types of land use—arable pasture plus range and grazing land. Arable pasture includes all those improved and rotation pasture areas that are considered plowable and that might easily be shifted into cropland use. Areas of this type are often interspersed with and sometimes rotated with croplands. These two overlapping uses are frequently treated together as arable farmland. The term *range*, in contrast, is ordinarily associated with the large, naturally vegetated, and often unfenced grazing areas found in the western states and in parts of the South. Some western range lands have a cropland potential, particularly if they can be irrigated, but most are best adapted to permanent grazing use because of their low rainfall, rough topography, or high altitude.

Forestland includes the areas used for commercial timber production together with noncommercial woodlands, farm woodlots, cutover lands with a timber growth potential, and some brushland areas. This classification occasionally overlaps certain other agricultural uses. Grazed woodlands, for example, may be treated as either grazing or forest land. Similarly, a number of tree crops have value for food as well as for timber production.

Aside from the residual class of barren and waste lands, most of the remaining types of land use may be grouped together as special-use areas. *Mineral lands* vary from open-pit sources of coal or iron ore to the much smaller surface areas required to support the operation of oil wells and underground mines. *Recreation lands* include parks, beaches, resort areas, racetracks, game preserves, and open space and scenic areas that are used largely for recreation and closely related purposes. *Transportation lands* include those areas used for highways, streets, alleys, parking purposes, railroads, airports, harbors, and wharves. The concept of *service areas* overlaps that of other special-use areas but applies specifically to uses such as military reservations, prisons, cemeteries, reservoirs, and hydropower sites.

Clear-cut classifications of land uses are often complicated by the complementary nature of many uses and by overlapping and multiple-use patterns. Tracts used primarily for single uses such as cropland, forestry, or industries often contain areas utilized for necessary complementary purposes such as roads, parking, or service uses. Sites also may be used simultaneously for more than one purpose. Large areas in the western United States, for example, are designated as forestlands but at the same time have important values for grazing, recreation, and watershed uses. Similarly, a site occupied by a hotel may be used primarily for residential purposes but at the same time provide commercial, recreation, transportation, and service facilities.

Land Use-Capacity

Land use-capacity involves the relative ability of a given unit of land resource to produce a surplus of returns and/or satisfactions above the costs of utilization. This concept, which is closely associated with that of land rent, measures the productive potential of units of land utilized for a given use at a given time with given technological and production conditions. The amount of net return or satisfactions secured provides an index of relative use-capacity. When these indices are compared for particular tracts or units of land resources, the concept of use-capacity provides a common measure of the quality or excellence of the units considered. For example, one might assume a comparison of three land areas of equal size, each of which is used for the same purpose. If the first area produces a net return of $50, the second $100, and the third $15, the second area naturally has the highest economic use-capacity.

Use-capacity has two major components—accessibility and resource quality. *Accessibility* involves the convenience, time, and transport cost savings associated with specific locations with respect to markets, shipping facilities, and other resources. It is concerned with optimizing transportation and communication costs and time-distance considerations. *Resource quality* involves the relative ability of a land resource to produce desired products, returns, or satisfactions. With agricultural lands, quality is usually viewed in terms of native fertility or fertility in combination with ability to respond to fertilizer inputs. Quality may reflect climatic advantages—favorable temperature and precipitation levels, low wind velocity, or infrequency of storms. It may involve aesthetic considerations such as scenery, presence of trees or water attractions, nearness to parks or open space, and access to schools and cultural opportunities. In urban areas, quality can involve items such as functional area planning, neighborhood attractiveness, architectural styling of buildings, and other conditions that affect the satisfactions and values people associate with properties.

The concept of use-capacity is used in land economics to distinguish between the comparative abilities of different units of land resources to provide net returns and other satisfactions. From an overall point of view, this concept involves all the factors that affect the ability of a unit of land resource to produce a net return as compared with some other unit. In practice, it is often employed with examples that involve a single criterion of accessibility or quality—all other factors being assumed as constant. With discussions of the productivity of farmlands, for example, use-capacity is often equated with differences in fertility. In discussions of site location advantages, it is frequently associated with transportation costs. Similarly, with examples involving urban location differences, use-capacity may be thought of in terms of the relative amounts of time and effort required to transport persons or things from particular sites to other places, such as the downtown business district.

Comparisons involving use-capacities assume a given instant of time. Observations based on particular comparisons can remain unchanged for long periods. Shifts take place, however, with changes in the resource base, changes in operator know-how, and changes in the uses made of land resources. Factors such as urban blight or the depletion of a mine can downgrade the use-capacity of land, whereas resource-development programs usually increase the use-capacity of particular resources. In similar fashion, new inventions, the building of railroads, and the development of new markets can raise the use-capacity of given sites—sometimes at the expense of others. Changing opportunities and the shifting of land areas to new uses, such as the movement of land from farming to suburban residential uses, can also have marked effects on the relative use-capacities of individual properties.

Highest and Best Use

Most land areas are suited for a variety of uses. The highly valued land found in most central business districts could be used for forestry, grazing, crop production, or residential purposes as well as for commercial uses. It is used as it is, however, because owners have an economic incentive to use their land resources for those purposes that promise them the highest return. In this respect they allocate their land resources in accordance with the concept of highest and best use.

Land resources are at their *highest and best use* when they are used in a manner that provides an optimum return to their operators or to society. Depending on the criteria used, this return may be measured in strictly monetary terms, in intangible and social values, or in some combination of these values. Real estate is ordinarily considered at its highest and best use when it is used for that purpose or that combination of purposes for which it has the highest comparative advantage or least comparative disadvantage relative to other uses.

The highest and best use of any particular site is often subject to change. Like the concept of use-capacity, it can shift with changes in the quality of the land resource, changes in technology, and changes in the demand picture. Sometimes it is affected by zoning ordinances and other public policies. Under most circumstances, a certain amount of shifting can also be expected to take place in response to the bidding and counterbidding that goes on between various operators.

In modern society, land resources usually earn a higher return when used for commercial or industrial purposes than for other types of use. As a result, these uses are usually able to outbid other uses for almost any site. Residential uses ordinarily have next priority, followed by various types of cropland, pasture, grazing, and forest uses. This simple ordering of land uses suggests a definite profile such as that depicted in Figure 1-2, in which the highest-value lands at the center of our cities are used for commercial purposes, while areas with successively lower values are used for residential, cropland, grazing, and forestry purposes, respectively.

Profiles of this type represent a generalized average. They are never as fixed or static as they may at first appear. Variations from this norm often occur because of the differences found in individual type of use classes and because of the tendency of some classes to overlap. Exceptions to this pattern also exist. Some industrial

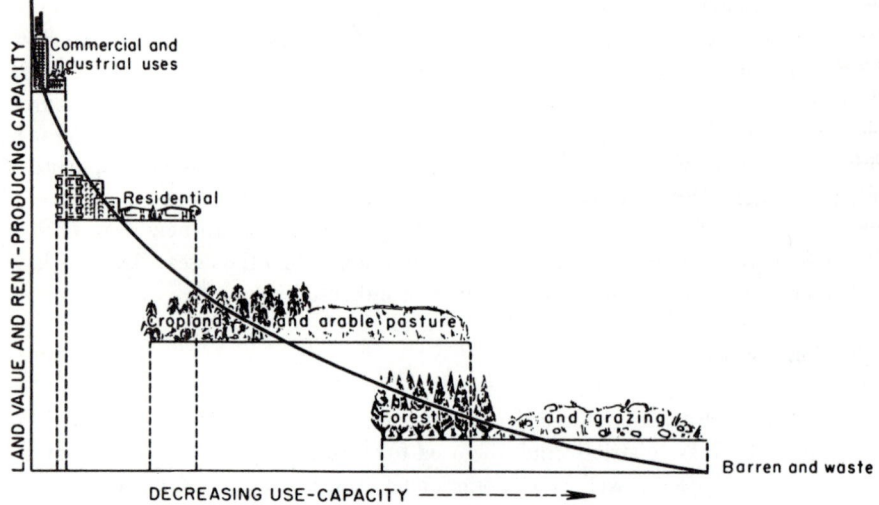

FIGURE 1-2.   Generalized profile of land uses showing the overlapping ranges within which selected uses may be regarded as the highest and best use.

and commercial uses deliberately seek low-cost sites, and residential uses such as apartment houses occasionally outbid commercial and industrial uses for particular sites. Similarly, in areas such as the oasis communities of the Sahara Desert, where the supply of arable land is extremely limited, residential quarters may be located on the edge of the desert while the arable land they could occupy is used for food production.

Considerable overlapping exists among the agricultural uses. This is particularly true when good management calls for the simultaneous operation of two or more enterprises. Typical farms accordingly usually have areas used for farmstead, cropland, pasture, and woodlot purposes. Overlapping is also common in cities where a lot may be used for residential purposes, as one's place of work, and as the site of a vegetable garden.

Wide variations exist with the priorities associated with special-use areas. Mining often claims top priority, particularly when the prospects for profitable operations are high. Recreation sites vary from low-priority wilderness or residual areas to urban parks that may be reclaimed or redeveloped at great expense. Lands used for transportation purposes may be low in value in some instances but involve substantial costs with street-widening projects, new parking lot developments, or new metropolitan airport facilities. A similar wide range of use priorities applies to service areas. Some of these land uses, such as watershed protection areas, reservoirs, and hydropower sites, are closely associated with special characteristics provided by nature. This gives these uses a first choice of particular sites. Competition for the use of these sites can also lead to high land values. Where less demand exists, site values may be low in spite of the high priorities claimed by particular uses.

Though the highest and best use of a given unit of land resource at any one time can usually be computed in monetary terms, differences of opinion frequently arise when weight is given to welfare considerations and nonmonetary satisfactions. One landowner may keep a forested area as a woodlot. A successor in ownership may choose to convert it into farmland, residential lots, or possibly a public park. An individual may feel that a lot should be retained for residential or commercial purposes, while a city may insist that part of it be given up for a street-widening project. A public-spirited group may insist that a virgin forest be maintained as a public park, while other groups argue that it be cut to provide timber for a growing nation. These examples indicate the need for differentiation between the marketplace, or *economic*, and the *social* concepts of highest and best use. Economic highest and best use is a meaningful measurable concept with numerous applications in the commercial world. Social highest and best use, in turn, is a less quantifiable concept that reflects the varying aspirations, goals, and value judgments of different individuals and groups.

—SELECTED READINGS

Ely, **Richard T.**, and **George S. Wehrwein**, *Land Economics* (Madison: University of Wisconsin Press, 1964), Preface and pp. 25–28. Originally published by Macmillan, 1940.

Johnson, V. Webster, and **Raleigh Barlowe**, *Land Problems and Policies* (New York: McGraw-Hill, 1954), chap. 1.

Renne, Roland R., *Land Economics*, 2nd ed. (New York: Harper & Row Publishers, Inc., 1958), chap. 1.

Salter, Leonard A., Jr., "The Content of Land Economics," *Journal of Farm Economics*, 24 (February, 1942), pp. 226–36.

2

AVAILABILITY OF LAND
FOR ECONOMIC USE

Humanity has enjoyed a mixed relationship with land resources throughout history. While the resource base has sometimes provided bountiful opportunities for growth and affluence, individuals and groups have frequently found their conditions of life hard and their prospects for the future bleak. Those who have been fortunate enough to live at productive sites and benefit from new knowhow have usually prospered, while those who have lived under adverse conditions have often experienced privation and mere survival.

Availability of technology more often than not has provided the key for successful use of the land resource base. The significance of this strategic input has been amply demonstrated during the last 300 years, the period of the so-called Agricultural and Industrial Revolutions. During this short segment of the world's history, operators have employed a steady flow of new technology to unlock and open the gates of resource productivity. The human race has benefited greatly from this bounty; and in enjoying it, many have found it possible to attain higher and higher levels of life while total population numbers have tripled and then tripled again.

Overall, the accomplishments of recent decades show that we have many reasons for being pleased with the productive opportunities offered by our store of land resources. But before anyone is lulled to a satisfied sense of security concerning our future prospects on this planet, it must be noted that numerous competent observers question the sufficiency of our land resource base to do all that we expect of it.

Individual operators experience this problem every time they find that they lack access to the types of land resources needed for their operations. Our primitive ancestors faced it when hunger forced them to abandon their hunting and gathering economy and settle down in areas where they could plant and harvest crops. Communities and nations have felt it when droughts and crop failures have caused famines and human privation. Concerns about the future adequacy of our land resource base have a persistent habit of refusing to go away. They provide an ever-present backdrop for much of our thinking about land and its use. In practice, these concerns involve three questions: How much land is available for economic use? How much demand is there for its use? Do we have sufficient land resources to care for our emerging needs? These topics are addressed in this chapter and in the two chapters that follow.

ADEQUACY OF OUR LAND RESOURCE BASE

Most people have a keen interest in their prospects for survival and for the future maintenance and improvement of their levels of life. With this interest, it is only natural that repeated questions are raised concerning the ability of our land resource base to supply the food and other materials needed for an expanding population. A gloomy answer to these questions was presented by the Reverend Thomas Robert Malthus when he published his theory of population growth two centuries ago. Since that time, our farmers and ranchers, miners, and industrialists have demonstrated time after time that the Malthusian conclusion was premature if not inaccurate.

The industrialization and urbanization trends of the 1800s and 1900s clearly indicate that the earth has been able to supply more and more food and other raw materials. These increases have come largely in response to new approaches and techniques of production. Additional improvements can be expected in the future. But does the fact that we have experienced two centuries of rising returns in response to our utilization efforts mean that we can expect indefinite continuation of this phenomenon? Or is it possible that we have gone through what people some centuries hence will view as a short-term production binge, during which we skimmed off much of the cream from the world's resource base?

Authorities frequently disagree in their answers to these questions. Many are optimistic about future prospects for increasing production.[1] Some even talk of inexhaustible resources, of the folly of conserving resources such as oil that may be worthless in another generation, and of expected abilities to secure needed resources from other planets.[2]

[1] For examples, see Harold J. Barnett and Chandler Morse, *Scarcity and Growth: The Economics of Natural Resource Availability* (Baltimore, Md.: Johns Hopkins University Press, 1963); Wilfred Beckerman, *Two Cheers for the Affluent Society* (New York: St. Martin's Press, 1974); Herman Kahn, William Brown, and Leon Martel, *The Next 200 Years* (New York: Morrow, 1976); Pierre Crosson, ed., *The Cropland Crisis: Myth or Reality?* (Baltimore, Md.: Johns Hopkins University Press, 1982); and Julian Simon, *The Ultimate Resource* (Princeton, N.J.: Princeton University Press, 1981).

[2] Cf. Eugene Holman, "Our Inexhaustible Resources," *Atlantic Monthly*, 189, no. 6 (June 1952), 29–32; Jacob Rosin and Max Eastman, *The Road to Abundance* (New York: McGraw-Hill, 1953); and Gerhard Anders, W. Philip Gramm, and S. Charles Maurice, *Does Conservation Pay?* (Los Angeles: International Institute for Economic Research, 1978).

Quite a different group, including several prominent economists and environmentalists, argues for a more conservative and pessimistic outlook. They see the earth's supply of resources as finite in quantity. In their view, we face a real danger of running out of resources and of losing the economic viability of many of our past investments in economic development.[3] The views of these thinkers are sometimes written off as the mouthings of gloomer-doomers. But the uncertainties of the future and an absence of absolute criteria that policy makers can apply in planning future courses of action suggests a need for caution. Wasteful uses of resources should surely be avoided, and serious thought should be given to the acceptance of policies that will bring the orderly, efficient, and effective use of land resources.

Secular law of diminishing returns. Long before the impact of technology on production was as well understood as it is today, economists of the early classical school postulated a theoretical answer to the land resource adequacy issue with their formulation of the secular law of diminishing returns. As stated by Alfred Marshall, this law proclaims that "whatever may be the future developments of the arts of agriculture, a continued increase in the application of capital and labor to land must ultimately result in a diminution of the extra produce which can be obtained by a given amount of capital and labor."[4]

As the diagrammatic presentation of this concept in Figure 2-1 indicates, over the long run, users of land resources must expect to eventually reach a point of overall diminishing returns. Technological advances can postpone this day of reckoning and make it possible for people to expand production capacity far beyond earlier potentials. During recent decades, workers in the more developed nations have been able to enjoy the rising standards of life that come with operations during periods of increasing returns. Present prospects indicate that important new increases in technology are probable and that all of the world's people can share in new opportunities for higher levels of living. But no one knows how long this process will continue. Our prophets of gloom may be unduly pessimistic in assuming that we are at a higher position on the long-run production curve shown in Figure 2-1 than we actually are. Yet regardless of where we are, simple logic suggests that we must eventually face the prospect of diminishing returns.

While the concept of secular diminishing returns seems to be logically consistent, some economists argue that it will not apply in our time because the prospects for new technology and new input-output combinations of resources for production

---

[3] Prominent examples of this point of view are provided by Barry Commoner, *The Closing Circle* (New York: Knopf, 1971); Edward Goldsmith *et al., A Blueprint for Survival* (Boston: Houghton Mifflin, 1972); Donella H. Meadows, Dennis Meadows, *et al., The Limits to Growth* (New York: Universe Books, 1972); Mihajlo D. Mesarovic and Eduardo E. Pestal, *Mankind at the Turning Point* (New York: Dutton/Reader's Digest, 1974); Robert L. Heilbroner, *An Inquiry into the Human Prospect* (New York: Norton, 1974); William Ophuls, *Ecology and the Politics of Scarcity* (San Francisco: W. H. Freeman & Company Publishers, 1977); Edward J. Misham, *The Economic Growth Debate* (London, U.K.: G. Allen & Unwin, 1977); U.S. Department of State and Council of Environmental Quality, *The Global 2000 Report to the President* (Washington, D.C.: Government Printing Office, 1981); and Lindsey Grant, *The Cornucopian Fallacies* (Washington: The Environmental Fund, 1984.)

[4] Alfred Marshall, *Principles of Economics*, 8th ed. (New York: Macmillan, 1938), p. 153.

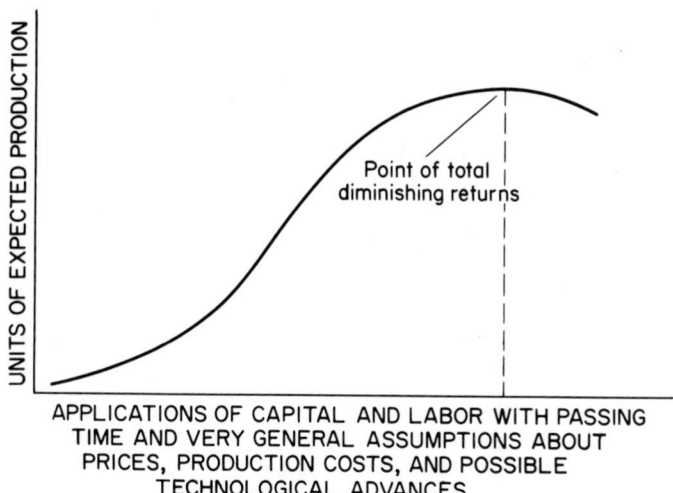

FIGURE 2-1.   Graphic illustration of the operation of the secular law
of diminishing returns.

provide almost infinite opportunities for increasing productivity. Barnett and Morse, for example, indicate that "a strong case can be made for the view that the cumulation of knowledge and technological progress is automatic and self-reproductive in modern economies, and obeys a law of increasing returns. Every cost-reducing innovation opens up possibilities of application in so many new directions that the stock of knowledge, so far from being depleted by new developments, may even expand geometrically."[5] Whether this situation can or will come about remains to be seen. Meanwhile, increasing production costs and decreasing returns per unit of cost outlay are being experienced in many parts of the world. They are exerting important impacts on operator efforts to develop and secure additional economic supplies of many land resources. Since the best and most accessible supplies of many types of resources have already been put to use, operators face an increasingly difficult task in drawing on residual resource bases which are dwindling both in quantity and quality. Larger outlays of labor and capital are neded in many instances to secure the same quantities of land products; and without major advances in production technology or a leveling off of demand, these problems may well become more rather than less serious.

FACTORS AFFECTING THE ADEQUACY
OF LAND RESOURCE SUPPLIES

Realistic assessments of the adequacy of our land resource base must start with an examination of the nature of the resources we have. Answers are needed to the questions of how much land we have, what are its use capabilities, what constraints

[5] Barnett and Morse, *op. cit.*, p. 236.

limit its potential for use, and what factors can unlock new doors for added productivity. The threefold framework suggests a logical approach for addressing these issues.

Physical and Biological Capabilities and Constraints

The physical and biological framework sets the basic parameters for land use. It tells us what resources we have, how much of each type of resource is available, where it is located, and quite often when and during what seasons it can best be used. Nature has provided us with a vast resource heritage. But the gifts of nature are not scattered evenly; and as often as not, they are cloaked with characteristics that tax human ingenuity. In their age-old effort to find, develop, and utilize new resources, people have found that they seldom operate in a Garden of Eden. Resources are available, but users have to cope with problems of uneven distribution and often unfavorable climate, topography, and location.

The physical and biological resources of different areas vary widely in their natural characteristics and in their general use-capacities. Some variations stem from differences in (1) sunlight and temperature, (2) precipitation and access to water supplies, (3) topography and drainage, (4) soil conditions, subsurface strata, and presence of minerals, and (5) physical location with respect to markets and transportation facilities. Others involve the fixity and renewability of the resources themselves.

Limitations involving these characteristics definitely limit the areas and classes of resources suited for particular uses. Fortunately, the purposes for which we use land vary almost as much in their need for particular land characteristics as do the characteristics of the land resource base. Thanks to this happy circumstance, most of the earth's surface has potential value for some use or group of uses, although the total area suited for any one is often quite limited.

Natural limits on the supply of land for agricultural uses. Almost all of the earth's surface enjoys sufficient access to sunlight to permit some type of crop, range, or forest use. But short growing seasons, unseasonable frosts, or year-round winter prevent the utilization of many otherwise available areas for these uses. Approximately one-fourth of the earth's land surface is too cold for wheat culture.[6] Much of this area has value for forestry purposes, as is attested by the northern forests of Alaska, Canada, the Scandinanvian countries, and the Soviet Union. Some of it also has commercial value as summer range and for the support of wildlife and reindeer. Yet large areas such as the icy expanses of the Arctic and Antarctic must be written off as waste land so far as agricultural use is concerned.

The supply of moisture for plant use also presents a varied picture. Only 34 percent of the world's surface land area enjoys both an adequate and reliable supply of

---

[6]Cf. O. E. Baker, "The Potential Supply of Wheat," *Economic Geography*, 1 (March 1925), 31; also Baker, "The Population Prospect in Relation to the World's Agricultural Resources," *Maryland*, alumni publication (College Park, University of Maryland Press, 1947).

rainfall for crop growth, whereas only an additional 1 percent benefits from irrigation.[7] Of the 41 million square miles with suitable temperatures for wheat culture, only 11 million square miles, 20 percent of the earth's land surface, have suitable temperature and moisture conditions.[8] Thirteen million are too wet and 17 million are too dry. Portions of this area, however, can be used for other purposes. Wetlands can be planted with water-loving crops such as rice or be used for forest culture. Summer fallowing practices can be applied in arid regions to store two years' precipitation for use during a single crop season. Irrigation can be used to bring water to areas with inadequate supplies of moisture for normal crop growth. And areas considered unsuited for cultivation can frequently be used for ranching or grazing.

Much of the area climatically suited for crop use is too mountainous, rough, or steep for successful cultivation. Pearson and Harper indicate that only 64 percent of the world's land has favorable topography for crop use, while Baker estimates that 4 of the 11 million square miles of earth surface climatically available for wheat culture are unfit for this use because of hilly or rough land.[9] Terracing is used in many rough land areas to augment the limited supply of agricultural land. Mountainsides in Southeast Asia are often covered with small fields that have been reclaimed in this manner. This type of development calls for large expenditures of capital, time, and effort. Where these expenditures are economically impractical, steep, rough, and mountainous areas often find a residual use in either forestry or grazing.

The soils that cover the earth's surface vary considerably in color, structure, texture, physical constitution, and chemical composition and in their other natural characteristics. They range from light-colored soils to black earth, from heavy clay to sand and gravel, from shallow soils to deep formations, from soils that tend to be acid to those that are alkaline, and from soils that provide plants with little more than space and foundation to soils of high inherent productive capacity. Agricultural uses also vary in their soil requirements, but most crops are responsive to fertile and productive soils. The same may be said of grazing and forest uses, even though these uses are often relegated to the less fertile and less desirable lands.

Pearson and Harper report that around 46 percent of the earth's surface is covered with "good soils" suitable for crop use.[10] This estimate may be taken as a general measure of the world's soil characteristics though it must be remembered that some types of crops make better use of low-quality soils than others and that deficiencies in soil quality can often be overcome with soil building, fertilization, and other practices. Overall, however, our supply of prime productive soils is definitely limited,

---

[7]Cf. Frank A. Pearson and Floyd A. Harper, *The World's Hunger* (Ithaca, N.Y.: Cornell University Press, 1945), pp. 27–28. They define an adequate and reliable supply of rainfall as a minimum of fifteen inches of precipitation annually in temperate areas (forty inches in equatorial areas of high evaporation), varying from year to year by less than 20 percent from the normal average. They found that 79 percent of Europe, 70 percent of South America, 38 percent of North America, 29 percent of Asia, 25 percent of Africa, and only 9 percent of Oceania enjoy both an adequate and a reliable supply of rainfall.

[8]Cf. Baker, *The Potential Supply of Wheat*, pp. 27, 31.

[9]Pearson and Harper, *op. cit.*, p. 42; and Baker, *ibid.*, pp. 28, 31.

[10]Pearson and Harper, *op. cit.*, p. 46.

and we face the prospect of having to draw on soils of lower and lower inherent productive potential.

Natural limits on supplies of land for nonagricultural uses. Climate, topography, and soils have widely different effects on the availability of land for nonagricultural uses. They have only a limited impact, for example, on decisions to use particular sites for mining purposes. The known or believed presence of commercial deposits of minerals is the necessary ingredient with this use; and when this condition is filled, mining or oil drilling activities will sometimes take place in otherwise distant and desolate locations.

Recreation areas, in contrast, frequently owe their attraction to natural features such as sunshine, climate, flora and fauna, and the presence of scenery, water resources, and a beach or mountain environment. High values are ascribed to such diverse features as desert sand dunes, underground caverns, mountain lakes, wilderness areas, fishing streams, hunting marshes, pleasant country vistas, warm ocean beaches, tropical splendors, winter wonderlands, pollen-free air during the summer, and dry desert air during the winter. But important recreation values may also be associated with sites of historical significance and with the man-made facilities, playgrounds, athletic fields, golf courses, swimming pools, and sports palaces found in or near most urban areas.

Urban growth usually requires a strong commercial, trade, or industrial base. This calls for locations near sources of raw materials and near the consumers who will eventually buy and consume the products processed in cities. Since agriculture has long provided a high proportion of the raw materials used in cities plus a market for processed and manufactured goods, it is only natural that most urban centers are found in or near areas also used for agricultural purposes. Sites around harbors, along navigable streams and railroads, and in other spots that offer natural trade advantages also provide favored locations for cities.

As human beings, most of us prefer to live in areas that are neither too hot nor too cold for agricultural use. From a strictly functional standpoint, however, urban areas are not subject to the same range of climatic controls as crops. Sunshine is desirable, but numerous urban functions are carried on with artificial light, heating, and air conditioning. Access to water supplies is necessary, but most cities could get along without torrential rains, snow, sleet, or fog. Well-drained sites and level lands offer advantages for urban developments, and swamps can be drained or filled and rough terrain can be leveled or bridged. Subsurface support is needed for buildings and other structures; but while fertile and permeable soils offer advantages for urban residents, cities can be established on barren sites.

Harbor and dock facilities are found where nature provides the best natural sites or where population and other pressures dictate that harbors should be built. Airports, railroads, streets, and highways are located where it is hoped they will facilitate commerce or convenience. Large bodies of water and rough terrain often make it expedient to direct roads over something other than the shortest or most direct route between points. But while barriers of this type complicate transportation developments, they seldom prevent the linking of cities and other centers. Rain, snow, sleet, and fog conditions can create transportation-use problems; but it is a rare occasion when these factors play more than a minor role in influencing the location of transportation routes.

Service areas call for a wide variety of natural characteristics. Urban-associated uses such as public buildings, cemeteries, and water-filtering facilities must usually be located near the populations they serve. These uses ordinarily call for well-drained locations. Low-lying sites and areas with rough terrain can often be used for city dumps; but toxic waste disposal facilities must be located at isolated sites that involve a minimum of human contact and zero possibilities for contamination of water supplies. Other service areas such as military reservations, reservoir sites, and watershed areas may call for sizable tracts and climatic, land-cover, or topographic conditions that narrow the number of potential sites available for their use.

Renewal and regeneration of productive capacity. The ability of a land resource base to provide products for human use over time is also affected by its self-renewal and regenerative qualities. Again, resources vary in their characteristics. Urban and farming areas often have high capacities for use, whereas wetlands and wilderness areas are usually fragile in the sense that they can sustain only limited amounts of use before they suffer losses of productive capacity. Some resources such as sunshine and precipitation are being continually replenished, whereas others such as copper and coal occur in relatively fixed quantities and have no prospects for renewal within our planning horizon. Some biological resources such as pests, rats, and weeds regenerate with ease even when concerted efforts are made to destroy them; others such as condors and whooping cranes seem to lose their biological urge to reproduce once their numbers are reduced and they become endangered species.

Land resources can be classified in a meaningful manner according to the relative fixity and exhaustibility of their supplies. The resources that occur with fixed or finite supplies in nature can be described as *fund resources*. Some fund resources (coal, oil, and other mineral fuels) are exhausted through use, while others (copper, iron, and lead) may be recycled for future uses if care is taken to prevent their loss or wasteful use. Another grouping known as *flow resources* (rainfall, tides, sunlight, and changing climate) occur in predictable flows over time, are self-renewing, and can seldom (except in the case of water) be stored or saved for later use.

Biological resources (plants and animals) are a leading example of a composite group of resources that have both fund and flow characteristics. At any given moment, the total supply of any species of these resources could be treated as a fund resource and be entirely exhausted through exploitive use. As long as seed stock is retained for future regeneration, however, these resources are renewable and can be viewed as a type of flow resource.

For long centuries, our forebearers were able to use land resources without giving much thought to their replacement. But this situation has changed. George Perkins Marsh observed in 1864 that mankind was fast acquiring the ability to reshape and even destroy its physical and biological environment.[11] This human power to build,

---

[11] Cf. George Perkins Marsh, *Man and Nature, Physical Geography as Modified by Human Action* (New York: Scribner's, 1864).

perpetuate, and destroy has been greatly expanded during the past century; and how we develop, manage, and conserve or exploit the fund, flow, and biological resources entrusted to our stewardship has greater implications for our future now than ever before.

Economic and Engineering Considerations

What operators do with the resources at their disposal depends in large measure on the economic and engineering or technological feasibility of their alternative use opportunities. In our assessment of the impacts these considerations have on the provision of land resource supplies for human use, it is important that we start by considering the nature of the concepts of supply and demand, after which we will examine some specific examples of the roles played by economic and engineering factors.

Concepts of supply and demand. Discussions of the adequacy of our land resource base call for frequent use of the terms *supply* and *demand*. Like many other terms used by economists, these words do double duty in the sense that they have more than one meaning. Economists frequently speak of the term *supply* as the schedule of amounts of a good or service sellers will offer on the market at different prices during a given time period, all other factors being equal. In similar fashion, the term *demand* is used to describe the schedule of amounts of a commodity buyers are willing to purchase at all possible prices during a given time period, all other factors being equal. These specialized concepts have their place in rigorous economic analysis. However, economists often revert to the popular usage of these terms (1) when they talk generally about amounts of goods or resources rather than specific supply or demand schedules, and (2) when they are concerned with supply or demand conditions over time or in periods during which other relevant conditions may change. Throughout our discussion we will follow popular usage in speaking of *supply* as the quantity of goods or resources available for use and of *demand* as the amounts of a commodity people want and are willing to buy.

As we examine the supply side of the land resource picture, two factors bear notice. The overall supply of land resources is limited to the sum total of the resources provided by the earth's surface. The amounts available for use by individual operators, nations, and humankind as a whole, however, vary with changing circumstances. This situation suggests the need for a distinction between the physical and the economic concepts of supply.

When we speak of the *physical supply* of land, we are concerned with the physical existence of land resources. This concept can be applied to particular resources, such as the physical supply of mineral fuels or the areas with selected soil types. It can also be applied to area units, as when we speak of the sum total of the land resources found in individual ownership units, counties, nations, or the entire world.

The *economic supply* of land involves only that portion of the total physical supply that people want and use. Land becomes a *resource* in an economic and technical sense when people begin to use it, compete with others for its use or control, put a price or value on it, or indicate willingness to undertake the costs associated with its development.[12] The economic supply of land is seldom fixed. It is responsive to price and demand factors. New technology and new demands can give values to areas and physical materials that had little or no economic value before. With changing demands, the economic supply of land for any use can expand (and sometimes contract) and is limited only by the total physical supply of land that can be made available for that use.

The concept of demand is analogous to that of supply. As a physical concept, demand is associated with desires, needs, and wants for commodities and services. *Resource requirements*—the amounts needed to provide average diets, housing facilities, or education or recreation opportunities of a given level of adequacy—represent a measure of physical demand. This concept can provide guidelines for program planning and public policy but has limited value in economic analysis. It is economic *effective demand*—the willingness and ability of people to buy—not the mere existence of unsatisfied needs or desires for land products that influences the determination of prices and the movement of products in the market.

Most demand for land resources involves a derived type of demand. Everyone demands products of land; but our demands are directed more at the quart of milk, pound of rice, barrel of oil, or ton of steel than at the land from which these products ultimately come. *Derived demands* are reflected back through a series of steps to demands for land itself. Most operators see ownership as a means to an end. They may have sentimental attachments to given parcels of real estate, but they are usually more interested in the potential values associated with its productivity, location, scenery, or ability to generate satisfactions than with the land itself.

Interaction of supply and demand.   Though supply and demand are often treated as separate factors, they normally operate in close conjunction with each other. It is the interaction of these factors that gives us the concept of a market. Under free market conditions, it is this interaction that sets the market price at which products sell.

This situation is illustrated by the model depicted in Figure 2-2. The supply curve SS' represents a schedule of the increasing quantities of a product sellers would offer in the market at a given time at a series of rising price levels. The demand curve DD' in turn represents the schedule of increasing quantities of the product buyers would take in the same market at the same time at a series of decreasing prices. With these supply and demand schedules, the only possible equilibrium price occurs at P, the point of intersection between SS' and DD'. At this price, the quantity of product offered and the quantity buyers are willing to purchase are equal. If the price were set at a higher point such as P', sellers would be willing to supply a larger quantity of product, but some buyers would take less and some would drop out of the market. With a lower price such as P'', the reverse situation would hold.

Under real market conditions, supply and demand factors tend to follow the

---

[12]Cf. Erich W. Zimmerman, *World Resources and Industries*, rev. ed. (New York: Harper & Row Publishers, Inc., 1951), p. 7.

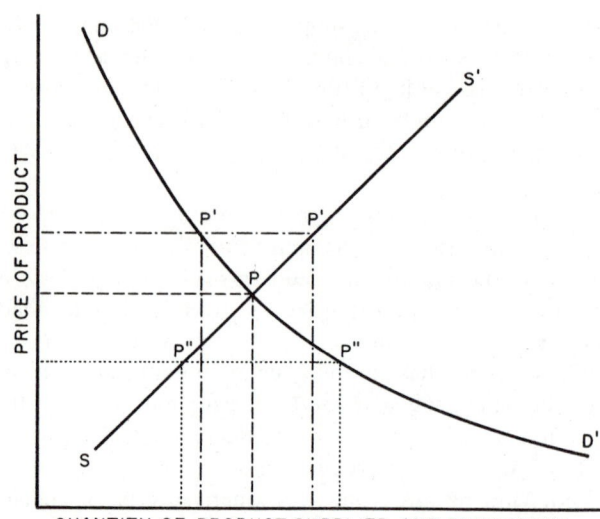

QUANTITY OF PRODUCT SUPPLIED AND PURCHASED

FIGURE 2-2.   Interaction of supply and demand factors in determining market prices under free market prices and free market conditions.

model of the perfect market. Variations occur with differences in the elasticity of supply or demand. A supply or demand is said to be *elastic* when a given change in price results in a more than proportionate change in the quantity of product supplied or purchased. When a given price change results in less than a proportionate change in quantity supplied or purchased, the supply or demand is *inelastic*. With the example depicted in Figure 2-2, the supply would be elastic if circumstances required a drawing of the *SS'* supply line at a lower angle, whereas it would be inelastic if *SS'* rose to a steeper angle. These concepts are most meaningful when applied to products such as wheat, coal, or office space that are bought and sold in given markets. The amounts of any given land resource supplied or demanded are usually responsive to price changes; but in most market situations, supply and demand responses tend to be inelastic.

Complications in the model can arise because of differences in the knowledge and market expectations of individual buyers and sellers and because of the linkage and interdependence of various factors in the economy. Significant deviations from this model can also occur when individual operators or firms use their power as buyers or sellers to set their own prices under the conditions of monopolistic or oligopolistic competition.

Importance of economic considerations.   Economic feasibility is an all-important requirement with the development and provision of new supplies of land resources. Business operators will not undertake resource developments or production programs unless they anticipate a surplus of benefits or economic returns above expected costs. It would be irrational for them to operate otherwise. In their decisions affecting the provision of resource supplies, they are cost-conscious because they have strong economic incentives for maximizing their monetary returns. They try to hold development and operating costs down. At the same time, they recognize that competition can affect cost and price levels and that market-determined prices play major roles in allocating products and materials between buyers and users. They also are keenly aware of the *opportunity costs*—the income that could have been secured by utilizing their resource inputs in their most productive alternative uses—associated with their resource allocation decisions. When they can do so, they frequently shift their investments and production inputs to other enterprises if the opportunity cost returns expected with them exceed those they can secure from their present enterprises.

Operators are naturally inclined to make first use of those land areas that have the highest use-capacities for their particular enterprises. As additional land is needed, they resort to the use of lower and lower grades of land. Higher product prices are needed to cover the higher production costs per output unit encountered on the lower-quality lands. Buyers who want more products bid market prices up to the levels necessary to bring forth the products they need; and these higher prices logically lead both to more intensive use of the existing supplies of developed land and to the bringing of new areas into use.

Most programs that add to the economic supplies of land call for the development of new, less productive, less favorably located, and harder-to-develop lands. When the pressure for new land development is strong enough, drainage, irrigation, terracing, and other practices are used to "create" agricultural lands, and leveling, filling, landscaping, and multistory building practices are used to "create" urban and recreation sites. Each of these types of development has its price tag, and its feasibility always depends on the willingness of buyers and of society to pay the higher market prices and possible subsidies associated with their cost.

Competition among individuals and among land uses also has its impacts on land resource supplies. In the bidding and counterbidding that takes place between individual operators and uses, resources normally go to those operators and those uses that offer the highest prices and enjoy the greatest prospects for their remunerative use. From an overall standpoint, the supply of land resources available to individual operators is limited only by their willingness and ability to pay the going prices asked for the resources they need. In practice, operators often use fewer and lower-quality resources than they desire because they lack adequate financing or because they have limited opportunities to make optimum use of the resources they could acquire through the bidding process.

Much the same situation applies in the competition between various land uses. Serious supply problems do not develop as long as each type of use can expand without impinging on the areas needed for other purposes. But complications arise when conflicting uses compete for the same land areas. At this point, the more highly valued and more economically productive uses usually take precedence, thus crowding the lower-priority uses to outlying or lower-quality areas. Once most of the land has been brought into use, any continued expansion of the areas utilized for urban, cropland, and other high-premium usage must come at the expense of the residual areas previously put to lower-priority uses. In this respect, continued expansion of high-priority uses inevitably leads to a diminution of the economic supplies of land available for lower uses.

Impact of technological factors. Most of the value we ascribe to land resources is directly related to our ability to use them. In this sense, our economic supplies of land always reflect current levels of technological development. Primitive man had no conception of the worth of iron, coal, and petroleum and accordingly placed little economic value on them. Fertile valleys were valued mostly as places for hunting until mankind learned the arts of agriculture. The flow of new technology that came with the opening of the doors of science during our modern industrial age has had tremendous effects in generating demands for land products and stimulating new land resource developments.

Numerous examples can be cited to illustrate the impact of these changes on land supplies. Development of the steam engine revolutionized industry, increased the demand for coal and other raw materials, and stimulated the growth and expansion of many cities. Railroad-building programs opened many remote areas for commercial cropland, forestry, and other uses. The cotton gin made cotton production economically practicable and prompted the widescale development of new areas for this purpose. Improved oil-drilling techniques paved the way for the rise of the automobile industry, and this development in turn prompted an increased demand for oil and mineral resources and for more and better highways. Substitution of steel girders for solid masonry construction permitted an expansion of the usable space in downtown commercial areas by making it possible for buildings to rise above a six- to ten-story maximum height. Air conditioning has added considerably to the comfort and attractiveness of life in tropical climates and in this way has contributed greatly to the economic growth potential of Sunbelt communities.

Technology, in meeting the problem of scarce supplies, has frequently provided substitutes superior to the products replaced. Use of coal and petroleum products has freed us from much of our earlier dependence on water power. Development of the synthetic dye industry during the late 1800s virtually wiped out the market for the 2 million acres that had been used to produce indigo and madder, the age-old sources for blue and red dyes. In similar fashion, our use of automobiles, trucks, and farm tractors has freed some 80 million acres in the United States that were once used to produce feed for horses and mules.

Institutional Arrangements Affecting Supplies

Institutional factors also have far-reaching effects on land resource supplies because of the roles they play in determining what practices are acceptable and what resources can be used. Economic analysis typically assumes that decisions are made by operators who have almost unlimited freedom of choice concerning how, when, and what resources to use in their production operations. In the real world, operators find this freedom of action restricted by the nature of the rights they hold in property. *Property* involves rights to the possession and use of economic objects. It involves a complex of rules and procedures which determines how land resources are owned, leased, mortgaged, and legally transferred to others. As such, it represents a powerful institutional constraint that guides the operation of the economic system. Operators can use and exclude others from the use of properties they own. When they lack ownership rights, they can buy or lease the rights of others; but without ownership rights they have no legal right of access to land regardless of how desirable or profitable their proposed uses may be.

Most of the more productive and more valuable land resources of the United States are held in private ownership, but private ownership accounts for only 58 percent of the total land area. This means that approximately 42 percent of the nation's surface land is owned, managed, and administered as public property by agencies of the federal, state, and local governments. An additional class of property known as *common property* has acquired particular significance in recent years. (See Chapter 12 for a more detailed discussion of this.) It involves resources such as ocean fisheries, public grazing areas, and wildlife, which were treated as free goods that anyone had a right to use in times past but whose use is now subject in most cases to public regulations.

Concerns about the sharing of property rights and responsibilities frequently lead to economic problems that call for public mediation and regulations.[13] Economic analysis typically assumes that operators bear all the costs and reap all the benefits from their activities. This situation applies to many decisions involving private uses of land resources. But individual actions, as when one plants trees, builds a road, or opens a restaurant, often provide benefits or *positive externalities* in the form of scenery, increased accessibility, or added facilities for others. Residents of the community are usually free riders in the enjoyment of these benefits. Their attitudes are apt to be quite different when the operator's activities result in *negative externalities* (side effects that have negative impacts on others) such as air or water pollution, the generation of chemical and toxic wastes, or the destruction of scenic resources.

Externalities arise because of the joint and multiple products and costs often associated with uses of land resources. Forests, for example, can be used for timber production, recreation, grazing, game production, watershed protection, bird watching, mushroom picking, and wilderness enjoyment. Private operators often find it to their economic advantage to manage forests for a single dominant use such as commercial timber production. But the social values associated with the expected generation of positive externalities and spillover values that bring benefits to others may justify programs that emphasize multiple-use management objectives.

As this example suggests, what is economically best for individual operators and what is socially best for communities often depends on who pays the costs and who

[13]Cf. Robert H. Haveman, "Efficiency and Equity in Natural Resource and Environmental Policy," *American Journal of Agricultural Economics*, 55 (December 1973), 868–78; and A. Allan Schmid, *Property, Power, and Public Choice: An Inquiry into Law and Economics* (New York: Praeger Special Studies, 1978).

gets the benefits. The forest owner who is persuaded to adopt a multiple-use rather than a dominant-use managerial program may be a victim of *disassociation of benefits and costs* if public sector assistance is not supplied to help generate the socially valuable joint products that are of little commercial value to the forest owner.

In similar fashion, private operators may receive inordinate benefits from programs financed by other operators or by the general public. Disassociation of benefits and costs can occur any time operators participate in enterprises in which part or all of the benefits are received by parties other than those who pay the costs or part or all of the costs are borne by parties other than those who receive the benefits. Classic examples occur with the activities of operators who exploit or overuse common property and environmental resources. Individual actions in these cases can easily deplete or undermine the productive capacities and values of important resources. However, in following the course they do, aggressive operators can often realize significant short-term private benefits while others bear the cost of regenerating or renewing the resource base.

Another institutional arrangement with important implications for land use involves the possession and use rights one can hold in the columns of developed air space associated with the upper floors of multi-storied buildings. These rights can be divided and held separately by different owners under *condominium* ownership. With this arrangement, owners of upper-floor apartments and office suites have rights of physical support for their properties, and all owners share common-use rights in certain building and ground facilities together with responsibilities for their care and maintenance. *Time sharing* arrangements are used with some condominium properties to provide seasonal owners with rights to their possession and use during specified time periods each year.

Other examples of institutional arrangements affecting land resource supplies are provided by the activities, programs, and regulations of government. A series of favorable land laws stimulated the early settlement and development of the lands along the western frontier of the United States and Canada. Reclamation projects, protective tariffs, farm and home credit programs, flood control, and urban renewal and highway construction programs are used to promote the development and expansion of certain types of land use. Zoning ordinances, building codes, and development permits are used in many communities to guide and control future land-use development. Still other measures involving restrictive legislation, repressive taxation, acreage controls, and court injunctions can be used to limit particular land-use practices and encourage the shifting of areas to other uses.

IMPORTANCE OF THE FIXED-LOCATION FACTOR

One of the most basic characteristics of land is its fixed location in space. Particular land resources such as mineral deposits, soil, forest products, and houses may be moved about; but land as space remains fixed, immobile, and indestructible. As Alfred Marshall has indicated;

. . . the fundamental attribute of land is its extension. . . . The area of the earth is fixed; the geographic relations in which any particular part of it stands to other parts are fixed. Man has no control over them; they are wholly unaffected by demand; they have no cost of production; there is no supply price at which they can be produced.[14]

This fixed-location characteristic has an important impact on the supply of land available for economic use. Evidence of this is seen in the manner in which it (1) affects human decisions regarding the value and use-capacity of various sites; (2) influences land-utilization practices; (3) facilitates private ownership and ties ownership conditions to the local environment; and (4) affects the legal description of properties.

Economic Location

Location and accessibility play decisive roles in determining the uses for which various tracts of land are suited. Operators usually prefer and ordinarily find it advantageous to concentrate their use efforts on particular sites. Individual decisions vary, but whenever groups of operators compete for the use of certain areas, the sites in question acquire economic significance. Preferences may be associated with natural or man-made advantages such as soil fertility or location near a harbor or water-power site, along a railroad, or near a good market area. But it is the fact that different operators recognize these advantages and are willing to compete for their use that gives them economic value. In this process, human choice is combined with physical location to create situs or economic location.[15]

The concept of *economic location* assumes that some areas enjoy locational advantages over others. This advantage often involves savings in transportation costs and time and stems partly from the fact that the law of diminishing returns makes it both physically impossible and economically impracticable for operators to produce all market goods at points adjacent to a central market. Location advantages also result from the higher productivity and lower production costs associated with particular sites.

Since most sites can accommodate only one use at a time, competition naturally exists for their possession and control. Under free market conditions, this control goes to those uses with the highest prospects for profit and the highest capacity to bid up market values. This bidding and counterbidding process goes on continuously. It assigns economic values to different areas and in so doing identifies some sites as prime spots for certain types of economic activity. In urban areas, these prime locations for commercial activity are often described as "100 percent" sites.

[14] Alfred Marshall, *Principles of Economics*, 8th ed. (New York: Macmillan, 1938), pp. 144–45.
[15] Cf. H. B. Dorau and A. G. Hinman, *Urban Land Economics* (New York: Macmillan, 1928), pp. 167–69; also Richard T. Ely and George S. Wehrwein, *Land Economics* (Madison: University of Wisconsin Press, 1964), p. 65. Originally published by Macmillan, 1940.

Effects of Location on Land Utilization

Fixity of location means that land resources must be used where they are and that enterprises involving movement or shipment of land products are located at sites where the cost, time, and trouble associated with necessary transportation can be minimized. Since production sites are invariably located at varying distances from processing plants and final markets, differential advantages are naturally enjoyed by some sites relative to others for almost every type of land use. These advantages may spring from savings of time and effort required for commuting or transporting products over varying distances or from savings of the monetary costs associated with the movement of people and products. In either case, they frequently favor more intensive use of sites located near markets than sites located farther away and also dictate the most profitable uses that can be made of different sites.

Modern technology has done much to free mankind from the tyranny of space that once greatly restricted the availability of sites for most productive uses and also the ability of various sites to compete for markets. By speeding up the transportation process and reducing transportation costs, these developments have facilitated the productive use of many areas where production would not have been economically practicable. These changes have tempered but not negated the impact of fixed location on land-utilization decisions. As Ely and Wehrwein have observed:

> No matter how much transportation is perfected it can never become instantaneous, effortless, or costless. There will always be a cost of overcoming friction, gravitation, and loss of time in moving goods and people. Farmers near a city will always have some advantage over those farther from the market who are raising the same crops and who have identical transportation facilities. . . . The distance from which people can commute comfortably is still a matter of time, convenience, and costs, complicated many times by the congestion caused by modern transportation.[16]

In addition to its effect on choices concerning land uses, the fixed location factor and the spread-out nature of some uses (farming, forestry, grazing) as compared with the smaller space requirements of most commercial and industrial uses can have important impacts on the economic organization of user enterprises. These factors influence decisions about relative intensity of use, optimum scale of operations, and an operator's ability to set product and input supply prices.

Effects on Ownership and Community Ties

The fixed location of land makes it easy for people to establish and exercise ownership rights over surface units of the earth. But in their exercise of these rights, they must use land where they find it. Some substitution of tracts is always possible in the production process. Operators may have a choice between several prospective

[16] Ely and Wehrwein, *op. cit.*, p. 71.

commercial sites, all of which fit their needs. But they do not have the option of moving low-value land to high-value sites. The fact that surface space may be idle and inexpensive in Nevada or in northern Michigan means very little to operators who compete for commercial space on Manhattan Island or in the Chicago Loop. Operators in these areas must either pay the going price for the space they need or move to other areas where real estate values are lower and where the land may be less suited for their intended purposes.

A singular characteristic of real estate is that it cannot be standardized. Most commercial products such as canned peas or specified automobile parts can be substituted for one another with ease. This degree of standardization is never possible with land, because even when tracts have the same size, shape, and soil characteristics, they always differ in location and have different neighbors and different spatial relationships with other properties.

The fixed-location factor also ties real estate values, uses, and ownership conditions to local environments. Landowners cannot disassociate themselves and their land-use operations from their local communities. They enjoy the same climate as their neighbors and usually the same marketing, social, and governmental services. Ordinarily, they carry on the types of farming for which their areas have some comparative advantage or find that their use of land for residential, commercial, and other purposes complements other land uses carried on in their areas.

When landowners want to sell property, they normally look for a market among local buyers. If the community has a high property tax levy, they suffer the consequences because they lack the option of picking up their land and moving elsewhere. Whole groups of owners are adversely affected when communities are hit by floods, storms, droughts, insect invasions, industrial stagnation, loss of established markets, or possible urban or rural blight.

Legal Description of Properties

In addition to its other effects on land use, the fixed-location factor makes it possible for one to describe the location of real estate holdings in very specific terms. These descriptions facilitate the legal registration of land titles and also the identification, location, and measurement of the surface extent of different properties. Three principal systems of land measurement are used in the United States. These systems involve measurement by metes and bounds, by rectangular survey, and by platting. All three of these systems were devised for the practical purpose of identifying the precise boundaries of land holdings. None of them ties into a truly national or global system of site description, and each lacks the specificity needed for computerized mass analysis and handling of remotely sensed geographic data. More-refined measurement techniques involving descriptions of specific sites in terms of latitude and longitude, state plane coordinates, and Universal Transverse Mercator (UTM) ticks have been developed for these purposes.[17]

---

[17]Cf. David Greenhood, *Mapping* (Chicago: University of Chicago Press, 1964).

In most of the older settled areas of the United States, properties are described by *metes and bounds*—that is, in terms of their location with respect to local landmarks and natural objects such as streams, rock formations, and trees. Metes-and-bounds descriptions ordinarily start with a reference to some carefully identified monument, such as a stone, tree, body of water, building, or piece of pipe driven into the earth. They then indicate the distance and direction to each boundary corner so that a surveyor might accurately locate the property boundaries. The property that later became Washington's Mount Vernon estate, for example, was described by metes and bounds in 1726 as

> . . . a moiete or half of five thousand acres formerly Lay'd Out for Collo Nicholas Spencer and the father of Capt. Lawrence Washington. Bounded as follows Beginning by the River Side at the Mouth of Little Hunting Creek according to the several courses and Meanders thereof nine hundred Eighty and Six Poles to a mark'd A Corner Tree standing on the West side of the South Branch being the main branch of said Hunting Creek. From there by a lyne of Mark'd trees west eighteen Degrees South across a Woods to the Dividing Lyne as formerly made Between Madam Francis Spencer and Captain Lawrence Washington and from hence W by the said Lyne to ye River and with the River and all the Courses and Meanders of the said River to the Mouth of the Creek afor'sd.[18]

Metes-and-bounds descriptions are sometimes quite involved, partly because of their references to local topographic features such as streams or mountain ridges. Despite their cumbersome nature, however, these descriptions fill a purpose as long as the various boundaries and corner monuments can be easily identified. Problems arise when the boundary descriptions are vague, when property owners mentioned in the description have been forgotten, when original monuments have been moved or destroyed, or when properties have been subdivided.[19] When complications of this type develop, the title clearance process becomes both time-consuming and expensive.

A second type of land measurement, the *rectangular-survey system*, applies in the thirty public domain states and in parts of several other states. With this system, principal meridians running north and south and base lines running east and west have been established in various parts of the country. Working out from the intersection of these two lines, additional meridians and parallels have been surveyed at six-mile intervals. The intersection of these lines suggest a huge gridiron, with each six-mile square representing a township of thirty-six square miles. Each township is numbered according to the number of ranges it is from the principal meridian and the number of tiers of townships it is from the baseline. The township with the cross-hatching of section lines in Figure 2-3 can be described as Township 2 North, Range 3 West of the Sixth Principal Meridian.

[18] Taken from court house records of 1726.

[19] The author once asked two surveyors with multistate experience to relate the most unusual metes-and-bounds descriptions they had encountered. One reported a Kentucky description which identified a property boundary as "south two hollers of a hound dog." The second recalled a Connecticut boundary described as "north to the place where the boy killed a bear."

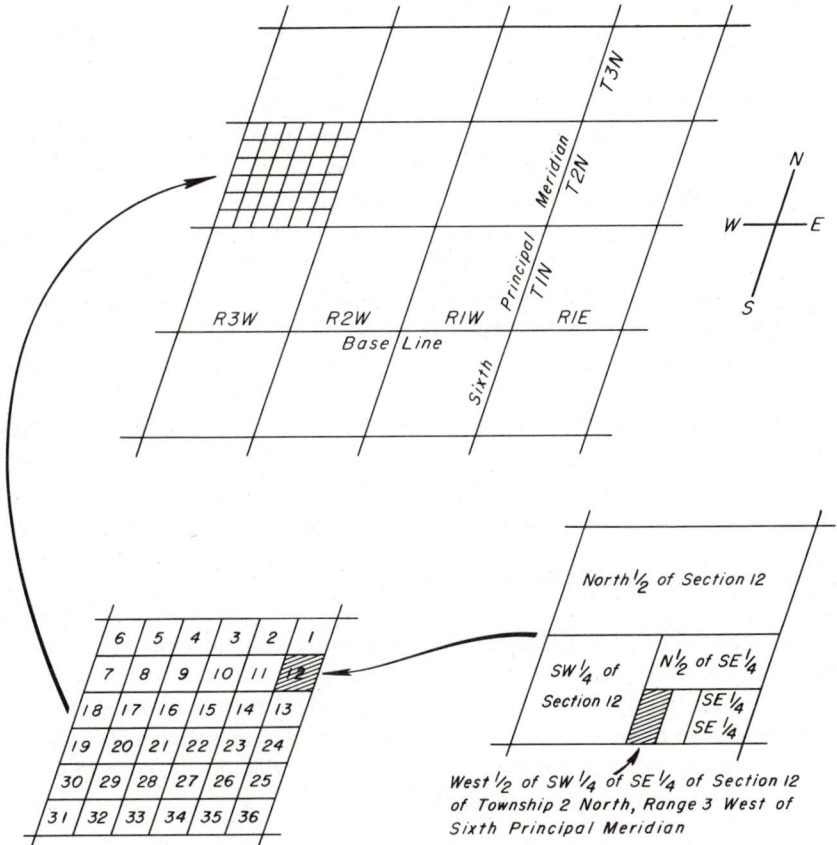

FIGURE 2-3. Legal description of areas under the rectangular-survey system.

Every full township is divided into thirty-six sections. These are a mile square in size and contain 640 acres. In the United States, these sections are numbered from 1 to 36 beginning in the northeast corner as in Figure 2-3. Each section can then be divided into quarters containing 160 acres, and each quarter section can be divided into "forties" and sometimes even smaller units.[20] The 20-acre parcel designated in Figure 2-3 has the following description: W½ of SW¼ of SE¼ of Section 12 of T 2 N, R 3 W of the Sixth Principal Meridian.

The rectangular-survey approach has provided a workable and systematic method of land measurement. In those parts of the United States and Canada where it has been applied, it has had a pronounced effect on the size and shape of rural land holdings. It has also affected the legal boundaries of local units of government and

[20]For other descriptions of the rectangular-survey system, cf. William M. Shenkel, *Modern Real Estate Principles* (Dallas, Tex.: Business Publications, Inc., 1977), chap. 7; Bruce Harwood, *Real Estate Principles*, 2nd ed. (Reston, Va.: Reston, 1980), chap. 2; and William G. Murray, *Farm Appraisal and Valuation*, 5th ed. (Ames: Iowa State University Press, 1969), chap. 4.

the siting of most public roads. Despite its numerous advantages, this approach has some weaknesses: (1) Frequent correction lines are needed to compensate for the curvature of the earth; (2) the presence of streams and bodies of water requires the designation of irregular lots as well as section descriptions; (3) section lines ignore natural property boundaries such as streams or mountain ridges; and (4) complications have resulted from errors in the original surveys.

A third system of land measurement known as *platting* is used in the legal description of most urban and suburban properties. The areas subdivided for these uses are first located according to their metes-and-bounds or rectangular-survey descriptions. Careful surveys are then made, corner monuments are established, and information concerning the size and location of each lot and the areas dedicated for streets and public purposes is recorded on a map that is filed with the proper local authorities. Thereafter, each tract of land may be legally described for tax and other purposes by lot number rather than by metes-and-bounds or rectangular-survey descriptions. A tract might thus be described as Lot 3, Block 5 of the Wellsley subdivision or as Lot 187 in Glencairn subdivision number 4.

THE PRESENT LAND-USE SITUATION

Almost all of the earth's surface has been explored, and most of its visible resources have been appropriated by individuals or governments for some type of use. The uses found in different areas range over a wide spectrum from wilderness and barren regions to densely populated urban centers. For an overall view of how land is used, we look first at the distribution of land uses found on the world scene and then at the land-use situation in the United States.

World Land-Use Picture

Altogether, the world has a total surface area of approximately 197 million square miles, of which 55 million square miles, or 35.7 billion acres, are land surface. When a deduction is made for the ice-covered wastes of Antarctica, it appears that the six major continents have a total surface area of slightly over 32.3 billion acres (Table 2-1). Approximately 3.6 billion acres, or 11.1 percent of this total area, can be classified as arable land. An additional 7.7 billion acres (23.8 percent) are used for meadows and pasture, while 10.1 billion acres (31.3 percent) are used as forestland. This leaves 10.9 billion acres (33.7 percent of the surface land) which are classified by the Food and Agriculture Organization of the United Nations as "built upon, unused but potentially productive, wasteland or other."

Land-use distribution patterns vary considerably by world regions and individual countries. More than a fourth of the surface area of Europe is classified as arable land. This compares with only 5.5 percent of Oceania (Australia, New Zealand, and the Pacific Islands) and only 6.1 percent of Africa. At the same time, 55.3 percent of Oceania is classified as meadows and pasture, while only 16.5 percent of North

TABLE 2-1. Major Land Uses by World Regions and Selected Nations

AREA	TOTAL AREA*	ARABLE CROPLAND	MEADOWS AND PASTURE	FOREST-LAND	BUILT UPON, UNUSED BUT POTENTIALLY PRODUCTIVE WASTELAND, AND OTHER
	Millions of acres		Percentages		
	33,092	11.1	23.8	31.3	33.7
World regions:					
Europe	1,204	29.8	18.3	32.7	19.1
North America	5,239	12.7	16.5	32.0	38.8
South America	4,403	7.2	25.6	53.8	13.4
Asia	6,814	17.0	22.5	20.4	40.1
Africa	7,495	6.1	26.4	23.5	44.0
Oceania	2,103	5.5	55.3	18.0	21.2
USSR	5,536	10.4	16.8	41.3	31.5
Selected nations:					
Australia	1,899	5.8	59.3	14.0	20.8
Brazil	2,103	7.3	18.8	68.0	5.9
Canada	2,465	4.8	2.7	37.0	59.8
China	2,371	10.7	23.6	12.5	53.2
Egypt	247	2.9	–	–	97.1
France	135	34.2	23.6	26.7	15.5
India	812	56.9	4.0	22.7	16.4
Italy	74	42.4	17.5	21.6	18.5
Japan	92	13.2	1.5	67.4	17.9
Mexico	487	12.1	38.7	25.2	23.9
Nigeria	228	33.4	22.9	16.4	27.3
Sweden	111	7.2	1.8	64.2	26.8
United Kingdom	60	29.0	47.5	8.7	14.8
United States	2,314	20.9	26.0	31.2	21.9

*Totals include reported area within boundaries, not total surface area.

Source: Food and Agriculture Organization, *1981 FAO Production Yearbook* (Rome: FAO, 1982), Table 1. The data reported by countries represent the latest official reports from each nation. All totals have been converted from hectares to acres. Percentage distributions by land uses are of total reported surface area. The world has a surface area of 32.3 billion acres.

America is in this classification. Some 53.8 percent of South America as compared with 18.0 percent of Oceania is classified as forestland.

Greater differences exist between individual countries. Canada and Egypt reported less than one-twentieth of their area as arable cropland as compared with more than two-fifths of India and Italy. Some nations reported little meadow or pasture, while 68.0 percent of Australia is used for this purpose. Egypt has very little forestland, while approximately two-thirds of Brazil, Japan, and Sweden are forested. The residual classification accounts for less than one-sixth of France and the United Kingdom but for more than half of the area of Canada, China, and Egypt.

Many reasons can be given for these wide differences between continents and between countries. Relatively favorable climatic, topographic, and soil conditions favor agricultural developments in Western Europe and the United States. In contrast, the use of land for crops is regimented by arid climates in Egypt and Australia, by a short growing season in the northern parts of Canada and the Soviet Union, by mountainous topography in parts of Latin America, and by tropical jungles in countries such as Brazil. Population pressure also has had its effects in India, Italy, and Japan in favoring the terracing and reclamation of areas that might pass as wasteland in other countries.

Use of Land in the United States

Slightly over one-fifth of the surface land area of the United States can be classified as arable cropland, more than one-third as pasture and grazing land, and almost one-third as forest land. These classifications overlap, and each involves a considerable area that is used for more than one purpose. When these multiple-use lands are arbitrarily assigned to a single major use, 20.7 percent can be classified as cropland; 55.3 percent as grazing, forest, and woodlands; 14.0 percent as special-use areas; and 10.0 percent as miscellaneous residual areas. (See Table 2-2).

As these data indicate, 76 percent of the nation's surface land was used for cropland, pasture, grazing, forestry, and other agricultural purposes in 1982. Of this total area, 1,028 million acres (45 percent of the total area) were included in farms.

Agricultural uses. U.S. Department of Agriculture reports indicate that the nation had 469 million acres of cropland within farms in 1982. This total included 383 million acres that were used for crop production plus 21 million acres of idle cropland and 65 million acres of cropland used for pasture. More than 99.8 percent of the cropland is found in the 48 contiguous states where it accounts for 24.6 percent of their total land area.

Around 820 million acres (36 percent of the nation's area) were used for pasture and grazing purposes. This total included 597 million acres of grassland pasture and open-range grazing areas, 158 million acres of grazed farm woodland and grazed forestlands, and the 65 million acres of cropland used as pasture. A substantial additional acreage of harvested cropland also was grazed after the principal crops had been removed. Approximately a third of the area used for grazing involves publicly-owned open range and grazed forestlands located in the western states.

Forestlands accounted for 655 million acres or 29 percent of the area of the United States. Not included in this total is an additional wooded area of more than 100 million acres which was designated as park and recreation lands. Around two-thirds of the acreage classified as forestland is commercial forestland. The remaining third has limited value for timber production though much of it produces fuelwood or is used for grazing, recreation, game cover, and other multiple-purpose uses. Approximately one third of the forestlands are publicly owned. Table 2-2 provides a general picture of the land use situation in the United States in 1982.

TABLE 2-2. Major Uses of Land in the United States, 1982

LAND USE	MILLIONS OF ACRES		PERCENTAGE OF TOTAL AREA
Total land area		2,265	100.0
Area in public ownership		885	39.1
Indian lands		51	2.2
Area in private ownership		1,329	58.7
Area in farms		1,028	45.5
Cropland			
Used for crops	383		16.9
Idle cropland	21		0.9
Used for pasture	65		2.9
Total cropland		469	20.7
Grazing, forest, and woodland			
Grassland pasture and range	597		26.4
Grazed forest and woodland	158		7.0
Nongrazed forest and woodland	497		21.9
Total wooded and grazing		1,252	55.3
Special uses			
Urban areas	47.3		2.1
Rural transportation	26.7		1.2
Rural parks and recreation areas	115.8		5.1
Wildlife refuges	95.2		4.2
Defense and national industrial areas	24.0		1.1
Farmsteads, farm roads	8.0		0.30
Total special areas		317	14.0
Miscellaneous lands*		227	10.0

*Includes areas not inventoried and areas of little use such as marshes, swamps, bare rock, deserts, and tundra.
Source: H. Thomas Frey, and Roger W. Hexem. *Major Uses of Land in the United States: 1982,* U.S. Department of Agriculture Economic Research Service, Agricultural Economics Report No. 535, 1985.

The land-use pattern varied a great deal, however, among regions. While some regions approached the national average, 61 percent of the area of the Corn Belt region and 55 percent of the Northern Plains area were used as cropland. This compared with 8 percent in the Mountain region and less than 1 percent of Alaska. Almost 56 percent of the Mountain region and 58 percent of the Southern Plains region were used for pasture and range as compared with less than 8 percent of the Alaska, Appalachian, Corn Belt, Delta States, Lake States, Northeastern, and Southeastern regions. Almost 65 percent of the Northeastern region and 60 and 58 percent of the Southeast and Appalachia were forested as compared with only 2 percent of the Northern Plains region. Wider variations occurred within regions, as some communities had mixed land-use patterns while others utilized almost all of their land for one or two specific uses.

As the trend data reported in Table 2-3 indicate, the agricultural land-utilization picture has changed considerably since 1880. Total area in farms more than doubled

TABLE 2-3. Agricultural Land-Utilization Trends, United States, 1880-1982*

YEAR	LAND IN FARMS	CROPLAND HARVESTED	OTHER CROPLAND	PASTURE AND GRAZING	FOREST AND WOODLAND	OTHER USES
			Millions of acres			
1880	536	166	22	935	628	153
1890	623	220	28	892	604	160
1900	841	283	36	831	579	175
1910	881	311	36	814	562	181
1920	959	349	53	750	567	185
1930	990	359	54	708	607	176
1940	1,065	321	79	723	602	180
1950	1,161	344	65	700	606	189
1959	1,124	311	71	699	728	455
1969	1,064	286	98	692	723	465
1978	1,030	321	74	663	703	503
1982	1,028	326	54	662	655	544

*Data for years prior to 1959 are for the 48 contiguous states. Alaska and Hawaii are counted for 1959 and later years.

Sources: Adapted from *U.S. Census of Agriculture, 1978* and *1982.* (Washington, D.C.: Government Printing Office), 1980 and 1983; Hugh H. Wooten and James R. Anderson, *Major Uses of Land in the United States*, U.S. Department of Agriculture Information Bulletin No. 168, 1957, pp. 36–37; H. Thomas Frey, *Major Uses of Land in the United States, Preliminary Estimates for 1974*, U.S. Department of Agriculture Economic Research Service, Agricultural Economics Report No. 487, 1982; and H. Thomas Frey and Roger W. Hexem, *Major Uses of Land in the United States: 1982*, U.S. Department of Agriculture Economic Research Service, Agricultural Economics Report No. 535, 1985.

between 1880 and 1950 and has since declined. The area of harvested cropland increased from 166 million acres in 1880 to 359 million acres in 1930 after which it declined and then rose again in response to changes in the nation's agricultural production policies.

Land clearing, drainage, and irrigation operations brought significant acreages of new cropland into use between 1930 and 1982. But these additions were more than balanced by the shifting of large areas of cropland to pasture, woodland, and non-farm uses. This shifting process has been going on for more than 200 years. Numerous areas reached their peaks in cropland use before the turn of the present century. New Hampshire, for example, reached its peak in 1860 when it had 2,367,000 acres of tillable land in farms as compared with a later low of 172,000 acres in 1974. Wooten has estimated that around 150 million acres in the eastern and southern states have been involved in a long-time rotation process in which forestlands have been cleared, cultivated for awhile, then allowed to revert to forestry, only to be recleared later.[21]

Most of the increase in cropland prior to 1920 came at the expense of grazing

[21] Hugh H. Wooten, *Major Uses of Land in the United States*, U.S. Department of Agriculture Technical Bulletin No. 1082, 1953, p. 5.

areas and to a lesser extent from the clearing of forested areas. The total area used for pasture and grazing purposes has fluctuated considerably in recent decades. In the case of forestlands, the reforestation movement and the tendency for some cleared lands to revert to forest use have generally balanced the clearing of forestlands for crop and pasture use. An upward trend in forest and woodland acreage started around 1910 and boosted the total forestland acreage to the 605-million-acre level, around which it remained from 1930 until the forested areas of Alaska and Hawaii were added to the total in 1959 and until a substantial acreage of forestlands were redesignated as park and recreation areas in 1978.

Nonagricultural uses. Tables 2-2 and 2-3 treat the areas used for nonagricultural uses as residual lands. These lands include those used for urban, transportation, park and recreation, and service area uses along with the waistlands not included in farms. Unlike the agricultural areas, few census data have been collected for these lands. Available estimates, however, indicate that the total area used for the principal nonagricultural uses increased from around 40 million in 1920 to 309 million acres in 1982 (See Table 2-4).

The acreage used for urban purposes increased more than three-fold from 15.0 to 47.3 million acres between 1945 and 1982. More than 50 million acres were

TABLE 2-4. Approximate Acreage of Lands in Principal Nonagricultural Land Uses, 1920, 1930, 1945, 1959, 1969, and 1982

TYPES OF USE	1920	1930	1945	1959	1969	1982
	Millions of acres					
Urban areas	10.0	12.0	15.0	27.2	31.0	47.3
Highways and roads	15.0	19.0	19.1	20.5	21.0	21.5
Railroad rights-of-way	4.0	4.0	3.4	3.4	3.2	3.0
Airports	*	*	1.3	1.4	1.8	
Rural park, recreation and wilderness areas	8.0	12.0	17.9	29.7	49.3	115.8
Wildlife areas	*	1.0	4.7	17.2	32.0	95.2
National defense areas	2.0	2.0	24.8	24.4	23.5	24.0
State and federal institutions and industrial lands	*	*	*	*	4.0	*
Total	39.0	50.0	86.2	123.8	165.8	309.0

*Not separately reported.

Sources: Data for 1920 from U.S. Department of Agriculture, *Yearbook of Agriculture, 1923* (Washington, D.C.: Government Printing Office, 1924); for 1930 from National Resources Board, *Land Planning Committee Report, 1934* (Washington, D.C.: Government Printing Office, 1934) for 1945 from Lawrence Ruess, Hugh H. Wooten, and F. J. Marschner, *Inventory of Major Land Uses,* U.S. Department of Agriculture Miscellaneous Publication No. 663, 1948; for 1959 from Hugh H. Wooten, Karl Gertel, and William Pendleton, *Major Uses of Land and Water in the United States: Summary for 1959,* U.S. Department of Agriculture Agricultural Economics Report No. 13, 1962, p. 10; for 1969 and 1978 from H. Thomas Frey and Roger W. Hexem, *Major Uses of Land in the United States: 1982,* U.S. Department of Agriculture Economic Research Service Agricultural Economics Report No. 535, 1985.

added to the holdings designated as rural park, recreation and wilderness areas and wildlife refuges during the 1950 and 1960s; and another 130 million acres, mostly publicly-owned lands in Alaska, were assigned to these uses during the 1970s. Meanwhile, the acreage used for highways, railroad rights-of-way, and airports increased slightly after 1945 to a total of 26.7 million acres in 1982 while the area used for national defense purposes remained relatively constant.

Nationwide data are not available on the acreage used for various purposes within urban areas. Three studies indicate, however, that while land-use patterns vary considerably between cities, around two-fifths of the developed area of the typical average city is used for residential purposes.[22] An additional third of the area is used for railroads, streets, and alleys; a tenth for industrial and commercial purposes; and a sixth for parks, playgrounds, and other public and semipublic uses.

Specific data are lacking on the extent of the areas utilized for several other uses. Millions of acres of forested and nonforested land are now used primarily for recreation purposes. Substantial areas are used for schools and public buildings, cemeteries, golf courses, storage areas, power sites, reservoirs, flowage and watershed protection areas, dumping grounds, and other service purposes. Large areas also are used as sand or gravel pits, stone quarries, open pit mines, and as the service areas for oil and gas wells and underground mines.

— SELECTED READINGS

Ely, **Richard T.**, and **George S. Wehrwein,** *Land Economics* (Madison: University of Wisconsin Press, 1964), chaps. 2 and 3. Originally published by Macmillan, 1940.

Frey, H. Thomas, and **Roger W. Hexem,** *Major Uses of Land in the United States: 1982*, U.S. Department of Agriculture, Agricultural Economics Report No. 535, 1985. *Von Thur*

Renne, Roland R., *Land Economics*, 2nd ed. (New York: Harper & Row Publishers, Inc., 1958), chap. 3.

Zimmerman, Erich W., *World Resources and Industries*, rev. ed. (New York: Harper & Row Publishers, Inc., 1951), chap. 7.

[22]Cf. Harland Bartholemew, *Land Uses in American Cities* (Cambridge, Mass.: Harvard University Press, 1955), Tables 3 and 7. Generally comparable distributions were reported from later studies by John H. Niedercorn and F. R. Hearle ("Recent Land Use Trends in Forty-eight Large American Cities," *Land Economics* 40, February 1964, pp. 105–110) and A. D. Manvel, ("Land Use in 106 Large Cities," The National Commission on Urban Problems, *Three Land Research Studies*, Research Report No. 12, Washington: Government Printing Office, 1968).

3

POPULATION PRESSURE AND THE DEMAND FOR LAND

When we speak of demand for land, we are concerned for the most part with a derived type of demand. People seldom desire real estate for its own sake. They want it because of what it produces. Its products vary from the food we eat, clothes we wear, and materials we use to the scenery we enjoy and the prestige and other satisfactions associated with land ownership.

Our overall demand for land resources finds its roots in the needs and aspirations of the many individuals who make up society. People have different wants and desires. Up to a point, everyone is primarily concerned with the physical need to secure sufficient food and other materials to sustain life. Beyond that, what people want of land is influenced by their knowledge of how land resources can be used, their cultural and educational backgrounds, incomes and spending power, individual tastes and personal goals, and by the changing attitudes that come with advancing age. Each of these factors helps to condition the overall demand picture. But the single most important factor that affects demand for land is population numbers. It is imperative, therefore, that we begin this discussion with some consideration of the problems of population pressure and population growth.

POPULATION TRENDS AND OUTLOOK

Widely different attitudes are held concerning increasing population and the question of what, if anything, should be done about it. In newly settled areas and rapidly developing economies, popular opinion often favors large families and an increasing population. Similar attitudes have won acceptance in areas where religious dogma or social prestige have favored large families or several sons. This attitude has also been fostered in military-minded nations by policies oriented toward the provision of more troops for the armed services.

At the other extreme, the effect of increasing population numbers on the sufficiency of food supplies and maintenance of the quality of local environments can become matters of critical concern. The occurrence of problems of this nature in times past has caused some cultures to accept population control measures such as sex taboos, delayed marriages, birth control, infanticide, and senicide.[1] Moral restraints now prevent popular endorsement of population control measures that involve the taking of human life; but strong support exists for the acceptance of practices that will check the rate of population increase.

Regardless of the position one takes on the controversial question of population control, it must be recognized that increasing population pressure has important impacts on the demand for land and its products. An understanding of these impacts calls for consideration of the population situation in various parts of the world, current trends in population growth, and the future outlook for population increase.

The World Population Picture

Population reports for the premodern era are both fragmentary and incomplete. Most demographers agree that the world's total population probably did not pass the 500-million mark until some time after A.D. 1500. The population problem during this early period was characterized by high birthrates, high mortality rates, and a relatively short average span of life.

Naturally there was some increase in population numbers during this period. Yet almost every upward surge resulting from the high birthrate was counterchecked by the doleful effects of famine, plague, or war.[2] Some 600 famines were recorded in Europe in the first eighteen centuries after Christ. Three times this number have been reported for China. Plagues such as the Black Death of the fourteenth century wiped out more than two-thirds of the population of some countries and probably claimed between a fifth and a fourth of all the lives in Europe. Intermittent warfare also provided a strong check against population increase. The Thirty Years' War, for example, reduced the populations of Bohemia and the German States to between a third and a half of their former numbers.

The overall importance of these checks against population increase may be illustrated by two simple comparisons. The average child born in the United States in 1980 had a life expectancy of 73.8 years. This is double that of children born during the 1780s and probably three times that of babies born in Europe during the

---

[1]Cf. Warren S. Thompson and David T. Lewis, *Population Problems*, 5th ed. (New York: McGraw-Hill, 1965), pp. 238–39.

[2]Cf. T. H. Hollingsworth, "Population Crises in the Past," in Bernard Benjamin, Peter R. Cox, and John Peel, eds., *Resources and Population* (New York: Academic Press, 1973).

Middle Ages. An average of 2.1 children per marriage is now sufficient to maintain the total population. During the plague-ridden years that followed the appearance of the Black Death, population maintenance called for 10 or 11 children per family.[3]

With the onset of the agricultural revolution in Western Europe around 1700, the population situation began to change. Increases in food production provided more sustenance for human life and paved the way for an upward trend in population numbers. This trend was accelerated by the industrial and sanitary-medical revolutions of the next two centuries. Increasing trade and commerce, the settlement of new areas, the trend toward industrialization and higher productivity per worker, and the success with which medical developments lowered mortality rates all opened the way for new increases in total population numbers.

Between 1650 and 1900 the population of the world almost tripled. It then almost tripled again during the next 80 years to a total of 4.4 billion in 1980. Numbers increased at a rate of 1.9 percent annually between 1950 and 1980. Continuation of this rate can give the world a population of more than 6 billion people by the end of the present century and more than 11 billion by 2050.

The general trend in world population growth by continents from 1650 to 1980 is summarized in Table 3-1. As this tabulation suggests, more than half of the world's people live in Asia, while less than a fifth live in Europe (including the USSR) and only a seventh live in the two Americas. The significance of this distribution may be visualized in terms of the individual prospects faced by the thousands of new babies born into the world each day. Only about 1 in 20 will be born in the United States and 1 in 17 in the Soviet Union. About 1 in 4 will be Chinese and 1 in every 7 will be born in India. Only 1 in 4 will be born in a so-called Christian country, and only 1 in 4 will be white.

TABLE 3-1.   World Population Growth, 1650–1980 (in Millions)*

AREA	1650	1750	1850	1900	1930	1950	1960	1970	1980
Africa	100	95	95	120	164	219	275	354	469
America	13	12	59	144	242	330	414	509	615
Asia†	330	479	749	937	1,120	1,380	1,683	2,091	2,558
Europe†	100	140	266	401	534	572	639	704	751
Oceania	2	2	2	6	10	13	16	19	23
World total	545	728	1,171	1,608	2,070	2,513	3,027	3,678	4,415

*Area totals vary from world totals because of rounding.

†The entire population of the Soviet Union is included in the total for Europe for the period since 1900.

Source: Estimates for 1650 through 1900 from A. M. Carr-Saunders, *World Population: Past Growth and Present Trends*, 1936, p. 42. By permission of the Clarendon Press, Oxford. Population estimates for 1930 through 1980 from *Demographic Yearbook, 1980* (New York: Statistical Office of the United Nations, 1982).

[3]Cf. J. C. Russell, "Demographic Patterns in History," *Population Studies*, 1 (March 1948), 393–94.

Distribution of Population in Relation to Land Resources

Most of the world's people live on a small proportion of its surface (see Figure 3-1). They are not distributed between regions or nations in accordance with either land area or population-carrying capacity. Some parts of the world are far better endowed to supply food and other land resources for human use than are others. Some are approaching the limits of their population-carrying capacity (assuming existing levels of technology), while others are capable of supplying high levels of living for somewhat larger populations than they now have.

Population density. The relationship between population numbers and land resources is often stated quantitatively in terms of *population density*, or the number of persons per square mile or other area unit. As Table 3-2 indicates, wide differences exist in the average population densities reported for various countries. Some populous areas, such as Japan, the Netherlands, and Taiwan, have densities of more than 300 people per square kilometer; while others, such as Australia and Canada, have averages of 1.9 and 2.4 persons per square kilometer, respectively.

Population distribution patterns differ as much within as between countries; and it is usually unwise to accept averages on population density as accurate representations of the actual relationships between people and their land resource base. With Egypt, for example, the average density figure of 42 persons per square kilometer suggests a population distribution similar to that found in the United States, which has an average density of 29.3 in its contiguous 48 states. The situation appears differently, however, when it is noted that 98 percent of Egypt's population is concentrated on about 3 percent of its land area and that the Nile Valley has what is probably the highest density of agricultural population in the world.

Comparable differences in population distribution patterns can be found in Algeria, Australia, Canada, and the Soviet Union. Population densities in the United States in 1980 ranged from a high of 902 persons per square mile in Rhode Island to 4.9 in Wyoming and 0.7 in Alaska. On a smaller unit basis, these densities ranged from the 62,317 persons per square mile reported on Manhattan Island to the zero densities found in the uninhabited portions of some western states.

Man-land ratios. A concept known as the *man-land ratio* is often used to express the qualitative relationship between people and their land resource base. This ratio deals with specified total populations or segments of populations and the supplies of particular types of land resources with which they work or on which they depend. As such, it often provides a more realistic measure of the relative dependence of people on the carrying capacity of their resource base than do simple measures of population density.

Man-land ratios can be used to describe the areas of cropland per farmer or the average areas of cropland equivalent per person. Data on average availability of food calories or nutrients per capita provide a general measure of the ability of an area to supply its residents with food resources up to an optimum level of consumption. Indices of average per capita consumption of energy resources provide another man-

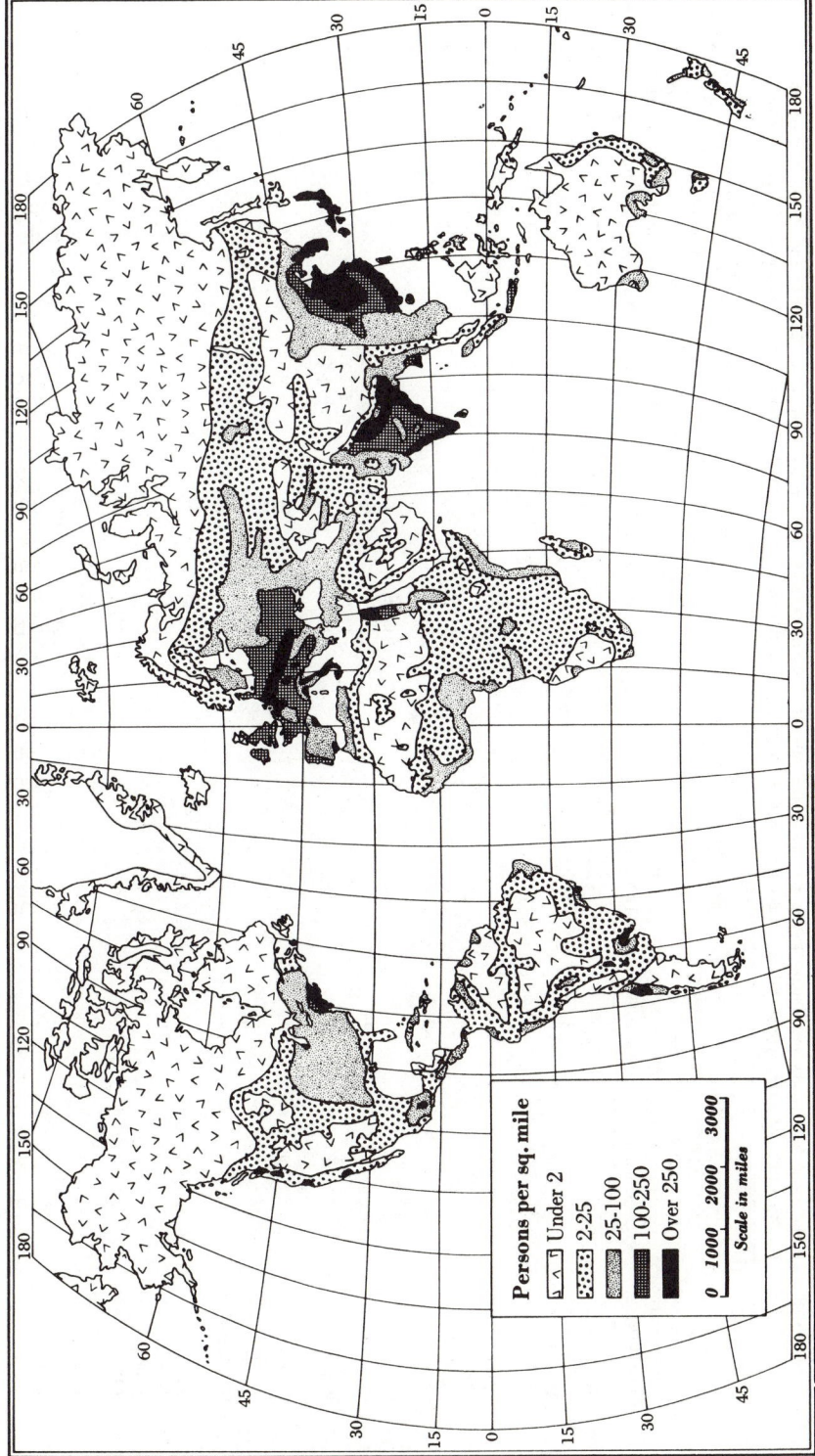

FIGURE 3-1. The distribution of world population. (From Oliver H. Heintzelman and Richard M. Highsmith, Jr., *World Regional Geography*, 4th ed. [Englewood Cliffs, N.J.: Prentice-Hall, 1973], p. 25.

TABLE 3-2. Population, Population per Square Kilometer, Birth and Mortality Rates, Average per Capita Supplies of Calories and Proteins, Average per Capita Consumption of Commercial Energy Resources, and Average per Capita Gross National Product for Forty-eight Selected Countries, 1980.

SELECTED COUNTRIES	MIDYEAR ESTIMATE OF POPULATION 1980 (millions)	POPULATION DENSITY PER SQUARE KILOMETER	CRUDE BIRTHRATE	MORTALITY RATE	PER CAPITA DAILY FOOD SUPPLIES OF Available Calories	PER CAPITA DAILY FOOD SUPPLIES OF Percent of FAO recommended level	PER CAPITA DAILY FOOD SUPPLIES OF Proteins (grams)	PER CAPITA ANNUAL CONSUMPTION OF ENERGY RESOURCES (kgms. of coal equiv.)	PER CAPITA SHARE OF GROSS NATIONAL PRODUCT IN U.S. DOLLARS
Algeria	18.6	8	48	15	2,406	100	62.1	597	1,920
Argentina	27.1	10	26	9	3,386	128	111.9	1,818	2,390
Australia	14.6	2	16	8	3,202	120	116.7	6,032	9,820
Brazil	123.0	14	37	9	2,517	105	59.3	761	2,050
Canada	23.9	2	15	8	3,358	126	97.8	10,241	10,130
Chile	11.1	15	21	7	2,732	91	74.0	1,014	2,160
China (Mainland)	956.8	100	26	9	2,472	105	65.4	619	290
China (Taiwan)	17.9	497	26	5	2,386*	101	62.5	n.a.	2,160
Colombia	27.5	24	34	9	2,363*	n.a.	49.1*	743	1,180
Costa Rica	2.2	44	31	5	2,630	117	60.3	589	1,730
Czechoslovakia	15.3	120	18	12	3,472	141	132.6	6,481	5,820
Denmark	5.1	119	12	10	3,495	130	166.4	5,225	12,950
Egypt	42.0	42	38	11	2,949	117	76.6	473	580
Finland	4.8	14	14	9	3,119	115	136.1	5,135	9,720
France	53.7	98	14	10	3,390	135	145.2	4,351	11,730
Germany, West	61.6	248	9	12	3,537	132	163.4	5,727	13,590
Greece	9.6	73	16	9	3,400*	n.a.	103.2	2,137	4,520
Haiti	5.0	180	43	17	1,882	83	44.5	52	270
Honduras	3.7	33	49	14	2,175	96	52.0	191	560
India	663.6	202	34	15	1,998	90	29.2	191	240
Indonesia	151.9	78	42	17	2,295	96	36.2	220	420
Iran	37.4	23	43	12	2,912	121	60.2	1,246	n.a.
Ireland	3.3	47	21	11	3,766	150	139.6	3,252	4,880
Israel	3.9	186	25	7	3,045	118	108.3	2,368	4,500

Italy	57.0	189	13	9	3,288	145	139.1	3,318	6,480
Japan	116.8	314	15	6	2,916	125	81.6	3,690	9,890
Kenya	16.4	28	51	14	2,055	89	56.8	109	420
Mexico	71.9	36	42	6	2,803	120	72.2	1,770	2,130
Netherlands	14.1	346	13	8	3,372	n.a.	88.5	6,183	11,470
New Zealand	3.1	12	17	8	3,511	n.a.	151.9	3,456	7,090
Nigeria	77.1	83	50	20	2,335	99	53.2	144	1,010
Pakistan	82.4	103	36	12	2,300	100	42.1	218	300
Peru	17.8	14	41	14	2,166	92	56.3	619	930
Philippines	48.4	161	41	10	2,315	102	32.7	328	720
Poland	35.6	114	19	9	3,520	134	128.0	5,590	3,900
Saudi Arabia	8.4	4	50	20	2,233	n.a.	63.0	1,677	11,260
South Africa	29.3	24	38	12	2,827	115	74.0	2,595	2,290
Spain	37.4	74	17	8	3,333	135	131.2	2.530	5,350
Sweden	8.3	18	11	11	3,157	117	147.9	5,269	13,520
Syria	9.0	48	47	14	2,863	115	76.2	1,047	1,340
Thailand	46.5	90	40	11	2,965	104	25.3	371	670
Tunisia	6.4	39	36	13	2,751	115	72.8	636	1,310
Turkey	45.4	58	40	15	2,029	118	64.7	737	1,460
USSR	265.5	12	18	10	3,460*	n.a.	103.4*	5,595	4,550
United Kingdom	55.9	229	12	12	3,316	138	149.1	4,835	7,920
United States	227.6	24	15	9	3,652	138	106.7	10,410	11,360
Venezuela	13.9	15	36	7	2,649	107	70.8	3,375	3,620
Yugoslavia	22.2	87	17	9	3,511	138	104.2	2,049	2,620

Sources: Estimates of midyear populations for 1980 for all countries from United Nations, *Demographic Yearbook, 1980* (New York: Statistical Office of the United Nations, 1982), Table 5. Population densities computed from 1980 population numbers and surface land areas reported in Food and Agriculture Organization, *Production Yearbook, 1980* (Rome: FAO, 1981). Data on average crude birth and mortality rates are from The Environmental Fund, *1980 Population Data Sheet* (Washington: The Environmental Fund, 1980). (Data used with permission of The Environmental Fund.) Data on average per capita daily available supplies of food calories and proteins are from Food and Agriculture Organization, *Production Yearbook, 1981* (Rome: FAO, 1982), Tables 97 and 98. Data marked with an asterisk are from Food and Agriculture Organization, *Food Balance Sheets and per Capita Food Supplies* (Rome: FAO, 1980). Percentages of FAO recommended levels of food calorie consumption calculated from data reported in Food and Agriculture Organization, *The State of Food and Agriculture, 1981* (Rome: FAO, 1982), Annex Table 16. Data on average annual energy consumption per capita are from United Nations, *Yearbook of Energy Statistics, 1980* (New York: United Nations, 1981), Table 6. Data on average annual per capita shares of gross national product for 1980 are from World Bank, *World Bank Atlas, 1981* (Washington: International Bank for Reconstruction and Development, 1981.)

resource ratio indicating the extent to which areas are supporting an industrialized society.

Ability to produce or otherwise provide citizens with adequate diets is a prime requisite for economic growth in the modern world. No nation can expect to make much progress in raising the living standards of its people until it meets this first requirement. Average levels of availability of food calories and other food nutrients can thus be treated as indicators of an area's relative ability to implement goals beyond the provision of minimum levels of living for its citizens.

United Nations agencies have recommended average daily per capita food consumption levels that range from 2,160 calories in warm countries such as Indonesia to 2,710 calories in Finland.[4] Examination of the data on average availability of food for consumption reported in Table 3-2 shows that all of the world's more-developed nations provided more than the recommended calorie levels for food supplies in 1978-1980. They also supplied averages of eighty or more grams of protein per capita per day, whereas several of the less-developed countries supplied less than the recommended food calories and less than fifty grams of protein per capita per day.

Dietary levels range between the lower limit required for survival and an upper limit set by the capacity of the human stomach. No complete index has been developed to provide a comparable measure of the relative amounts of nonfood resources people require in their daily lives. A partial measure of their abilities to move beyond a stage of primary emphasis on food production, however, is provided by the data on average per capita consumption of energy resources. As Table 3-2 indicates, average persons in most of the more industrialized nations used more than 5,000 kilogram-equivalent units of energy in 1980. Average consumption levels of less than one-twentieth of this amount were reported for several developing nations.

Area Differences in Population Growth

In addition to wide differences in man-land ratios, notable differences also exist in the net rates of population growth experienced in different areas. World regional comparisons (see Table 3-1) show that all the continents have experienced rapid rates of population growth since 1900. Some countries, such as Mexico, Nigeria, and Thailand, have experienced spectacular rates of growth since 1950, whereas others, such as France, Germany, and Sweden, enjoyed their periods of major population increase somewhat earlier. Projections based on the recent experiences of the rapid-growth nations can suggest the dire prospect of overpopulation and "standing room only." Analyses of the experiences of those nations that have reduced their rates of population increase, on the other hand, indicate hopeful possibilities for stabilizing population numbers at levels our land resource base can support and provide with opportunities for high levels of life.

[4]Cf. Food and Agricultural Organization, and World Health Organization, *Energy and Protein Requirements*, Joint FAO/WHO Ad Hoc Expert Committee Report (Rome: FAO, 1973).

How much demand for land resources we must prepare for in the future depends in large measure on the answers to the questions of when and at what level will the world's population stabilize. Will this level permit improvements in the living standards of the great mass of the world's people, or will the pressure of population numbers force a vast majority to live at a subsistence level?

Malthusian doctrine. A gloomy answer to this question is suggested by the now-famous Malthusian doctrine of population growth. This doctrine, first introduced by the Reverend Thomas Robert Malthus in 1789, asserts that "population invariably increases where the means of subsistence increases . . . unless prevented by some very powerful and objective checks." Malthus was deeply concerned with human welfare, but he was not optimistic about humanity's future prospects in the world he saw. His fears are illustrated in shocking fashion by his assumption that population tends to increase in a geometric ratio—1, 2, 4, 8, 16—while the means of subsistence tend to increase in arithmetic ratio—1, 2, 3, 4, 5. The experience of the American colonies led Malthus to believe that population would double every twenty-five years were it not held back by lack of subsistence. Over a two-century period, this trend would bring a 256-fold increase in population numbers but only a 9-fold increase in food supplies.

As Professor Warren S. Thompson has observed:

Clearly, this is an impossible situation, but Malthus explained that the fact that man's numbers did not generally show this rapid increase was due to vice and misery, which operated to produce a high death rate. He recognized from the very first, however, two types of checks to population growth—the positive and the preventive. The only form of the latter that he ever contemplated was abstention either temporary or permanent from marriage. He did not believe, however, that this would so reduce the rate of increase of any people that the former—that is, the positive checks—would be rendered inoperative; consequently, he did not believe that schemes for improving human institutions, of themselves, were ever likely to do much to increase man's happiness. It seemed to him that, however good human institutions became, they would serve to mitigate the positive checks for only a short time, until man's numbers grew up to his productive capacity under his new institutions. Then he would suffer from hunger, disease, war, and a host of other ills as he had always done in the past.[5]

Malthus modified some of his views in later editions of his essay. His basic concept of the tendency of mankind to breed to the limit provided by the means of subsistence still stands, however, as one widely accepted answer to the question of how far population will increase.

The Malthusian doctrine provided a reasonable interpretation of population trends 200 years ago and has since demonstrated itself time after time in many

---

[5] By permission from *Population Problems*, by Warren S. Thompson, Copyright 1942. McGraw-Hill Book Co., Inc., pp. 21–22.

parts of the world. But while this doctrine has a certain logic in theory, it has not always held up in practice. Throughout much of the world, application of new technology has made it possible for people to increase their productivity. A true Malthusian would argue that this victory of production over increasing population numbers is temporary and that in years to come population will eventually outrun production and again force the mass of mankind down to a subsistence level. This view is subject to one important fault. Experience shows that where people have been able to raise themselves above subsistence levels, they have also usually shown a willingness to reduce their birthrate to lower levels.

Cultural and social approach. This observation suggests a cultural and social approach to the problem of population growth that stands in marked contrast to the naturalistic approach accepted by Malthus and his followers. It indicates that there is no need for passive or fatalistic acceptance of the inevitability of a Malthusian balance between population numbers and the means of subsistence. Recent history shows that the rising levels of life and reduced mortality rates found in industrialized areas have usually led to a gradual decline in birthrates. This decline normally lags some years behind the drop in mortality rates and thus provides a significant potential for population increase. But as long as birthrates continue to follow a downward trend, there is hope that whole populations can be moved from low Malthusian levels of existence to higher plateaus where better levels of life can be maintained through relative balance between reduced birth and mortality rates.

Part of the reason for this decline in birthrates may be attributed to the fact that fewer births are needed for replacement purposes when mortality rates are low than when they are high. Another important reason involves changing attitudes regarding optimum family size and the place of children in modern society. As families become accustomed to the advantages of modern life, they have little desire to return to subsistence levels of existence. Instead they often willingly accept family limitation as a means of insuring continued high living standards for themselves and their children. This attitude may be prompted to some extent by the fact that children are less and less the economic asset they have been in times past when child labor was more common; but it stems mostly from the desires of modern families to improve rather than lower their levels of living.

The presence of children in many families is now regarded as a deliberate matter of choice and planning. With these families, children are no longer accepted as a matter of fate but instead are treated as an integral part of the family's standard of living. Children are added to the family because they are desired for their own sake and because they add to the satisfactions and fullness of life. Yet from a cost standpoint, they compete with other items that also constitute parts of the family's standard of living. Family size in these cases depends not only on the ability of parents to have children and their willingness to limit family size but also on budgetary considerations and the attitudes of prospective parents concerning the effect children might have on their living standards both now and in the future. With depressed economic conditions, competition between wants for babies and wants for goods may favor family limitation. During more prosperous times, couples may choose family limitation as a means for expanding their opportunities to secure benefits such as more goods and services, better housing, a more pleasing environment, fewer hours of labor, "earlier marriage, somewhat more living children per family, and greater ease in old age."[6]

Classification of nations by population-growth potential. One of the biggest problems demographers face in forecasting world population trends is that of estimating how fast various peoples will shift from a Malthusian balance of high birthrates and high mortality rates to a more rational balance of low birthrates and low mortality rates. Rapid shifts from one level to the other can involve relatively small increases in population numbers. Tremendous increases in population can result, however, when the decline in birthrates lags far behind the reduction in mortality rates.

Demographic forecasts of future population levels are usually based on careful analyses of the age-group profiles of the present population and interpolation of known or assumed mortality, fertility, and immigration rates into the future. In this process, particular consideration is usually given to numbers of prospective mothers and to fertility rates—the number of children under five years of age for every thousand women of child-bearing age (usually all women between fifteen and forty-four) —because these rates provide a better means for predicting trends in births than the birthrate, which measures the number of live births per thousand people of all ages in the population. This approach is meaningful when applied to individual countries. Its application on a worldwide basis, however, is complicated by the wide range of average fertility and mortality rates found in different countries and by widespread differences in cultural attitudes regarding family size.

Analysis of the prospects for world population increase calls for classification of the status of different nations into the five stages depicted in Figure 3–2. The first of these involves a situation in which high mortality rates counterbalance high birthrates and provide a *Malthusian balance* in population numbers. Stage 2 involves a *high-growth-potential* group of nations that have high birthrates but which are experiencing a decline in mortality rates. Stage 3 is characterized by a *transitional-growth* situation in which considerable population growth can be expected before the now declining birthrates drop to the same plateau as the already lowered mortality rates. A fourth stage involves an *incipient-decline* prospect with mortality rates at a low level while birthrates have declined to a point at which population stability could be attained rather easily. Stage 5 involves a *cultural balance* at which population numbers are relatively stable and births and deaths are approximately in balance at the same low level.

At the end of World War II, three-fifths of the world's people were in the high-growth-potential group, one-fifth in the transitional-growth group, and a final fifth in the incipient-decline group.[7] In the period that followed, some incipient-decline areas such as the United States temporarily shifted back to the transitional-growth classification. More recently, the trend toward lower mortality and birthrates has

---

[6] Joseph S. Davis, "Our Changed Population Outlook and Its Significance," *American Economic Review*, 42 (June 1952), 318.

[7] Cf. Frank W. Notestein, "Population—The Long View," *Food for the World* (Chicago: University of Chicago Press, 1945), pp. 42–57.

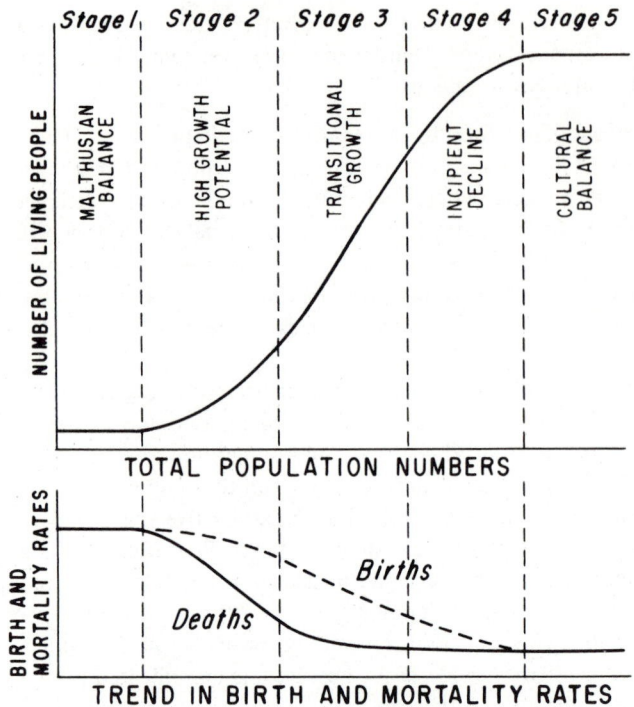

FIGURE 3-2.   The five stages of population growth.

brought the shift of transitional-growth countries such as China, Japan, and the Soviet Union to the incipient-decline group and the shift of numerous high-growth-potential areas to the transitional-group category. By 1980, no nation was operating in the Malthusian balance stage, and only a small number could be counted as high-growth-potential areas. Most developing nations had shifted to the transitional-growth or early incipient-decline stages, while several European nations had moved to the cultural-balance stage.

The relevance of the five stages for future world population growth may be illustrated with the examples reported in Table 3-3. High-growth-potential nations such as Ethiopia, Malawi, Nepal, and Yemen still have high birth and mortality rates. For them, the Malthusian devil is still unchained. Mortality rates can be expected to drop, but tremendous population growth is probable during the time it will take for birthrates to drop to a comparably lower level. Transitional-growth countries such as Honduras, Iraq, Kenya, and Mexico have already enjoyed the benefits of medical services that have helped them to reduce their mortality rates. Population will continue to increase rapidly in these nations until their birthrates also decline. With the incipient-decline nations, the period of rapid population increase is hopefully over. Birthrates have declined to levels slightly above mortality rates, and population stability can be achieved within a short time if attainment of this goal appears desirable.

TABLE 3–3. Comparison of Birth, Mortality, and Population Growth Rates for Four Groups of Nations, 1980

SELECTED NATIONS	BIRTHRATE	MORTALITY RATE	GROWTH RATE
High-growth-potential areas			
Ethiopia	50	25	2.5
Malawi	51	27	2.4
Nepal	44	23	2.1
Yemen	49	26	2.3
Transitional-growth areas			
Honduras	49	14	3.5
Iraq	47	15	3.2
Kenya	51	14	3.7
Mexico	42	6	3.6
Incipient-decline areas			
France	14	10	0.4
Italy	13	9	0.4
USSR	18	10	0.8
United States	15	9	0.7
Cultural-balance areas			
Austria	11	13	−0.2
Belgium	12	12	0.0
Germany, West	9	12	−0.3
Sweden	11	11	0.0
United Kingdom	12	12	0.0

Source: Birth, mortality, and growth rate data from *World Population Estimates, 1980* (Washington, 1980). (Used with permission of The Environmental Fund.)

The extent to which world population numbers will increase in the future depends on the progress individual countries make in shifting from the second, third, and fourth to the fifth stage of population growth. Tremendous progress has been realized in bringing medical knowledge and expertise to most of the world's people and in lowering mortality rates. This trend has been followed by sizable reductions in the birth and fertility rates of most nations and can logically lead to additional reductions in the future. Without further rapid reductions of mortality and birthrates, it is questionable whether the people of some high-growth-potential and transitional-growth nations will ever be able to raise themselves much above a subsistence level of living.

Population projections. Popular interest in the subject of population growth has prompted numerous predictions of future population levels. One prominent planner has suggested a world population of 20 billion by 2050.[8] A British physicist has calculated that the world's population can double every 37 years to a total of 400 billion in 260 years and then rise to an ultimate peak of 60 million billion in

[8] Cf. Constaninos Doxiadis, "Water and Human Environment," *Proceedings of International Conference on Water for Peace* (Washington, D.C.: Government Printing Office, 1967), Vol. 1, p. 36.

another 890 years.[9] Neither of these predictions is apt to come about for the simple reason that people with their ability to reason and plan can be expected to take such actions as are necessary to keep total numbers from rising to these levels.

A study of world population trends prepared for the United Nations in 1981 projects world totals of 7.2, 8.2, and 9.1 billion people by 2025 with the respective assumptions of low, medium, and high variant rates of increase. Table 3-4 lists population projections at the medium variant rate for the world's major regions in 2000 and 2025. These projections signal a strong prospect for continued rapid increase in world population numbers. Most of the expected increase will come in the developing nations of South Asia, Africa, and Latin America. In contrast with the more developed nations where birthrates are already approaching a replacement level, population is increasing at a 2.5 to 3.0 percent annual rate in many of these countries and is not expected to drop to the 1.0 percent level until sometime after 2025.

Population growth in the United States. The United States provides an excellent example of a nation that has experienced tremendous population growth. In 1800 it had a population of 5.3 million. Between then and 1850 this total increased

TABLE 3-4.   **Expected World Population Numbers with a Medium Variant Rate of Increase (in Millions)**

	1980	2000	2025
Less developed regions			
Africa	470	853	1,542
East Asia	1,058	1,346	1,581
South Asia	1,404	2,075	2,819
Latin America	364	566	865
Others	5	8	11
Total	3,301	4,847	6,818
More developed regions			
North America	248	299	343
Western Europe	374	391	391
Eastern Europe and USSR	375	431	486
Japan	117	129	131
Oceania	18	22	25
Total	1,131	1,272	1,377
World total	4,432	6,119	8,195

Source: World Population Prospects as Assessed in 1980, United Nations Department of Economic and Social Affairs Population Studies No. 78 (New York: United Nations, 1981).

[9]Cf. John M. Fremlin, "How Many People Can the World Support?" *New Scientist*, October 29, 1964, pp. 285–87.

more than fourfold to 23.2 million. It then more than tripled to 76.2 million in 1900, almost doubled in the next half century to a total of 151.3 in 1950, and rose to 226.5 million in 1980.

Much of this rapid rate of increase can be credited to the land settlement and development opportunities associated with the nation's bountiful land resource base. It also was facilitated by high birthrates, early progress in the reduction of mortality rates, and immigration policies favorable to settlement. Between 1830 and 1980, 49.1 million immigrants officially entered the country; and, except for the 1930s, this flood of immigrants passed the 1 million mark in every decade between 1841 and 1980.

As Table 3-5 indicates, population increases of between 23 and 26 percent were reported for every decade from 1790 to 1890. This rate of increase declined after 1890 and reached a low of 7.3 percent during the 1930s (see Figure 3-3). The low growth rate of the 1930s can be credited largely to adverse economic conditions which caused major reductions in marriage, birth, and immigration rates. This situation changed with the population explosion that followed World War II. The birthrate rose to 25.7 per thousand people in 1946 and remained at a high level until the middle 1960s. High births in combination with a low mortality rate and a resurgence of immigration brought an increase of 28.0 million people, an 18.5 percent rate of increase, during the 1950s.

The population boom tapered off in the 1960s as the small baby crop of the 1930s came into their child-bearing years. The birthrate dropped below 20 in 1965 and then to a low of 14.7 births per thousand in 1975 and 1976. By 1983, many of the children born in the early post–World War II period were married and ready to have families of their own. Yet the birthrate remained low, largely because of popular acceptance of family-planning measures and widespread concern over the problems of population increase.

Numerous projections have been made of population trends for the United States. Some of these have been high, some low, some on target. Abraham Lincoln was overly optimistic when he extended the eightfold national growth rate of the 1790–1860 period to project a population of 250 million for 1930.[10] Lyndon B. Johnson made a similar error in projecting the growth rate of the 1950s to a national total of 400 million by 2000. In contrast, a National Resource Committee projection in 1935 was unduly pessimistic in assuming that the nation's population would peak at 158 million around 1980 and thereafter decline.

An authoritative projection based on known and expected demographic trends was published by the U.S. Bureau of the Census in 1982.[11] It indicates the nation's

---

[10]Cf. Abraham Lincoln, *Message to Congress of December 3, 1861*, Senate Executive Documents, 37th Cong., 2nd sess., Vol. 1, p. 20.

[11]*Projections of the Population of the United States: 1982 to 2050*, U.S. Bureau of the Census, Current Population Reports Series P-25, No. 992, 1982, p. 6.

TABLE 3-5. Population Trends in the United States, 1790-1980

YEAR	TOTAL POPULATION IN MILLIONS	AMOUNT OF INCREASE IN MILLIONS	PERCENT OF INCREASE	IMMIGRATION BY DECADES* (thousands)	PERCENTAGE DISTRIBUTION OF POPULATION			PERCENTAGE DISTRIBUTION		
					Native white	Foreign-born white	Non-white	Urban	Rural nonfarm	Rural farm
1790	3.9	–	–	–	80.7		19.3	5.1		94.9
1800	5.3	1.4	35.1	–	81.1		18.9	6.1		93.9
1810	7.2	1.9	36.4	–	81.0		19.0	7.3		92.7
1820	9.6	2.4	33.1	–	81.6		18.4	7.2		92.8
1830	12.9	3.2	33.5	152	81.9		18.1	8.8		91.2
1840	17.1	4.2	32.7	599	83.2		16.8	10.8		89.2
1850	23.2	6.1	35.9	1,713	74.6	9.7	15.7	15.3		84.7
1860	31.4	8.3	35.6	2,598	72.6	13.0	14.4	19.8		80.2
1870	38.6	7.1	22.6	2,315	72.8	14.3	12.9	25.7		74.3
1880	50.2	11.6	30.2	2,812	73.5	13.1	13.5	28.2		71.8
1890	62.9	12.8	25.5	5,247	73.0	14.5	12.5	35.1		64.9
1900	76.2	13.2	21.0	3,688	74.5	13.4	12.1	39.7		60.3
1910	92.2	16.0	21.0	8,795	74.4	14.5	11.1	45.7		54.3
1920	106.0	13.8	15.0	5,736	76.7	13.0	10.3	51.2	18.7	29.9
1930	123.2	17.2	16.2	4,107	78.4	11.4	10.2	56.2	19.0	24.8
1940	132.2	9.0	7.3	528	81.2	8.6	10.2	56.5	20.3	23.2
1950	151.3	19.2	14.5	1,035	82.8	6.7	10.5	64.0†	20.7	15.3
1960	179.3	28.0	18.5	2,515	83.4	5.2	11.4	69.9	21.4	8.7
1970	203.2	23.9	13.3	3,322	83.4	4.3	12.3	73.5	21.7	4.8
1980	226.5	23.3	11.5	3,962	83.2		16.8	73.7	23.6	2.7

*Immigration data for 1830 cover period from October 1, 1819, to September 30, 1830; data for 1840 cover period from October 1, 1830, to December 21, 1840; data for 1850 and 1860 cover calendar years; data for 1870 cover January 1, 1860, to June 30, 1870; and data for the decades since then cover periods beginning on July 1 and ending June 30.

†Part of the increase in the urban percentage for 1950 resulted from a change in definition which included much urbanized unincorporated area, previously classed as rural nonfarm, in the urban classification.

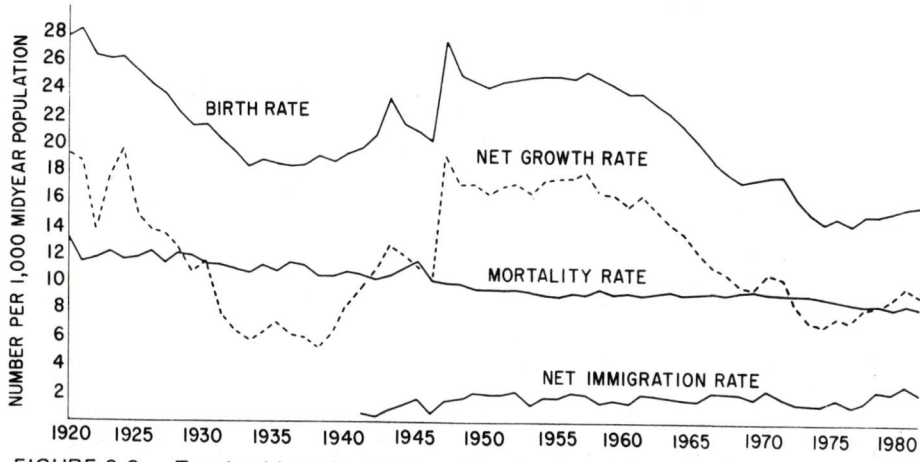

FIGURE 3-3.   Trends with births, deaths, net immigration, and net population growth in the United States, 1920-1982.

probable future population assuming averages of 1.6, 1.9, and 2.3 children per woman. The populations projected at these levels are:

Year	Low series	Middle series	High series
1982	231.8 million	231.8 million	231.8 million
1990	245.5	249.7	254.7
2000	255.6	268.0	282.3
2020	261.6	296.3	341.9
2050	230.8	308.9	428.7

Other assumptions include average life expectancies by 2050 of 76.7, 79.6, and 83.3 with the low, middle, and high series and assumed average annual immigration levels of 250,000, 400,000, and 750,000 with the three respective series. All three series project larger proportionate increases for blacks than for the white segment of the total population.

The nation's population was increasing at approximately the middle series rate in 1983. Continuation of this rate will bring a national total of around 268 million people in 2000. It will also bring a stabilizing of the total number of whites in the population before 2050 and zero population growth sometime after 2050. A reduction in the assumed annual level of 400,000 immigrants can lead to zero population growth at an earlier date, while increases in immigration or in average birth rates can lead to higher rates of population growth.

Changing Characteristics of the Population

While population numbers provide an important index of the demand for land and its products, the nature of the demand for specific types of land often reflects the makeup of the population and changes from one period to another in population characteristics. The significance of changing population characteristics is amply illustrated by the example of the United States. The nation's resource base now supports a larger and more urbanized population than it has in the past. Households are more numerous and smaller in average size. People are more urban-oriented in their work and thinking. A higher proportion of the women work outside their homes. Most adults have had more years of education, are more skilled in their training, enjoy higher real incomes, and are more mobile and communication-minded. The average person is taller, heavier, healthier, and has a longer life expectancy. A significantly larger proportion of the population is in the 65 and older age group; and the population mix, which was dominated by individuals with white, Anglo-Saxon, and Protestant backgrounds until the mid-1900s, now contains increasing proportions of people with Black, Hispanic, and Asian roots.

Urbanization. In 1790 only about one person in twenty in the newly constituted United States lived in an urban community (see Table 3-5). By 1980, three out of every four Americans lived in urban places, while only one in thirty-seven still lived on a farm. As a result of this change, the thinking of most people is no longer as farm-oriented as it has been in times past. Urbanization has divorced the interests of most city folks, particularly those who have not grown up on farms, from intimate use of agricultural land resources. As consumers, they are interested in food, fibers, fuels, building materials, and the prices at which these products retail in their processed form. But they usually tend to associate milk with the sanitary container in which it is purchased or fresh vegetables with the well-lit greens department of the local supermarket rather than with the cow or truck garden from which they came. So long as there is no threat to their supply of food and other land products, they have only an indirect demand for agricultural land. Their direct interests are expressed more in the demand for housing, commercial and industrial developments, parking space, and recreation areas.

Urbanization has also brought changes in the cities. Most cities grew in size between 1940 and 1970, but this growth was associated with considerable suburbanization as thousands of residents, businesses and industries moved to outlying suburban locations. Residents with above average incomes were among the first to leave the central cities. Meanwhile, the proportion of Blacks in the central cities increased from 12.3 percent in 1950 to 22.6 percent in 1980, while the proportion of Hispanics and others not counted as Blacks or whites rose from 1.9 percent in 1970 to 8.2 percent in 1980.

Number and size of households. Table 3-6 shows that the number of households has increased at a significantly faster rate in the United States than the total population since 1890. The number of households more than doubled between 1900 and 1940, and again between then and 1980. The decreasing average size of households has been an important factor contributing to this trend. The average number of people per household dropped from 4.9 in 1890 to 3.8 in 1940 and 2.75 in 1980. These trends have had an important impact on the demand for land resources because each new household unit brings with it an independent demand for

TABLE 3-6. Trend in Number and Size of Households, Median Age of Population,
and Proportion of Population under Fifteen and over Sixty-Five, United States 1890-1980

YEAR	NUMBER OF HOUSEHOLDS (millions)	AVERAGE NUMBER OF PERSONS PER HOUSEHOLD	TOTAL POPULATION (millions)	MEDIAN AGE OF POPULATION	PERCENTAGE OF POPULATION Under 15	Over 65
1890	12.7	4.9	62.9	22.0	33.8	3.9
1900	16.0	4.8	76.0	22.9	34.4	4.1
1910	20.3	4.5	92.0	24.1	32.1	4.3
1920	24.4	4.3	105.7	25.3	31.8	4.7
1930	29.9	4.1	122.8	26.4	29.4	5.4
1940	34.9	3.8	131.7	29.0	25.0	6.8
1950	42.9	3.4	151.3	30.2	26.9	8.1
1960	52.8	3.3	179.3	29.5	31.1	9.2
1970	63.4	3.1	203.2	28.0	28.5	9.9
1980	80.8	2.76	226.5	30.0	22.6	15.7

Sources: U.S. Bureau of the Census, *Statistical Abstract of the United States, 1982-1983* (Washington: Government Printing Office, 1982). Proportions of population under fifteen and over sixty-five calculated from population data reported in U.S. Bureau of the Census, *Historical Statistics of the United States from Colonial Times to 1970* (Washington: Government Printing Office, 1975), Part I, Series A-119-A123, 133.

additional living quarters, furnishings, and other products. A fourfold increase in the number of nonfamily households between 1950 and 1980 (from 4.7 to 20.7 million as compared with from 38.8 to 58.4 million family households) also signaled increasing demands for new types of household facilities.

As in decades past, urban households tend to be adult-oriented and have less than their proportionate share of persons under the age of twenty. Prior to 1950, central cities looked to the nation's farms for much of their continuing demand for potential urban workers. This situation changed with the decline in farm population numbers. They now find their principal reservoir of potential workers in suburban communities where large numbers of urban workers live with their families. A Bureau of the Census survey for 1981 shows that only 33.5 percent of the husband-wife families with children who resided in the nation's standard metropolitan statistical areas (SMSAs) lived in central cities as compared with 66.5 percent in the outlying ring portions of the SMSAs. Some 64.9 percent of the Black families in the SMSAs lived in the central cities as compared with 18.9 percent of the white families.[12]

Changing age distribution of population.   More of the nation's people are living to adulthood and to old age than was the situation earlier. Average life expectancies at birth increased from 59.1 years for males in 1920 to 69.9 years in 1980 and from 54.6 to 77.6 years for females. Reductions in the mortality rate made these gains possible and explain the fact that far larger proportions of the people born in the United States now live through infancy and childhood to become productive workers than was once the case. This trend has added to the size and productivity of the nation's working force. From an economic and social point of view, it has greatly enhanced the returns the nation can expect from its investment in the nurture and training of the young. It also means that almost every baby born now will consume and use land products as a child and as a youth, then as a working adult and parent, and probably later as a retiree.

The median age of the population rose from 16.7 years in 1820 to 30.2 in 1950, then dropped temporarily with the upsurge of new births before climbing to a high of 30.3 in 1981. The proportion of the total population represented by children under 15 dropped from 34.4 percent in 1900 to 22.6 percent in 1980. Meanwhile, the proportion of people in the 65 and over age bracket increased from 4.1 percent in 1900 to 15.7 percent in 1980. This aging of the population has called for new programs to deal with the health, housing, and recreation of older citizens. Retirement benefit programs also have made it possible for larger numbers of retirees to relocate to places where they want to be without regard for employment opportunities.

Age-sex pyramids such as those shown for 1900, 1950, and 1980 in Figure 3–4 can be used to illustrate the changing age distribution of the population. The distribution of males and females by successive age brackets in the United States in 1900 suggested an almost perfect pyramid. New developments, including a tendency

---

[12] Cf. *Household and Family Characteristics, 1981*, U.S. Bureau of the Census, Current Population Report Series P-20, No. 371, 1981, pp. 129-33.

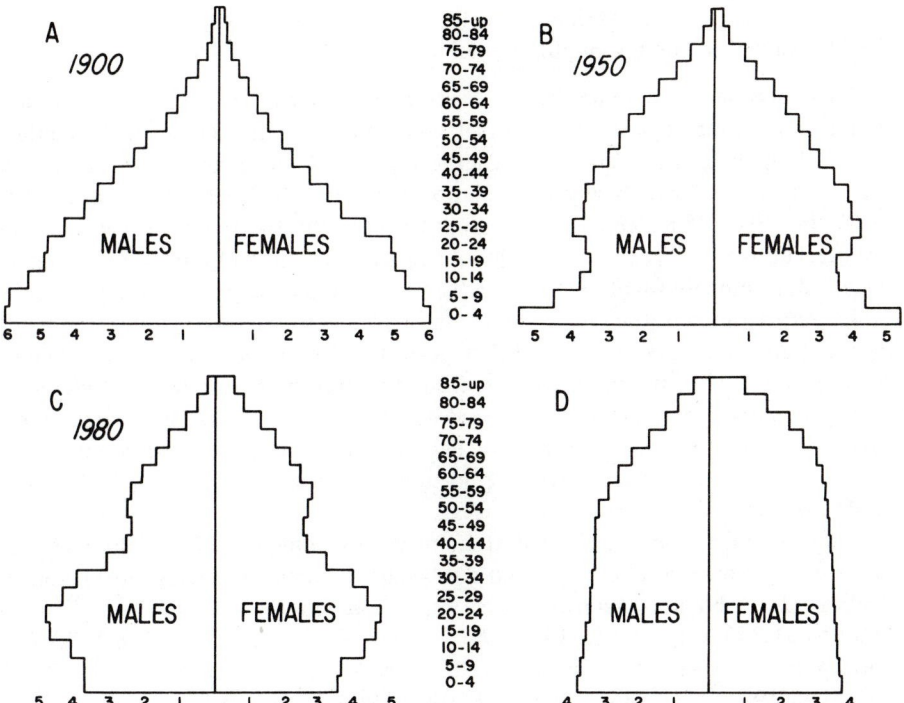

FIGURE 3-4. Comparison of age-sex distributions of the population of the United States for 1900, 1950, 1980 and a projection to 2050, which assumes growth after 1980 at a replacement level and no net immigration.

for more people to live beyond the ages of fifty and sixty-five, an upsurge in birthrates during the 1920s, a short crop of persons born during the 1930s, and the beginning of a "baby boom" at the end of World War II, explain the changed age-sex distribution pictured by the pyramid for 1950. With the passing of time, the continuation of the "baby boom" through the 1950s and early 1960s, and the downturn of birthrates in the later 1960s, the population pyramid for 1980 took the form shown in Figure 3-4C.

These age distribution patterns have a considerable potential for impact on the demand for land and its products, on the nation's work force, and on population trends. The large number of children born in the post–World War II period guarantees a large and growing demand for food, fibers, housing, and other land products for years to come. It also brought a large influx of new workers into the labor force during the 1970s and early 1980s at a time when the nation's economy was not well-geared to meet the challenge of providing all of its workers with employment opportunities. The population bulge provided by the surplus births between 1946 and 1965 can also be a source of future imbalance, for just as the surplus babies born in the 1920s became parents in the 1940s and 1950s and as the smaller number of children born during the 1930s contributed to the decline in birthrates in the 1960s, the large number of young people in the fifteen to thirty-five age bracket in

1980 portends a future avalanche of births if family-planning techniques are not used to control the rate of population increase.

Education and income levels. It was common practice until the 1920s for most young Americans to end their formal education on completion of elementary school. Emphasis is now given to higher levels of educational attainment, and more than half of the nation's adult population have been high school graduates since 1966. In 1980, 85.6 percent of the population in the twenty-five to twenty-nine age bracket had four years of high school and 22.5 percent had four years of college training, as compared with 38.1 and 5.9 percent of the people in this age group in 1940. Approximately the same proportion of young people now graduate from college as graduated from high school fifty years ago. This trend has delayed the entry of large numbers of potential workers into the labor market. It has also increased their productive potential. It has opened new horizons in awakening desires for enjoyment of the material and amenity benefits of modern life. High among these benefits are those associated with opportunities to use, own, and control various land resources and land products.

Emphasis on higher educational attainment and employee skill development has led to rising worker productivity and a substantial increase in average real incomes. Average disposable personal incomes per capita rose from $676 in 1929 and $362 in 1933 to $1,355 in 1950, $1,934 in 1960, $3,348 in 1970, and $8,012 in 1980. Calculated in constant 1972 dollars, these averages rose from $1,887 in 1929 and $1,351 in 1933 to $2,386 in 1950, $2,700 in 1960, $3,620 in 1970, and $4,473 in 1980.[13]

This rise in real incomes has been associated with a reduction in the average workweek from 60.2 hours per week in 1900 and 49.7 hours per week in 1920 to approximately 40 hours per week since 1950.[14] Paid vacations have become a common phenomenon with most jobs since 1946. The combination of higher real incomes with increased leisure has made it possible for most American families to care for their subsistence needs and still have additional income for better housing, household luxuries, a second car, sometimes a second home, vacation trips and travel, college educations, and investments and savings. These additional expenditures and most particularly the demands for better housing, recreation opportunities, and improved highways have had direct impacts on the demand for land resources and land products.

Changes in mobility. Few changes in population characteristics have had more impact on land-use decisions than changes in individual, family, and operator mobility. Land-use decisions were definitely limited in 1900 by transportation constraints.

---

[13] Cf. U.S. Bureau of the Census, *Historical Statistics of the United States: Colonial Times to 1970* (Washington: Government Printing Office, 1975), Part 1, p. 297, col. 197; and U.S. Bureau of the Census, *Statistical Abstract of the United States, 1981* (Washington: Government Printing Office, 1981), p. 423.

[14] Cf. J. Frederic Dewhurst and Associates, *America's Needs and Resources* (New York: Twentieth Century Fund, 1955), p. 40.

The steam engine and the railroad had brought transportation miracles, but industries were tied to sites adjacent to sources of power and next to railroad sidings and wharves that permitted rail or water transportation of raw materials and finished products. Most workers also were tied to housing locations along streetcar lines or within walking distances of their places of employment.

The relaxation of the transportation constraint that came with the development of electric power and widespread acceptance of automobiles and trucks have permitted the location of industries at sites far away from railroads and waterways. It has favored worker acquisition of suburban and rural homesites, the sprawling outward growth of cities, and numerous land-development decisions that have contributed to the decline of central cities.

Americans love their cars and are greatly dependent on them. A 1979 study of 54.1 million workers showed that they travel a mean distance of 11.9 miles to work each day and that 86.4 percent traveled by automobile, truck, or motorcycle, while 5.9 percent used public buses, subways, railroads, or taxis; 5.0 percent walked or road bicycles, and 2.3 percent worked at home.[15] Widescale ownership and use of automobiles, trucks, and other new transportation and communication developments have greatly expanded the areas within which people operate. It has made it practicable for operators to develop and use sites that were previously relatively inaccessible. It has revolutionized the recreation and travel industry. It has prompted demands for highway, street, and airport improvements and also for the production of large quantities of portable motor fuels.

INTERRELATION WITH OTHER DEMAND FACTORS

Every new birth means a new mouth to feed, a new body to clothe and house, and a new person whose health and happiness call for the use of land and land products. It is little wonder that most discussions of the demand for land emphasize the significance of population numbers. But the question of how much land is needed to fill these needs varies with the productivity of land, levels of technological development, operator incomes and purchasing power, and the consumption and buying habits of people. These factors have far-reaching impacts on the actual areas needed for producing food, feed, forest, and mineral products and for providing sites for nonagricultural uses.

Demand for Agricultural Lands

Agricultural land requirements are affected by three principal factors—population numbers, nutritional and other consumption standards, and land productivity. The impact of the last two of these factors makes it inappropriate to generalize directly

[15] *The Journey to Work in the United States, 1979*, U.S. Bureau of the Census, Current Population Reports Special Studies, Series P-23, No. 122, 1979, pp. 2, 5.

from indications of population increase to assumptions of increasing agricultural land requirements. Increases in population ordinarily call for more production. But the question of how much more production depends on food consumption levels and land-use practices and how much land is needed to provide this production depends on trends in crop and livestock yields.

Consumption and nutritional standards. Human stomachs have approximately the same average capacity all over the world. From the standpoint of quantity of food consumed, there are no great differences between people or nations. Pearson and Harper report a world average per capita consumption of 558 pounds of food (dry weight) each year, with the difference between continents ranging from a low of 543 pounds per capita in Asia to a high of 587 pounds per capita in Europe. Pearson and Harper list the average per capita consumption for North America as 576 pounds a year (dry weight). Between 1909 and 1980 the average annual per capita consumption of food products in the United States varied from a low of 1,414 pounds (retail weight equivalent) in 1963 to 1,651 pounds in 1945. Except for the 1942–1947 period, this average never exceeded 1,600 pounds. It has averaged less than 1,450 pounds annually since 1958.[16]

Greater differences exist in the quality of diets. More emphasis is placed on the consumption of livestock products, fruits, and vegetables in the Western nations than in most parts of the world. The people of these countries consume more calories and ordinarily have diets that are better balanced from a nutritional standpoint than is the case with areas where people rely largely on the consumption of cereals and pulses with limited use of animal products.

Dietary tastes and nutritional differences have a considerable effect on the amounts of land needed for food-production purposes. Some types of food provide large outputs of food nutrients from relatively small areas. Others call for the extensive—and sometimes luxurious—use of considerably larger areas. The sugar produced on a sixth of an acre planted to sugar beets can provide more than enough calories to meet the energy equivalent needs of a moderately active man for a year. This same energy requirement can be met by slightly less than an acre used to produce apples, wheat, or beans, while larger acreages are required with livestock products. It takes around 7.5 acres for feed crops plus 2.3 acres of pasture to produce enough dressed beef to provide the annual food-energy equivalent requirements of a moderately active man.[17] This makes beef one of the most expensive foods from the standpoint of land requirements. Among the cereals, wheat is the most expensive. This explains why wheat is often displaced in heavily populated areas by rice, potatoes, or other crops such as oats and rye, which yield more grain per acre, particularly on less fertile lands.

[16] Cf. Frank A. Pearson and Floyd A. Harper, *The World's Hunger* (Ithaca, N.Y.: Cornell University Press, 1945), p. 12; and U.S. Department of Agriculture, *Agricultural Statistics, 1967 and 1981* (Washington: Government Printing Office, 1967 and 1981), p. 693 and p. 552.

[17] Cf. Raymond P. Christensen, *Efficient Use of Food Resources in the United States,* U.S. Department of Agriculture Technical Bulletin No. 963, October 1948.

The heavy emphasis the developed nations place on the consumption of livestock products means that large areas must be used for grain and feed crops, pasture, and range. In areas of high population pressure, this could represent a wasteful practice because crops fed to livestock lose 80 to 90 percent of their food value before they re-emerge as meat, milk, or eggs. Observations of this type suggest that the United States could support three or more times as many people with Asiatic dietary standards as with its present standards. Clark indicates that the average American-type diet requires three times as much land per capita as the average Japanese standard. He calculates that by use of all the world's surface land at the average American level, the world could support 47 billion people. At the Japanese level, this projected carrying capacity rises to 157 billion people.[18] It must be noted, however, that nutritional values are associated with diets rich in livestock products. Also, to a considerable extent animal and crop production are supplementary rather than competitive; and under the climatic and other conditions prevailing in many parts of the world, a system of mixed farming based in part on grazing or grass production gives best results for both crops and livestock products. In some parts of the world animals constitute the main form of draft power, in the absence of which crop production would seriously suffer.[19]

Improvements in dietary standards in nations that suffer from inadequate food supplies call for cultivation of more land and more intensive use of the areas in current production. Capital investments, acceptance of improved production practices, and market and trade developments are often needed for the attainment of this goal. Even with these adjustments, the possible production of additional food, fibers, forests and other materials does not necessarily mean that average consumers will benefit from more nutritious diets or the availability of more products. Products must be provided at prices buyers are willing and able to pay. When prices rise relative to buying power, consumers do without many products and do the best they can with those they can afford.

Changes in land productivity. Although the demand for agricultural land and its products ordinarily increases with population numbers, the amount of land actually needed to meet demand always reflects levels of productivity. Decreasing yields resulting from past exploitation practices can lead to higher land requirements and in some cases may force a tightening of a nation's belt. Increasing productivity, in turn, may make it possible for a nation to supply a growing population with more products per capita without increases in the area used in production.

The United States provides an excellent example of the relationship among productivity trends, land requirements, and total food supplies. Total farm output almost doubled between 1900 and 1950 and then increased another 67 percent by 1980. The nation's farms produced 52 percent more produce from 20.8 percent less

---

[18] Colin Clark, *Population Growth and Land Use* (New York: St. Martin's Press, 1968), pp. 152–53.

[19] Food and Agriculture Organization, *The State of Food and Agriculture: Review and Outlook, 1952* (Rome: FAO, 1952), p. 36.

harvested cropland (75 million acres less) in 1969 than in 1944. Some of this increase resulted from the displacement of horses and mules on farms, but most of it can be credited to higher crop yields and greater efficiency in the conversion of feed into livestock products.

This same situation applies with the production of fibers and other nonfood crops. Only 41 percent as much land was used for cotton production in 1980 as in 1910. Total production was down slightly because of changes in export and per capita demands, but average yields were up from 176 pounds per acre in 1910 to 498 pounds for 1979-1981.

Nonagricultural Land Resource Needs

Higher total population numbers mean greater demand for nonagricultural as well as agricultural lands. With increases in population, nations need millions of new houses, automobiles, television sets, refrigerators, and other consumer goods. They also need schools, factories, shopping centers, streets, and parks. These needs call for building materials, minerals, energy resources, and additional water supplies. They also call for new residential, commercial, and industrial developments and for land areas that can be used for recreation, transportation, and service purposes.

Although overall demand for nonagricultural land is strongly influenced by population trends, it is also conditioned by per capita consumption rates and by our ability to better utilize the resources we have at our disposal. Technological developments have tremendous impacts on demands. Primitive man had little appreciation or use for many of the energy and mineral resources we now hold in great esteem. But as mankind has acquired technical knowhow, new ways have been found to convert resources into goods people can use in their daily lives, and new demands have emerged for many types of raw materials. New inventions such as the steam engine, the automobile, and the electric light bulb provide good examples of new developments that have led to substantial increases in the demand for mineral and energy resources.

Increases in labor productivity have brought higher per capita consumption of nonagricultural products. Low labor productivity frequently means that average workers must spend most of their time producing food and other subsistence items. With increasing productivity, the production of subsistence items can be handled by smaller portions of the total labor force, and increasing numbers of workers are freed to produce the nonsubsistence items and services associated with high levels of life.

Higher worker productivity also means higher real incomes and increased consumer purchasing power. Much of the increased purchasing power associated with increasing labor productivity in this country has gone for nonfood items—for new houses, automobiles, household goods, recreation, and travel. These expenditures, together with our exposure to a continuous barrage of advertising, have stimulated widespread desire for more goods and for still higher living standards. In this sense, we have come to regard yesterday's luxuries as the necessities of today.

Effects of increasing urbanization. Aside from the growing need for mineral and material resources, most of our increasing demand for nonagricultural resources is directly associated with the phenomenon of increasing urbanization. During the past century, the United States has been transformed from a predominantly rural, farm-minded country into an urban-oriented nation. This trend has brought major demands for the conversion of once open areas into sites now used for residential, commercial, industrial, and other urban-associated purposes.

Until recent times, only small proportions of the populations of most countries have lived in cities; most cities were small; and only limited areas were needed for urban developments. Factors such as enclosure within fortified walls, movement of traffic at an ox-cart pace, and the simple requirement that most people walk to and from markets and their places of residence and work, placed high premiums on space. Crowded quarters were typical in most cities. At the height of its grandeur, Rome had many more palaces, temples, public squares, and baths than the average city. Yet for every villa or private house there were twenty-six blocks of teeming *insulae* or apartment houses.[20] Because of their shoddy construction, the *insulae* often had a tendency to crumble and collapse. In one of the first building codes on record, Augustus Caesar decreed a maximum height of around sixty feet for these buildings and specified certain minimum standards in construction. *Insulae* were located along dark, crooked, and narrow streets (sometimes only a few feet across) and frequently stood six or seven stories in height. Inside these structures the bulk of the Roman populace found themselves quartered in small apartments or single rooms with only primitive lighting, heating, and sanitary facilities. Another regulation prohibited vehicular traffic during daylight hours on most Roman streets. This regulation relieved traffic congestion during the day but made it necessary for tradespeople to transport their supplies at night.

The crowded conditions of ancient Rome have been duplicated time after time after time in the older sections of many cities. Transportation developments have freed cities from this need for congestion. In so doing, however, they have contributed to patterns of sprawling outward growth and to new demands for the shifting of rural areas to urban uses. Approximately 870,000 acres of rural lands shifted to urban use each year in the United States between 1959 and 1982.

Urbanization and increasing citizen affluence also are contributing significantly to new demands for recreation, transportation, and service lands. With more leisure time and higher real incomes, urban residents demand opportunities for outdoor recreation seldom dreamed of by average workers in earlier times. Parks, playing fields, golf courses, and other user-oriented facilities are needed in urban areas, while millions of acres outside cities must be set aside and developed for parks and recreation uses. Urbanization also has brought new demands for highway and airport facilities and other service areas. Important among the service areas are those required for the provision of municipal and industrial water supplies. These supplies

---

[20]Cf. Jerome Carcopino, *Daily Life in Ancient Rome* (New Haven, Conn.: Yale University Press, 1940), p. 23.

often call for the construction of reservoirs, canals, and conduits as well as pumping stations and treatment plants, for the development of watershed protection systems, and sometimes for diversion of water resources now used for other purposes. Significant demands also have emerged for the provision of adequate and hazard-free sites for the processing and disposal of urban and industrial wastes.

Mineral and energy needs. Much of the industrial progress the world has experienced in recent decades has been associated with increasing demands for mineral and energy resources. Heavy reliance has been placed on the use of fund resource deposits; and as known deposits have been used, market demands have called for the discovery of new sources of supply and for extension of mining and oil drilling activities to new areas. Increases in demand and in provision costs have brought higher market prices for many mineral and energy resources. These higher prices have provided economic incentives for additional production and also for consumer economizing in the use of these resources.

The prospect of higher energy and mineral prices comes as a cruel blow to millions of poor people and to the economic development aspirations of many developing countries. Both would like to benefit from the opportunities associated with low-cost energy. Without subsidies or miracle answers, both must accept the problems that come with the higher cost of using these resources.

Competition between Land Uses

With continued population growth and the increasing material requirements of modern life, market demand is bound to increase for almost every type of land use. It would be convenient if every use could expand without impinging on lands needed or used for other purposes. Unfortunately, the fixed nature of the world's land resource base makes this an idle dream. Accordingly, each new upward spurt in the demand for land may be expected to contribute further to the competition and possible conflicts between existing and emerging land uses.

Competition typically occurs both between uses and users for the utilization of land. In the competition between uses, decisions must be made concerning whether specific tracts and sites should be used as they were in nature, be treated as forest or grazing areas, be cultivated as farms, or be developed for various urban purposes. Within these uses, decisions must be made concerning whether a field should be utilized as a pasture, grow crops such as wheat or soybeans, or support an orchard. Competition between urban uses calls for decisions that may lead to the use of specific sites for streets, parks, housing, or commercial or industrial purposes. Competition between users involves the activities and bidding of various operators, both owners and tenants, as they seek rights of access and control over the possession and use of land.

Under completely free market conditions, the control of real estate resources normally goes to the highest bidders. Each new demand for land is rated in comparison with all competing demands, and the tract or site is utilized for its highest and best use. With this concept at work, wild lands have shifted to agricultural uses, farmlands to residential subdivisions, and houses have been torn down to make way for commercial developments. The displaced uses in each case have been forced to move to sites with lower use-capacities.

This concept of land always moving to its highest and best use is generally operative; but it does not operate as smoothly and perfectly as it might. Many land uses are not compatible with others; and operators frequently have different combinations of interests and objectives that cause them to assign different weights to the private and social benefits associated with alternative land uses. Conflicts of interests are a frequent result. These conflicts may involve the rival interests of two or more operators who have different plans for the same tract. The one who gets the site will be able to carry out his or her plan, while the losers must look elsewhere for suitable sites of equivalent use-capacity, possibly move to other communities, or operate their present holdings at less than optimum scale.

Conflicts of interest may also involve conflicts between the money-making opportunities of individual operators and the broader interests of society. Examples occur when one wants to open a mining operation in a natural area preserve, clear-cut a forest on a scenic mountainside, develop and sell building sites in a prime agricultural area, or operate a junk yard or a porno shop in a residential neighborhood. A not infrequent variation of this problem occurs when groups of citizens fight against the location of publicly needed but locally unwanted land uses in their communities. Few people want to live near waste disposal sites, prisons, oil refineries, or other similar facilities; but it is obvious that places must be found for the siting of these facilities.

Most of the adjustments necessitated by competition between new and existing demands for land can be handled by the normal operations of the marketplace. Exploitive practices and consumptive use encroachments on lands that could be held for better long-run production purposes are often tolerated, particularly if communities have plentiful supplies of land. Over the longer run and as continuing competition affects larger and larger proportions of available land resource supplies, however, communities will bear more of the ultimate responsibility for deciding how limited supplies of land shall be used. Their willingness to assume this responsibility will be dictated in part by a need for positive measures to minimize and mediate private and public conflicts of interest concerning land use. It will also be affected by a growing awareness of the effects that questionable resource-use practices have on the environment, on costs that must be borne by society, and on needed uses of land resources.

— SELECTED READINGS

Bogue, Donald J., *Principles of Demography* (New York: John Wiley, 1969).

Brown, Lester R., *World Population Trends: Signs of Hope, Signs of Stress* (Washington, D.C.: Worldwatch Institute, 1976).

Clark, Colin, *Population Growth and Land Use* (New York: St. Martin's Press, 1968).

Davis, Joseph S., "Our Changed Population Outlook," *American Economic Review*, 42 (June 1952).

Johnson, V. Webster, and Raleigh Barlowe, *Land Problems and Policies* (New York: McGraw-Hill, 1954), chap. 8.

Ratcliff, Richard U., *Urban Land Economics* (New York: McGraw-Hill, 1949), chaps. 3–5.

Robinson, Harry, *Population and Resources* (New York: St. Martin's Press, 1981).

Thompson, Warren S., and David T. Lewis, *Population Problems*, 5th ed. (New York: McGraw-Hill, 1965).

United Nations, *Demographic Yearbook*, current years (New York: United Nations).

——, *World Population Prospects as Assessed in 1980*, United Nations Department of International Economic and Social Affairs, Population Studies No. 78 (New York: United Nations, 1981).

World Population Growth and Responses 1965–1975: A Decade of Global Action (Washington, D.C.: Population Reference Bureau, 1976).

4

LAND RESOURCE REQUIREMENTS

Questions concerning the adequacy of our land resource supply necessarily go beyond the present situation. The world obviously has sufficient land and productive capacity to do the job it is doing in providing living space, food, fibers, and other land products for its present population. But is its resource base adequate to provide high levels of life for all the world's people? Can it care for the needs of an expanding population? And can it meet the new demands that will come with continued technological advance?

These questions can be answered only in a general way. We have far too little information about the expected future land requirements of different peoples to more than guess at the world's future needs. Considerable thought has been given to this subject in the United States, however; and projections of the future land resource requirements of this country are presented here to illustrate the nature of the problem. This presentation is followed with a discussion of the prospects all countries face in meeting their future needs.

LAND RESOURCE REQUIREMENTS
IN THE UNITED STATES

Wide differences exist between the present and future land resource needs of different peoples. Average per capita needs may be relatively simple and change slowly in tradition-oriented societies. With the acceptance of new marketing and technological developments, most consumers in areas such as the United States, however, demand diets rich in fruits, vegetables, and livestock products. They demand more and better housing and use more forest products, more mineral and energy resources, more water, and more land for recreation and transportation purposes than the people of less developed areas. With the present rate of population increase, it is realistic to talk of setting additional plates at the nation's table and building additional houses for millions of new citizens. Increasing affluence has brought higher levels of per capita food consumption and demands for better housing, more opportunities for outdoor recreation, and greater awareness of environmental concerns. This combination of a growing population and rising consumption levels poses a real challenge for those who work with future resource needs. It calls for more productive use of current stocks of land resources and for additions to the supply of land resources already in use.

Realistic projections of the land resource requirements associated with any particular use and any consuming group calls for careful consideration of several variables. Regardless of whether one is concerned with land requirements for food, fibers, minerals, water, housing, recreation, or some other land product, one's calculations involve assumptions about (1) the size of the population to be served, (2) levels of consumption or resource use, (3) production trends and costs, (4) allowances for imports and exports, and (5) possible adjustments in production and demand. Other often unstated assumptions such as continuation of favorable climatic and weather conditions, continued availability of low-cost energy resources, peace with other nations, and an absence of catastrophic events also underlie most projections.

Crop and Pasture Land Requirements

History shows that people have always been more or less concerned with the adequacy of their food supplies. Frequent experience with the pangs of involuntary hunger has often made them food-conscious—a condition well expressed by the first request in the Lord's Prayer: "Give us this day our daily bread." This concern over whether we have food enough has been tempered somewhat in the Western world during modern times. But it still remains a chronic problem of persistent importance in many parts of the world. It can be likened in some ways to a smoldering volcano. On many occasions, it lies almost dormant. Sometimes its fires flare up in localized areas. And at times, particularly during and after wars, it erupts in almost epidemic proportions as a problem of worldwide concern.

The problem of food supplies has never been as serious in the United States as in many countries. The nation has faced a bigger challenge in this century in dealing with food surpluses than with shortages. Its experiences during and immediately after World Wars I and II and again during the 1970s in supplying food for deficit-producing nations have had their sobering effect, however, and have given rise to numerous questions concerning long-run crop and pasture land needs.

Numerous estimates have been made of the approximate cropland acreage the United States will need for future production purposes. An example is provided by the crude illustrative projection of cropland requirements reported in Table 4-1.

TABLE 4-1. Illustrative Projection of Cropland Requirements in the United States for a Future Population of 275 Million Persons (All Figures in Millions)

ITEM	1947–49	1967–69	LEVELS OF EXPORTS*		
			Low	1980	High
Average total population	147	223	275	275	275
Cropland requirements:[†]					
Cropland for domestic consumption of food and nonfood products[‡]	286	222	274	274	274
Export crops	46	117	60	138	200
Feed for horses and mules [§]	24	4	5	5	5
Total cropland requirements	356	343	339	417	479
Total cropland requirements assuming a 15 percent increase in productivity			303	381	443
Total cropland requirements, assuming a 30 percent increase in productivity			275	353	415

*The 138 million acres used to produce export crops in 1980 is treated here as the medium level. The low level of 60 million represents less than half this total, whereas the high level of 200 million assumes a substantial increase in export activity.

[†] Cropland requirements are measured in acres.

[‡] This acreage represents the area reported by the U.S. Department of Agriculture as planted for the production of food and nonfood crops for domestic consumption.

[§] Around 91 million acres were needed in 1915 to produce feed for the 26 million horses and mules on farms plus horses found in nonfarm areas. The U.S. Census of Agriculture has not counted numbers of horses and mules since 1964, but it is assumed that their numbers are increasing.

This example assumes a future population of 275 million—a total that the nation will probably reach shortly after 2000. As a first step in the computation, the average acreage of cropland used per person in the 1977-1979 period is used to indicate the cropland area that would be needed to support the assumed population at 1977-1979 food and fiber average consumption levels. Higher or lower levels of consumption could, of course, be assumed. Adjustments must then be made to provide for the acreages that will be needed to supply agricultural products for export and feed for horses and mules. Adjustments could also be made for projected changes in acreages planted to soil-improvement crops and for the cropland equivalent of pasture and grazing lands but are omitted here as this projection deals only with cropland requirements.

The acreage equivalent used to produce agricultural products for export more than doubled between 1947-1949 and 1980 and by some estimates could call for the harvest from 200 million acres by 2000.[1] Table 4-1 recognizes the significance

[1] Patrick M. O'Brien, "Global Prospects for Agriculture," *Agriculture Food Policy Review*, U.S. Department of Agriculture Economics and Statistical Service AFPR-4, 1981, reports that exports, which had accounted for only 13 percent of the value of the nation's farm marketings in 1960 and 25 percent in 1980, could rise to 43 percent by 2000.

of this variable and assumes three alternative export levels: a low level requiring less than half of the 138 million acres used for this purpose in 1980, the 1980 level, and a high level requiring the use of 200 million acres of cropland for exports.

Table 4-1 shows that provision of sufficient food and nonfood crops to supply a population of 275 million people at 1977–1979 consumption levels will call for the use of 274 million acres if production and yields remain at their 1977–1979 levels. When the areas needed to supply products for export and feed for horses and mules are added, it appears that the nation could provide for its domestic consumption needs with the same area used in 1977–1979 if it cuts back to the low level of exports. Maintenance or increasing the 1980 level of exports would call for bringing additional cropland into use.

This conclusion assumes no changes in production resulting from improved technology. In practice, substantial increases have been realized in recent decades and additional increases can be anticipated in the future (see Table 4-2). The U.S. Department of Agriculture's index of farm productivity increased from 52 in 1930 (1967 = 100) to 73 in 1950 and to 134 in 1980. Its index of total farm output doubled between 1940 and 1980 (see Figure 4-1), while its indices of total crop production and total livestock product production rose 100 and 53 percent, respectively, in the thirty-one years between 1950 and 1981.

Naturally, it is difficult to predict the net effect new technology and improved production practices will have on crop and livestock yields in the future. However, if productivity increases 15 percent on the acreage used to produce for domestic consumption, the nation would be able to handle its total cropland needs plus a 1980 level of exports with 381 million acres as compared with the 343 million acres used for this purpose in 1977–1979. With a 30 percent increase in productivity, the nation can maintain the 1980 level of exports with 353 million acres of cropland, but an increase in exports would call for bringing additional land into cropland use.

This simple projection of future cropland requirements is presented for illustra-

TABLE 4-2. Average Yields per Acre of Selected Crops in the United States, 1937–1939, 1979–1981, with Two Projections to 2000

CROP	1937–1939	1979–1981	PROJECTIONS TO 2000 CARD	RCA
Corn, bushels	28.6	103.5	110.4	107
Cotton, pounds	253	498.0	815	525
Hay, tons	1.28	2.33	4.43	3.2
Soybeans, bushels	19.7	29.6	40.6	36
Wheat, bushels	13.6	34.0	40.4	35

Sources: U.S. Department of Agriculture, *Agricultural Statistics, 1969* and *1982* (Washington, D.C.: USDA, 1970 and 1983); Anton D. Meister, Earl O. Heady, Kenneth J. Nicol, and Roger W. Strohbehn, *U.S. Agricultural Production in Relation to Alternative Water, Environmental, and Export Policies,* Iowa State University Center for Agricultural and Rural Development CARD Report 65, 1976, pp. 152–53; and U.S. Department of Agriculture, *Soil, Water, and Related Resources of the United States: Analysis of Resource Trends,* RCA 1980 Appraisal, Part II, August 1981, p. 68.

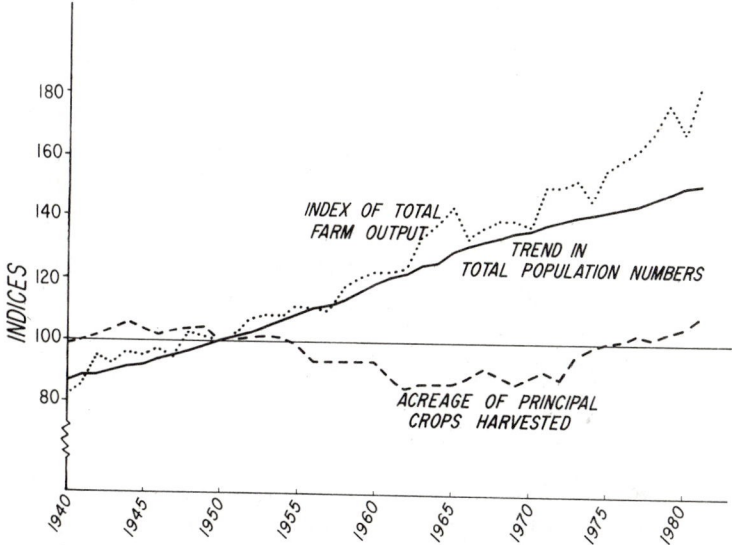

FIGURE 4-1. Comparison of trends of indices of total population growth, total farm output, and acreage of principal crops harvested, United States, 1940–1981 (1950 = 100).

tive purposes only. As an indicator of actual future cropland requirements, its value is definitely limited by its implicit assumption that agricultural lands will be distributed between specific crop and livestock uses in approximately the same way in the future as during the 1977-1979 period. Important shifts will undoubtedly take place as the demand pattern for agricultural products responds to changes in the nation's eating and other consumption habits and as improved production practices affect some types of land use more than others. Changing trends in food consumption provide a dramatic example of a factor that must be considered in these models. The average per capita annual consumption of meat, poultry, and fish in the United States rose from 177.9 pounds in 1909 to 237 pounds in 1981. Meanwhile, average consumption of potatoes and sweet potatoes declined from 211.9 pounds in 1909 to 79.4 pounds in 1981. Average consumption of flour and cereal products dropped from 295 pounds in 1910 to 140 pounds in 1973 and then rose to 151 pounds in 1981.[2]

A more sophisticated analysis would call for setting up aggregate models that would give detailed consideration to the impact of these and other factors on cropland requirements. Important among the other factors are adjustments for changes in incomes; buying patterns; availability of additional croplands; losses associated with the shifting of production areas to other land uses; effects of soil erosion and conservation practices on land productivity; and availability and cost of energy, fertilizer, water, and other necessary inputs.

[2]Cf. U.S. Department of Agriculture, *Agricultural Statistics, 1972* and *1982* (Washington, D.C.: USDA, 1972 and 1983), p. 690 and p. 518.

Examples of these more comprehensive analyses are provided by two recent studies. An analysis conducted by Resources for the Future projects a need for 417 million acres of harvested cropland in 2010.[3] A linear programming model developed by the U.S. Department of Agriculture projects cropland requirement levels of 306, 390, and 457 million acres in 2030 with the assumptions of production for domestic use only, domestic use plus a 1975-1977 average level of exports, and domestic use plus an increasing level of exports, respectively.[4] Assuming peacetime conditions, normal weather, and no major catastrophes, both models indicate that the nation faces no imminent danger of running out of food. But both also recognize that important production areas are shifting out of agriculture to other uses, that continued soil erosion poses a threat to future production capacity, and that provision of sufficient cropland to meet the 2010 and 2030 projections will require bringing additional land areas into agricultural use.

Forestland and Range Land Requirements

Because of the long-term nature of the forest production process, decisions affecting future production must be made years in advance. The quantity and quality of the areas reserved for forest use are influenced by current estimates of future supply and demand conditions. On the demand side, every forestland owner speculates with future price levels. Prospects for success are affected by possible changes in the need for forest products, by the threat of increasing competition with substitute materials, and by possible developments of new uses and new markets for wood products. On the supply side, owners must consider prospective forest yields, probable production costs, and the effects of imports and exports on their competitive position. They must also respond to pressures to shift their forestland to other alternative uses and to a variety of managerial and risk factors that affect the willingness of operators to engage in long-term forestry.

The United States used approximately 13.3 billion cubic feet of roundwood and 62.5 billion board feet of sawtimber in 1976. Assuming rising relative prices for timber products, it is expected that the nation will have a market demand for 22.7 billion cubic feet of roundwood and 94.2 billion board feet of sawtimber in 2000. More than three-fourths of these products involved softwoods in 1976, and will come from softwoods rather than hardwoods in 2000. It is also expected that 728 million acres in 2000 and 718 million in 2030 will be classified as forested as compared with 737 million acres in 1977.[5]

[3] Cf. Pierre R. Crosson and Sterling Brubaker, *Resource and Environmental Effects of U.S. Agriculture* (Washington, D.C.: Resources for the Future, 1982), pp. 56–66.

[4] Cf. U.S. Department of Agriculture, *Soil, Water, and Related Resources of the United States: Analysis of Resource Trends*, RCA 1980 Appraisal, Part II, August, 1981, p. 70; also Robert F. Boxley, "Competition for Agricultural Land to the Year 2000," in *Agricultural Land Availability: Papers on Supply and Demand for Agricultural Lands in the United States*, Senate Committee on Agriculture, Nutrition and Forestry, 97th Cong., 1st sess., Committee Print, 1981.

[5] Cf. U.S. Forest Service, *An Assessment of the Forest and Range Land Situation in the United States*, FS–345 (Washington, D.C.: U.S. Department of Agriculture, 1980), pp. 32 and 337–38.

Sufficient young and maturing timber is now growing in the United States to meet most of the projected future needs. Market demand, however, is expected to exceed domestic production if timber prices continue at their 1950–1975 price trend levels. Higher prices can depress demand and thereby bring supply and demand equilibrium. But increases in production can also be secured by reservation of sufficient areas for forest culture, measures to provide protection against long-term uncertainties, and the devising of market incentives.

Significant increases in forest product production can be secured with the wide-spread adoption of improved management practice. Clawson has suggested that the nation can meet its future forest production needs with the application of intensive management practices on only 200 million acres of its highest quality commercial forestlands.[6] But several factors are working against the attainment of this objective. Intensive management in many cases could call for large investments, as much of the nation's forestland is sparsely stocked or is stocked with low-value species. Furthermore, intensive management can be controversial because programs designed to increase commercial timber production are often incompatible with competing multiple-use goals such as the promotion of recreation, game, grazing, watershed protection and the preservation of scenic resources. Also, much of the privately owned forestland is held in small, hard-to-manage units and is subject to indifferent management, because owners show as much or more interest in diverting their lands to service areas, farming, recreation, or residential uses as in retaining them for forest production.

In addition to its croplands and forestlands, the nation had 587 million acres of grassland pasture and range land in 1978. These lands are not cultivated, have limited tree cover, are often mountainous or rough, and usually receive too little rainfall to be used for crop culture. They are frequently viewed as residual areas that can be diverted for reclamation projects, recreational, or other uses. Yet they do provide considerable grass and browse for livestock, sheep, and wildlife. U.S. Forest Service data indicate that the demand for average months of animal units of grazing can be expected to increase 35 percent betweeen 1976 and 2000 and 41 percent between 1976 and 2030. With this expectation, grazing land totals (including cropland used for pasture and forestland grazed) of 796 and 764 million acres are projected for 2000 and 2030, respectively, as compared with an estimated 820 million acres used for these purposes in 1977.[7]

Housing and Residential Site Requirements

Recent population trends provide an excellent indication of the nation's future housing needs. Calculations concerning the children born since 1950 show that the nation can expect a significant increase in its number of households. Every child represents a potential demand for at least one-half of a residential unit. Several will

---

[6]Marion Clawson, ed., *Forest Policy for the Future* (Washington, D.C.: Resources for the Future, 1974), pp. 179–80.

[7]U.S. Forest Service, *op. cit.*, pp. 32, 52.

be financially able to own or lease additional units which they will use on an occasional or seasonal basis. Pitkin and Mesnick have projected the following increases in housing demand by types of units needed. All figures are given in the thousands.[8]

	1980	1990	2000
1-family units	53,913	63,558	70,199
2- to 4-family units	9,870	11,747	12,327
5-family or more units	12,029	14,796	16,057
Mobile units	4,780	7,512	9,390
Total households	80,592	97,613	107,973

Another projection of housing needs is reported in Table 4-3. This tabulation shows that the United States had 88.2 million residential units, 80.1 million of them occupied, in 1980. This total will increase considerably in the decades ahead. Average annual demand for new housing is expected to reach 2.59 million and 2.24 million during the 1980s and 1990s. Most of the expected demand will call for the construction of new single-family residential units, but there will also be a substantial need for multifamily units and additions to the stock of mobile homes. More unoccupied units also are expected by 2000. Some of these will be truly vacant, but most will be second homes used for vacation and weekend recreation purposes.

Several factors can affect the attainment of these housing objectives. A slowing of the rate of population increase could reduce the need for new housing construction, and population stability could call for new construction for replacement purposes only. Because of their impact on effective demand, business conditions will also have a strong determining influence on the amount of new housing that will be needed. Full-employment, high disposable incomes, low interest rates, and readily available credit will bring demands for considerable new housing; the reverse situation can have a dampening effect.

Many of the new housing units constructed in the next decade will be built on vacant lands now within established city limits. Thousands of other units will come with the redevelopment of older urban areas. Most new construction, however, will likely come in suburban areas, currently unincorporated subdivisions, and in areas now classified as open country. This trend will call for the shifting of extensive areas of rural land to urban and suburban uses. An average of around 870,000 acres of rural land a year was taken for new housing and other urban purposes during the 1960s and 1970s. Resources for the Future, Inc., has projected a further taking of 25 to 30 million acres of cropland for urban uses between 1977 and 2010, and the U.S. Department of Agriculture has projected a loss of 44 million acres of cropland for this and other purposes by 2030.[9]

---

[8] Cf. John Pitkin and George Mesnick, *Projection of Housing Consumption in the United States, 1980 to 2000, by a Cohort Method*, U.S. Department of Housing and Urban Development, Annual Housing Survey Studies No. 9, 1980.

[9] Cf. Crosson and Brubaker, *op cit.*, p. 58; and RCA 1980 Appraisal, *op cit.*, p. 70.

TABLE 4-3. Number of Residential Units and Average Annual Demand for Housing in the United States, 1920–1980, with Projections to 2019

| YEAR | TOTAL NUMBER OF RESIDENTIAL UNITS | NUMBER OF OCCUPIED UNITS | DECADE | AVERAGE ANNUAL DEMAND FOR NEW HOUSING UNITS | AVERAGE ANNUAL DEMAND FOR CONVENTIONAL UNITS | | | AVERAGE ANNUAL DEMAND FOR MOBILE HOUSING UNITS* | |
					Total units	One-family	Multifamily	Primary residences	Others
1920	24,552	24,353			*thousands of residential units*				
1930	32,495	29,905	1920–29	803.4	803	527	276	–	–
1940	37,439	34,855	1930–39	365.1	365	304	61	–	–
1950	46,137	42,826	1940–49	809.0	780	657	123	29	–
1960	58,326	53,024	1950–59	1,572.4	1,509	1,276	233	41	22
1970	68,657	63,450	1960–69	1,648.7	1,443	929	514	164	41
1980	88,441	80,390	1970–79	2,140.0	1,774	1,143	631	292	74
1990			1980–89	2,590.0	2,250	1,680	570	270	70
2000			1990–99	2,240.0	1,930	1,540	390	250	60
			2000–09	2,300.0	1,960	1,410	550	270	70
			2010–19	2,270.0	1,930	1,390	540	270	70

*All new mobile units are counted in the totals for average annual demand for new housing units through 1979. Only those mobile units that will be used as primary residences are included in the projections of average annual demand for later decades.

Sources: Number of residential units and occupied units from U.S. Bureau of the Census, *Historical Statistics of the United States* (Washington: Government Printing Office, 1975) p. 636 and from U.S. Censuses of Housing, 1920 through 1980. Data on average annual demand for housing from U.S. Forest Service, *An Assessment of the Forest and Range Land Situation in the United States*, FS–345 (Washington, USDA 1980), p. 319.

Industrial and Commercial Site Needs

Commercial and industrial sites have an importance that far exceeds their actual area requirements. Only 10 to 16 percent of the area of most cities is used for these purposes. With continued expansion of the national economy and of the areas used for urban and suburban residential purposes, more and more land probably will be needed for commercial and industrial sites. Some of these sites will be provided by redevelopment of urban areas now used for residential purposes. But unless there is a reversal of recent trends, most of the new commercial and industrial development will come in suburban and nearby rural areas. The lure of lower land values, hopes for more space, and desire to escape the problems of the central cities will cause more and more operators to join in the flight to the suburbs.

Increasing suburbanization has favored the development of hundreds of suburban shopping centers in recent years. These centers range in size from small neighborhood centers that cover a few acres and serve up to 10,000 people to large regional shopping centers that cover 100 acres or more, have over 1 million square feet of building area, and serve more than 200,000 people. Most centers are laid out on a single-floor level and many are enclosed. Substantial open-space areas are needed for wide sidewalks, streets, adequate parking, buffer areas, and possible landscaping.

Industrial space needs also reflect the trend toward decentralization. No longer is it as necessary as it once was for industries to locate next to railroads or navigable waters or to seek sites that provide natural water power. The widespread use of motor trucks for transport purposes and the use of electricity and mineral fuel for power and energy have freed industry from an earlier "tyranny of space." These changes together with the relative scarcity of vacant industrial sites within urban centers have made it practicable for many industries to locate in suburban and rural communities where they can spread out over sizable areas.

Land Requirements for Recreation

Recreation land requirements was a subject of secondary interest in the United States until the middle 1950s. True, the nation had its parks, scenic drives, and recreation areas at the local, state, and national levels. It had a dedicated group of park and recreation workers and also its share of outdoor recreation devotees. Yet the subject of recreation land resource requirements was usually viewed as a miscellaneous residual need. This situation suddenly changed when an affluent burgeoning population found that it had more disposable income and more leisure time but lacked adequate opportunities for outdoor recreation.[10] Fire was added to the demand for additional outdoor recreation areas in 1959 when Marion Clawson argued that rising population numbers, higher family incomes, increases in leisure time, and transportation improvements could call for a five to fifteen-fold increase in the demand for outdoor recreation by the end of the century.[11] An Outdoor

[10] Cf. Michael Chubb and Holly R. Chubb, *One Third of Our Time* (New York: John Wiley, 1981), pp. 34–39.

[11] Cf. Marion Clawson, "The Crisis in Outdoor Recreation," *American Forests*, 65 (March 1959), pp. 22–31, 40–41.

Recreation Resource Review Commission was appointed to study the problem, and public programs were soon initiated to acquire and develop more lands for public recreation use.

The problem of future recreation land requirements is somewhat elusive. Individuals differ in their recreation likes and dislikes. Some choose to take practically all of their recreation in their homes, night clubs, taverns, and theaters. The per capita recreation land area requirements of this group are low. Others make considerable use of city parks, beaches, and golf courses. Their per capita land requirements also are relatively small but involve intensively used properties of high value. At the other extreme, many people take camping, fishing, and hunting trips or go on extensive tours that give them opportunities to "rough it," commune with nature, and visit wilderness areas.

Recreation lands also vary over a wide range. The intensively used totlots, parks, playing fields, and golf courses found in and around cities may be described as user-oriented facilities.[12] Wilderness areas, national parks, some state parks, and the recreation areas in national forests, on the other hand, are resource-based in the sense that they feature natural wonders along with the flora and fauna of nature. These facilities are ordinarily used on an infrequent vacation basis and often entail considerable travel by the users. Between these two classes is an intermediate classification of parks and recreation areas, which involves natural and man-made developments located within a few hours' driving time of most of their users. Emphasis has been given to the development of all three of these types of facilities. It is obvious, however, that there is a limited supply of wonders of nature and that much of the future acquisition and development of parks and recreation areas should come at sites located within easy commuting distance of the using populations.

The question of how much land should be reserved for recreation uses depends mostly on the needs of the urban population. A rule of thumb on municipal recreation land requirements was suggested by the National Recreation Association around 1923 when it recommended that cities of 10,000 or more have ten acres of recreation land for every 1,000 people. Most park administrators regard this standard as inadequate; and some recommend twenty acres of city and county parks and outdoor recreation facilities plus an additional 65 acres of state and 150 acres of federal parks and outdoor recreation areas for every 1,000 people.[13]

Acreage standards can be meaningful with stable populations but can also be unrealistic in rapidly growing metropolitan areas. The fact that acreage standards rise with population growth and at a time when prospective park sites become both scarce and more expensive means that these standards often go unfilled. Goals stated as a percentage of an urban region's total area provide a more workable standard over time for public planning purposes. The standard suggested by the Na-

[12]Cf. Marion Clawson, *ibid.*, pp. 30 and 40; and Marion Clawson and Jack Knetsch, *The Economics of Outdoor Recreation* (Baltimore, Md.: Johns Hopkins University Press, 1966), pp. 36–40.

[13]Cf. Charles E. Doell and Louis F. Twardzik, *Elements of Park and Recreation Administration*, 4th ed. (Minneapolis, Minn.: Burgess, 1979), pp. 96–110.

tional Recreation Association calls for larger holdings of municipal and county or regional recreation lands than the nation now has. Reports of the Bureau of Outdoor Recreation indicate that cities had 805,336 acres of parks and recreation areas plus 96,965 of school recreation lands in 1965.[14] The nation had an additional 211 million acres of rural parks, wilderness, and wildlife refuge areas in 1982. Doell and Twardzik indicate that around 234 million acres of public lands are available for recreation of which 100 million are probably used for intensive and extensive recreational activities.[15] An additional 257 million acres of privately owned corporate and noncorporate land holdings also have been opened for general public recreational uses.[16]

Considerable headway has been realized since 1960 by federal, state, and local agencies in reserving and acquiring public wilderness, recreation, and open-space lands. But additional demands for the recreational opportunities associated with public park and recreation areas and privately owned hiking and hunting areas can be expected in the future. The U.S. Forest Service has projected a 35 percent increase in the demand for developed outdoor recreation between 1977 and 2000 and a 105 percent increase by 2030 and a 23 percent increase in the demand for dispersed types of outdoor recreation (hiking and hunting) by 2000 and a 71 percent increase by 2030.[17]

Transportation and Service Area Needs

Approximately 39 million acres were used for transportation purposes in 1982.[18] This estimate includes 21.5 million acres in rural highways and roads; 3.0 million acres in railroad rural rights of way; 1.9 million acres in farm roads and lanes; 2.3 million acres in airport sites; and 10.5 million acres used for streets, alleys, parking areas, and railroad lands in cities, villages, and unincorporated subdivisions.

By and large, the boom period of railroad and local highway building in the United States has passed. Significant areas will be needed in the future, however, for the construction of additional limited-access highways and turnpikes; and rights of way will be required for highway improvement projects. These projects will require upward of 35 to 40 acres of land per mile of new highway construction. Substantial additional areas will also be used in the construction of new and wider city streets and for the provision of off-street parking facilities in urban places.

Finding adequate space for parking has become a major problem for cities. Most projects for this purpose involve an extensive type of urban redevelopment. Yet

[14] Bureau of Outdoor Recreation, *Recreation and Parks Yearbook, 1966* (Washington, D.C.: National Park and Recreation Assn., 1967).

[15] Doell and Twardzik, *op. cit.*

[16] U.S. Department of Interior, Heritage Conservation and Recreation Service, *The Third Nationwide Outdoor Recreation Plan*, Appendix IV (Washington, D.C.: Government Printing Office, 1980), p. 49.

[17] U.S. Forest Service, *A Recommended Renewable Resource Program, 1980 Update*, (Washington, D.C.: U.S. Forest Service, 1980), p. 36

[18] Cf. H. Thomas Frey and Roger W. Hexam, *Major Uses of Land in the United States, 1982*, U.S. Department of Agriculture Economic Research Service, Agricultural Economics Report No. 535, 1985.

merchants and city residents often support this type of action in the hope that it will attract new business and prevent migration of present trade to suburban shopping centers. Comparable problems are associated with the provision of improved airport facilities. With the development of large commercial planes, most cities have felt a need for enlarging their airport facilities. Fulfillment of these needs usually calls for tracts of substantial size located within reasonable commuting distance and accordingly results in expensive land-acquisition programs and the frequent displacement of other desired uses.

Land requirements for service areas. Service areas tend to fall into two groups—those that exist primarily for single-service purposes and those that fit into multiple-use patterns. Watershed protection areas and large storage reservoirs provide leading examples of multiple-purpose service areas. The need for these types of service areas will in all probability increase with population numbers and with society's need for additional water supplies.

Military reservations rank most important areawise among the single-purpose uses (24 million acres in 1982). Other important examples include the sites used for cemeteries, schools and public buildings, water-filtering operations, power plants, and public dumping grounds. The area requirements for most of these uses will probably increase with population growth. Some of the most critical problems with single-purpose service areas will concern the provision of acceptable and appropriate sites for locally unwanted land uses.[19] One of the most controversial of these involves the location of disposal and storage sites for chemical, nuclear, and toxic wastes.

Water Resource Needs

Water resources and water quality have become subjects of widespread public concern since the late 1950s. Interest in the increasing water resource needs of an expanding economy prompted the appointment of a U.S. Senate Select Committee on National Water Resources in 1959. This committee issued a series of reports, which projected a 2.5-fold increase in municipal water supply needs and an 8-fold increase in industrial water supply needs between 1954 and 2000. These reports were followed by the creation of a U.S. Water Resources Council and the launching and expansion of several public programs for the development and treatment of water supplies, improvement of desalinization processes, and enhancement and protection of water quality.

In its second national assessment of the nation's water resources, the Water Resources Council reported major increases in water withdrawals and water consumption between 1965 and 1975 but an expected slowing down of these rates of increase in the future.[20] (See Table 4-4.) The nation withdrew 338.2 billion gallons

---

[19]Cf. Frank J. Popper, "The Environmentalist and the LULU," *Environment*, vol. 27, March 1985, pp. 7–11 and 37–40.

[20]Cf. U.S. Water Resources Council, *The Nation's Water Resources, 1975–2000, Second National Water Assessment*, Vol. 1, Summary (Washington, D.C.: Government Printing Office, 1978).

TABLE 4-4. Withdrawals and Consumption of Water in the United States in 1975 with Projections of Water Usage for 1985 and 2000*

TYPE OF USE	TOTAL WITHDRAWALS			TOTAL CONSUMPTION		
	1975	1985	2000	1975	1985	2000
Fresh water uses						
Domestic						
Municipal	21,164	23,983	27,918	4,976	5,665	6,638
Rural	2,092	2,320	2,400	1,292	1,408	1,436
Commercial	5,530	6,048	6,732	1,109	1,216	1,369
Manufacturing	51,222	23,687	19,669	6,059	8,903	14,699
Agriculture						
Irrigation	159,743	166,252	153,846	86,391	92,820	92,506
Livestock	1,912	2,233	2,551	1,912	2,233	2,551
Steam electric						
generation	88,916	94,858	79,492	1,419	4,062	10,541
Minerals industry	7,055	8,832	11,328	2,196	2,777	3,609
Public lands and						
others	1,866	2,162	2,461	1,236	1,461	1,731
Total fresh water	338,500	330,375	306,397	106,590	120,545	135,080
Saline water	59,737	91,236	118,815			
Total water	398,237	421,611	425,212			

*Figures are in millions of gallons per day.
Source: U.S. Water Resources Council, *The Nation's Water Resources; 1975-2000, Second National Water Assessment*, Vol. 1, Summary (Washington, D.C.: Government Printing Office, 1978), p. 29.

daily and consumed 106.6 billion gallons daily in 1975. Almost 90 percent of the water withdrawn was used for irrigation, thermal power generation, and industrial uses; while agricultural irrigation accounted for more than 80 percent of the consumptive use. Most of the water withdrawn is used and then returned to streams, lakes, or the ocean for possible reuse.

The nation has sufficient access to water supplies to meet its projected demands. Unfortunately, however, water supplies are not always found at the places where they are needed at the specific times and in the quantities and quality desired. In a very practical sense, most of the nation's water-use plans must be built around the supplies that are normally available at specific sites under the operations of the hydrologic cycle. Additional supplies can be secured by pumping from greater depths, conducting waters through canals or conduits from greater distances, transforming saline waters into fresh water supplies, and treating used waters for reuse. These practices involve added costs which when passed on to water users can discourage the use of high-cost water supplies for low-value uses.

Water resource and watershed management practices are needed to protect water supplies, minimize damage from floods, facilitate retention of water supplies for use in low-flow periods, minimize pollution problems, and enhance quality maintenance. Like other supplies of free natural resources, which have appeared more than adequate in times past, water supplies must now be stretched in some cases to meet the needs of larger numbers of users. This stretching process calls for discipline in water use, new regulations affecting individual use rights, water quality maintenance programs, and emphasis on recycling practices that permit the use of the same water resource many times.

Energy and Mineral Requirements

Great quantities of energy and mineral inputs are needed to keep the wheels of modern industry moving. As one of the industrial powers of the present world, the United States uses considerably more energy and mineral resources on an average per capita basis than the average country. Our consumption of most minerals has been "expanding at compound rates and thus is pressing harder and harder against resources which, whatever else they may be doing, are not similarly increasing."[21] Projections of expected future demands show that the nation will need substantially larger annual supplies of these resources in the years ahead.

Table 4-5 reports the amounts of selected energy and mineral resources required to fill primary demands in the United States in 1980 together with the rates of demand projected for 1990 and 2000. These projections indicate that a steady increase in demand can be expected for all energy and mineral resources that are not expected to be in short supply. Rising prices and finite limits on total supplies pose major problems in filling these demands.

Needed supplies will come from both domestic and world sources. An indication of the extent of this dependence on foreign sources is provided by comparison of the information reported in Table 4-5 on quantities of production and amounts demanded in the United States in 1979-1980. Even when allowances are made for the processing of some minerals for sales of finished products abroad, it is apparent that the nation is dependent on foreign sources for much of its needed supplies of key minerals such as copper, lead, petroleum, and zinc. It also draws heavily on foreign sources for its supplies of bauxite, iron ore, and many other minerals.

A major problem associated with the filling of future demands is that of providing new and additional supplies of minerals. Minerals are scattered in fairly good supply throughout the earth's crust. But most of the surface sources have been discovered and exploited. This means that we must look harder and go deeper to find new economic sources of supply. With steadily increasing rates of drain against known reserves, this search process is becoming more and more critical and may logically lead to higher costs and some material shortages.

Minerals are a fund resource. Their supplies are not renewable, and although metallic minerals are not destroyed by use, as is the case with mineral fuels, their recovery for future use involves time, trouble, and a certain amount of conscious planning. Continuation of our present rates of use can easily bring exhaustion of the world's low-cost reserves of most minerals within a few decades. But this does not mean that the world will run out of minerals. In practice, we are dealing with a

[21] President's Materials Policy Commission, *Resources for Freedom*, Vol. 1, Summary (Washington, D.C.: Government Printing Office, 1952), p. 2.

TABLE 4-5. Production and Primary Demand for Selected Energy and Mineral Resources in the United States in 1979-1980 with Projection of Demands for 1990 and 2000.*

ENERGY AND MINERAL RESOURCES	PRODUCTION IN 1979-1980	PRIMARY DEMAND IN 1979-1980	PROJECTION OF PRIMARY DEMAND AT MEDIUM ASSUMPTION LEVEL FOR	
			1990	2000
Petroleum (millions of gallons per day)	10.2	16.9	14.8	14.0
Natural gas (quadrillion British thermal units per year)	19.8	20.8	18.8	18.2
Nuclear (quadrillion British thermal units per year)	2.7	2.7	7.6	10.5
Aluminum (thousand short tons)	5,023	6,032	11,400	17,200
Coal (million short tons)	781	681	989	1,475
Copper (thousand metric tons)	1,441	2,408	3,400	4,600
Iron ore (million short tons)	135.9	133.3	104	120
Lead (thousand metric tons)	526	1,404	1,900	2,300
Phosphate rock (thousand metric tons)	51,000	38,000	50,000	63,300
Potash (thousand metric tons)	2,245	6,916	8,000	10,100
Sand and gravel (million short tons)	979	978	1,085	1,200
Zinc (thousand metric tons)	267	1,009	1,400	1,800

*Production and current primary demand data are for 1979 for minerals and for 1980 for petroleum, natural gas, and nuclear resources.

Sources: U.S. Energy Information Administration, Annual Report to Congress, vol. 3, Energy Projections, (Washington: Government Printing Office, 1982); and Mineral Facts and Problems, 1980, U.S. Bureau of Mines Bulletin No. 671, 1981.

two-dimensional supply situation (see Figure 4–2). As we satisfy our demands for minerals, we will add to our known reserves by seeking and tapping new sources now viewed as probable and possible reserves. More effort and higher cost outlays per unit of product recovered can also be expected as we move to utilize lower grades of minerals. The expected higher costs associated with these processes will call for higher prices. Higher prices in turn will help to correct the shortage problem by generating three types of activities. They will provide (1) increased incentives for the discovery and development of new mineral sources, (2) opportunities for the profitable processing of lower grades of ores, and (3) pressures for shifts to possible alternative resources that could not compete with the minerals in question when their prices were lower.

It is true that the United States has skimmed the cream from its stock of mineral resources insofar as it has been able to find them. There has been nothing unnatural in this process. Bounteous supplies made it economically feasible for the nation's copper producers to pass up ores with concentrations of less than 5 percent copper around 1900. Ores were being smelted in 1968 with concentrations as low as 0.4 percent copper, and new processing efficiencies can make the mining of even lower concentrations economically feasible in the future.[22] On this point, Harrison Brown has observed that "if at some future time the average concentration of copper in copper ore were to drop to 0.01 percent, and if there were still an acute need for copper, there would be little question but that the metal could be extracted in high yield. . . . Given the brainpower and the energy, the people of the world could, if

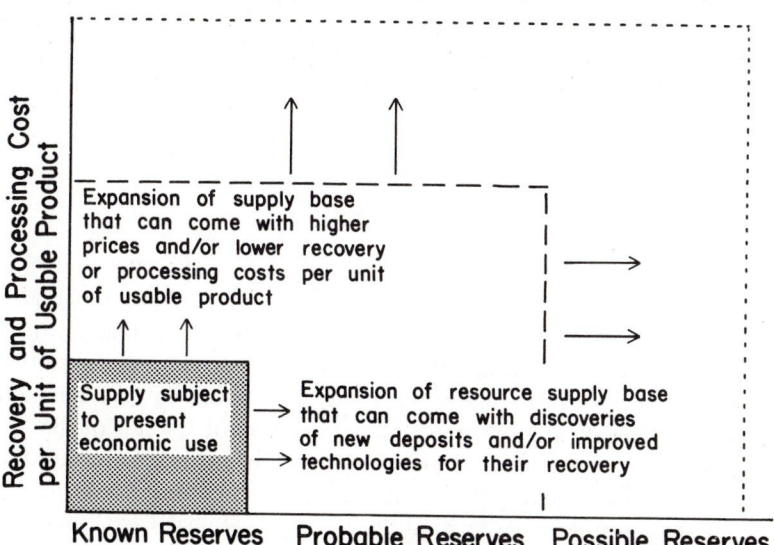

FIGURE 4–2.   Opportunities for increasing our economic supplies of mineral resources.

[22]Cf. Thomas S. Lovering, "Mineral Resources from the Land," National Academy of Sciences–National Research Council Committee on Resources and Man, *Resources and Man* (San Francisco: W. H. Freeman & Company Publishers, 1969), p. 111.

need be, support themselves entirely with the leanest of ores, the waters of the oceans, the rocks of the earth's crust, and the very air around them."[23] This outlook justifies a certain amount of optimism. But a note of caution is in order. While higher prices and new technological developments can justify the search for new mineral deposits and the mining of lower-grade and harder-to-recover deposits, raising prices is not a cure-all answer to the long-run threat of mineral or energy famine. This is particularly true of mineral fuels.

Modern society is heavily dependent on the energy inputs currently supplied by the burning of fossil fuels. The supplies of these resources are both finite and exhaustible. The world still has tremendous proven, prospective, and possible reserves of petroleum, natural gas, coal, and uranium, most of which can be claimed for human use through good management, applications of improved know-how, and the exploitation incentives afforded by opportunities for economic profits. But the total recoverable amounts are limited. With increasing worldwide demands for and dependence on energy, one must accept the inescapable conclusion that the world will run out of its commercially available supplies of each of these resources some x years in the future.

A sane energy resource-management approach dictates that steps be taken to (1) prevent wasteful uses of these energy resources; (2) facilitate and encourage the development and substitution of new practicable sources of energy; and (3) accept reservation and allocation policies that will husband substantial reserves of specific resources such as oil and natural gas for essential uses (e.g., petrochemicals and fertilizers) that cannot be supported by alternative resources. Fossil fuels have provided the cheap energy needed to generate and fuel our present industrial society. Over time, plans must be made to shift from our dependence on this source of energy to biological resources such as wood and plant cellulose and to flow resources such as wind, water, geothermal, nuclear, and solar power.

PROSPECTS FOR MEETING FUTURE NEEDS

Few problems provide a bigger challenge than that of meeting the world's future requirements for food and other land resources. With food, as with other land products, the problem of increasing supplies is always one of expanding production where the people live, transporting food from where it can be produced to the people, or moving people to the food supply. Problems arise with each of these alternatives. Factors such as a limited resource base, operation of the law of diminishing returns, and present production and ownership patterns may discourage efforts to increase production in the areas where people live. High transportation costs, limited consumer purchasing power, trade barriers, and international exchange problems frequently have similar effects on the possible use of areas located away from markets. The movement of people to areas of potential surplus production in turn can

[23]Harrison Brown, James Bonnet, and John Weir, *The Next Hundred Years* (New York: Viking, 1957), pp. 90 and 92.

be stymied by immigration barriers, migration costs and restrictions, or the reluctance of people to move.

Opportunities for Better Man-Land Relationships

Nations or communities that desire a better or more rational balance between their supplies of land products and their demands for these products can pursue three principal types of programs for this purpose. They can (1) enlarge their resource base and bring new lands into use. They can (2) make more intensive use of the lands they have available. Lower man-land ratios can also be secured by (3) reducing population numbers or by accepting downward adjustments in dietary and product consumption standards.

Bringing new resources into use. Almost every country has areas that can be brought into agricultural or other uses should the price and demand situation so warrant. Areas used for cropland can be supplemented by the clearing, draining, irrigation, or terracing of new lands. Sometimes these lands represent areas of high potential productivity whose development has been delayed because of imperfect human knowledge, lack of accessibility to market, or high costs of development. Changed demand and price situations may also justify the cultivation of lower-grade lands, building of dikes, development of pasture and grazing areas, or planting of trees in cutover and wasteland areas.

Because of their higher than average values, commercial and industrial, residential, and other urban-oriented uses can secure additional land by bidding it away from agriculture and other open-space uses and in some cases by leveling rough areas and filling in low and submerged sites. In similar fashion, new areas become available for mining purposes once mineral resources are discovered and developed.

Individuals and nations have long since found that it is possible to extend their controls over new resources without going through the land development process. Individuals can buy the already developed resources of others. Nations have at times used brute force and military power to acquire control and sovereignty over additional resources. This method is frowned upon in the modern world, but it has been used quite extensively in times past by most of the world's great powers. World War II was precipitated by the expansionist policies of the Axis Powers; and tensions in international politics still spring from national ambitions for control, exploitation, and use of the resources of other countries.

Large investments in resource developments can tie an area's economic future to the economy of the nation from which the investment funds have come. Industrialization and trade policies can also be used by nations to enlarge the economic base on which they depend. Nations such as Great Britain, Belgium, and Japan have found it profitable to market the products of their labor and manufacturing skill in a world market. In this process they import a high proportion of their supplies and raw materials. By processing these materials at home, selling much or most of their finished products abroad, and then reinvesting in raw materials, they have found it possible to expand their accessible resource bases far beyond their geographic and political boundaries.

Intensifying the use of available land resources. Higher production can come from more intensive utilization of lands currently in use as well as from the development of new land areas. Land resources can frequently be used more intensively than they have been. This is particularly true in periods of rising prices and when new technological developments offer a prospect of increasing production or reducing costs.

Numerous examples of intensification practices may be listed. With croplands, these may involve the use of improved cultural practices, better seed, pesticides, more fertilizer, irrigation water, and possibly some double cropping. Pasture and grazing lands can often be planted to improved grass species, fertilized, irrigated, and stocked with better livestock. Forestland use can often be intensified through the use of timber-stand improvement and selective cutting practices. Skyscrapers and multiple-storied structures can contribute to the more intensive use of urban sites, and leveling, filling, and bridging operations can be used to make sites more available for urban developments.

Intensification practices also have an important impact on the use of mineral resources. Higher prices and technical improvements frequently make it profitable for operators to capture a higher proportion of their known reserves of ores, coal, and oil. This condition leads to more intensive mining operations and in so doing reduces the amount of physical waste by favoring use of the lower-grade or less-accessible deposits, which might otherwise be discarded or bypassed. Minerals are also used more intensively when means are developed to make them work harder and longer than they have in the past and when scrap materials are reclaimed for reprocessing and reuse.

More effective land use can also be secured by actions that overcome or offset barriers to intensive land use. Highways, railroads, and other transportation and communication developments can make areas more accessible. Canals and aqueducts bring water to thirsty cities and fields. Sewers and drainage ditches take away unwanted wastes and surplus waters. Flood control programs provide protection against possible flood damage. Commercial fertilizer compensates for soil-nutrient deficiencies.

Downward adjustments in demand. Less heroic approaches for improving the balance between the demands of people and their land resource bases can involve measures for reducing the size of the "man" factor in the man-land ratio. This can be accomplished by stabilizing or reducing total population numbers and by stabilizing or reducing average per capita levels of consumption.

Population stabilization programs can involve measures that encourage the migration of segments of the population to other areas. They may also call for population control programs such as China's policy for limiting births to one child per family. Ireland used a combination of these programs to stabilize its population numbers during the century following the potato famine of the 1840s.

The forced or voluntary rationing of food and other products can also affect man-land ratios by reducing individual consumption to levels that can be served by currently available supplies. Over long periods of time, forced rationing can bring significant changes in consumer attitudes. Societies that once looked to livestock products for significant portions of their food, for example, could turn to primary dependence on food crops that involve less luxurious uses of land.

Outlook for New Land Development

A considerable portion of the earth's land surface can still be developed for agricultural and other uses. The Food and Agriculture Organization has classified 3.6 billion acres, 11.0 percent of the world's land surface, as arable cropland, and approximately a billion additional acres as unused but potentially arable land. Substantial areas now used for pastures, meadows, and forests might also be shifted to cropland uses. Large tracts of both arable and nonarable land can be drawn on for urban, transportation, recreation, and service-area uses. The area used for mining will increase as new deposits are discovered and opened up for exploitation.

Most of the world's easy-to-develop fertile soils have already been brought into agricultural use. Soil classification and climatological studies, however, show that the earth has 7.86 billion acres of potentially arable land, while an additional 9.02 billion acres has a potential for grazing.[24] Most of the land with an arable use potential but not now used for that purpose is found in Africa and South America. But almost 1 billion acres are found in North America, Australia, New Zealand, and smaller areas in Europe and Asia.[25] Irrigation is needed with 850 million of the 7.86 billion acres, while land-clearing, irrigation, drainage, and other land-development practices are necessary for optimum use of much of the remaining area.

More land can be brought into cropland use even in the world's great "bread basket" areas. A natural resources inventory conducted by the U.S. Soil Conservation Service in 1982 showed that the nation had 421.4 million acres used as cropland. In addition, it had a reserve of 35.3 million acres with a high potential and 117.6

[24] Cf. President's Science Advisory Committee, Panel on the World Food Supply, *The World Food Problem* (Washington, D.C.: Government Printing Office, 1967), Vol. 2, p. 423. Without irrigation, multiple cropping could increase the gross cropped-area (the cultivated area times the number of crops) to 9.8 billion acres annually, or about three times the world's present cultivated area. Use of irrigation and double and triple cropping in those areas where it is feasible would permit a maximum gross cropped-area of 16.3 billion acres. (Cf. *ibid.*, p. 434.)

[25] These totals are somewhat higher than earlier estimates of the world's gross area of potential cropland. During the 1940s, for example, Charles E. Kellogg ("Food Production Potentialities and Problems," *Journal of Farm Economics*, 31 [February 1949], 251–62; and "World Food Prospects and Potentials: A Long-run Look," *Alternatives for Balancing World Food Production and Needs* [Ames: Iowa State University Press, 1967], pp. 98–111) estimated that 1.3 billion acres (1 billion acres of tropical soils and 300 million acres of temperate area podzol soils) could be shifted into food production. In 1964, he reported a soil classification study that showed the world had 6.59 billion acres of potential arable land. (Cf. Kellogg, "Potentials for Food Production," *Farmer's World: The Yearbook of Agriculture, 1964* [Washington: GPO, 1964], pp. 57–69.)

million acres with a medium potential for conversion to cropland use. Almost 71 percent of this cropland potential was currently used for pasture or range, 27 percent for forest production, and the remainder for other uses.[26]

Problems with new land development. A false sense of security often comes with knowledge that physical supplies of land are available for cropland use. Before these "available acres" acquire economic value, they must be "produced." This calls for favorable economic, social, political, and institutional conditions. Cost outlays are necessary, and justification of these expenditures may call for much higher food prices than now exist.

In areas such as the tropics, for example, the development of new croplands will call for more than just land-clearing, drainage, and terracing activities. Suitable transportation, communication, sanitation and health, and housing facilities must be provided. Consideration must be given to the nature of local social services, to marketing and international trade problems, to local tax systems, to possible governmental restrictions, and to the problems of growing nationalism. Attention must also be given to tropical soil management problems; to the financing of land settlement and development projects; to relationships between outsiders and native populations; to the impact commercial agriculture will have on subsistence and barter economies; and to adjustments that must be made to local problems of poverty, disease, and illiteracy. The scope of these problems suggests the risks and headaches often associated with new land developments. These factors, together with the problem of economic uncertainty, provide effective barriers to the development of many new areas.

Losses of land area to other uses. The problem of resource adequacy is also affected by the shifting of land areas to competing uses. This phenomenon affects all uses but has its most critical impact on those uses that cannot compete with others in the economic bidding process. This situation is well illustrated by the example of cropland in the United States where approximately 20 million acres, much of it used earlier for farming purposes, shifted from rural to urban and built-up uses between 1959 and 1982. More shifts of this nature can be expected in the future. This shifting process can have negative effects on the nation's ability to supply the food, feed, and forest products it needs, because unlike shifts of land back and forth between cropland, pasture, and forests, movements of land from agricultural to urban uses are for all practical purposes nonreversible. Each acre shifted is an acre lost and a net subtraction from the nation's cropland reserve.

Public interest in the loss of farmlands issue was fanned during the early 1980s by an assertion that the nation was losing 2.9 million acres of agricultural land each year.[27] Continued losses at this rate could have caused the nation to run out of its

[26] Cf. U.S. Soil Conservation Service, *Natural Resources Inventory, 1982*, National Summary, Tables 2a, 31a, 32a, and 33a.

[27] Cf. National Agricultural Lands Study, *Final Report* (Washington, D.C.: Government Printing Office, 1981); and Michael F. Brewer and Robert F. Boxley, "Agricultural Land: Adequacy of Acres, Concepts and Information" and comment by Philip M. Raup, *American Journal of Agricultural Economics*, 63, (1981), 879-93. The National Agricultural Lands Study took a

reserve of potential croplands in less than forty years. More careful analysis of the changing land-use situation showed that rural land was shifting to urban uses at a much slower rate and that average losses of cropland to urban uses probably did not exceed 750,000 a year during the 1970s. With this slower diversion rate, the nation's cropland reserve could last for a longer time; but the eventual problem of using up the reserve of potential cropland still remains.

The seriousness of the farmland loss problem is complicated by three additional factors. First, land developers have shown a distinct preference for taking prime farmlands, because the flat and fertile areas used for farms are usually easier to work with and are better supplied with utilities, roads, and public services than the alternative rougher sites that could often be used just as well for urban developments. Second, the area generally viewed as potential cropland is composed mostly of pasture and forestlands that are also needed to sustain future grazing and forest production needs.[28] And finally, the continued productivity of the nation's present farmlands is threatened by soil erosion problems that are expected to have an impact on total production potential equal to the loss of 23 million acres over the next fifty years.[29]

Crop surpluses belied any threats of an imminent cropland crisis during the early 1980s. Indeed, the emphasis on increasing production for export markets posed a greater potential for causing the nation to run out of farmlands than did the shifting of rural areas to urban-oriented uses.[30] Yet cropland was still shifting out of agriculture, and the makings of a possible future crisis were at work. In addressing this issue, it must be remembered that some shifting of rural lands to urban uses is both necessary and normal. With adequate planning, positive steps can be taken to minimize the wastage of resources often associated with the shifting process. Measures are also needed to retain large areas of productive farmlands in their present uses so that technology will not have to work harder to secure comparable future production responses from lower-grade replacement lands. Policies designed to deal in a forthright manner with these issues and with the problem of continued soil erosion would seem to provide the best means for protecting the nation's agricultural production potential.

---

pessimistic view of the farmland loss situation, as did W. Wendell Fletcher and Charles E. Little, *The American Cropland Crisis* (Bethesda, Md.: American Land Forum, 1982); and R. Neil Sampson, *Farmland or Wasteland: A Time to Choose* (Emmaus, Pa.: Rodale Press, 1981). Another group argued that the nation had an adequate supply of cropland. Cf. Pierre R. Crosson, ed., *The Cropland Crisis: Myth or Reality?* (Baltimore, Md.: Johns Hopkins University Press, 1982); and Julian L. Simon, "Are We Losing Our Farmland?" *The Public Interest*, 67 (1983), 1–14.

[28] Cf. Robert G. Healy, "Land in the South; Is There Enough to Satisfy Demands?" *Conservation Foundation Newsletter*, September 1982.

[29] Cf. National Agricultural Lands Study, *Soil Degradation: Effects on Agricultural Productivity*, Interim Report No. 4 (Washington, D.C.: Government Printing Office, 1980).

[30] Cf. Philip M. Raup, "Competition for Land and the Future of American Agriculture" in Sandra S. Batie and Robert G. Healy, eds., *The Future of American Agriculture as a Strategic Resource* (Washington, D.C.: The Conservation Foundation, 1980), pp. 41–77.

The Promise of Technology

Science and new technology have played prominent roles in helping nations to feed themselves. They have also contributed to the rise of the modern city, provided better housing and living conditions for most people, facilitated the manufacture of new types of products, and brought more effective exploitation and use of the world's energy and mineral resources.

Because of this help in times past, many people have come to regard the flow of new technology as endless. Without doubt, new technological developments can go a long way in solving our future land-requirement problems. Yet the assumption of continuous technological improvement is a tenuous one. One cannot be certain that new technology will unlock the doors to new resource potentialities as fast in the future as in recent decades. Serious questions can thus be asked concerning the extent to which policymakers should depend on hoped-for answers to emerging resource needs.

Outlook for new technology. Developments with important implications for how we use land resources have come in a steady stream since our entry into the age of science. This continuing bounty would seem to justify future optimism. As we look ahead, however, we must admit that our outlook is far from clear. Our perception of what will happen is colored in part by a haunting fear that the prospect of diminishing returns will limit the flow of new technology in the same way it limits individual operators in their production activities. It is also clouded by our inability to foresee future events that may affect production and by concerns about the risks associated with the generation and adoption of new technologies.

Many scientists argue that we are entering an age of tremendous technological change that will bring major benefits to humanity. Wittwer, for example, argues that "far from reaching its scientific and biological limits, the world has only begun to explore the possibilities for increasing food production."[31] Insofar as this is the case, science can help us meet our increasing requirements for agricultural, building, energy, and other materials. It can promote further utilization of our air and water resources as well as those of surface land and teach us how to use these resources without impairment of environmental values. Science may help us synthesize new materials and most certainly can open the way to new opportunities for resource use and development.

From a hard-headed practical point of view, however, it is important that we keep our balance and not be carried away by unrealistic hopes for the future. It is easy to talk of a world of "inexhaustible resources" and a "chemistic society" in

---

[31] Sylvan Wittwer in National Academy of Science, *Long Range Environmental Outlook* (Washington, D.C.: National Academy of Science, 1980), p. 67. Cf. also Tao-chi Lu and Leroy Quance, *Agricultural Productivity: Expanding the Limits,* U.S. Department of Agriculture, Economics, Statistics and Cooperatives Service Agricultural Information Bulletin No. 431, 1979; and Earl Heady, "The Adequacy of Agricultural Land: A Demand-Supply Perspective" in Crosson and Brubaker, eds. *op. cit.,* pp. 23–61.

which food is synthesized and plants are used for decorative purposes only.[32] Unfortunately, visions such as these must be regarded as fanciful dreams until they are demonstrated in reality. In practice, we must recognize that diminishing returns are a physical and economic fact of life. Indeed, there is evidence that advances in some scientific areas are already coming at a decreasing rate. Improved technology brought average annual increases of 0.75 percent in U.S. agricultural production during the 1970s as compared with annual averages of 1.6 percent for the 1960s.[33] This continued increase at a decreasing rate was caused at least in part by expansion of crops to less productive lands and rising costs for energy, fertilizer, machinery, and water inputs. But it has caused some scientists such as Norman Berg to suggest "that available agricultural technology may have reached a point of diminishing returns."[34] James H. Anderson argues that unlike the situation at the end of World War II when increases in agricultural production were favored by a large "reservoir of science and technology waiting for practical applications," this "reservoir of unused technology had been depleted" by the 1970s.[35] Also we must note that while technology can open doors to a more abundant future, careful monitoring is needed to control its potential for bringing horrendous consequences. Generation of toxic substances, genetic engineering, and nuclear developments provide fitting examples of scientific Pandora boxes that should be opened with utmost care.

New technologies will be necessary if this planet is to provide its expanding human population with opportunities for high levels of life. Continued financial and moral support for scientific research and development is definitely needed for this purpose. Emphasis should clearly be given to attainment of the following tasks of technology: (1) foster new techniques of discovery, (2) utilization of materials that have thus far evaded our efforts, (3) promotion of resource recycling techniques, (4) greater utilization of low concentrations of useful materials, (5) economic development and utilization of naturally renewable resources, and (6) substitution of plentiful resources for those in scarce supply.[36]

While promoting new technology, we must not forget that we are working with a finite and somewhat fragile resource base. This base can provide opportunities for "a good life" for a stabilized population for centuries to come. But to do so, it must be managed in accordance with sound ecological principles. Demands for the products of land must be kept in line with the earth's sustained capacity to produce,

---

[32]Cf. Jacob Rosin and Max Eastman, *The Road to Abundance* (New York: McGraw-Hill, 1953); and Eugene Holman, "Our Inexhaustible Resources," *Atlantic Monthly*, 189, no. 6 (June 1952), 29–32.

[33]National Agricultural Lands Study, *Final Report* (Washington, D.C.: Government Printing Office, 1981), p. 58.

[34]Norman Berg quoted in American Land Forum, Inc., *Land and Food: The Preservation of U.S. Farmland*, American Land Forum Report No. 1 (Washington, D.C.: American Land Forum, Inc., 1979), p. 25.

[35]James H. Anderson, "The Effects of Structure, Organization and Orientation on Decision Making and Prioritization in Agricultural Research," paper presented at American Association for the Advancement of Science Symposium, May 1984.

[36]President's Materials Policy Commission, *op. cit.*, vol. 1, pp. 132–39.

and exhaustible resources once used cannot be reused. Until the promise of a better future offered by new technology and improved resource management becomes a reality, society must face the issues squarely and take care not to "count its chicks before they are hatched."

Challenge of the future. When one considers the problems that lie ahead in meeting the expanding land resource needs of a growing population, it is hard not to be impressed by the divergent views expressed on this subject. On the one extreme, writers of the neo-Malthusian school speak of the future with dire forebodings. To them the threat of overpopulation is real and suggests future shortages, famine, and a "road to survival."[37] On the opposite extreme, some optimists predict an age of industrial synthesis that will transform the world into a new Garden of Eden. In their enthusiasm, they talk almost glibly of a world in which "notions like 'hunger,' 'food problems' and 'over-population' will become things belonging to a past barbaric age."[38]

In all probability, the world will follow a middle course between these extreme points of view. The problem of supplying the world's growing need for food, minerals, and other products will continue to loom large on the horizon. But it is not a hopeless problem. Countries such as the United States and Canada face no imminent danger of famine or starvation. Indeed, barring emergencies or a major catastrophe, their problem during the next decade is more apt to involve food surpluses than shortages. For many overweight individuals in these countries, the major nutritional problem will continue to be that of excessive food consumption in relation to the decreasing muscular-energy requirements of a pushbutton civilization.

Happy as the food situation may be in some areas, the problem of want and hunger remains acute on the world front. Properly used, the world's present resource base can suffice to provide increasing supplies of food and other materials for its present population. New land development and the adoption of new technology can do much in providing for the needs of a larger population. But there are limits to the carrying capacity of the world's resource base. No one knows where these precise limits are. Yet it should be obvious that mankind cannot enjoy the fuller life promised by science on a standing-room-only basis. This means that over the long run, population numbers must level off or society will suffer the squeeze of diminishing returns.

The real problem is not "Can the world produce enough?" but rather "At what cost?" and "With what adjustments?" As we face up to our land resource and raw material supply problems, more and more attention must be given to economic and institutional considerations and to the adoption of sound long-run resource management programs. On the world front, this will require better planning and resource

---

[37]Cf. footnote 3 in Chapter 2, p. 18. For other examples of this point of view, cf. Georg Borgstrom, *The Hungry Planet* (New York: Macmillan, 1965); Paul E. Ehrlich, *The Population Bomb* (New York: Ballantine, 1968); William Vogt, *Road to Survival* (New York: Wm. Sloan Associates, 1948); Fairfield Osborn, *Our Plundered Planet* (Boston: Little, Brown, 1948); and Edward M. East, *Mankind at the Crossroads* (New York: Scribner's, 1923).

[38]Rosin and Eastman, *op. cit.*, p. 57.

management, and it may call for new policies affecting economic development, trade, and immigration. These adjustments sound simple; but in the final analysis, they may require sweeping changes in national attitudes and policies. They almost inevitably will involve sacrifices on the part of some people together with some re-allocations of the world's resources and wealth.

Finally, we must plan for the future in terms of our available resources. In this process, we must recognize the need for maintaining and conserving our present resource base. We must do what we can to prevent waste and discourage exploitive practices that undermine future productive capacity. Much as we might wish it, we cannot depend on technology to come to our aid if we dissipate and waste our resource heritage. If we insist on strangling the goose that lays our golden egg, we must expect to find ourselves reduced to a state of poverty.

— SELECTED READINGS

Brown, Lester R. *et al.*, *State of the World 1985* (Washington: Worldwatch Institute, 1985).

Clawson, Marion, R. Burnell Held, and **Charles H. Stoddard**, *Land for the Future* (Baltimore, Md.: Johns Hopkins University Press, 1960).

Crosson, Pierre R., and **Sterling Brubaker**, *Resource and Environmental Effects of U.S. Agriculture* (Washington, D.C.: Resources for the Future, 1982).

Fisher, Anthony, C., *Resource and Environmental Economics* (Cambridge; Cambridge University Press, 1981), chap. 4.

Howe, Charles W., *Natural Resource Economics* (New York: John Wiley, 1979), chaps. 1–7.

National Commission on Materials Policy, *Material Needs and the Environment Today and Tomorrow* (Washington, D.C.: Government Printing Office, 1973).

Outdoor Recreation Resources Review Commission, *Outdoor Recreation in America* (Washington, D.C.: Government Printing Office, 1962).

U.S. Department of Agriculture, *Soil, Water, and Related Resources of the United States: Analysis of Resource Trends*, RCA 1980 Appraisal, Part II (Washington, D.C.: U.S. Department of Agriculture, 1981).

U.S. Forest Service, *An Assessment of the Forest and Range Land Situation in the United States*, FS-345 (Washington, D.C.: U.S. Department of Agriculture, Forest Service, 1980).

U.S. Water Resources Council, *The Nation's Water Resources: 1975–2000, Second National Assessment*, Vol. 1, Summary (Washington, D.C.: Government Printing Office, 1978).

5

INPUT-OUTPUT RELATIONSHIPS AFFECTING LAND USE

Land economics deals with our attitudes, behavior, and decisions concerning the use of real estate resources. Some of its most significant aspects involve the workings of the economic framework within which land use takes place. This framework can be described largely in terms of the principles that affect, condition, and control (1) the response of land as a factor of production to varying input combinations of capital, labor, and management; (2) the economic returns that accrue to land in the production process; (3) the factors that influence land resource development and resource conservation; (4) the location considerations that affect land use; and (5) the concept of real estate value and its measurement. Emphasis is given to these economic aspects of land resource utilization in this chapter and in the five chapters that follow.

BASIC ASSUMPTIONS OF ECONOMIC ANALYSIS

Realistic economic analysis calls for a broad understanding of the many factors that influence economic behavior. In their search for explanations, economic theorists ordinarily employ an inductive approach. They recognize that limitations of the human mind complicate the analysis of interrelations involving more than a few variables at any one time. With this fact in mind, they identify and isolate the factors that appear to have the most significant impacts on behavior. They then assume idealized situations in which the operation of all other factors can be ignored or held constant. These assumptions permit the construction and manipulation of economic models that highlight the interaction of limited numbers of factors on one another and in so doing facilitate the development of meaningful theories and explanations of the relationships that affect economic behavior.

This approach is basic to fundamental economic analysis. As a method of analysis, it has the advantage of focusing attention on important relationships that would otherwise be hidden by the simultaneous operation of the maze of variables that complicate everyday life. The validity of this approach is always conditioned by the nature of the assumptions on which it is premised. Unrealistic assumptions give rise to unrealistic theories. Because of this situation, theories should always be tested in the light of reality and their worth appraised in terms of their usefulness in explaining real world problems.

Like most theoretical concepts, the body of economic thought that has been developed to explain the patterns and processes of land utilization rests on a number of basic assumptions. The most important of these is the assumption that operators are rational beings who behave in a logical and reasonable manner. This is probably the most basic assumption in all economic analysis. As such, it underlies a number of other important assumptions. Two of the most important of these involve the assumption that operators normally try to maximize their self-interests and that prices tend to allocate resources.

Before turning to these two basic concepts, it should be noted that economic analysis is usually based on simple cause-and-effect reasoning with the assumption of "other things being equal." The far-reaching nature of this assumption can be illustrated by the simple statement that "other things being equal, people will buy more of a good at a lower price than at a higher price." The condition of "other things being equal" in this case assumes: (1) no appreciable change in consumer income or consumer tastes, (2) no change in the price of other goods, (3) no anticipation on the part of buyers of further price reductions, (4) no new substitute for the good in the market, and (5) no complications of prestige value, which may lead buyers to purchase products simply because they are high in price.[1]

Frequent use is made of economic models and examples that assume conditions of perfect competition with perfect knowledge on the part of buyers and sellers, perfect mobility of goods and productive factors, and a perfectly elastic supply of productive factors. These assumptions can be criticized as being unrealistic. But they have value because they make it possible for analysts to ignore the operation of many factors while they focus attention on the interaction of specified factors on economic behavior.

Maximization of Individual Self Interests

The so-called economic man who is often visualized as the prime mover in economic society, is motivated by a desire to maximize net economic returns. This economic person has an uncanny awareness of alternative opportunities and of what can be logically expected to happen under varying production, price, and cost situations. The assumption of perfect knowledge gives this economic person a tremendous advantage over the operators found in the real world. The principle of individual

[1] Albert L. Meyers, *Elements of Modern Economics*, 3rd ed. (Englewood Cliffs, N.J.: Prentice-Hall, 1948), pp. 3–4.

self-interest, however, is generally realistic. Up to a certain point, it is descriptive of most economic activity. Business operators are interested in pushing plant production to its optimum level; farmers attempt to combine their productive factors in such a way as to enjoy maximum income from their activities; workers demand the highest wages they can get for their labor.

All rational individuals try to optimize their value returns and the satisfactions they derive from life. But wide differences exist between individuals in the extent to which they measure their satisfactions in monetary returns. Some people place a high value on profits and the maximization of monetary returns *per se*. Most, however, regard monetary returns as an intermediate rather than a final goal. For them, money is a means to the attainment of more ultimate ends. When profit maximization conflicts with other goals, they often settle for less money and more leisure, more security, or more of some other goal. Recognition of this factor is important, because it explains why landowners frequently fail to behave in a strictly economic manner, even when it might be clearly to their financial interest to use real estate resources in a different manner than they do.

The challenge operators face in maximizing returns would be much simpler if they did possess perfect foresight and knowledge, if they could predict in advance the exact combination of factors and economic decisions that would prove most profitable. Unfortunately, this is not a human trait. Operators seldom enjoy perfect security in their economic expectations. Uncertainties often cause even the best-laid plans to go astray. As a result, almost everyone must operate at least partially in the dark. Even when they are well supplied with information, they make numerous decisions with no certainty concerning the net outcomes. Yet in making these decisions, they and their families must bear the consequences, whether good or bad.

People differ in their reactions to this situation. Some are inclined to take calculated risks or perhaps even gamble on an occasional long shot. Others place much more emphasis on security. This basic difference in willingness to take risks often has important effects on economic behavior. Some operators who take long chances succeed in parlaying their earnings into substantial fortunes; others may end up without their shirts. On the other extreme, security-minded people often act too conservatively. They may muddle along with old, tested, and proved practices, while more aggressive operators profit by accepting new techniques and by readily adapting themselves to new and changing situations.

Successful operators in the modern world must make decisions and bear the consequences of their decisions. There is no simple formula for success in the decision-making process. Even the most skilled operators expect an occasional setback. Experience shows that most successful operators find that they can best maximize their returns by plotting a course between the extremes of overconservatism and outright gambling. These operators are interested in profits; but they also accept the concept of the "minimax."[2] They seek to maximize their returns but at the same time attempt to minimize potential losses.

[2] Cf. John McDonald, *Strategy in Poker, Business and War* (New York: Norton, 1950); also John von Neumann and Oskar Morganstern, *Theory of Games and Economic Behavior* (Princeton, N.J.: Princeton University Press, 1944).

In following this policy, they take a forward view in their decisions and carefully consider the available facts before they act. Their knowledge and willingness to take risks steer them beyond a policy of inaction. At the same time, intuition may cause them to back away from long-shot decisions that could lead to fortune but will more likely result in failure. This middle road often leads to less spectacular economic action; but it provides operators with a reasonable return and saves many from possible financial ruin.

Prices and the Allocation of Resources

Economics is sometimes described as the science that deals with the allocation of scarce resources. In this sense, it deals with prices, because in our society the amounts people are willing to pay usually determine who gets what.

Rights for the use of land resources ordinarily go to those buyers who can bid up and pay the highest prices. Land resources also tend to gravitate to those uses that command the highest market prices and offer the highest net returns. They seldom move into production without some promise of a suitable market for their products. Rising price levels often favor the bringing of additional land into use and the more intensive use of areas already in use. Declining price levels, on the other hand, can force retrenchment policies, shifts to lower uses, and sometimes land abandonment.

Though prices tend to allocate resources under free market conditions, there are occasions when other factors interfere. Factors such as haste, ignorance of the facts, custom, conspicuous consumption, or the maximizing of other than monetary returns often prevent prices from playing their normal role in the allocation of resources. Failures of prices to allocate resources in accordance with accepted concepts of distributive justice may lead to ameliorative measures. Merchants sometimes ration their sales of scarce or price-leader commodities. Governments follow similar policies when they institute price control and rationing programs and assign priorities for the purchase and use of vital materials. Community chest, charity, and public welfare programs also use other resource-allocation criteria when they make resources available to people who have difficulties in commanding their use in competitive open markets.

THE CONCEPT OF PROPORTIONALITY

Most of our economic thinking concerning effective land utilization starts with the concept of *proportionality*. This concept involves the rationale operators use and the skills they show in combining and proportioning their various production inputs in their effort to maximize production goals. In the normal production process, operators must always add inputs of capital, labor, and management to their land if they are to secure production of the products they desire. As Ely and Wehrwein have observed, "Land in itself is not productive. It yields wheat, forest products, or office space only when labor and capital are applied to it."[3]

---

[3] Richard T. Ely and George S. Wehrwein, *Land Economics* (Madison: University of Wisconsin Press, 1964), p. 50. Originally published by Macmillan, 1940.

Regardless of the purposes for which land is used, operators always find it necessary to combine capital and labor with real estate resources in the economic production process. The amounts of capital and labor needed vary with different uses of land. Large inputs of capital and labor per unit of land area are ordinarily needed with commerical and industrial developments and with many farming activities. Forest and grazing operators, in contrast, may benefit from the fact that trees and grass can grow without human assistance.

Producers ordinarily try to secure the most economically productive combinations of their available input factors. Their decisions are guided by a natural desire to maximize their net returns. Ignorance of facts, lack of knowhow, or shortages of capital or labor can dampen their success in attaining this goal. By working with the information and resources they have, they usually try to make input-output decisions that provide the highest surplus of returns above costs. Their success in this regard ordinarily calls for awareness and appreciation of the workings of the law of diminishing returns plus ability to recognize and adjust to the strategic roles limiting factors play in the production process.

The Law of Diminishing Returns

It has long been observed that whenever successive inputs of a productive factor are added to a limited fixed factor, a point is soon reached after which the additional or marginal output of product per unit of input decreases and eventually becomes a negative quantity. This principle is known as the *law of diminishing returns*. By its very nature, it is one of the most important factors that affect people in their use of land. Without the operation of this principle, operators could concentrate all of their production activities on one spot. They could raise the world's entire food supply in a flowerpot. Everyone could be housed on a single building site.

The concept of diminishing returns can best be illustrated by the use of an example such as that reported in Table 5-1. This table assumes a single unit of land as the fixed input factor (column 1), with composite homogeneous units of capital and labor treated as variable input factors (column 2). Up to a certain point (the fourteenth variable input unit), the addition of each successive input of capital-labor to the fixed factor results in an increase in total output (column 3). This total is called the *total physical product*.

The average yield or output of product per variable input unit is known as the *average physical product* (column 4). This measure is determined by simply dividing the total physical product by the number of variable inputs used in its production. For example, the use of the eighth variable input unit in Table 5-1 brings a total physical product of 78 and thus results in an average product of 9.75 (78 divided by 8 = 9.75). The point of highest average return in this illustration comes with the tenth variable input unit.

In addition to the concepts of total and average physical product, operators are also interested in the amount of output associated with the use of each successive additional input unit. This concept is known as the *marginal physical product*

TABLE 5-1. Illustration of Operation of Law of Diminishing Returns

INPUTS OF FIXED FACTOR (LAND)	INPUTS OF VARIABLE FACTOR (CAPITAL-LABOR)	UNITS OF TOTAL OUTPUT (TOTAL PHYSICAL PRODUCT)	AVERAGE UNITS OF OUTPUT PER VARIABLE INPUT UNIT (AVERAGE PHYSICAL PRODUCT)	INCREASE IN OUTPUT PER ADDITIONAL VARIABLE INPUT (MARGINAL PHYSICAL PRODUCT)	VALUE OF MARGINAL PRODUCT AT: 50¢ per unit	80¢ per unit	$1.20 per unit
1	1	2	2	2	$1.00	$ 1.60	$ 2.40
1	2	6	3	4	2.00	3.20	4.80
1	3	13	4.333	7	3.50	5.60	8.40
1	4	23	5.75	10	5.00	8.00	12.00
1	5	35	7	12	6.00	9.60	14.40
1	6	49	8.167	14	7.00	11.20	16.80
1	7	64	9.143	15	7.50	12.50	18.00
1	8	78	9.75	14	7.00	11.20	16.80
1	9	91	10.111	13	6.50	10.40	15.60
1	10	102	10.2	11	5.50	8.80	13.20
1	11	111	10.091	9	4.50	7.20	10.80
1	12	118	9.833	7	3.50	5.60	8.40
1	13	122	9.385	4	2.00	3.20	4.80
1	14	123	8.786	1	.50	.80	1.20
1	15	121	8.07	-2	-1.00	-1.60	-2.40

(column 5). In the example, the use of six variable inputs results in a total physical product of 49, while a seventh input pushes total production up to 64. The difference between these two totals (64 - 49 = 15) represents the additional yield or marginal physical product associated with the use of the seventh variable input.

The concepts of total, average, and marginal physical product are often symbolized by the letters *TPP*, *APP*, and *MPP*, respectively. These concepts may be depicted graphically as in Figure 5-1. As this diagram indicates, the changes in total production associated with the use of each successive variable input unit suggests a series of steps that go up to a peak level and then start down again. For analytical purposes, these steps are usually smoothed out and depicted by production curves. The *TPP* curve in Figure 5-1 shows the cumulative increase in total physical product (measured on the vertical axis) that comes with the addition of each successive input of variable factor (measured on the horizontal axis). Whenever an input-output relationship can be described by a continuous curve of this type, it may be described as a *production function.*

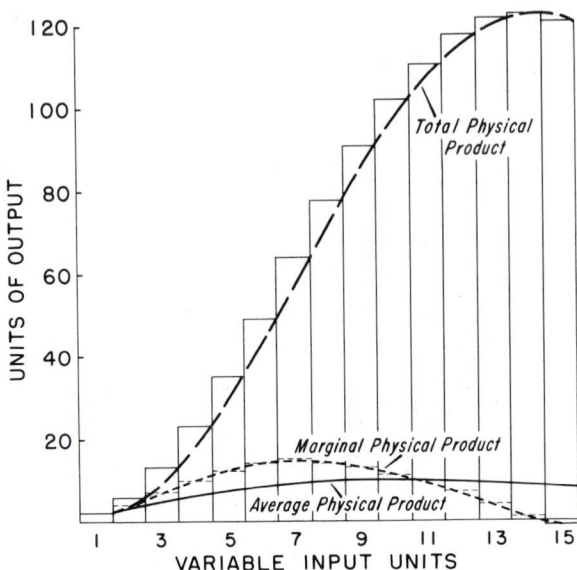

FIGURE 5-1.   Illustration of production function and input-output relationships involved in operation of physical law of diminishing returns.

As Figure 5-1 indicates, every production function involves three points of diminishing return. Total production increases at an increasing rate until the *MPP* curve reaches its peak. From this point on, the marginal physical product diminishes while the total physical product continues to increase at a decreasing rate. The *TPP* curve reaches its highest level—the point of total diminishing returns—at the same point on the variable input scale as that at which the *MPP* curve intersects the baseline and becomes zero. Any additional application of variable inputs beyond this point results in both a decrease in total physical product and a negative marginal return.

The *APP* curve always reaches its maximum height at the point at which it intersects the declining *MPP* curve. Beyond this point, the average physical product gradually decreases. An operator could continue to add variable inputs until the added inputs reduce both the total and the average physical products to zero. It does not make economic sense, however, to add inputs beyond the point of highest physical production.

Economic Law of Diminishing Returns

Emphasis has been given thus far to the physical law of diminishing returns. This concept is basic in production, but most operators want to view their input-output prospects in monetary terms. They think not only of physical inputs and outputs but also of the costs and returns associated with these units. They are concerned not so much with the maximization of physical output as with maximization of economic net returns. In short, they are concerned with the economic law of diminishing returns.[4]

Transition from the physical to the economic concept of diminishing returns can be achieved simply by assigning a cost to each of the input factors and a market value or price to each unit of output produced. With this adjustment, one can speak of the total, average, and marginal returns secured in the production process and of the total, average, and marginal costs associated with these respective measures of return. With this transition, operators find it most profitable to push production to the point at which the value of their marginal physical product equals or just exceeds the cost associated with its production. This is the *point of diminishing economic returns*. As long as operators combine their variable input factors around their scarce or limiting factors, they can always expect their highest net return at this point.[5]

Most economic analysis involving production problems builds on the economic concept of diminishing returns. In their applications of this concept, economists sometimes find it desirable to calculate costs and returns on an input-unit basis. On other occasions they find it more appropriate to deal with the costs and returns associated with units of output. Both approaches have merit; both have numerous applications in land economics analysis; and both involve applications of the same basic principle.

Input-unit approach. As long as costs and returns are computed on an input-unit basis, one can shift from the physical to the economic concept of diminishing returns simply by assigning a value to each unit of physical product and a production cost to each unit of variable input factor. With this adjustment, the value of the marginal physical product may be described as the marginal return per input unit, or more simply as the *marginal value product* (*MVP*). Similarly, the concepts of *total value product* (*TVP*) and *average value product* (*AVP*) are used to describe the value of the total and average physical products, respectively.

On the cost side, the term *factor cost* may be used to describe the costs associated with the variable factor inputs. The additional cost associated with the application of each last successive variable input thus becomes the *marginal factor cost* (*MFC*), and the average cost per input unit is known as the *average factor cost* (*AFC*).

One can illustrate the *value product* or input-unit approach for explaining eco-

---

[4] Different terms are frequently used in discussions of this concept. Some authors describe the physical concept as the *law of diminishing productivity* or as the *law of diminishing physical outputs*. The economic concept is often referred to as the *law of diminishing returns*, the *law of variable proportions*, or the *law of proportionality*. Some writers limit the concept of diminishing returns to combinations in which land is treated as the central or limiting factor and use the concept of variable proportions to signify other types of combinations.

[5] The difference between an operator's total production costs and the value of his or her total product is referred to here as *net return*. Depending on one's assumptions concerning the nature of the fixed factor and the items treated as variable inputs, this net above cost can also be described as an economic surplus, as land rent, or as operator's profit. As explained in the next chapter, this surplus is land rent in those instances in which land resources are treated as the fixed factor. It is operator's profit only when management is treated as the fixed factor.

nomic input-output relationships simply by assuming a value of $1 for each of the physical output and variable input units listed in Table 5-1. With this assumption and accepting land as the fixed factor, an operator will find it profitable to push production to the thirteenth input unit. At this point, there is a marginal value product of $4, a marginal factor cost of $1, and a realized net return of $3 for the last variable input unit. If a fourteenth input is added, the marginal value product of $1 just equals its marginal factor cost, and there is no gain nor loss. Our operator could choose to add the fourteenth input if the cost of the last variable input includes a fair allowance for the operator's management, capital and labor but may also choose to withhold this last input because it contributes nothing to net returns above variable input costs. Should a fifteenth input be added, the marginal factor cost would still be $1, and the marginal value product would become a negative quantity.

If the operator in this example thinks in terms of total net returns, he or she may ask why it is necessary to push production to the point at which marginal value product equals or just exceeds the cost of the last variable input? Why not stop at some point of higher marginal return? A few simple calculations with the cost assumptions above and the production data reported in Table 5-1 show why it is most profitable to push production to the point at which $MFC = MVP$. If the operator stops with the tenth input, there will be a net return of $92 (total physical product valued at $102 less $10 in variable input costs). Use of an eleventh input brings the net return up to $100. Total net returns rise to $106 with the twelfth variable input and $109 with the thirteenth input. The fourteenth input barely pays for itself, and total net returns drop to $106 if a fifteenth input is applied.

The precise number of variable inputs one should combine with the fixed factor always depends on cost and price relationships. When variable inputs cost $4.75 each and the units of product are priced at 50 cents each (see the last three columns of Table 5-1), it pays to push production only to the tenth input. At a price of 80 cents, it pays to push production to the twelfth input; and at a price of $1.20, it is most profitable to apply thirteen inputs of variable factor. If the unit cost of the variable inputs is raised, operators will reduce the number of variable inputs they use. Conversely, if variable input-unit costs are lowered, it will usually pay to apply additional inputs.

Value product analysis highlights the choices available to operators as they make decisions concerning combinations of successive numbers of variable inputs with their fixed factor. Operators want to maximize their net returns. They can accomplish this goal by using the number of variable inputs that permits the maximum spread between their total value product and their total factor cost. As Figure 5-2A shows, this maximum spread always comes to the point at which MVP equals or just exceeds MFC. This mathematical truism favors the focusing of economic analysis on the relationships between marginal value products and marginal factor costs and between average value products and average factor costs.

Figure 5-2B provides a diagrammatic example of the simplest form of value product analysis. In common with similar value product or input unit diagrams, it

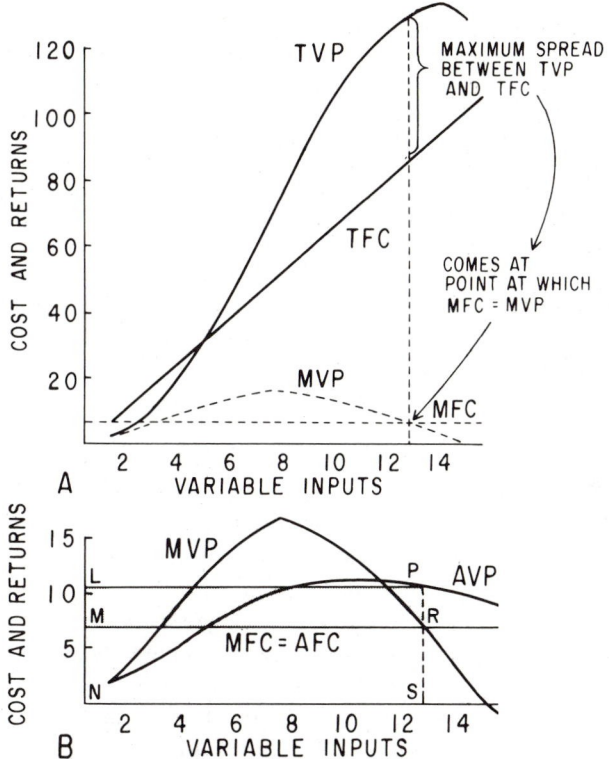

FIGURE 5-2.   Use of value product curves to determine the net return to a fixed input factor (land) at most profitable point of operations.

assumes a single fixed factor (land), a single type of homogeneous variable input, a fixed price per unit of product, and a fixed cost per variable input.[6]

The example in Figure 5-2 assumes the production function reported in Table 5-1, an average factor cost of $7 per variable input unit, and an average market value of $1.10 for each unit of product. With this combination of factors, one would find it advantageous to stop with the twelfth variable input, because this is the last logical breaking point before $MFC = MVP$. At this point, total value product (AVP times the number of variable inputs used) is represented by the large rectangle $LNSP$, while total factor cost (AFC times the number of variable inputs used) is

[6]Modification of these assumptions can bring significant changes in the basic value product diagram. When two fixed factors are assumed, for example, it is sometimes expedient to assign fixed or overhead costs to one of them. Average overhead or fixed cost curves can then be added to the diagrams as they frequently are in cost curve analyses. Assumption of two types of variable inputs with different production functions complicate the analysis by requiring three dimensional diagrams, whereas assumptions involving three or more types of variable inputs call for elaborate mathematical models such as those used in linear programming. An assumption of declining product prices with increased production requires adjustments in the slopes of the MVP and AVP curves. Similarly, an assumption that factor costs will increase as more factors are used in production calls for separate MFC and AVC curves.

represented by the lower rectangle *MNSR*. The rectangle *LMRP* represents the share of the total value product above factor costs that may be credited to the fixed factor.

Computed in arithmetic terms, the operator secures a total value product of 10.817×12 (*AVP* $\times$ number of variable inputs used) = $129.80. Total factor costs are 7×12 (*AFC* $\times$ number of variable inputs used) = $84, and the net return is $129.80 - $84.00 = $45.80. With only eleven variable inputs, the net return would be $45.10; with thirteen, it would be $43.20.

Transition to cost curves. Economists make frequent use of the input-unit or value product approach, particularly when they deal with production and management situations. They also have need at times for viewing production in terms of cost and returns per output unit. Output-unit or *cost curve analysis* is used for this purpose. Like the input-unit approach, this approach has its own set of production concepts.

When product values are computed on an output-unit basis, they are called *returns* or *revenue*. The value of the total physical product thus becomes the *total return*, and the average value associated with each output unit is the *average return*. In similar fashion, the value of the marginal or last additional unit of production is called the *marginal return*. With this approach and the assumption of a uniform price for all product units, the concepts of average return and marginal return can be depicted in diagrams by a horizontal line, which represents the price level.

Three cost concepts—total cost, average cost, and marginal cost—also play significant roles in cost curve analysis. *Total cost* represents a sum of all the production costs incurred at any given point in the production process. With the example assumed in Table 5-2, total cost is equal to total factor cost and involves the cost associated with the use of x number of variable inputs. The term *average cost* (or *average total unit cost*) is used to describe the proration of total costs among the various units of output (total costs divided by the number of units of output).[7]

[7]It is common practice in production analysis to distinguish between *variable costs* and *fixed costs* as components of the concepts of total and average costs. When this practice is followed, total cost represents the sum of the total variable and fixed costs, and average cost represents the sum of the average variable and fixed costs. *Variable cost* in this case represents the aggregate cost of the variable inputs used in production. When plotted in a diagram, average variable costs decline with the increasing output secured from use of the initial inputs of variable outputs and reach their lowest level at the point of highest physical product per input unit. Beyond this low-cost level, they rise with the higher marginal costs per unit of output associated with the decreasing physical product secured by each successive additional variable input.

Fixed costs involve costs associated with overhead and other outlays, which are viewed as fixed throughout the production process. These costs are fixed for the production period and are just as high when one variable input is used to produce two units of product as when fourteen inputs are used to produce 123 units. Average fixed costs decline steadily as production mounts to its maximum level.

Overhead and fixed cost assumptions involve identification of two or more fixed factors in production. This approach is frequently used when some factor other than land is treated as the fixed factor, in which case an allowance of a fixed payment for land and taxes is often viewed as an overhead cost. No assumption of fixed costs is used in Table 5-2 or Figures 5-2 or 5-3, for the simple reason that land is treated here as the single fixed factor in the analysis.

TABLE 5-2. Illustration of Economic Costs and Returns Calculated on an Input-Unit and on an Output-Unit Basis Assuming the Production Function Reported in Table 5-1, a Standard Price of $1.10 per Unit of Output, and a Uniform Cost of $7.00 per Variable Input-Unit

NUMBER OF VARIABLE INPUTS USED WITH FIXED FACTOR	TOTAL UNITS OF PHYSICAL PRODUCT PRODUCED (TPP)	MARGINAL PHYSICAL PRODUCT (MPP)	STANDARD PRICE PER UNIT OF OUTPUT (AR and TR)	VALUE OF TOTAL PHYSICAL PRODUCT (TVP and TR)	MARGINAL VALUE PRODUCT (MVP)	AVERAGE VALUE PRODUCT (AVP)	UNIFORM COST OF VARIABLE INPUT (AFC and MFC)	TOTAL COST OF VARIABLE INPUTS USED (TFC and TC)	AVERAGE COST PER UNIT OF OUTPUT (ATUC or AC)	MARGINAL COST PER OUTPUT UNIT (MC)	NET RETURN (TVP - TFC or TR - TC)
1	2	2	$1.10	$ 2.20	$ 2.20	$ 2.20	$7.00	$ 7.00	$3.50	$3.50	($4.80)
2	6	4	1.10	6.60	4.40	3.30	7.00	14.00	2.333	1.75	(7.40)
3	13	7	1.10	14.30	7.70	4.77	7.00	21.00	1.615	1.00	(6.70)
4	23	10	1.10	25.30	11.00	6.33	7.00	28.00	1.217	.70	(2.70)
5	35	12	1.10	38.50	13.20	7.70	7.00	35.00	1.00	.583	3.50
6	49	14	1.10	53.90	15.40	8.98	7.00	42.00	.857	.50	11.90
7	64	15	1.10	70.40	16.50	10.06	7.00	49.00	.766	.467	21.40
8	78	14	1.10	85.80	15.40	10.73	7.00	56.00	.718	.50	29.80
9	91	13	1.10	100.10	14.30	11.12	7.00	63.00	.692	.538	37.10
10	102	11	1.10	112.20	12.10	11.22	7.00	70.00	.686	.636	42.10
11	111	9	1.10	122.10	9.90	11.10	7.00	77.00	.694	.778	45.10
12	118	7	1.10	129.80	7.70	10.82	7.00	84.00	.712	1.00	45.80
13	122	4	1.10	134.20	4.40	10.32	7.00	91.00	.746	1.75	43.20
14	123	1	1.10	135.20	1.10	9.66	7.00	98.00	.797	7.00	37.30
15	121	–1	1.10	133.10	–1.10	8.87	7.00	105.00	.868	—	28.10

Marginal cost represents the addition to total costs associated with the production of each last additional unit of output (cost of the last variable input divided by the marginal physical product).[8]

With these concepts of returns and costs and the basic production data reported in Table 5-2, one can illustrate input-output relationships with cost curve diagrams such as those presented in Figures 5-3 and 5-4B. Again the operator seeks to maximize net returns and finds that this can be done by pushing production to the point at which the maximum spread exists between total returns and total costs, (see Figure 5-4B). This point corresponds with the number of units of output secured in Figure 5-3 at the point at which $MC = MR$. This is the *point of diminishing economic returns* and as such it corresponds with the point at which $MFC = MVP$ in Figures 5-2 and 5-4A.

With the cost curve diagram shown in Figure 5-3, the operator should produce to the 118th unit of output because this is the last unit of output that can be produced at a cost less than its product value. The 118th unit has a value of \$1.10 and a marginal cost of \$1.00; the 119th unit would cost \$1.75. The operator's total returns (total units produced times the average return per unit) are represented in the diagram by the large rectangle *LNSP* in Figure 5-2. Production cost (total units produced times the average cost per unit) is represented by the lower rectangle *MNSR*, and the return to the fixed factor is represented by the rectangle *LMRP*. This net return may also be calculated by subtracting average costs from average return and multiplying the difference by the number of output units. Measured in this way

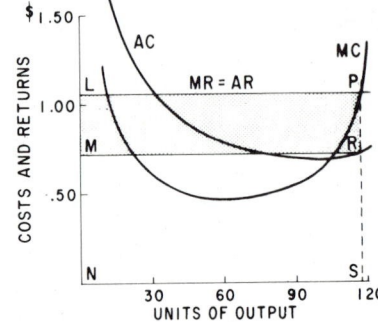

FIGURE 5-3   Use of cost curves to determine the net return to a fixed input factor at the most profitable point of operations.

[8]The marginal cost of each successive output unit, when plotted in a diagram as in Figure 5-3, declines rapidly with the increasing production per unit experienced at the beginning of the production process. The *MC* curve then levels off, reaches its lowest level at the point of highest marginal physical product, and then rises—gradually at first and then at an accelerating rate—until it goes straight up at the point of total diminishing returns. In its upward swing the marginal cost curve intersects both the average variable cost curve (if computed separately) and the average cost curve at their lowest points.

The marginal cost curves shown in most diagrams suggest continuous input-output relationships. In practice, a discontinuous relationship normally exists with whole groups of output units being associated with the cost of successive input factors. Marginal cost in these cases represent the average cost of each last group of output units.

($1.10 - 0.7119 = 0.3881 \times 118 = 45.80), the net return in Figure 5-3 corresponds with that reported with the value product diagram in Figure 5-2B.

SOME APPLICATIONS OF PROPORTIONALITY

At this point, one might well ask how much consideration average operators give to the concept of proportionality. Is it a fancy theory or does it have real-life significance? In addressing this question, first consideration should be given to the impact the proportionality concept has on managerial decisions. Other examples of its worth include the recognition it gives to limiting factors in production and its implications for public resource policy.

Use in Managerial Decisions

The central goal with proportionality is to combine the inputs used in production to provide a maximum net return. All producers are concerned to a greater or lesser degree with this objective. True, many have never heard of "proportionality," and some still pattern their practices on those of their fathers. But most successful operators have a definite feel for this concept and realize that their success depends in large part on the skill with which they proportion the production factors they have at their disposal. Economic efficiency is attained when operators proportion their inputs in a manner that maximizes their net returns; and maximization of net returns is the end objective of most typical short-run business operations. It must be noted, however, that this objective gives no weight to distributional, social justice, or social well-being objectives. Moreover, the smooth path to attainment of an operator's economic efficiency goals can be disrupted by market imperfections and instability. It can also be adversely affected by monopolistic and oligopolistic pressures and by the unequal access different operators have to knowledge and needed resource inputs.[9]

Examples of day-to-day applications of proportionality occur with almost every type of land use. Industrialists employ this concept when they decide how much raw material to use, how many workers they should hire, and what adjustments they should make for changing costs or prices. Commercial businesspeople apply it when they consider how much floor space they will use, how much they should spend on advertising, and what types of goods and services they will offer. Farmers employ it when they decide how much seed or fertilizer to use or how much grain they should feed their livestock.

Foresters consider proportionality when they decide how much they can afford to spend on tree planting and stand-improvement measures and whether they

[9]Cf. Alan Randall, *Resource Economics: An Economic Approach to Natural Resources and Environmental Policy* (New York: John Wiley, 1981), chap. 8; and Richard H. Day, "Adaptive Economics and Natural Resources Policy," *American Journal of Agricultural Economics*, 60 (1978), 276–83.

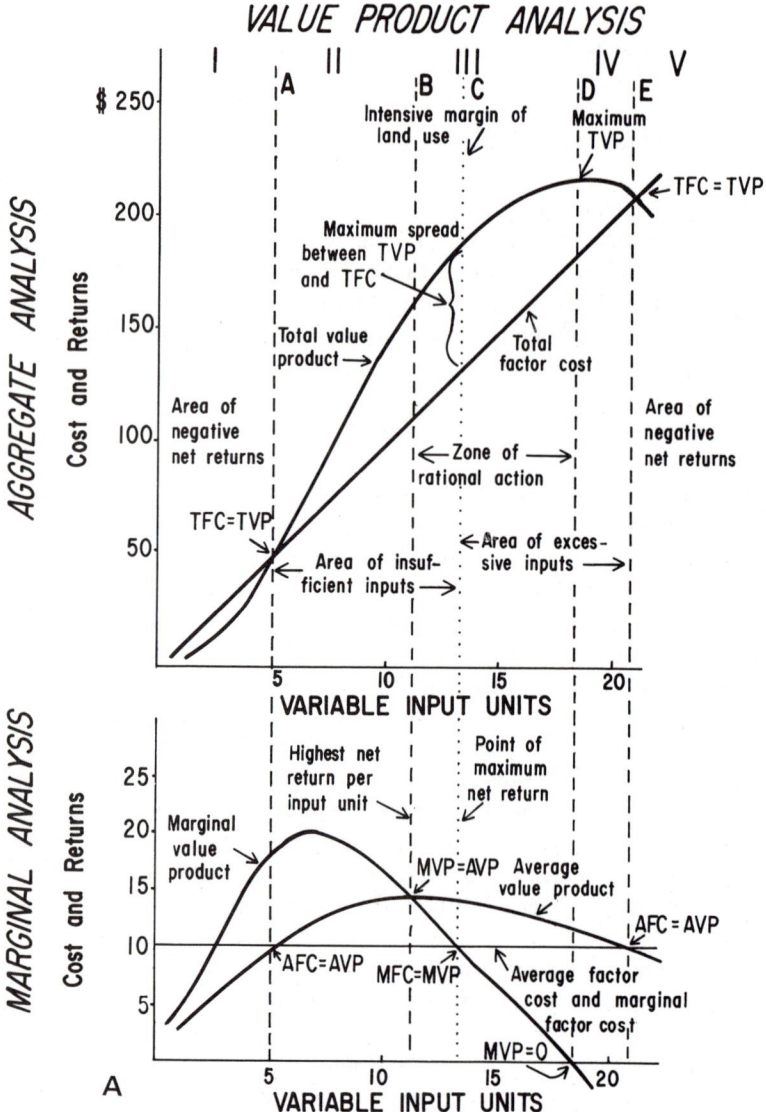

FIGURE 5-4. Comparison of the value product and cost curve approaches to input-output productivity analysis.

should cut their trees for pulp or hold them for sawtimber. Proportionality affects the decisions of real estate developers who want to know the optimum size, height, and layout of an office building they plan to build or the amount and nature of the site improvements they provide for a shopping center.

Proportionality also affects operators who are motivated, at least in part, by non-economic goals. Architects and planners use this concept when they attempt to

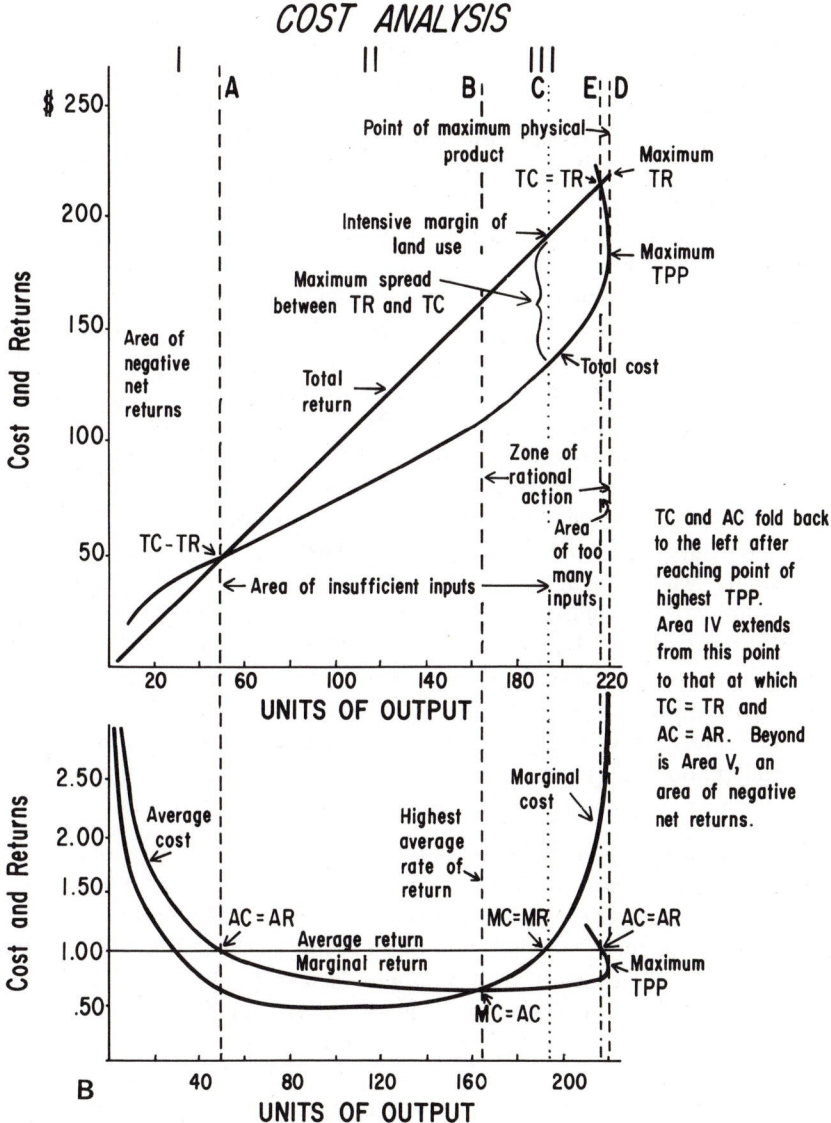

combine the factors in their plans to achieve the highest standards of quality possible within the limits of their budget specifications. Public administrators and other program or project managers use this approach when they determine the optimum allocation of the funds placed at their disposal.

The success with which operators apply the proportionality concept is conditioned both by the clarity of their reasoning and by their response to the problems of uncertainty and imperfect knowledge. With perfect knowledge and foresight, average operators would find it relatively easy to use static input-output models in proportioning inputs to the exact point of maximum returns. But these assumptions seldom apply in the real world.

Real-life operators seldom have the ability to predict in advance the exact combinations of output that will bring them the highest net return. New enterprises call for experiments with various input combinations and identification of the points of highest economic and physical return. Even when one has the necessary experience or data to determine the optimum combinations of input factors, it is necessary to cope with vagaries of climate and nature and with uncertainties occasioned by changing cost, price, market supply, and consumer demand relationships. With this complicated situation, one might logically ask how an operator can apply the proportionality concept in the decision-making process. How can one use static economic models with rigid assumptions such as those shown in Figures 5-2 and 5-3 as guides when the operator lacks perfect knowledge and foresight, must be concerned with unpredictable changes in input costs and product prices and yields, lacks the luxury of dealing with a single production function, and has to operate with a specified or limited supply of variable input factors? What does one do when land is not fixed in supply, when it is not the truly limiting factor in production, when important inputs are large and indivisible, or when it is possible to substitute other inputs for those in short supply? Real world operators make adjustments in their proportioning decisions as they relax these static model assumptions. Some of their most important adjustments involve (1) efforts to operate within a zone of rational action, (2) reformulation of planning models to recognize changing production situations, (3) adaptations for multiple-production functions, and (4) acceptance of the equi-marginal principle in operations involving two or more enterprises.

Zone of rational action. Most successful business operations involve what is known as the *zone of rational action*. This zone covers the range of input-output combinations with any given production function within which producers can best expect to maximize their economic returns. Lacking perfect foresight, operators seldom find it possible to gauge their inputs to the exact point of diminishing economic returns. But by following economic input-output models for particular enterprises, they can push production to points near the economic optimum. In this sense, one can visualize the production point at which $MFC = MVP$ or $MC = MR$ as the bull's-eye on a target. In aiming at this target, operators may sometimes overshoot this mark, sometimes undershoot it. As long as operations are kept within a reasonable range, with producers using their feel for proportionality to consistently hit the target area, economic returns can be optimized even though operators may seldom hit the bull's-eye.

Shown graphically in Figure 5-4A, the zone of rational action can be visualized quite simply as the distance between B and D or between the points at which $MVP = AVP$ and $MVP =$ zero in the marginal value product diagram (Figure 5-4A). With cost curve analysis (Figure 5-4B), this zone again occurs between B and D or the

corresponding points at which $MC = AC$ (the lowest point on the average cost curve) and the point of maximum physical product. The overall rationale for designating this area as a zone of rational action is illustrated by the value product curves shown in Figure 5-4A.[10]

The production map depicted in Figure 5-4A has several subareas. Areas I and V are areas of negative net returns, in which average factor costs exceed average value product and total factor costs exceed total value product. No rational producer would stop operations in area I or continue operations into area V.

The area between points A and E in Figure 5-4A can be divided into two, three, or four production zones. The zone between A and C may be described as an "area of insufficient inputs" and that between C and E as an "area of excessive inputs." C comes at the point at which $MFC = MVP$ and represents the production point that provides operators with the highest possible net economic return. Within the area of insufficient inputs, operators have a strong economic incentive to push production as close as possible to C and at least as far as point B, at which they secure their highest average return per variable input. Likewise, in the area of excessive inputs between C and E, it is best to stop production near C and if possible avoid the rapid decline in net returns associated with operations near and beyond point D, the point of maximum physical product at which $MVP = 0$.

Figure 5-4A designates the portion of the area of insufficient inputs between A and B as area II and the portion of the area of excessive inputs between D and E as area IV. The most productive portions of the areas of insufficient inputs and excessive inputs lie between points B and C and C and D, respectively. These two areas are shown together in Figure 5-4A as area III and represent the zone of rational action.[11]

The rational action zone can be thought of as the larger target at which operators

[10] The zone of rational action is usually illustrated with value product curves. As Figure 5-4B indicates, however, it can also be shown with cost curves. In this diagram, area I, with its prospect of negative net returns, extends to the point at which $TC = TR$ and $AC = AR$. Area II extends from this point (point A) to the point of lowest average cost per output unit at B. Area III is the zone of rational action and extends from B to D, which represents the point of maximum physical product and highest total revenue. Once the production of output units reaches its maximum level at D, the average and total cost curves fold back to the left. This reversal of direction occurs because additional variable inputs are being applied beyond point C where $MPP = 0$. Further applications of variable inputs in area IV between D and E are not economically rational, even though a steadily decreasing amount of net return will exist until one reaches point E, at which the total cost and average cost curves again cross the total return and average return curves. Area V beyond point E is a second area of negative net returns.

[11] An added rationale for setting the boundaries of this zone at B and D is found in the economic reasoning operators would be expected to accept with changing assumptions concerning their factor costs. With rising factor costs, one could afford to operate as long as average factor costs do not rise above the highest point on the average value product curve at B. Operations at higher cost levels would be irrational, because total factor costs would then exceed total value product. Similarly, one could operate to advantage with lower costs but would not expect to operate beyond D, because this would be the point of total physical diminishing returns at which MVP becomes a negative quantity.

aim in their production decisions and which they hopefully will hit as they combine their inputs in ways they hope will maximize their net returns above production costs. Business operators have strong economic incentives for operating within the zone of rational action. Yet even the best managers sometimes undershoot or overshoot this goal. Their failure to operate in the rational stage can be attributed to factors such as faulty foresight, inadequate knowledge about production relationships, lack of know-how or managerial ability, or poor allocation of resources. Unexpected changes in factor costs and sudden changes in market prices can cause a firm to produce for a time at an uneconomic level. Natural catastrophes such as floods, drought, or forest fires can also have this effect on operations.

Adjustments to dynamic conditions. In their attempt to stay within the zone of rational action, operators must always cope with the problem of uncertainty. Unlike the "economic man," whose perfect knowledge permits work with a certainty model, operators in real life must always be ready to adjust their decisions to the changing conditions of a dynamic world. Decisions made at any given time are based partly on one's understanding of certain known facts and partly on expectations regarding future cost, price, and yield conditions.

With the passing of time and unfolding of the production process, an operator's original expectations may or may not materialize. When they materialize as expected, production can continue with the assumptions of the operator's original static planning model. When they change, alert operators substitute updated planning models that contain adjustments for changed circumstances. These new models are forward-looking. Past inputs that lack salvage values are regarded as fixed, as of historical value only, and emphasis is focused on maximizing operator returns from the present time on.

With an improved market price outlook for a given product, the new planning model may favor additional applications of those types of input (e.g., payments for overtime, bonuses for added work effort, applications of additional fertilizer to crops) that can contribute to higher production. Declining market prices, on the other hand, may cause manufacturers to scuttle production programs, sell existing inventories at bargain prices, or try to salvage what value they can from alternative uses of their product. Farmers whose income prospects are dashed midway through a crop season by a hailstorm or a collapse of market prices may adjust to this situation by harvesting a crop for what it is worth, converting it into cattle feed, or plowing it under as a soil-building crop. Operators must always plan ahead, since they do not have the option of stepping back in time to reclaim the value of inputs already applied.

The necessity for treating past inputs as fixed naturally narrows the alternatives available to operators as they approach the end of the production process. The significance of this situation may be illustrated by the example of a speculative house builder. Before the building process begins, the operator can choose between several options such as style and type of residential structure. Once a foundation is laid and a house is framed in, however, many of these options for change are foreclosed by the nature of already vested inputs. The builder can still modify details, substitute some types of inputs for others, and within limits either downgrade or upgrade a house to meet particular levels of market demand. But as the production process proceeds, there are fewer and fewer opportunities for changing the final product, and the builder will be less and less willing to backtrack and make major changes in work already completed.

Multiple-production functions. Throughout the discussion above it has been possible to think in terms of simple input-output or resource-product relationships that involve the application of successive inputs of some standardized factor to a fixed factor of land. This centering of attention on a single production function has facilitated isolation and recognition of the principles involved with proportionality. A look at the real world, however, shows that very few operators deal with single-production functions.

Production normally involves the combination of a variety of nonhomogeneous inputs (different types of raw materials, machinery, and labor) with one's fixed factor. Sometimes these inputs are complementary and must be used together; sometimes they may be regarded as substitutes for one another. Sometimes they are indivisible and must be used as whole units; at other times they can be divided for smaller applications. As a rule, they are applied separately at different times and often in different sequences, have different costs, and can be combined differently with other factors with different results.

Each of the several types of variable inputs used in production has its own production function. Some of these parallel or complement the production functions of other necessary inputs. But with typical production processes calling for the concomitant use of many types of inputs and each type responding in its own way to combinations with other types of inputs, operators face the prospect of choosing among hundreds of possible input combinations. How can they make the right decisions in proportioning their inputs to optimize production? In times past, people experimented on a hit-or-miss basis until they found production patterns that worked. These patterns were often copied by others, sometimes acquired customary status, and were then passed on from generation to generation as approved and expected ways of doing things. In more recent times, scientists, inventors, and others have questioned these ways of the past and have experimented with new combination possibilities. In so doing, they have opened up vast new opportunities for increases in production, and have highlighted promising frontiers for additional research.

Individual operators respond to the multiple-production function problem in different ways.[12] Some cling to customary approaches because they cannot afford to

---

[12] Various techniques have been developed in production economic theory for the analysis of multiple-production functions. Simple cases involving one or two variables can be depicted in geometric diagrams and can usually be handled with simple reasoning. Any further increase in the number of variables taxes the reasoning capacity of the human mind. Cases involving four variables, for example, call for five-dimensional diagrams. Mathematical and computer models can be used in the analysis of multiple-production functions, but relatively few decision makers think in these terms.

assume the risk of failure that accompanies experiments with new approaches. Others are willing to accept conventional practices and the recommendations of production specialists. Still others experiment with new combinations and from time to time discover new and improved ways of doing things. Despite their differences, operators usually base their decisions on what they regard as reasonable judgment. Lacking perfect knowledge and foresight, they use their feel for proportionality in what they regard as a reasonable manner as they seek workable and profitable combinations of their resource factors.

Equi-marginal principle. Up to this point, we have assumed that the average operator deals primarily with one enterprise and has a plentiful supply of variable input factors to combine with a single fixed factor. With these two assumptions, one might logically expect the operator to push production to the point of diminishing economic returns. Under practical conditions, however, most average operators have limited supplies of productive factors and can usually put these factors to several alternative uses. This situation calls for modifications of operator production goals. Instead of always pushing production to the point at which $MFC = MVP$ (or $MC = MR$), operators with limited resources tend to push production with any particular enterprise only to the point at which their marginal value product equals or promises to drop below the return they could secure by using their marginal variable inputs in some recognized alternative use.

In this equalizing process, operators with limited resources apply the *equimarginal principle*. This principle asserts that maximum profit can be secured only when each input of land, capital, labor, or management is used in such a way as to add the most to total return and when the various resources used in any one enterprise produce a marginal value product at least equal to that which they could secure from their best alternative use.

This principle encourages operators to shift to those enterprises that promise the most net returns and to allocate their inputs among enterprises and among competing units of the same enterprises in such a way as to maximize their expected total net returns. A land-use illustration of this concept is reported in Table 5-3, which assumes an operator with thirty units of variable input costing $3 each, which are applied to three tracts of land, each of which has a different production function. With no limitations on the supply of variable inputs, it would pay to push production to the point at which $MFC = MVP$, or to the sixteenth input on the first tract, the thirteenth input on the second tract, and the ninth input on the third tract. Production at these points would call for thirty-eight inputs but only thirty are available. The operator could choose to apply an equal number (ten inputs) to each tract. This would provide a total value product of $95 + $80 + $59 = $234. This obviously is not the best combination, because the tenth input on the third tract does not pay for itself. Shifting this last input from the third tract to the first tract brings a marginal value product of $9 instead of $2 and thereby increases total value product to $104 + $80 + $57 = $241. Further experimentation shows that the operator secures the highest total value product and the maximum net return from the three tracts

TABLE 5-3. Illustration of Application of the Equi-Marginal Principle in the Allocation of Thirty Variable Inputs Costing $3 Each to Three Tracts of Land with Different Production Functions When the Product Has a Market Value of $1 per Unit

NUMBER OF VARIABLE INPUTS	FIRST TRACT		SECOND TRACT		THIRD TRACT	
	TVP	MVP	TVP	MVP	TVP	MVP
6	$ 47	$11	$45	$10	$42	$9
7	59	12	56	11	49	7
8	72	13	65	9	54	5
9	84	12	73	8	57	3
10	95	11	80	7	59	2
11	104	9	86	6	60	1
12	112	8	90	4	60	0
13	119	7	93	3		
14	124	5	95	2		
15	128	4	95	0		
16	131	3				
17	133	2				

when the variable inputs are proportioned between the tracts in a manner that provides approximately equal marginal value products from each tract. By using thirteen inputs on the first tract, ten on the second, and seven on the third the operator receives marginal value products of 7 from each tract and secures a maximum total value product of $119 + $80 + $49 = $248 and a total net return of $248 − $90 = $158.

Fixed and Variable Nature of Land Costs

Our discussion to this point has assumed production under short-run conditions— *short run* being defined as any production period during which an operator is limited by the fixed supply of particular inputs. Under short-run conditions, land resources are almost always regarded as a fixed production factor; and the costs associated with the acquisition and holding of land resources—property taxes, insurance payments, and commitments to pay cash rent—are classified as fixed costs. This situation stems from the fixed location of real estate resources, the ownership rights concepts we apply to them, and the tendency of most enterprises to be tied to particular sites throughout their production periods.

Individual producers normally operate in a series of short-run periods during which they are limited by various fixed factors such as land. The duration of these short-run periods in the case of land resources reflects the conditions under which land is owned or leased and may involve the time needed to grow a crop, complete a production cycle, or justify certain operating expenditures. When these periods are treated together as part of the long-run situation, all fixed factors become variables. The supply of land resources available to individual operators over this longer run can change with individual adjustments. Industries secure new plant locations; commercial operators relocate, remodel, or rebuild their establishments; tenants renegotiate their leases; and operators reduce or add to their real estate holdings. Changes also occur in the ownership, size, and value of holdings and in the rental rates, tax, insurance, and other charges associated with the holding of land resource units.

As one shifts from this long-run situation to the shorter-run periods under which production takes place, most variable inputs become fixed. At one stage of this process—when the operator decides on the size and location of the property unit—the land unit is a key variable. At this point, consideration is given to the possible advantages associated with alternative choices and the operator presumably chooses a site that promises to more than pay for its cost. Once this decision is made, the site and size of the land factor are fixed and the cost of acquiring and holding this unit becomes a fixed or "sunk" cost.

Economies of size and scale. The long-run variable nature of land makes it possible for operators to adjust the size and scale of their enterprises to an optimum level at which they can secure the highest possible net return from their combinations of outputs. But what is the optimum economic size of a factory, an apartment house, or a farm? And assuming that such an optimum can be determined, how might an operator move to this optimum? To determine the optimum size of enterprises, operators should consider the effects different scales of operation can have on both their costs and returns. If they can secure more than proportionate increases in total output by using additional inputs in their present production combinations, they are said to enjoy increasing returns to scale. With this situation and even with constant returns to scale, operators have an economic incentive to increase the size of their enterprises. Once they experience a decreasing rate of return in response to increasing scale, they have passed their optimum size and find it desirable to reduce their scale of operations.

Operators are keenly aware of the cost outlays and expenditures they must make for additional land, labor, and materials if they are to increase their scales of operation. Insofar as their resources permit, they are often willing to expand their scale as long as this action leads to production efficiencies and to a lowering of average production costs per unit of output. In this respect, they are concerned with the possible economies and diseconomies of scale associated with enterprises of varying sizes.

Cost economies occur whenever operators find that a larger scale of operations facilitates more effective use of their managerial abilities or better utilization of the underused capacity of particular factors such as specialized equipment. Similar economies result when expanded operations permit job specialization, work-simplification techniques, increased use of labor-saving machinery, and the savings that can come with bulk purchase of materials and supplies. In addition to these internal economies, operators may also benefit from external economies. A producer may benefit, for example, from the improved processing or marketing facilities attracted by a larger volume of production. In similar fashion, a large subdivision and apartment development may benefit from the improved bus, school, and community shopping facilities it may attract.

In addition to the cost economies that can come with increasing scale, diseconomies are also associated with the enlargement of enterprises. Some of these result from operator delegation of managerial responsibilities to individuals of lesser ability. Others may arise when an operator substitutes impersonal dealings with hundreds of employees for earlier personal contacts with a small labor force; when communication and transport problems develop because enterprises are spread over larger areas; or when larger scale entails greater risks from disease, fire, and other hazards. External diseconomies also develop when the expanded needs of an enterprise force it to bid up the prices it pays for the factors it uses in production.

Over the long run, operators should seek the size or scale level at which cost economies most exceed possible diseconomies of scale. They should visualize a planning curve (Figure 5-5) that connects the cost curves associated with the optimum proportioning of their productive factors in a series of enterprises of increasing size.[13] With this model in mind, they can then adjust the size of their factories, farms, or businesses to those scales with the lowest average cost curves. Operations at this point permit them to enjoy the maximum economies of scale.

Most operators who use land resources work with production combinations of less than optimum size. Some shift to larger enterprises over time, whereas others seem to be only slightly motivated by the comparative economies and diseconomies of scale. Several reasons may be advanced for this situation. Some lack opportunities

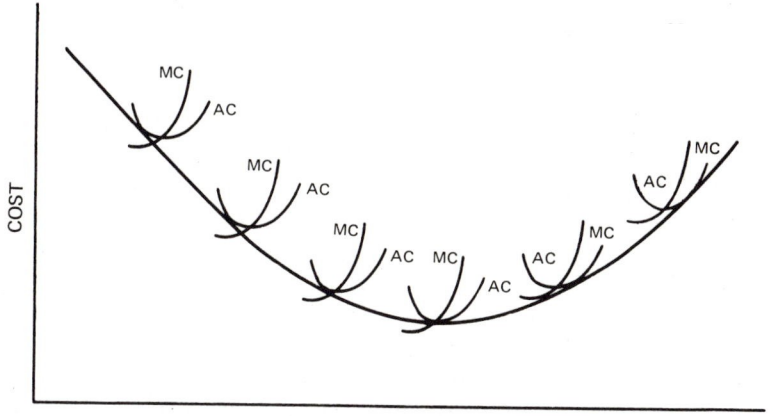

UNITS OF OUTPUT FOR A SERIES OF ENTERPRISES
OF INCREASING SIZE

FIGURE 5-5.   Use of a planning curve to indicate effects of economies and diseconomies of scale upon the optimum operating size of an enterprise.

[13] Cf. W. David Maxwell, *Price Theory and Applications in Business Administration* (Pacific Palisades, Calif.: Goodyear Publishing Company, Inc., 1970), pp. 129-34.

to acquire additional land in their communities or lack the initial capital outlay needed to finance such acquisitions. Some are held back by lack of imagination, initiative, or ability or because their personal goals call for the maximization of satisfactions that are not compatible with the operation of larger enterprises.

Importance of Limiting and Strategic Factors

Another important problem with proportionality involves identifying and making adjustments for limiting and strategic factors. This problem arises because of the scarce and frequently indivisible nature of the resources available to individual producers. Every operator deals with a limited supply of productive resources and ordinarily treats some one limiting factor, such as land, as the fixed factor around which other inputs are combined. Some operators have access to all the variable inputs needed for optimum input combinations with their fixed factor. But the scarce supply or strategic nature of particular inputs frequently causes them to stand out as "bottleneck" factors that prevent the normal functioning of the production process.

All things considered, lack of knowhow is probably the most limiting factor with which we must deal. Most operators try to make the best use they can of the knowledge they have in working out productive combinations of their available resource factors. In this process, they must often make adjustments for the strategic roles played by particular inputs and note that some factors may play strategic roles at particular moments and yet be of no more than routine importance on other occasions. An adequate supply of water or moisture for industrial or crop use, for example, may be more or less taken for granted under ordinary circumstances. It may suddenly loom as a critical factor, however, if a well runs dry, a water main bursts, or an area is affected by drought.

Successful operators must be able to identify their limiting factors and be ready to shape their production decisions around these factors. When the supply of some particular factor such as water is scarce, it may be necessary to ration the use of this resource and try to combine the other inputs in order to secure the highest return to this critical factor. Similarly, when supplies of capital, labor, or management are limited, it may make more economic sense to treat these factors rather than land as the fixed factors around which one proportions productive factors.

Indivisible inputs. An important example of limiting and strategic factors in production is provided by the indivisible inputs found in most production combinations. Because of their indivisible nature, operators must often choose between using larger supplies of a resource than they need or being content with a smaller supply than that required for most effective use. Many resource inputs are highly divisible. Productive factors such as chemicals, fertilizer, or water may be used in minute quantities or may be added by the ton. Some resources, however, are indivisible or, if divisible, come in large units. Labor, for example, may be calculated in hours and minutes; but a hired man or a skilled laborer must usually be treated as a unit whose services are sold by the day, week, or month. Similarly, it may be possible for two industrialists to share a computer center or drop forge or for two farmers to share a grain combine. Neither may need an entire unit, but desire for ownership control may cause both to consider the purchase of whole units even though this forces each operator to choose between the diseconomies of operating with insufficient or inadequate equipment and the unused capacity that could come with full ownership of the needed equipment.

Land is often thought of as an easily divided factor, and it is true that fields and lots can be divided and easily added to. Yet farms, lots, and buildings are usually sold as units, not as ten-acre tracts, so many square feet, or separate rooms. Because of this factor, industrialists and retailers often content themselves with cramped quarters because no adjacent space is available for expansion. Farmers may also content themselves with farms of uneconomic size because they cannot enlarge their present units or because they would have to acquire entire farms in the enlargement process.

Resource substitution in production. Producers frequently find that they can use different combinations of input factors to secure approximately the same net return. In this process, they can often adjust to their limiting factors by substituting other resources for those in short supply. Most producers are quite mindful of their opportunities for substituting resource factors for one another. They are intensely interested in ways and means for securing more production at less cost. They constantly seek new materials and processes they can use to cut costs or increase production, and they have a natural inclination to favor resource substitution whenever it promises higher net returns.

On an individual-operator basis, the relative scarcity of any particular input factor is usually gauged by the operator's opportunity to find a satisfactory substitute. When the price of a resource input increases relative to the price of a possible substitute, and when the substitution process involves only nominal cost and trouble, operators ordinarily shift to use of substitutes. Thus, an increase in labor costs relative to machinery costs will often cause operators to consider installations of automated equipment. Similarly, the high value of land relative to other input costs in the land-hungry areas of the world often favors the substitution of capital and labor for land in the production process.

Opportunities for resource substitution also play an important role in favoring technological development. With the progress of the Industrial Revolution, man has found it possible to substitute many new materials and devices for other factors in production. The steam engine and the gasoline engine have displaced millions of units of animal and human labor. Mass-production techniques in industry have been widely substituted for the less efficient use of labor in cottage industries. The combine harvester and other types of farm machinery have saved great quantities of farm labor for other uses. These examples merely suggest the tremendous impact new technology has had on resource substitution in production. Additional developments may be expected in the future as we acquire and utilize new ways of dealing with limiting factors in production.

INTENSITY OF LAND USE

Much of our economic theory regarding land use is rooted in the concept of proportionality. This central concept, with its emphasis on marginal analysis, input-output relationships, and considerations that affect operator decisions concerning the proportioning of resource factors, provides the keystone for production economic theory. As such, it helps explain why we use land resources as we do in production and also a number of other land economic concepts such as rent, land values, highest and best use, and allocation of land resources among competing uses. Another direct application of proportionality concerns the intensity with which land resources are used and the intensive and extensive margins of land use. When applied to land use, the term *intensity* refers to the relative amounts of capital and labor combined with units of land in the productive process.[14] People speak of those types of land use that involve high ratios of capital and labor inputs per land unit as intensive uses. Those enterprises involving large land areas relative to the amounts of capital and labor used are described as extensive uses.

Land areas differ a great deal in the intensity with which they are used. Urban lands, particularly those found in central commercial districts, are usually subject to very intensive use, farmlands are ordinarily the subject of somewhat less intensive use, whereas forest and grazing lands are used on a less intensive basis.

Intensive and Extensive Margins of Land Use

In contrast to the concept of intensity of use, the *intensive margin of land use* occurs at the point with any use of a given tract of land at which the marginal or last variable inputs of capital, labor, or management barely pay their cost.[15] This concept is applicable to all types of land use. As shown in Figures 5-4 and 5-6, the intensive margin is reached with the last successive variable input that can be applied before $MFC = MVP$ or before MC rises above MR in a cost curve diagram.

In contrast to the intensive margin which can be reached with any productive use of land for any specified purpose, the *extensive margin of land use* applies only with the lowest grade of land or the least accessible site that can be used to economic advantage for any given use. This margin occurs when typical operators who are applying their variable inputs to the intensive margin for a given use of land find that they are using the lowest grade of land of decreasing use-capacity they can afford to operate. At the extensive margin (area C in Figure 5-6A and B), the average value product is at its highest level and equals the average factor cost of the variable inputs ($AFC = AVP$ at the point at which $MFC = MVP$), the average cost per output unit is at its lowest level and equals the average return per output unit ($AC = AR$ at the point at which $MC = MR$), the total value product equals total factor costs ($TFC = TFC$ and $TC = TR$), and there is no economic surplus above production costs.

[14]Cf. Arthur C. Bunce, *The Economics of Soil Conservation* (Ames: Iowa State College Press, 1945), p. 29.

[15]John Ise, *Economics* (New York: Harper & Row Publishers, Inc., 1946), p. 440.

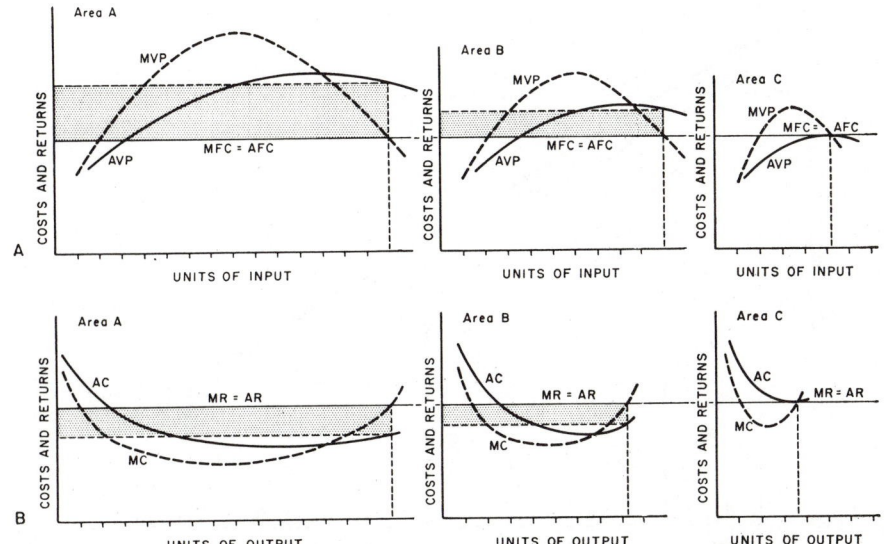

FIGURE 5-6.   Use of value product and cost curves to illustrate location of intensive and extensive margins of land use on three areas of differing use-capacities.

It is irrational for operators to apply variable inputs beyond the extensive margin, because sufficient returns to pay for the cost of their inputs can no longer be secured.

The intensitve and extensive margins of land use shown in Figure 5–6 assume the same enterprise on three land areas of different use-capacities. Area *A* has the ability or *economic capacity* to absorb fifteen variable input units to advantage, whereas area *B* can absorb ten inputs and the operator in area *C* barely breaks even with an optimum combination of five variable inputs with the fixed factor. With this difference in economic capacities of the three sites, the intensive margin with area *A* comes with the application of fifteen variable inputs, ten with area *B*, and five with area *C*. With area *C*, the value of the product produced just equals its cost of production. No area of lower use-capacity can produce enough value product to cover its production costs.

The situation portrayed in Figure 5–6 may also be visualized as a continuum such as that pictured in Figure 5–7. In this illustration, the horizontal axis measures decreasing use-capacity, and the vertical axis indicates the economic capacity or number of variable inputs that can be used to economic advantage with each successive grade of land. Note that area *A* uses fifteen inputs at its intensive margin, whereas *B* uses ten inputs and *C* uses five inputs. Other land areas with use-capacities between *O* and *R* could also be located along *OR* and would find their intensive margins along *MN*.

The line *NR* may be identified in this example as the *no-rent* or *extensive margin* of land use. This line comes at the no-rent margin because it intersects those points on the horizontal axis and on the *MN* line at which no economic surplus above pro-

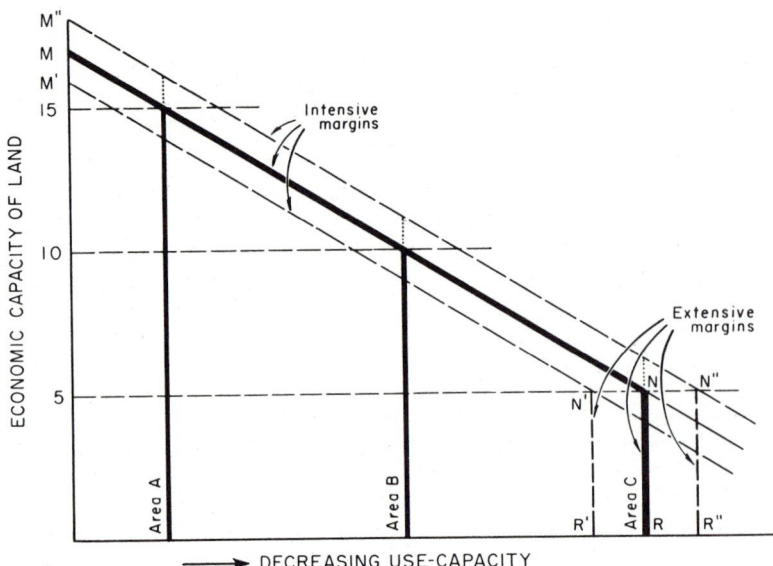

FIGURE 5-7. Illustration of intensive and extensive margins of land use.

duction cost occurs and beyond which it does not pay to bring new units of land into use. In practice, the intensive margin represents the economic point with each grade of land beyond which it does not pay to apply additional variable inputs. The extensive margin represents the point in a continuum of land areas with decreasing use-capacities beyond which it does not pay to bring additional land into production.

Changing price and cost conditions often bring shifts in the location of both the intensive and extensive margins. If production costs increase or if product prices drop, it may no longer pay the operator on area A to add the fifteenth input. In such an event, it may be economic to apply only fourteen or thirteen inputs. The operator on area B may find it best to stop with nine or eight inputs and the operator on area C would stop production entirely. With this situation, the intensive margin would drop to $M'N'$, while the extensive margin would shrink back to $N'R'$. A drop in production costs or an increase in product prices could have an opposite effect in encouraging A to add a sixteenth input, B to add an eleventh input, and C to go to six inputs. Under this circumstance, the intensive margin would rise to $M''N''$, and the extensive margin would move out to $N''R''$.

Marginal and submarginal land. Economists frequently speak of some land areas as being marginal or submarginal for particular types of use. The usual inference with these statements is that the areas fall either at or below the no-rent or extensive margins for the particular uses considered.

Past experience shows that operators sometimes exercise bad judgment in bringing new lands into production or in shifting already developed lands from lower to higher uses only to find they are not economically suited for the uses contemplated. These examples often involve a malallocation of resources. Once this fact is established, the submarginal uses are usually abandoned and the lands revert to lower types of use.

Changing price conditions provide a second major cause of submarginality. During the Great Depression of the 1930s, large areas that had paid their way in production under more favorable price and business conditions suddenly became submarginal when lower product prices forced a leftward shift of the extensive or no-rent margin. These conditions forced many operators out of production. In numerous instances, however, operators continued to produce at a financial loss. Sometimes they maintained themselves by drawing on personal or family reserves, by accepting a lower labor income and thus reducing their level of living, by borrowing funds from others, or by accepting financial aid or subsidies from public and private agencies.

With the passing of the 1930s, less was heard about marginal and submarginal lands. The reason was simple. With higher price levels and better business conditions, the extensive margin shifted to the right; and it again became profitable to bring many of the afflicted lands of the 1930s back into production. This experience has far-reaching implications for the future, because it indicates the effect higher prices can have in fostering the extension of land uses to areas now considered submarginal for these purposes.

Some Factors Affecting Intensity of Use

Levels of use intensity are often dictated by the types of use to which land resources are put. Land areas used for commercial and industrial purposes, for example, ordinarily call for far larger applications of capital, labor, and management per unit of land area than if used for residential, farming, grazing, or forestry purposes. With any given type of land use, levels of use intensity also are affected by factors such as land use and land area characteristics, changing supply and demand conditions, one's mix of productive factors, and operator attitudes.

Rising prices and increases in market demand make it profitable for operators to intensify their use of already developed lands and to bring new lands into use at the extensive margin. Depressed product prices can have a reverse effect. Changes in production and marketing costs can also affect land-use intensity. With higher costs and no change in price, operators ordinarily find it logical to cut back production. A reduction in production or marketing costs with no change in price, on the other hand, usually favors the more intensive use that comes with applications of additional inputs up to the new point at which marginal costs equal marginal revenues.

Limiting factors in production frequently have significant effects on the intensity with which land is used. When the supply of land or space used by an industrial establishment, a commercial shop, or a farm is the limiting factor, operators have every economic incentive to push their operations to the intensive margin. But when some nonland resource such as the operator's managerial capacity, an insufficient supply of operating capital, or a fixed labor force is the limiting factor, operators often find it more profitable to proportion their factors around their scarce resources, even though this may result in less intensive land use.

Family and operator attitudes also have important impacts on intensification practices. Certain immigrant groups and religious communities have at times displayed a willingness to accept hard labor and low levels of living. With this set of values, these operators have often found it possible to push production further than most of the operators with whom they compete. This willingness to accept the lower marginal returns to labor and management that have come with their intensive land-use practices has frequently made it possible for these operators to outbid other prospective buyers in the purchase of land.

All things considered, the intensity with which land is used always involves the interrelationship of several contributing factors. Areas of high use-capacity can ordinarily be used more intensively than areas of lower productive potential. Whether this relationship follows in actual practice depends on the impact of other conditioning factors such as population pressure, the stage of economic development, availability of capital and labor, and the attitudes and goals of those who own and operate the land. Differences involving these factors sometimes result in the intensive use of areas of limited use-capacity while nearby areas of greater productive potential remain underdeveloped or underutilized.[16]

— SELECTED READINGS

Black, John D., *An Introduction to Production Economics* (New York: Holt, Rinehart & Winston, 1926), chaps. 11–13.

Clough, Donald C., *Concepts in Management Science* (Englewood Cliffs, N.J.: Prentice-Hall, 1963), chap. 8.

Cramer, Gail L., and Clarence W. Jensen, *Agricultural Economics and Agribusiness* (New York: John Wiley, 1979), chaps. 4 and 6.

Doll, John P., and Frank Orazem, *Production Economics: Theory with Applications* (Columbus, Ohio: Grid, Inc., 1978), chaps. 2–6.

Heady, Earl O., *Economics of Agricultural Production and Resource Use* (Englewood Cliffs, N.J.: Prentice-Hall, 1952), chaps. 2–7.

Leftwich, Richard H., and Ross D. Eckert, *The Price System and Resource Allocation* 8th ed. (Chicago: Dryden Press, 1982), chaps. 7 and 8.

Nemmers, Erwin E., *Managerial Economics* (New York: John Wiley, 1962), Part III.

Roy, Ewell P., Floyd L. Corty, and Gene D. Sullivan, *Economics: Applications to Agriculture and Agribusiness*, 3rd ed. (Danville, Ill.: Interstate Printers and Publishers, Inc., 1981), chaps. 12 and 13.

---

[16] Cf. C. H. Hammar and J. H. Muntzell, "Intensity of Land Use and Resettlement Problems," *Journal of Farm Economics*, 17 (August 1935), 409–22; and Conrad Hammar, "Intensity and Land Rent," *Journal of Farm Economics*, 20 (November 1938), 776–91.

6

ECONOMIC RETURNS TO LAND RESOURCES

Rental payments have been made for the use of land resources since the beginnings of organized land settlement. These payments represent the economic return that goes to real estate resources for their use in production. This return was described in the last chapter as the net return or surplus of total value product above the operator's total factor costs, which should be credited to the fixed factor. Hereafter, it will be referred to simply as *land rent*.

Land rent is the key concept in land economic theory. It provides a theoretical base for explaining the value we place on real estate resources and much of the incentive we have for their ownership. It influences the allocation of real estate resources between individuals and between competing uses. It also affects leasing arrangements, taxation policies, the economics of land development and several other aspects of land resource use.

THE NATURE OF LAND RENT

The term *rent* is another of those common words for which economists have a specialized meaning. In their day-to-day use of this term, most people think of the payments made to property owners for use of their land and buildings. They speak of house rent, room rent, and the rent paid for commercial sites and farms. Like other people, economists often use the term *rent* in its popular sense. When they think in economic terms and, more particularly, of the economic returns to land resources, they find it appropriate to differentiate between three concepts of rent—contract rent, land rent, and economic rent.

Contract rent refers to the actual payments tenants make for use of the properties of others. The amount of these payments is normally agreed to by the landlord and tenant in advance of the period of property use and thus stems from mutual contractual arrangements. This concept is generally synonymous with the popular meaning ascribed to the term *rent*.

Land rent is a more specialized concept. It represents the theoretical earnings of land resources and may be defined simply as the economic return that accrues or should accrue to land for its use in production. This concept applies to the theoretical earnings of land and, as used here, applies to the combined earnings of land sites and their improvements. When distinctions are made between classes of land rent, it is sometimes expedient to distinguish between ground rents or site rents—the returns associated with building sites, bare ground, and raw land—and the improvement rents associated with buildings and other man-made real estate improvements.[1] Distinctions also can be made between location rents or rents that arise because of the favorable location of a particular tract of land and fertility or site-quality rents.

Economic rent is also a specialized economic concept. For over a hundred years, this term was used by economists to describe the economic earnings of land and had a meaning more or less synonymous with the present concept of land rent. With the refinements in economic thinking that have come during the past century, the focusing of more and more economic discussions on topics other than land, and the frequent tendency of economists to view real estate investments as a type of capital, a new meaning has been ascribed to the term *economic rent*. It is now defined as the surplus of income above the minimum supply price it takes to bring a factor into production.[2]

As now defined, economic rent is treated by most economists as a short-run economic surplus that a productive factor or an operator can earn because of unexpected demand or supply conditions. Over longer time periods, the supply and demand conditions affecting the factor are expected to come into balance, and the

[1] Some economists favor a narrower concept of land rent than that employed here and prefer to think in terms of what is here called *ground rent*. The broader concept of land rent is more meaningful and is accepted here because (1) nearly all land sites have benefited from man-made improvements, (2) it is difficult to distinguish between the portions of land rent that should go to raw land as compared with improvements, and (3) the focus of this study is on a broad concept of land and real estate resources, which includes both land sites and the improvements legally attached to them.

[2] Cf. Joan Robinson, *The Economics of Imperfect Competition* (New York: Macmillan, 1933), p. 102; and Kenneth E. Boulding, *Economics Analysis*, 4th ed. (New York: Harper & Row, Pub., 1966), Vol. I, 265. This modern concept of economic rent is an outgrowth of Marshall's views on "quasi-rents." Cf. Alfred Marshall, *Principles of Economics*, 8th ed. (New York: Macmillan, 1938), pp. 421–27.

It can be argued that economic rent, as defined above, encompasses the concept of land rent as long as one treats land as a free gift of nature. With this assumption, all of the earnings of land above necessary allowances for property taxes and insurance can be classified as land rent and also as economic rent. Two problems affect the acceptance of this approach. First, few owners are willing to accept the view that the minimum supply price of their land is zero. In practice, they usually equate their minimum supply prices with current contract rental values or with a given rate of return on the market value of their investments. From the standpoint of society, land can be viewed as a nature-given resource; but from the view of individual owners, there is no more justification for assuming a minimum short-run supply price of zero for land than for labor or any other input. A second difficulty comes with the theoretical assumption

phenomenon of economic rent disappears.[3] A real estate resource such as an apartment house, for example, can earn an economic rent above its normal land and contract rents when a sudden increase in demand for housing makes it possible for the owner to collect additional contract rent. This temporary advantage disappears over time as the bulge in demand disappears or as new housing is produced. Elements of economic rent may also be associated with the returns received by capital, labor, and management.[4]

Land rent and contract rent are the most important rent concepts used in land economics. These concepts differ from each other in one significant respect. Contract rent involves an actual payment to the property owner. This payment may exceed or fall below the amount of land rent theoretically earned by the property. When it exceeds the amount that should be paid as land rent, the tenant contributes the difference from returns that should go to his or her capital, labor, or managerial inputs. When it falls below this amount, the tenant is able to pocket part of the land rent.

Land Rent as an Economic Surplus

Land rent can be viewed as a residual economic surplus, as that portion of the total value product or total returns that remains after payment is made for total factor costs or total costs, respectively.[5] As illustrated with the value product and cost curves presented in Figure 6-1, land rent is the surplus depicted by the shaded rectangles *LMRP* that remains after the cost of the variable inputs (lower rectangles *MNSR*) is subtracted from the total value of product produced, which is represented

---

that a balancing of supply and demand conditions over time will bring the disappearance of economic rent. This assumption makes little sense when land rent is viewed as part of economic rent, for the simple reason that the differences in land use-capacity that provide the basis for land rent do not disappear with a balancing of supply and demand conditions. Cf. Robert H. Wessel, "A Note on Economic Rent," *American Economic Review*, 57 (December 1967), 1221-26; and Joseph S. Keiper *et al.*, *Theory and Measurement of Rent* (Radnor, Pa.: Chilton, 1961), pp. 108-13.

[3] Cf. Richard H. Leftwich, *The Price System and Resource Allocation*, rev. ed. (New York: Holt, Rinehart & Winston, 1963), pp. 294-96.

[4] Wages include an element of economic rent when they exceed the minimum supply price at which workers are willing to sell their services. Thus if one is willing to work for $250 a week but is offered a weekly wage of $300, the $50 of surplus represents an economic rent to labor. Other economic rents appear when an investor is willing to loan money at 9 percent but can get 14 percent (an economic rent of 5 percent), or a basketball star is willing to play professional ball for $20,000 but finds that he can collect $500,000. People are pleasantly surprised when they find that they can market their labor, capital, or special talents at higher prices than they expected. The resulting economic rents soon disappear, however, as they raise their minimum supply prices to match market levels.

[5] Malthus recognized this relationship during the early 1800s when he defined *rent* as "the excess value of the whole produce, or if estimated in money, the excess price of the whole produce, above what is necessary to pay the wages of labor and the profits of capital employed in cultivation." Cf. Thomas Robert Malthus, *Principles of Political Economy* (London: William Pickering, 1836), p. 136. Malthus outlined this same concept earlier in his 1815 essay "The Nature and Progress of Rent."

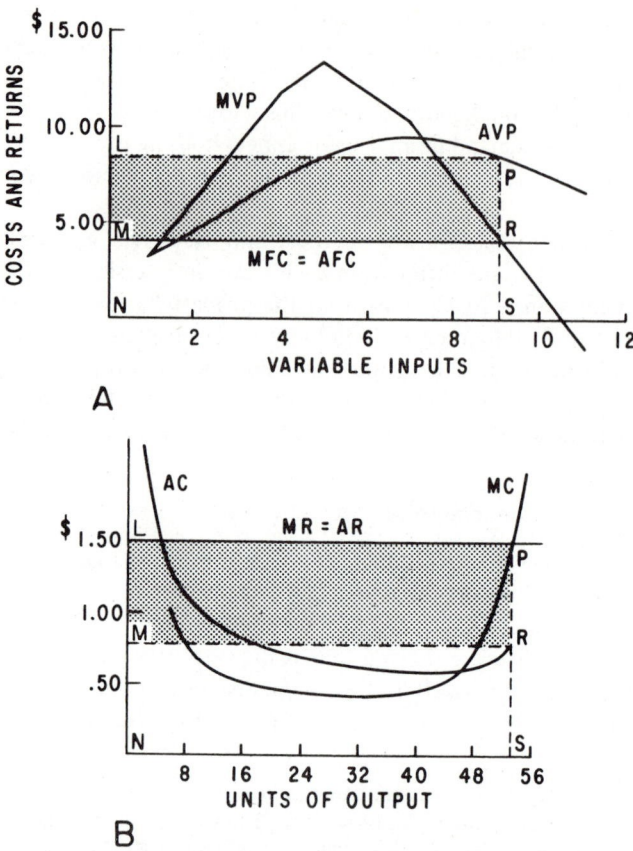

FIGURE 6-1.   Use of value product and cost curve diagrams to illustrate the concept of land rent as a residual economic surplus that remains after the payment of production costs.

by the large rectangles *LNSP*. With the value product diagram (Figure 6-1A), land rent is equal to *AVP − AFC* times the units of variable input applied. With the cost curve diagram (Figure 6-1B), it is equal to *AR − AC* times the units of output produced.

This simple formulation is both flexible and all-inclusive in its consideration of the factors that influence rental levels. Value product and cost curves can be used to explain differences in the amounts of land rent that accrue on different grades of land (See Figure 5-6). With the cost curves shown in Figure 6-2, for example, the units of output secured from three different grades of land can be assumed to have the same market value. Meanwhile, the average cost of production per output unit is lower on the grade *A* tract than on the grade *B* and *C* tracts, because the total production cost is spread over more units. With these differences in average unit production costs, the grade *A* tract yields considerable land rent, the grade *B* tract a smaller amount of rent, while the grade *C* tract is at the extensive or no-rent margin and produces barely enough to pay its production costs. The amount of rent

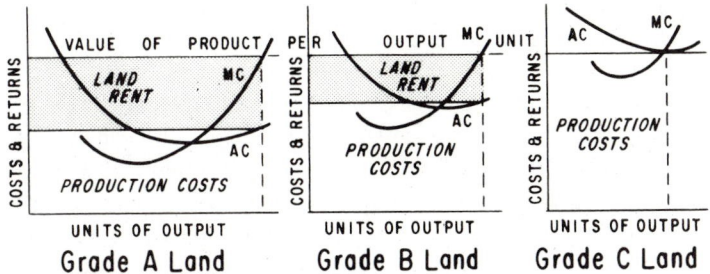

FIGURE 6-2.   Illustration of the effects differences in land quality have upon the amounts of land rent that accrue to three grades of land.

that accrues on each grade of land depends on the relationship between price levels and costs. With higher prices or lower costs, rents rise all along the line—even on the grade *C* land. Lower prices or higher costs, in turn, would lower the rents secured on the *A* and *B* tracts and force the grade *C* land out of use.

A similar comparative approach may be used to illustrate the effects of differences in location on the land rent produced on tracts of comparable quality. The left-hand diagram in Figure 6-3 indicates the amount of land rent that can be expected on a grade *A* site located at the market. Lands located at greater distances must pay a shipping cost to get their products to market. Since this cost is proportional to the number of units of output sold, it may be treated as a price-depressing factor that lowers the net price per output unit received at outlying production points.

As Figure 6-3 indicates, the lower net price received by producers located 250 and 500 miles from market has a considerable effect in reducing the amount of land rent received at these locations. These areas are just as productive as those located at the market; but with a transportation-cost handicap, the operators located at these sites must gear their production to a lower net price level. The lower land rents associated with the less advantageous locations may be attributed both to the

NET VALUE OF PRICE PER UNIT OF OUTPUT (AFTER DEDUCTION OF SHIPPING COSTS) AT POINT OF PRODUCTION

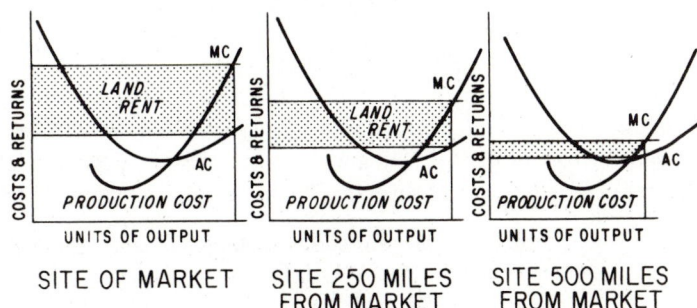

FIGURE 6-3.   Illustration of the effects of differences in accessibility upon the amounts of land rent that accrue to three tracts of land of comparable quality located at different distances from market.

lower net prices received by the operators at these locations and to the effect these lower net prices have in cutting back the number of variable inputs operators can profitably employ in production.

Viewed as presented in Figures 6-2 and 6-3, land rent may be considered as an economic leveler. Operators, who are free to move and who can apply their managerial talents equally well to different tracts of land, would realize no advantage in using one tract of land as compared with another. The returns to their nonland inputs would be the same in any case. This situation highlights a weakness of this type of analysis. In its treatment of land rent as the residual surplus that remains after payment is made for all other factors of production, marginal-productivity analysis assumes that the return to the other factors can be determined with some precision. This assumption is often unwarranted. The returns attributed to nonland factors in calculations of land rent are usually arrived at through an accounting process. They may represent the actual cash payments made for these factors, the going rates for these payments, or estimates of what a fair or normal return should be. Each of these methods gives value figures that bear the same relationship to actual productivity as contract rent bears to land rent.

The marginal-productivity approach can be used to measure the economic return attributable to nonland factors as well as to land. When it is used for this purpose, it is usual practice to impute a fair return to land just as fair returns are imputed to nonland factors when this approach is used to measure land rent. The accuracy of the final answers secured by this approach depends on the accuracy of the basic data used in the calculations. With superior management, the returns attributed to land may easily be too high if the return to management is calculated at an average going rate. Conversely, with mediocre combinations of productive factors, too little return may be attributed to land if all the other factors are compensated at their going rates.

It should also be recognized that land rent is not always viewed as a residual surplus. Rent may be regarded in this light from the standpoint of society and some individuals. Under real-life conditions, however, many individuals—and most particularly those who make contract rental payments—see rent as a fixed item among their production costs. Management or labor is treated as the fixed factor in these cases, and the residual surplus is viewed as either profit or labor income.

Classical Formulations of Rent Theory

Little consideration was given to economic explanations of the nature of land rent until relatively recent times. Sir William Petty made some pertinent observations concerning rent in 1662, as did several other writers in the next 150 years.[6] The

---

[6] Other important contributors to the early conceptualization of rent theory include Richard Cantillon (1730), Francois Quesnay (1756), A. R. J. Turgot (1770), Adam Smith (1776), James Anderson (1777), and James Mill (1804). Anderson was the only one of these writers to present a reasonably complete theory of rent. Cf. Keiper *et al., op. cit.*, pp. 4–21; and Edmund Whittaker, *A History of Economic Ideas* (New York: Longman, 1940), pp. 487–99.

beginnings of classical rent theory are usually associated, however, with the writings of a group of English economists at the conclusion of the Napoleonic Wars. At that time, the British Parliament was considering the controversial Corn Laws, and the attention given to this issue prompted several writers to publish essays regarding the nature of rent.[7] Three writers of this early post-Napoleonic period—Thomas Robert Malthus, David Ricardo, and Johann Heinrich von Thunen—made significant contributions to present land rent theory. Malthus outlined a residual surplus concept, which was largely ignored at the time but which foreshadowed the marginal-productivity concept of rent described in the last section. Ricardo attributed rent to differences in fertility and presented his views with such force and clarity that they were soon widely accepted as the basis for the classical concept of rent. Von Thunen authored an independently developed and complementary theory, which explained rent in terms of differences in location with respect to a central market.

Ricardo's emphasis upon differences in fertility. Ricardo was concerned with the problem of agricultural rents. He started his analysis by assuming a newly settled country with "an abundance of rich and fertile land, a very small proportion of which is required to be cultivated for the support of the actual population."[8] He then argued that only the most fertile lands would be brought into cultivation and that no payment of rent would be associated with their use. Rents arise on these lands only when increases in the demand for land justify the bringing of less fertile lands into use. In Ricardo's words,

> If all land had the same properties, if it were unlimited in quantity, and uniform in quality, no charge could be made for its use, unless where it possessed peculiar advantages of situation. It is only, then, because land is not unlimited in quantity and uniform in quality, and because, in the progress of population, land of an inferior quality, or less advantageously situated, is called into cultivation, that rent is ever paid for the use of it. When, in the progress of society, land of the second degree of fertility is taken into cultivation, rent immediately commences on that of the first quality, and the amount of that rent will depend on the difference in the quality of these two portions of land.
>
> When land of the third quality is taken into cultivation, rent immediately commences on the second, and it is regulated as before by the differences in their productive powers. At the same time, the rent of the first quality will rise, for that must always be above the rent of the second by the difference between the produce which they yield with a given quantity of capital and labour. With every step in the progress of population, which shall oblige a

---

[7]Thomas R. Malthus, David Ricardo, Robert Torrens, and James West published important pamphlets on rent in 1814 and 1815. Cf. Keiper *et al., op. cit.*, pp. 21–34; and Whittaker, *op. cit.*, pp. 499–503.

[8]David Ricardo, *The Principles of Political Economy and Taxation* (London: 1817: Everyman's edition, London: J. M. Dent, 1911), p. 34.

country to have recourse to land of a worse quality, to enable it to raise its supply of food, rent, on all the more fertile land, will rise.[9]

Ricardo's theory of rent determination may be illustrated by an example such as that portrayed in Figure 6-4. This example assumes four grades of land with yield capacities of 50, 40, 30, and 25 units of product, respectively, for a given input of capital and labor costing $100. With these assumptions, it costs $2.00 to produce each unit of output on the grade *A* land, $2.50 on the grade *B* land, $3.33 on the grade *C* land, and $4.00 on the grade *D* land. As long as there is enough grade *A* land to provide all the needed output, the market price of the product would correspond with the $2.00 per unit cost of production. No rent needs to be paid, because every land user is able to bring equally fertile areas into use, and any operator who attempts to raise the product price would be undersold by other producers.

This situation changes when grade *B* lands must be brought into use to provide products for the growing population. At this point, product prices must rise to the $2.50 level to cover the cost of production at the new extensive margin of

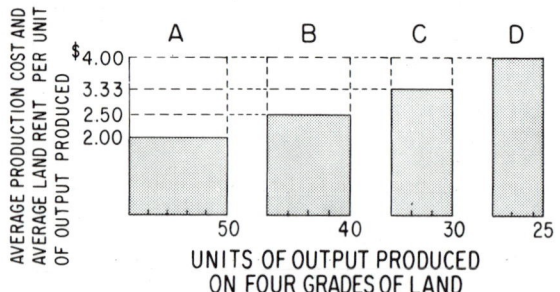

FIGURE 6-4.   Ricardo's explanation of land rent.

[9] *Ibid.*, p. 35. James Anderson expressed generally the same concept in 1777 when he wrote;

> In every country there are various soils that are endured with different degrees of fertility, and hence it must happen, that the farmer who cultivates the most fertile of these can afford to bring his corn to market at a lower price than others who cultivate poorer fields. But if the corn that grows on these fertile spots be not sufficient fully to supply the market, the price will naturally be raised to such a height as to indemnify others for the expense of cultivating poorer soils. The farmer, however, who cultivates the rich spots, will be able to sell his corn at the same rate with those who occupy poorer fields; he will consequently, receive more than the intrinsic value of the corn he raises. Many persons will, therefore, be desirous of obtaining possession of these fertile fields; being content to give a certain premium for an exclusive privilege to cultivate them, varying, of course, according to the more or less fertility of the soil. It is this premium which we now call *rent*; a medium by which the expense of cultivating soils of different degrees of fertility is reduced to a perfect equality.

From James Anderson's essay, "Observations on the Means of Exciting a Spirit of Industry," reported by J. R. McCulloch, *Principles of Political Economy* (Edinburgh, Scotland: 1843), p. 442.

cultivation. The higher product price, which encourages operators to cultivate the grade *B* lands, provides an economic surplus of 50 cents per output unit to operators of the *A* lands. This surplus is not needed to ensure continued production from the *A* lands, but since it exists, it goes as land rent to the owners of these lands.

Product prices must rise to the $3.33 level if the grade *C* lands are to be cultivated. This price increase provides a land rent of 83 cents for every unit of output produced on the *B* lands and an additional land rent of 83 cents for each unit of output produced on the *A* lands. A price of $4.00 per unit of output is needed to bring the *D* lands into use. At this point, a land rent of 67 cents for each unit of output arises on the *C* lands, and additional land rents arise on the *B* and *A* lands.

Ricardo believed that farm product prices are determined by the production costs associated with the highest-cost portions of the total supply needed by society. His theory assumes that prices are set by production costs at the intensive and extensive margins of cultivation. He recognized that product prices must rise with the outward shift of the extensive margin of cultivation and that these higher prices at the same time raise the intensive margin on the more fertile lands and thus favor their more intensive use. In his words:

> It often, and, indeed, commonly happens, that before No. 2, 3, 4, or 5, or the inferior lands are cultivated, capital can be employed more productively on those lands which are already in cultivation. It may perhaps be found that by doubling the original capital employed on No. 1, though the produce will not be doubled, will not be increased by 100 quarters, it may be increased by 85 quarters, and that this quantity exceeds what could be obtained by employing the same capital on land No. 3.
>
> In such case, capital will be preferably employed on the old land, and will equally create a rent; for rent is always the difference between the produce obtained by the employment of two equal quantities of capital and labour.[10]

Ricardo's thesis can best be illustrated by the use of cost curves as in Figure 6-5. When only the *A* lands are used, operators find it profitable to produce *p* units of output. At this point no rent arises. When the *B* lands come into use, the operators of these lands find it profitable to produce *j* units of output, at which point they experience an extensive no-rent margin. Their production at this price level creates a land rent for the owners of the *A* lands on the *p* units of output and also makes it profitable to increase production to a new intensive margin at *q* units of output. Similarly, when the *C* lands are brought under cultivation, the operators of the *B* lands secure a land rent on *j* units of output and find it profitable to increase production to the *k* units of output level, whereas the operators of the *A* lands secure larger land rents at the *p* and *q* levels and find it profitable to push to their new intensive margin with *r* units of output. A similar chain reaction takes place when the *D* lands are brought into use and the operators of the more fertile lands find

[10]*Ibid.*, p. 36.

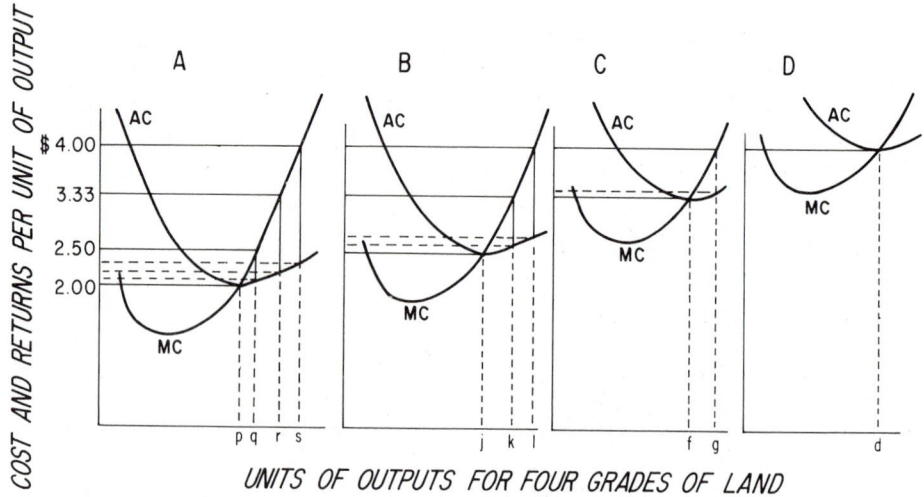

FIGURE 6–5.   Effects of bringing lower grades of land into use on intensive margins of use of better grades of land with Ricardo's concept of land rent.

that they can add to their rents by pushing their production to new intensive margins that yield additional units of output.

Ricardo held that rent was price-determined and "does not and cannot enter in the least degree as a component part of its price."[11] As he saw it,

> . . . raw produce rises in comparative value . . . because more labour is employed in the production of the last portion obtained, and not because a rent is paid to the landlord. The value of corn is regulated by the quantity of labour bestowed on its production on that quality of land, or with that portion of capital, which pays no rent. Corn is not high because a rent is paid, but a rent is paid because corn is high; and it has been justly observed that no reduction would take place in the price of corn although landlords should forego the whole of their rent. Such a measure would only enable some farmers to live like gentlemen, but would not diminish the quantity of labour necessary to raise raw produce on the least productive land in cultivation.[12]

[11]*Ibid*., p. 41. Critics sometimes argue that Ricardo's explanation of land rent is faulty because market prices are determined by the interaction of supply and demand factors, not by production costs at the extensive margin. This criticism is valid under short-run market conditions, because production uncertainties make it impossible for producers to predict in advance the exact points at which their extensive margins will occur. Consumer taste factors, acceptance of customary rental rates, and rental market rigidities imposed by monopoly pricing favor a short-run supply and demand explanation of urban and residential rents. Viewed in a longer run perspective, however, the prices determined through the interaction of supply and demand must reflect production costs at the extensive margin. When prices and rents are low, new areas are not brought into production at the extensive margin, and areas already in use often shift out of production. Rising prices and rents make it economically expedient to bring higher cost-of-production sites into use.

[12]*Ibid*., pp. 38–39.

Rent arising from location. In contrast to Ricardo's emphasis on differences in land quality, Petty (*A Treatise on Taxes and Contributions*, 1662) and von Thunen (*Der isolierte Staat*, 1826) stressed the fact that land rent also arises because of differences in location. Von Thunen observed that when crops produced for a central city market are grown on lands of like fertility, the lands located nearest the city enjoy a definite rent advantage over those located at greater distance. The extent of this advantage corresponds with the differences in the transportation costs that arise in shipments of products from the two areas to market.

In the days of ox-cart and wagon transportation, shipping costs definitely restricted the areas within which products could be produced at a profit. The cost of transporting cereal crops such as wheat, for example, was often so high that commercial production of this crop was not expedient at distances of more than twenty-five to forty miles from market. Technological developments have brought tremendous changes in this situation. But transportation costs still have significant effects on rent-paying capacity and the extent of the areas within which many products can be profitably produced.

The importance of the transportation costs associated with different locations may be illustrated by the example of a heavy and bulky product such as sugar beets. If sugar beets are worth $15 a ton delivered at a factory and can be produced at an average cost of $13.80 a ton (including loading costs and a fair return on the operator's capital, labor, and management), a surplus of $1.20 a ton will be available as land rent on those lands located at the market. With an average yield of ten tons per acre, this would result in a land rent of $12 per acre at this location.

Fields located at greater distances from market naturally have higher transportation costs and thus produce less land rent. With an average transportation cost of 3 cents per ton-mile, the amount of land rent drops 3 cents a ton, or 30 cents an acre with every additional mile between the location of the production site and the factory (Figure 6-6). This means that there can be only $6 in land rent per acre at

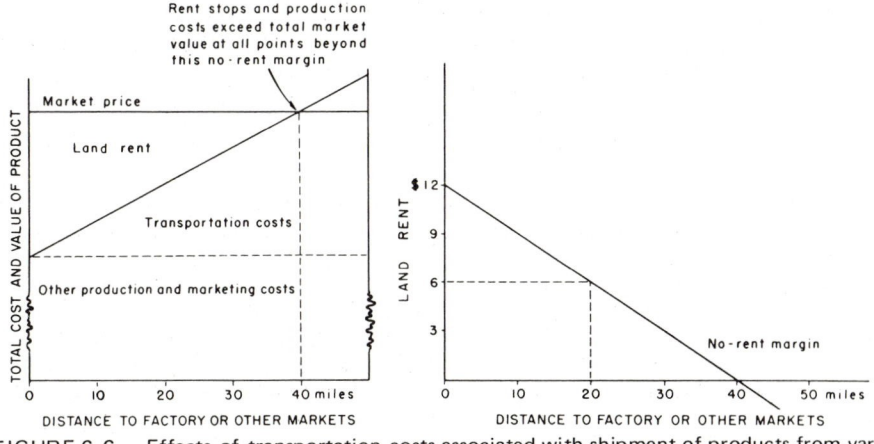

FIGURE 6-6. Effects of transportation costs associated with shipment of products from various locations to market on land rent.

locations twenty miles from the factory and that a no-rent point is reached forty miles from the factory. Any production carried on beyond the no-rent point would call for reduction of the payments that normally go to labor and management.

Fortunately, the no-rent points for different land-use enterprises occur at different distances from market. This factor, together with the multitude of markets found in modern society, gives almost every tract of land some rent-producing capacity. Lands located near the market or near the 100 percent spot of a central business district usually have a high income- and rent-producing capacity for any of several alternative uses. Lands located at more distant sites are usually beyond the no-rent margins for many uses. The operators of these lands have fewer choices of enterprises and frequently find it most profitable to concentrate on extensive land-use operations such as ranching.

Use-capacity and rent-paying ability. Differences in rent-paying capacity are often explained in terms of variations in either soil fertility or location. By themselves, neither of these factors provides a completely satisfactory explanation of the ability of land to pay rent; and even when the two are considered together, they can leave significant aspects of rent-paying capacity unexplained. Land rents can reflect levels of property improvement or amenity considerations such as a desirable neighborhood, a pleasing view, ready access to water supplies, or nearness to educational and recreation facilities. Convenience of access and possible savings in time-distance of travel can also influence the rent potential of various sites.

The cumulative impact of the various factors including soil fertility that affect land quality and of the items including location that affect accessibility is measured by the concept of land use-capacity. This concept has particular value because it permits relatively complete comparisons of the income-producing potential of various sites. Those areas with the highest use-capacity ordinarily have the highest value, the greatest production potential, and yield the most land rent.[13] The relationship between land use-capacity and the appearance of land rent is shown in Figure 6–7. This diagram assumes a continuum of lands of decreasing use-capacity ranging from areas of highest use-capacity at *A* to lands of much lower use-potential at *D*. As society resorts to the use of the lower-grade, less-productive, and less advantageously situated lands—as the extensive margin of land use shifts to the right—unit production costs gradually increase, and the net price per output unit rises enough to command whatever additional production is needed. When only the lands between *A* and *B* are used, prices are pegged at the *LR* level and the surplus above production costs available for land rent is small. When the extensive margin of

---

[13]Circuitous reasoning is involved with the assertion that land rents are associated with levels of use-capacity when *use-capacity* is defined simply as the relative ability of units of land to produce a surplus of returns above costs of utilization. This problem disappears when one remembers that differences in use-capacity stem from differences in land quality and accessibility. Lands of above-average fertility produce larger rents than less-favored lands because of their greater physical and economic productivity, and sites located near their markets produce more rent than less-favored sites, because their operators have lower costs for transporting products to market.

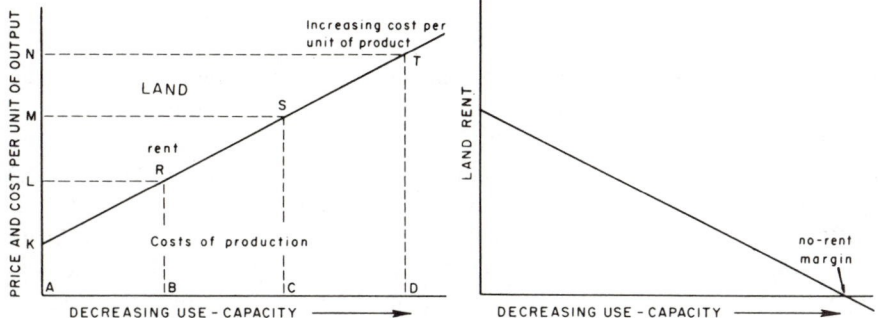

FIGURE 6-7. Illustration of relationship between use-capacity of land resources, production costs, and the appearance of land rent.

land use shifts from *B* to *D*, prices climb to *NT*, and the total volume of land rent increases from the area included within the triangle *KLR* to the area included within the triangle *KNT*.

Figure 6-7A assumes a continuum of lands of diminishing use-capacity and shows the quantities of land rent that arise at different sites as prices rise to meet the cost of utilizing lands at the extensive margin for some given use. The triangles *KLR* at price *LR, KMS* at price *MS*, and *KNT* at price *NT* indicate the amounts of land rent produced as the assumed use is extended out from *A* to points *B, C*, and *D*, respectively. For analytical purposes, these land rent triangles can be detached from the remainder of the diagram, turned over, and shown as land rent triangles such as that pictured in Figure 6-7B.[14] Land rent triangles or profiles of this type can be utilized to show the relationship between decreasing use-capacity and the amounts of rent produced for any land use.

This analysis assumes some abstraction from reality. The concept of a continuum of lands of decreasing use-capacity assumes that (1) society proceeds in its use of land resources from sites and areas of highest use-capacity to those of lower potential, and (2) areas with various levels of use-capacity are distributed more or less uniformly along the horizontal axis. Neither of these assumptions is completely valid. This situation naturally complicates the use of this approach. As long as these limitations are recognized, however, this approach can be used to provide a meaningful explanation of the effect of various levels of use-capacity on land rent. More important, it supplies an analytical basis for the concept of the land rent triangle.

[14]Each land rent triangle can be visualized as a cross-section of a land rent cone that surrounds a central market point (the site of highest use-capacity). Various terms have been used to describe the surface of the land rent cone, which is shown in Figure 6-7B simply as the hypotenuse or slope of the land rent triangle. Edgar M. Hoover, *Location Theory and the Shoe and Leather Industries* (Cambridge, Mass.: Harvard University Press, 1937), p. 23, uses the term *rent surface*. Edgar S. Dunn, Jr., *The Location of Agricultural Production* (Gainesville: University of Florida Press, 1954), p. 34, and Walter Isard, *Location and Space-Economy* (New York: John Wiley, 1956), pp. 194–95, speak of a "rent function." William Alonso, *Location and Land Use* (Cambridge, Mass.: Harvard University Press, 1964), pp. 40–41, describes it as a "bid rent function."

Other Views Concerning Rent

Although widely accepted, the marginal value productivity concept of land rent has had its critics. Some have attacked Ricardo's assumptions concerning the order of new land development, have challenged his assertion that rent does not enter into the determination of price, and have attributed rents to customary arrangements or the monopoly positions held by land owners.[15] These criticisms can be ignored as not affecting the long-run operation of the land rent concept. Two alternate views of the nature of land rent, however, deserve special attention. These are the concepts of rent as a return on capital investment and rent as an unearned increment.

Rent as a return on investment.   Mention has been made of the fact that real estate resources are often viewed differently from the standpoint of society than from the perspective of individual operators. Society sees land as a nature-given resource that was supplied in its original form at no cost to its users. Most investors, owners, and tenants, in contrast, see land rent as a return on their real estate investments. These operators are not particularly concerned with the fact that real estate involves intermixtures of nature-given land and man-made improvements. To their way of thinking, the real estate resources they work with are a type of capital. For them, the development of these resources calls for sizable investments in time, effort, and money, whereas the acquisition and use of already developed properties call for purchase or leasing arrangements. They see real estate as a capital good that can be bought, sold, or leased, and land rent as a return on the market value of land resources. Tenants typically view contract rental payments as an operating cost, not as a residual economic surplus due landowners because of the particular income-producing advantages associated with their properties. Landlords and owners in turn think of contract rent as a return on the capital value of their investments and compare these returns with those they could receive from alternative capital investments.

Rent as an unearned increment.   Ricardo treated rent as an economic surplus, as a payment to landowners that is not required to keep land in production. With this approach, it was an easy step for later observers to conclude that rent is an unearned increment or windfall return for which landowners do nothing and that they receive only because of their favored "monopoly" position.

This view of rent was accepted by three important nineteenth-century economists. John Stuart Mill regarded rent in this light and suggested that this unearned increment be subject to taxation. Henry George used it as the basis for his crusade favoring the single tax. Karl Marx saw rent from land as an unearned and unjusti-

---

[15]Cf. Keiper *et al., op. cit.,* pp. 35–52; Whittaker, *op cit.,* pp. 503–11; and Richard A. Walker, "Contentious Issues in Marxian Value and Rent Theory: A Second and Longer Look," *Antipode,* 7 (1976), pp. 31–53. Several Marxian economists support a concept of absolute rent which attributes land rent to the monopoly position of landlords and ignores the effect differences in land use-capacity have in generating rents.

fied monopoly return which owners could claim because of the institution of private property.

There is no necessary conflict between land rent and the concept of rent as an unearned increment. In some instances, rent may be regarded as something akin to a monopoly income. This is particularly true in areas where vestiges of feudal land-ownership systems persist, a high proportion of the land is controlled by a few families, or factors such as tradition or prestige of ownership discourage market transactions. In an economic sense, real estate ownership can provide owners with important differential advantages, but the presence of other owners holding similar ownership rights prevents the existence of a true monopoly.

Rent can be viewed as an unearned increment any time it arises from the mere holding of land. Whenever property owners enjoy an increase in land rent from the acts of others and not because of their own improvement efforts, the increase can be described as an unearned increment. Unearned increments of this type are associated with the returns to most factors of production and are often hard to identify. This identification problem is complicated with real estate resources by the frequent tendency of owners to make property improvements—a situation that gives them logical claim to part of any increase in rent—and by frequent property sales.

Unearned increments are ordinarily capitalized into the selling prices of properties at the time of sale, and their value thus goes to the seller. The new owners then start with properties that supposedly are worth no more than their purchase price. When these owners use savings from past sweat and toil along with current earnings to acquire properties, it is normal for them to regard land rent as a fair return on their investment rather than as an unearned income for which they have done nothing.

SIGNIFICANCE OF LAND RENT

Theoretical concepts such as land rent have little importance in and of themselves. Their real significance arises because of their value as tools of analysis that can be used in explaining real-life situations. Land rent is significant in this sense because it provides a key for explaining some of our most basic behavior regarding land resources. Four of the more important of these applications involve its relationship to contract rental arrangements, to property values, to land resource development and investment decisions, and to the allocation of land resources among competing uses.

Effects on Rental Arrangements

No rental arrangement is complete without some agreement on the amount of rent a tenant will pay for use of a landlord's property. How the concept of land rent affects the contract rental rate agreed upon is best illustrated by the workings of the rental bargaining process. Under ideal bargaining conditions, both parties should have an accurate understanding of the fair amount that should be paid as contract rent. In a hypothetical example, both parties may know that $300 represents a fair rental. Landlords in this case would find it to their advantage to demand the full $300 plus any additional payment they can get. Tenants in turn would refuse to pay more than $300 and would naturally favor payments of less than this amount. Should tenants agree to a higher payment, they would find it necessary to give up portions of the return that should go to their labor, management, or capital, while any willingness of landlords to accept a lower payment would redound to the benefit of the tenants.

With the ideal conditions assumed in this example, the contract rental rates agreed upon would closely approximate the theoretical land rent. Deviations from this model are common, however, when landlords and tenants operate with differing assumptions and knowledge concerning the rent-producing capacity of the land or when the two parties do not bargain as equals. And even when they enjoy equal knowledge and equal bargaining power, contract rental rates may differ from the theoretical land rent because of the failure of future production and income to match the conditions anticipated at the time of the rental agreement.

The rental bargaining process may involve sharp negotiations in which each party argues his or her position, or it may involve placid acceptance of terms already determined by the landlord. In either case, the problem of inadequate knowledge causes many landlords and tenants to guess at what is a fair rental rate. This guessing process can easily result in inequitable arrangements. Partly because of this situation, landlords and tenants long ago started modeling their rental arrangements on practices that had proved satisfactory in their areas. Acceptance of these precedents has often given rise to customary rental systems such as the "half and half" sharecropping system that prevailed for several decades in the South and the somewhat standardized housing rental rates accepted in many residential areas.

Customary rental arrangements often start with payments that correspond closely with the theoretical land rent. As these customary systems continue and spread, they may be applied under conditions that no longer fit the original assumptions. This situation can result in definite inequities for either the landlord or the tenant. Once established, these systems often resist change or modification. But adjustments can be made for changing supply and demand conditions. Landlords make rental concessions during periods when the supply of tenants is low. When an opposite set of conditions prevails, landlords frequently demand more rent. Tenants who need land may assist them by bidding up contract rental levels.

Numerous examples may be cited to illustrate these two extremes. At the time of the Black Death in England, serfs frequently used their strong bargaining position to secure more desirable tenure conditions and to work out favorable long-term leasing arrangements. Comparable benefits are sometimes enjoyed by residential tenants. During serious business recessions, apartment owners often reduce their rental rates or use special inducements, such as periods of free rent or agreements to redecorate, to attract new tenants. Increased competition between tenants in turn can lead to landlord demands for higher rental rates. The housing shortage experienced in many urban areas during World War II, for example, provided numerous landlords with opportunities to increase rents. Similar demand pressures have been a common phenomenon in the land-hungry areas of the world and have often contributed to peasant unrest. Counterbidding between tenants together with landlord greed gave rise to the famous "rack rents" of nineteenth-century Ireland. Exorbitant rental arrangements of a comparable nature also persisted for long periods in countries such as Egypt and India.

Short-run changes in supply and demand conditions can result in wide disparities between contract and land rents. Over the long run, however, contract and land rental levels ordinarily move in the same direction. When contract rents decline because of a decrease in the relative demand for land resources, land rents also tend to decline because of the lower income attributable to land resources. With the reverse situation, land rent tends to rise. These adjustments may be attributed in part to changes in the income-producing capacity of land, but they also involve the relative bargaining position of landlords and tenants.

The effects of different landlord-tenant bargaining conditions on the relationship between land rent and contract rent may be illustrated by the example of two areas with land of comparable productivity shown in Figure 6-8. Tenants in the area of Figure 6-8A enjoy a wide choice of alternative employment opportunities that allow them to insist on high returns to their labor and management inputs should they lease land. The area in Figure 6-8B, in contrast, offers few outside employment opportunities and has numerous potential tenants bidding against one another for a limited number of leasing openings. With their poor bargaining position, the tenants in this area are willing to sacrifice much of the otherwise normal return to their labor and management above its subsistence cost just to have access to land. As a result, landlords in the area in Figure 6-8B can claim a far larger proportion of

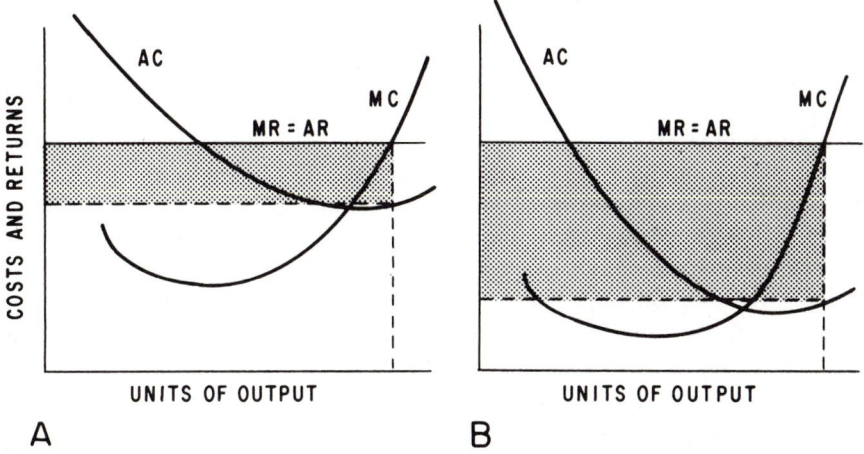

FIGURE 6-8.   Illustration of effects of high and low tenant bargaining power on land rental levels.

the total returns on their holdings as land rent and will receive considerably higher contract rental rates than is possible in the other area.

Relation of Land Rent to Land Values

Some types of land resources such as farms or forestland may be viewed as productive factors with almost unlimited productive lives. Others such as housing and office buildings have more limited economic lives but can be utilized over extended periods. Both types produce predictable future flows of reoccurring land rents that owners and investors consider as they visualize the current market values they associate with landed properties.

From a theoretical view, real estate resources have a current market value equal to the present value of their expected future land rents. The process used in determining these values usually calls for estimates of the value of the future flows of land rent expected from given properties and application of a procedure known as *discounting* to determine the present worth of these expected rental returns.

Discounting recognizes the fact that operators typically have a strong preference for receiving income now rather than waiting for its receipt at some future date. A guaranteed income of $10,000 fifty years, ten years, or even one year from now is never worth as much as an income of this amount now. Its current market value equals the worth of a current investment, which when held at an acceptable interest rate would equal the worth of the expected income at the time it is to be received. Discounting represents a negative premium on waiting, and with this process at work, properties that can produce endless flows of land rents into the future may have current market values that do not exceed the sum of the land rents expected in the next ten to twenty years.

In illustrating the discounting concept, one may assume a tract of land that is expected to produce net rental returns of $1,000 annually for x years into the future. The expected rental return for next year and for each year thereafter has a current market value of something less than $1,000 for the simple reason that the operator must wait to receive it. If the operator tried to sell or borrow money against this expected rental return for a given year or series of years in the future, the potential buyer or lender would calculate the present value of each year of expected return in terms of the amount of money it would take when invested now at an acceptable compound interest rate to yield $1,000 in the year in which the rental return would be realized. When discounted at 5 percent, an expected rental return of $1,000 one year hence has a current market value of $952.40, a return due in ten years has a current market value of $613.90, and a return due in twenty years a current market value of $376.90. With a 5 percent discount rate, an expected flow of net rents of $1,000 annually would have the currrent values shown in the shaded portion of Figure 6-9.[16]

---

[16]The present value of the future flow of returns can be charted infinitely into the future. However, as Figure 6-9 indicates, the present value of future returns becomes very small in a matter of forty to fifty years when discounted at 5 percent. Values decline more rapidly when higher interest rates are used and more slowly with lower rates.

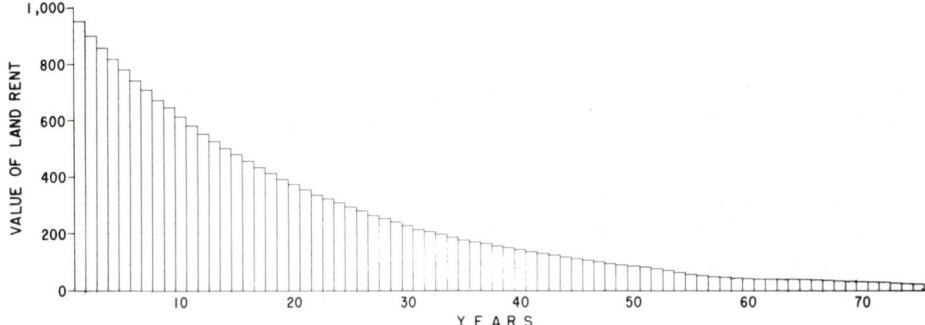

FIGURE 6-9. Present values of an expected flow of land rents of $1,000 annually for seventy-five years, discounted at 5 percent.

Summation of the discounted present values of a flow of annual land rents of uniform size that are expected to continue indefinitely into the future can be expressed by the formula $V = a/r$, in which V is the value of the property, a is the expected average annual land rent, and r is the capitalization interest rate.[17]

By way of illustration, if one assumes an expected average annual land rent of $1,000 and a 5 percent capitalization rate, the land resource in question has a value of $1,000 divided by 0.05, or $20,000. Similarly, with a 4 percent capitalization rate, the property has a value of $25,000, and with a 10 percent capitalization rate, a value of $10,000.

Despite its theoretical soundness, this income capitalization approach is not free from problems. These problems are considered in more detail in Chapter 10 along with some alternative approaches that are used in the valuation of real estate.

Applications to Land Resource Development Decisions

Expected flows of land rents also provide a guide for the decisions operators make concerning prospective investments in new or existing land resource developments. Investors expect a future payoff and will not commit funds to projects unless they anticipate a flow of sufficient land rent to more than repay their investment outlays within what they regard as reasonable time periods.

In projecting their expectations concerning future costs and returns, operators make whatever assumptions seem appropriate. They can assume that production,

[17]The term *capitalization rate* may be defined as the rate needed to convert a given periodic payment into a given cash value. As used here, it may be noted that the concepts of compound interest, discounting, and capitalization often involve different applications of the same interest rate. With compound interest, one starts with a present value and works toward a larger future value as accumulating interest payments are added to an initial amount of principal. Discounting involves a reversal of this process as one works back from an expected future value or return to a calcuation of its present worth. The same interest rate can be used in the capitalization process to indicate the current market value of a property that promises to provide a given flow of future land rents.

product prices, operating costs, and land rents will rise, remain constant, or decline. With an investment in an agricultural land development expected to yield a continuing flow of land rents and profits at some given level for an indefinite time period, they may assume a model such as that shown in Figure 6-10. The amount of land rent and profit represented by the difference between total returns (AR') and total operating cost (DR) would then be a key factor influencing their decisions as to whether they should proceed with the contemplated development.

Land resource developments frequently involve investments with limited economic futures such as that portrayed in Figure 6-10B. The investor in this instance may visualize considerable land rent and profits in the immediate future with a reduction and the eventual disappearance of these returns as the enterprise approaches the end of its economic life. Whether a prospective investor will proceed with this type of investment depends in large measure on his or her evaluation of the relationship between expected investment costs and the amount of land rent and profits expected during the economic life of the development. Discounting procedures may be used in both cases to indicate the present worth of the operator's expected investment opportunity.

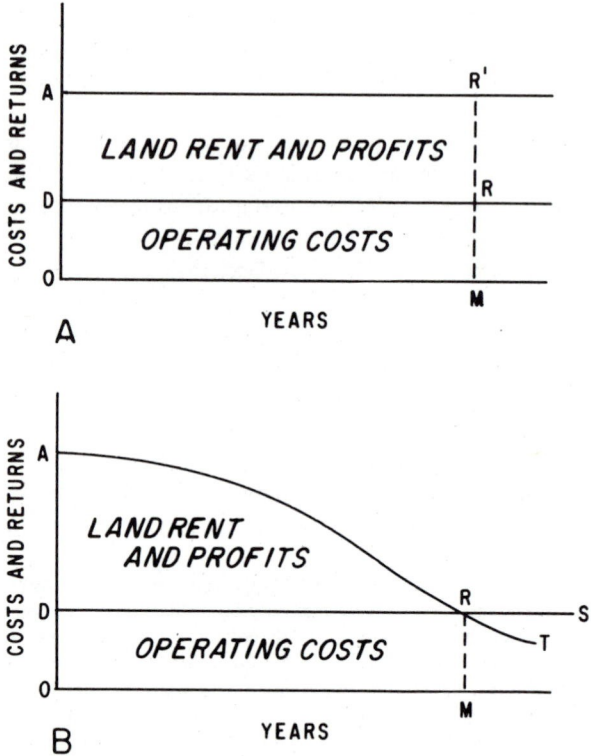

FIGURE 6-10.   Illustration of effects of expected costs, returns, and net land rents on resource development and investment decisions.

Effects on Land-Use Allocation

Our discussion of the economic returns to land resources to this point has assumed a single type of enterprise or land use. In practice, operators usually choose among a number of alternatives. Most operators tend to concentrate on those uses that will maximize their returns at their particular locations and with their combinations of productive factors. But they may also work with complementary enterprises, divide their attention among a variety of enterprises, or emphasize enterprises that correspond with their personal aptitudes and preferences.

In their choice of enterprises, operators are ordinarily interested in comparisons of the income-producing potentials of their various alternatives. These comparisons may be based on general observations or may involve calculations of the probable economic returns to land and management they can expect from each alternative. From an economic standpoint, comparisons of this type, particularly those involving both uses and locations, may be thought of in terms of overlapping rent triangles. The individual rent triangles for different land uses vary considerably in size and shape for different land uses. With the example used in Figure 6-11, they range from the high narrow triangle *EOP'*, which depicts the land rent secured from use *A*, to the low broad triangle *HOT*, which represents use *D*.

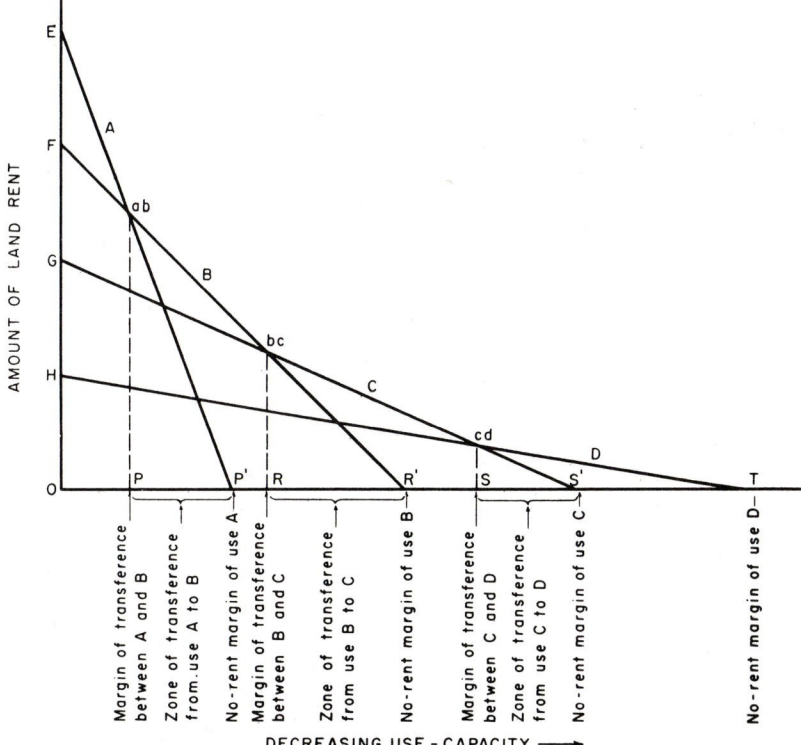

FIGURE 6-11.   Illustration of relationship between land rent and allocation of land resources between competing uses.

The four land rent triangles pictured in Figure 6-11 (*EOP'*, *FOR'*, *GOS'*, and *HOT*) may be used to describe the competition between four types of land use. Considered from an overall standpoint, they may represent commercial uses, residential uses, arable farming, and forestry or grazing, respectively. Under urban conditions, they may apply to a large department store, a light industry, a used-car lot, and the use of land for moderate-cost homes.

With each of these examples, the uses that produce the highest land rents ordinarily have first claim on the areas of highest use-capacity. Lower uses can always be carried on to advantage on the better lands, near the market, or at the 100 percent spot. But their lower comparative rent-producing capacity makes it impossible for them to compete with the more productive uses. As a result, they are crowded toward the outskirts to those locations where they can compete successfully with other uses. At any one location, some use can always return a higher land rent than any other and accordingly can be treated as the highest and best use of the site.

The hypotenuse of each of the four land rent triangles in Figure 6-11 represents the intensive margin for a particular use. The intensive margin for use A follows the line EP', and the intensive margins for uses B, C, and D follow the lines FR', GS', and HT, respectively. The points at which these intensive margins intersect are known as *margins of transference*. The intensive margins for uses A and B intersect at *ab* (point P on the horizontal axis). At this point it is more profitable to shift to use B than to continue with use A. Other significant margins of transference occur at points *bc* or R, where it becomes more profitable to shift to use C than to continue with use B, and at *cd* or S, where it becomes more profitable to shift to use D than to continue with use C. One can continue to operate in each case beyond the margin of transference to the extensive or no-rent margin and still receive land rent. Those who operate between their margins of transference and their no-rent margins (between P and P' in this case of use A; R and R' with use B; and S and S' with use C) are said to operate within their *zones of transference*. Operations carried on within these zones are profitable, but they are never as profitable as they could be if operators shifted to their highest and best use.

As this example suggests, the concepts of land rent and highest and best use can be used to explain both the competition between land uses and the resulting allocation of land resources between uses. This competition continues as a never-ending process, and its effects are observable in the continual allocation and reallocation of land resources that takes place between various uses and users.

Applications of the margin-of-transference approach.   The margin-of-transference approach provides a meaningful technique for explaining the allocation of land areas between uses with different rent-paying capacities. Diagrams such as Figure 6-12 can be used to illustrate the allocation process that takes place in and around typical cities. When cities are small, the triangles for use A (commercial) and use B (urban residential) may be reasonably small. As cities grow in size, the triangles expand both in height and in width, with the result that some sites used for residential purposes shift to commercial uses while farmlands around the city shift to residential uses.

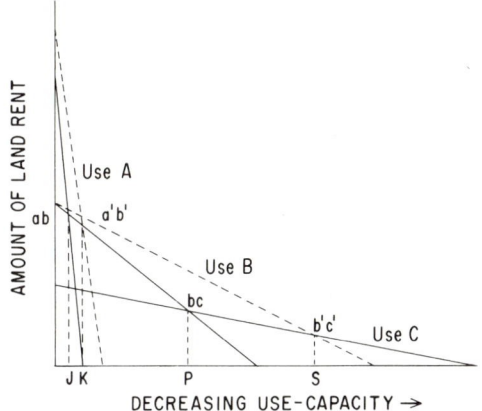

FIGURE 6-12 Example of applications of the margin-of-transference approach.

Figure 6-12 depicts two problem situations involving the use of land in urban areas. This first situation concerns the margin of transference *ab* at point *J*. This is the present margin between commercial and residential uses. Property owners just to the right of point *J* may assume that urban growth will soon push the margin to *a'b'* at point *K*. Anticipating this emerging higher use, they hold back on plans to remodel, repair, or rebuild properties in the transition zone between *J* and *K*. Should the expected emerging use develop, this decision may prove financially wise.

Unfortunately, the demand for the expected higher uses does not always materialize, or, if it does, it may not come until years after it was first anticipated. Without strong positive measures, the result is often a circle of spreading blight, slums, and urban decay around downtown commercial centers. Individual owners frequently sacrifice rental returns and satisfactions they could have had, although they may salvage their situations by acting as slumlords. The real tragedy in these cases involves the transference of social costs to the public, for it is society and the urban community that usually bear the major losses.

A second land-use problem centers on the margin of transference between residential and agricultural uses. In an earlier time period, when most urban residents lacked automobiles, the edge of the urban residential area occurred at point *P* in Figure 6-12. An increase in city size would have called for more intensive use of residential lands around the city's edge. With the relaxation of transportation constraints that has come with the widespread ownership of automobiles and the building of improved streets and highways, urban workers can now commute to work from point *S* in less time and with less effort than their grandparents expended in traveling from sites located to the left of point *P*.

Relaxation of the transportation constraint has facilitated a suburbanization and residential scatteration trend. It has made it possible for urban families to enjoy the advantages and amenities of suburban and rural living, but it has also greatly complicated the continued agricultural use of lands located in the vicinity of large cities.

This problem is most serious when occasional tracts between *P* and *S* are acquired for residential developments while large areas are expected to remain in agricultural and open-space uses. The higher rent-bid prices offered by residential property developers may call for only a small portion of the total land area, but they affect land prices, tax-assessment values, and the expectations of owners throughout the entire area. The new urban-oriented residents who move into the area often demand local public services not previously provided and add to the population that must be educated and protected. Farmers and other rural land users may feel that they are being squeezed out by rising property taxes and the larger investments required for any expansion of their business operating units; and speculators, attracted by the prospect of capital gains, may acquire lands that are often allowed to lie idle. Society again suffers as large areas become blighted for agricultural and other rural uses before a genuine need develops for their use for residential or other urban purposes.

Another application of the margin-of-transference approach can be visualized with public and private decisions concerning choices between single purpose and multiple-use alternatives in resource management. A public forest management agency may identify several competing uses such as commercial forest production, public recreation, or game management that could be emphasized as dominant uses in its management programs. It may also consider possible multiple-use management practices that give joint emphasis to two or possibly all three uses.

Realistic comparisons of the relative benefits associated with these management alternatives call for examination of the economic and social costs and returns associated with each management option. For multiple-use management to receive top emphasis, the sum of the economic and social land rents associated with this approach should exceed those attainable when managerial emphasis is given to any single dominant use. Once the economic and social land rents are determined, an agency may rationally decide, for example, that public recreation should be the dominant use of area A, multiple-use management should be applied to area B, and commercial forest production and game production should be emphasized in areas C and D, respectively.

Relationship between land rent and intensity of use. Land rents are often correlated with the relative intensity with which land is used. In practice, however, the two concepts are quite different.[18] Land rent represents the economic return land receives for its use in production, and intensity of use refers to the relative amounts of human and capital resources used in association with a given unit of land resources. These two concepts parallel each other because intensive use practices are often associated with high land rents; but it would be a mistake to assume that this situation always holds. Intensive use practices can be used to overcome the inherent deficiencies of low-rent sites. Businesspeople with poorly located sites sometimes use costly advertising programs to attract customers to their places of business. Farmers with soil of low natural fertility often use large inputs of fertilizer to increase the productivity of their lands. In similar fashion, peasant operators and

---

[18]Cf. Conrad Hammar, "Intensity and Land Rent," *Journal of Farm Economics*, 20 (November 1938), 776–91.

workers in cottage industries often find that lavish inputs of family labor are needed if they are to eke out a livelihood from their limited land resources.

Similarly, the fact that a site commands a high land rent does not necessarily mean that it is subject to intensive use. Low-rent housing facilities are usually subject to more intensive human use than high-rent luxury apartments. Low-rent commercial and industrial sites are sometimes used just as intensively as the high-rent locations found in downtown areas. And small family farms in low-rent areas are frequently used more intensively on an acre-to-acre basis than the larger commercial units found in areas of higher productive potential.

−SELECTED READINGS

Alonso, William, *Location and Land Use* (Cambridge, Mass.: Harvard University Press, 1964), chaps. 3–5.

Bober, Mandell M., *Intermediate Price and Income Theory* (New York: Norton, 1962), chap. 14.

Bye, C. R., *Developments and Issues in the Theory of Rent* (New York: Columbia University Press, 1940).

Due, John F., *Intermediate Economic Analysis*, 5th ed. (Homewood, Ill.: Richard D. Irwin, 1966), chap. 18.

Hall, Peter, ed. *Von Thunen's Isolated State* (London: Pergamon Press, 1966).

Keiper, Joseph S., Ernest Kurnow, Clifford D. Clark, and Henry H. Segal, *Theory and Measurement of Rent* (New York: Chilton, 1961), Part I.

Ricardo, David, *The Principles of Political Economy and Taxation* (London: J. M. Dent, 1911), chap. 2.

Whittaker, Edmund, *A History of Economic Ideas* (New York: Longman, 1940), chap. 20.

7

LAND RESOURCE
DEVELOPMENT DECISIONS

Land resource development decisions are much concerned with the economic productivity that can be expected from resources over time. Decisions frequently freeze the uses made of land for periods up to and often beyond the expected economic lives of the developments. They call for long-term commitments of an operator's capital, labor, and management resources. Their prospects for success call for proper timing, for making certain that a ready market exists for their expected products. They also call for balanced analyses of prospective costs and returns, efforts to keep costs within bounds, and realistic projections of the expected flows of land rents needed to justify investment expenditures.

Emphasis is given in this chapter to some leading factors that influence the land and water resource development decision-making process. Consideration is focused first on the concept of succession in land use and the economic rationale for resource development. This is followed with an examination of the principal costs associated with land resource development and finally with a discussion of the benefit-cost analysis techniques used in formal project evaluations.

SUCCESSION IN LAND USE

Land resources tend to move to those operators who bid the most for their control and to those uses that offer the highest return for their utilization. This concept operates with rural and urban lands alike. Its general operation suggests a phenomenon known as *succession in land use*. According to this principle, whenever changes in effective demand for different uses of land lead to changes in the use-capacities of the areas available for these uses, the real estate resources in question tend to shift to their highest and best economic uses unless prevented by institutional barriers, contrary goals, or operator inertia.

The history of man's use of land resources has been a long story of succession in land use. Some land resources have remained in their natural state. But most real estate resources have been modified and developed by human action. This development process has never been a once-for-all-time affair. Time after time, properties have been developed for particular uses only to be redeveloped within a few months or years for other uses that promise a higher net benefit or net return.

Examples of succession in land use appear all around us. Through the land development process, the forest primeval has given way to productive farms. Lands once written off as the Great American Desert have been utilized for irrigation, dry farming, and grazing purposes. Isolated wonders of nature have become summertime meccas for thousands of tourists.

This succession process finds its most vivid illustrations in the rise of our larger cities. The central business districts of most of the nation's cities were wilderness areas a scant two centuries ago. From this beginning, they became the sites of frontier trading posts and humble agricultural settlements, later the hubs of thriving business communities, and finally the commercial cores of expanding metropolitan centers.[1] In this succession process, the moccasined tread of the Indian and the frontier settler has yielded to the tumultuous traffic of downtown business areas. The crude shelters of early settlers have given way to the great banks, stores, and skyscrapers that now occupy the 100 percent locations. Lands that the government sometimes found difficult to sell for $1.25 an acre now support market values of thousands of dollars per front foot.

The succession process is dynamic. It calls for adjustments to changing demands and changing technology. As cities develop, houses and stores are built in the cow pastures and cornfields of yesteryear; private wells and primitive sanitation facilities give way to public water and sewerage systems; utilities are provided; new streets are built. Urban expansion and growth call for costly redevelopments. Streets that were suitable for horse and buggy traffic must be widened and relaid; sewers are dug up and enlarged; houses are torn down to make way for new commercial developments; and occasional areas are redeveloped to provide parks and open spaces.

Succession in land use often requires far-reaching decisions. Most resource developments call for substantial cash outlays. They call for careful investment calculations—for a weighing of expected benefits against the various costs that may arise with new developments. They frequently call for choices between alternatives—choices between different plans for development, between projects of differing size and scale, and between projects that promise to maximize personal profits and projects that accent community and social goals.

Emphasis is given here to three important types of decisions that arise with land resource developments: (1) the basic reasons for land resource development, (2) the

[1]Cf. Earl S. Johnson, *The Natural History of the Central Business District with Particular Reference to Chicago* (Chicago: University of Chicago Press, 1944).

rationale for redevelopment decisions, and (3) the problem of private versus social priorities in resource development.

Reasons for Land Resource Development

Land resource development is a necessary first step for securing most of the products we reap from land. Our basic motivation for resource development stems from the human urge for survival and from our desire to secure something more out of life than the food and shelter needed for subsistence. We develop resources because we must and because the products of development can add substantially to the utilities and satisfactions we secure from life.

These satisfactions are frequently measured in monetary terms, as when one clears farmland, lays out a subdivision, or erects an office building with the expectation of securing a larger net income. They may also be nonmonetary in character. They may involve spiritual or aesthetic values, a feeling of accomplishment, or any of many other individual or social goals. Thus a pyramid may be built as the resting place for one's soul, a garden as a place of beauty, a public building as a monument to one's philanthropy, or a system of fortifications as a line of military defense.

Regardless of whether they emphasize profits or nonmonetary goals, rational and knowledgeable operators do not proceed with a land resource development without certain basic assumptions. They first anticipate (1) an excess of expected benefits and returns above costs, (2) an unexpected flow of land rents and returns that continues into the future, (3) sufficient rental returns to justify their investment outlays within predictable time periods, and (4) development opportunities that rank high among their alternative investment options. Their evaluation techniques may be relatively precise or quite fuzzy. They may compare expected costs and returns both in the immediate and more distant future, place realistic values on expected future returns, and discount certain values back to the present. They may also involve faulty calculations of benefits and costs or inadequate allowances for risks and uncertainties. But well-planned or not, almost every land resource development involves some operator's attempt to maximize incomes and satisfactions.

The rationale of the business-minded operator who starts out to develop a land resource may be illustrated by the three planning models pictured in Figure 7-1.[2] With each of the examples in the figure, the operator starts with a given land resource base and a specific plan of development. The estimated gross receipts expected with each succeeding year are depicted by the line AT. Figure 7-1A assumes a development for a use—such as farming—that justifies an assumption of relatively constant gross returns over time. Figure 7-1B assumes a building investment that will yield a high annual return at first but lower and lower gross returns in future

[2]Decision-planning models depict the costs and returns visualized by operators at the time they make investment decisions. The models are forward-looking and indicative of what the operator thinks or hopes will happen. Operators do not really live into these models. New models must be formulated every time changing circumstances call for a reevaluation of one's investment opportunities.

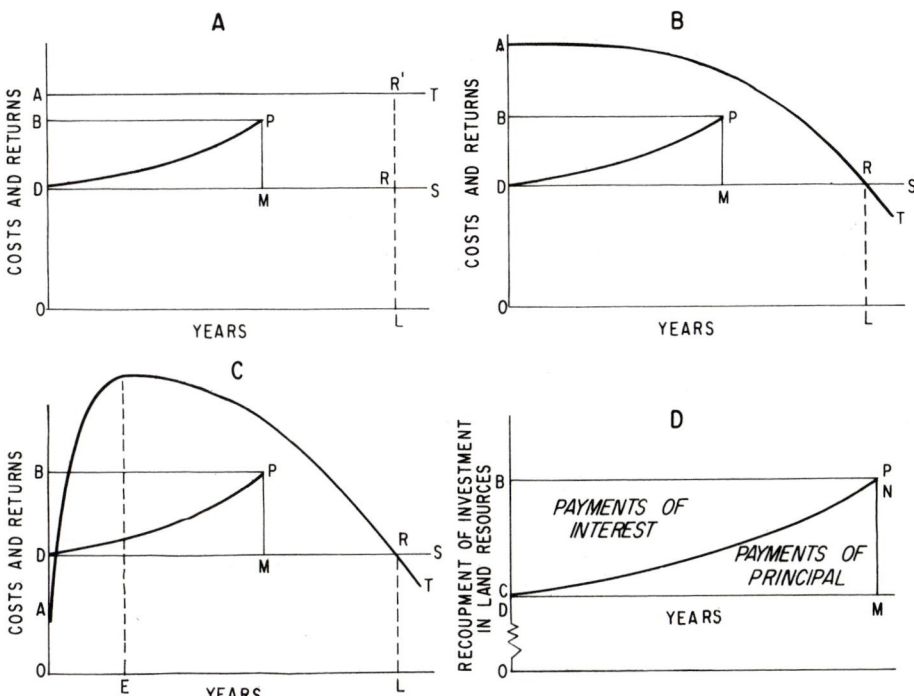

FIGURE 7-1.   Effects of expectations concerning future costs and returns on investment decisions involving land resource developments.

years as its economic life runs out and depreciation takes its toll. Figure 7–1C assumes the development of a commercial property that will not reach its highest earning capacity until year E and that will have a period of decreasing gross returns as its value depreciates.

With each of these examples, the operator must calculate an expected annual cost of business operation. A constant average level of operating costs may be assumed as with the cost line DS in each of the three diagrams; or it may be assumed that these costs will rise or fall with passing time. Attention also should be given to the size of the operator's investment outlay and to the prospects for its recoupment during the life of the development.

Operators who own their land should include their outlays for labor, materials, maintenance, taxes, insurance, and interest on loans for production purposes in their production costs. The surplus above these costs ($ADRR'$ and ADR for I years) represents a return of rent to land and profits to management. Of these two types of returns, the expected future flows of land rent can be quantified through use of the capitalization process.[3] Operators can thus treat the capitalized market values

[3]Capitalization of expected land rents to determine current market values is possible because flows of land rent are usually predictable. Similar quantification of the profit returns to management are less meaningful because of the unpredictable life spans of individual managers.

of their land rents as a fixed cost in their planning models and view any residual surplus above this amount as a profit return to their management.

With this approach, operators can visualize a block of land rental returns such as that symbolized by the rectangle *BDMP* in Figure 7-1D. This rectangle represents the portion of the expected future flow of land rents and profits needed to reimburse operators for the present market investment value of their proposed developments together with an opportunity cost allowance for interest on their equity funds and any use they make of borrowed capital during the assumed cost recoupment period. The block of returns symbolized by *BDMP* can be divided into two parts. The lower polygon, *CDMN*, in Figure 7-1D represents the net returns needed to permit full recovery of the current market value of the investment. This amount corresponds with the present discounted value of the expected future flow of land rents and in most cases can be closely identified with the operator's development and investment costs at the time commitments are made to proceed with a possible development or investment. The higher polygon, *BCNP*, represents the opportunity cost of using the operator's equity funds together with possible interest on borrowed capital and can be likened to the interest payments a borrower would make on a 100 percent loan of the value of the investment amortized over several years.

Once an operator has assembled and considered the necessary data on expected returns, investment costs, and operating costs required for a realistic evaluation of the prospects of a proposed development, he or she may feel free to proceed with the resource-development plan if the expected receipts (*AOLR'* in Figure 7-1A and *AOLR* in Figure 7-1B and C) exceed the expected costs (*DOLR* and *BDMP*, respectively). Serious consideration should be given at this point to the possible alternative uses that could be made of the investment inputs. Also if the decision is to go ahead with the development, plans will be made to limit operations to the time period (*L* years in Figure 7-1B and C) during which gross returns are expected to exceed operating costs.

These examples illustrate the operator's normal tendency to maximize land rents and profits. Similar types of reasoning are involved every time a family builds a home, plants a garden, or landscapes a yard and every time a city develops a civic center, lays a new street, or creates a public park. The emphasis in these examples is often placed on nonmonetary considerations—on personal, family, community, and social satisfactions and welfare. These values cannot always be expressed in monetary terms but nevertheless play important roles in the justification of land resource developments.

Rationale for Redevelopment of Land Resources

Most of the land resources with which people work have already experienced some measure of development. Decisions to further develop them for existing uses or to convert them to other uses call for their redevelopment. Redevelopment decisions are responsive to market pressures; and, as was the case in Figure 7-1, reflect the business calculations of individual operators. The average operator "is constantly alert to the possibility that some new use for his land may yield a greater return than the continued operation of the present use."[4] When opportunities of this type present themselves, the market values of properties are often bid up to the levels justified by the going estimates of their value for conversion to the new uses and their operators tend to shift to the higher and better uses.

The rationale for the redevelopment of existing resource developments may be illustrated graphically as in Figure 7-2. An owner visualizes current operations at the level suggested by the solid lines in the lower portion of Figure 7-1A or B. With the approach of year E, it appears that changing market conditions support a demand for the new type of development shown by the dashed lines. The proposed new development calls for higher annual operating costs ($D'S'$) than the existing development and also significantly higher levels of total returns, at least in the immediate future. The expected surplus of returns above costs promises higher levels of land rents, which again can be quantified in the rectangles $B'D'M'P'$, plus higher returns to management.

With these prospects, most operators who enjoy adequate financing would soon shift to the new higher and better use. They would capitalize on their future opportunities, even though this calls for writing off part of their existing investment. Some operators, however, may hesitate to shift to the new use because of pessi-

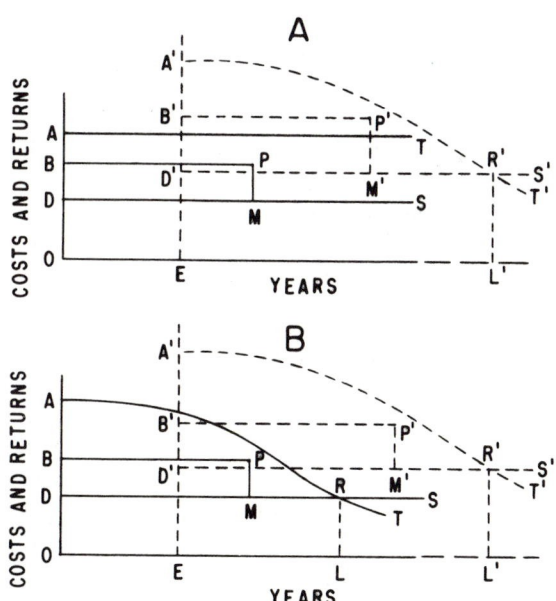

FIGURE 7-2. Effects of expectations concerning future costs and returns on investment decisions calling for the redevelopment of land resources.

[4] Richard U. Ratcliff, *Urban Land Economics* (New York: McGraw-Hill, 1949), p. 403.

mistic appraisals of their future opportunities, because of personal inertia, or because of concerns about supersession costs.

Should the new opportunity offer higher gross receipts with a net reduction of development and operating costs, only a foolish or overly conservative operator would refuse to shift. With a prospect of high development and operating costs, however, operators could find that they lack the financial backing needed to redevelop their properties. Some also could hold back until they have paid off their existing capital investment costs or hesitate to write off the current market value of existing developments.

The prospect of shifting to a higher use can pose quite a different problem with Figure 7-2A than with Figure 7-2B. The operator in Figure 7-2A will definitely lose income that otherwise could be received if he or she refuses to redevelop the property. Yet the shape of the present gross return curve (AT) provides a continuing prospect of profitable operations for as long as the present use continues. This expectation can be upset if higher property taxes or other rising operating costs (an upward tilt of the operating cost curve) narrow the expected margin of land rents and profits.

An operator faced with the situation pictured in Figure 7-2B may be content to continue with the current pattern of operations, even though a larger net return could be secured by redeveloping the property. It would not be economically wise, however, to postpone the redevelopment beyond year L. By then, the expected gross return will have declined to the level of the operating costs; and the operator must choose between redeveloping the property or losing money.

As was the case with Figure 7-1, the rationale for resource redevelopment has been viewed in terms of rent and profit maximization. This rationale can again be broadened to include various psychic returns, personal satisfactions, and social values. Families that remodel houses usually justify their actions more in terms of added family comfort and satisfactions than net increases in the market values of their properties. In similar fashion, cities redevelop streets and water and sewerage systems to increase the social utilities of these resources. Slum clearance and urban renewal projects are also undertaken for social as well as economic reasons.

Abandonment and shifts to "lower" uses. Real estate developers are perennially optimistic and in their optimism sometimes overestimate prospective returns, underestimate costs, or both. Miscalculations of this type frequently lead to developments that do not pay off as expected. When this situation arises, operators can charge off all or part of their past investments as sunk costs and continue to work with their developments, even though they provide fewer profits (or satisfactions) than expected. They can sell them for whatever price they can get and write off their investment losses. They may abandon them entirely if they cannot operate them at a profit or shift to some lower use, as in Figure 7-3, which calls for fewer variable inputs but which is really a higher and better use for the operator because it promises a larger net return above the expected lower costs of operation. A variation of this situation occurs in the blighted areas of many cities with the case of the

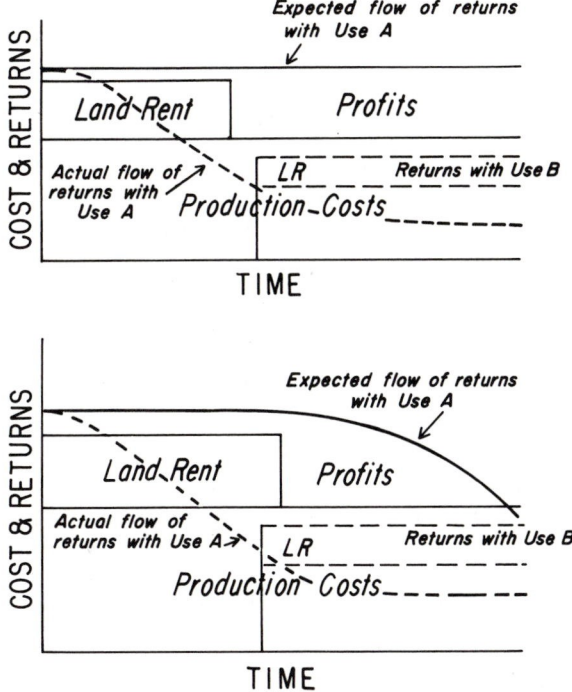

FIGURE 7-3. Economic rationale for decisions to shift to a "lower" use of land resources.

slumlord. Property owners in these areas often hope for emerging demands that will justify the redevelopment of their properties for higher uses. Quite often they have already reached the point *RL* shown in Figure 7-2B. Rather than abandon their properties, slumlords may exploit their remaining productive value by overcrowding and leasing for any price they can get while holding their expenditures for operation and maintenance costs to a minimum.

What happens to the unprofitable resource development depends largely on the location of the curves of its gross returns, operating costs, and capital investment costs. As long as gross returns exceed total costs, operators may be inclined to continue their planned use of the development. If gross returns exceed operating costs but are insufficient to cover capital investment costs, the use can continue with the writing off of all or part of the capital investment costs. If gross returns drop below operating costs, operators usually find it best to discontinue operations and abandon their development unless some means can be found to subsidize the cost of continued operations.

The history of land settlement and land resource development in the United States is replete with examples of resource developments that have not turned out as expected. Thousands of settlers have tried valiantly to develop productive farms in areas not physically suited for this purpose. Some of these settlements thrived for a time but gave out because of fertility exhaustion or changed market conditions. Some have survived because operators have subsidized them with large investments of outside capital and family labor. Many have been abandoned as unprofitable.

Much the same story applies with numerous other resource developments. Many building projects undertaken during boom periods have remained in operation in later, less prosperous years only because large portions of their initial investment costs were written off. Similarly, many large reclamation, public housing, and urban area renewal projects have depended heavily on public subsidies.

Resource developers and owners who have misjudged an expected cost and return situation should concentrate on maximizing their returns and satisfactions from this point on. Naturally they would like to recoup their losses on past investments, but their chief concern should be with the future, not with "crying over spilt milk" or moaning about costs already sunk. They should seek alternatives they can pursue to increase future net returns. They should consider converting underutilized commercial buildings into warehouses or shifting cultivated fields to grass or forest crops if the expected relationship between costs and returns promises a higher net return than can be secured from their existing development.

Land speculation.   Land speculation has been a common phenomenon in American history particularly along the frontier and more recently in fringe areas where rural lands are shifting to urban and recreation uses. As the term suggests, speculation involves the investment of monies at a risk with the hope of gain. Ventures in land speculation have brought moderate to substantial financial gains to some investors and losses to others.

From an economic point of view, *land speculation* may be defined as the holding of land resources, often in something less than their highest and best use, with primary managerial emphasis on resale at a capital gain rather than on profitable use in current production.[5] Traditionally, land speculators have shown little interest in the returns they could secure from current uses of real estate resources. They have tended to regard landed property as a commodity that should be bought and then sold at a profit. They sometimes invest in improvements that upgrade properties. But emphasis is placed on quick sales, on profitable and rapid turnover of capital investment, not on the continued holding and operation of properties. With this emphasis, they often leave their land holdings idle and unused while they look to profits from the sale of real estate for their principal source of income.

Land speculation often flourishes during periods of rising prices. With more stable

[5] Various shades of meaning are associated with the term *land speculation*. Horace Greeley is credited with describing the land speculator as "anyone who claimed or purchased raw land with no intent to farm it or who acquired more land than he could expect to develop." (Cf. Robert P. Swierenga, *Pioneers and Profits* [Ames: Iowa State University, 1968], p. 6.) This concept was generally accepted along the frontier, although typical frontier settlers usually distinguished between their own speculative landholdings and those of absentee owners.

Owners who hold land with the hope of later selling it at a profit are sometimes called *speculators*. Practically all American landowners are speculators under this definition.

market conditions, speculators sometimes find that properties must be held for months or years before they can be sold at a profit. Yet even with this outlook, they may hesitate to develop rural lands or build on city lots if they feel that the possible profit to be gained by waiting for an anticipated higher use will more than pay for the accumulated holding costs.

A different situation exists when high interest charges on borrowed capital, increasing outlays for property taxes and insurance, or other pressures discourage the playing of a waiting game. With these conditions, speculators may sell for whatever price they can get. They may give up their speculative dreams and develop or redevelop properties for the highest use justified by existing market conditions. They can also compromise by shifting resources to extensive-type uses such as urban parking lots, golf courses, or drive-in theaters, which promise sufficient returns to cover current holding costs while keeping open the option for shifting to some anticipated higher use such as a shopping center at a later date.

Private versus Social Priorities in Land Resource Development

Conflicts frequently arise between private goals in land resource development and the social interests of the community. Whose interests should govern in these instances? The answers to this question vary with time and circumstances. During some time periods and with some sets of circumstances, individuals are free to develop their land resources in almost any way they wish. On other occasions, they may find their opportunities definitely limited by social controls and regulations.

In facing up to the conflicts that occasionally arise between private and social priorities in land resource developments, it is important to note the reasons for these differences. The succession that takes place in private land resource use often reflects the bidding and counterbidding that takes place in the market. No problems arise as long as the owner or top bidder puts the property to some socially acceptable use. Conflicts may develop, however, if an operator decides to maximize personal profits or satisfactions by shifting to a use that damages or exploits the interests of neighbors or the community at large.

These exploited interests are often extramarket in character. They may involve the continued right of the public to enjoy the scenery provided by the trees, lakes, streams, or geological formations found on private lands or their opportunities to make recreational use of these resources. They may also include possible adverse impacts on the quality of the natural environment, as when an operator's activities lead to unsightly land developments, air or water pollution, or excessive noise or glare. These impacts can involve important negative externalities and social costs. They are hard to quantify in the usual market sense, and accordingly may have little impact on the economic calculus of individual operators. Group action calling for the exercise of social controls over private land-use practices is often recommended as an appropriate means for defending and protecting community interests in these instances. (See Chapter 18 for a more detailed discussion of these controls.)

LAND DEVELOPMENT COSTS

Land resources are sometimes described as a free gift of nature, but in their natural state they are seldom ready for immediate use as productive or consumption goods. Before they acquire much economic value, they must usually be processed or developed. They must be made accessible for use and in most cases must be modified through applications of capital and labor.

Cost considerations play a significant role in decisions concerning resource development. They guide operators in their choice of "what to produce, where and when to produce it, and by which of the available processes to produce it. Through a comparison of costs and prices with the productivity of the services hired, cost analysis coerces and lures the businessman into experimentation, into observation, into invention, with a view of finding out more economical ways of producing the things he is going to produce and more profitable things to produce."[6] Along with the prospect of benefits, cost considerations help to dictate the purposes for which land resources are developed and the timing of their development.

Several types of costs are involved in the development of real estate resources. First and most important among these are the actual outlays of cash and human effort required to bring new land resources into use and to qualify partly developed resources for higher uses. Other significant costs include the social costs associated with individual and group sacrifices; time costs, which arise because of the time it takes to bring resource developments into use; and the supersession costs associated with the scrapping of existing developments to make way for new resource developments. Reoccurring ownership and operating costs are also important once a development is brought into use.

Direct Outlays for Land Development

Land developments typically call for direct outlays of capital and labor. The extent and nature of these outlays vary both with the type of development undertaken and the period during which it takes place. Virgin forests may be opened for economic use through the building of logging trails. Modern skyscraper developments, in contrast, call for tremendous expenditures of capital and labor.

Much of the farm making that took place along the American frontier was accomplished with relatively low initial cash outlays. Settlers in Illinois in 1835 could buy 320 acres of prairie or woodland from the government for $400 and were able to supply themselves with a cabin, corncrib, and stable and to hire others to break the prairie sod and fence 160 acres for cultivation for an additional $745.[7] Many minimized their costs by clearing their own land, breaking the sod, and erecting their own buildings. This process required long hours of hard work, particularly in forested areas where it often took years for settlers to cut down trees, remove stumps and stones, erect buildings and fences, and expand their initial clearings into

---

[6] Jacob Viner, "The Role of Costs in a System of Economic Liberalism," *Wage Determination and the Economics of Liberalism* (Washington, D.C.: U.S. Chamber of Commerce, 1947), pp. 19–20.

[7] Cf. John Mason Peck, *A New Guide for Emigrants* (Boston, 1836), p. 313.

productive farms. By 1982 the average purchase price of a developed 320-acre farm in Illinois had risen to $600,800. Land-clearing costs, the installation of tile drainage systems in wet soil areas, and the provision of supplemental irrigation facilities often cost more than $1,000 per acre. Significantly higher costs also are associated with the provision of large-scale irrigation, drainage, and farmland development projects.

High land-development costs are expected with the development of land for most nonagricultural uses. The process of plotting and subdividing raw land for residential, recreation, and other urban-oriented uses usually involves substantial outlays for surveys; provision of roads, sewers, utilities, drainage facilities; and site improvements. These can easily exceed $10,000 per lot.

Higher price tags also are associated with the construction of residences, office buildings, shopping malls, and industrial facilities. Urban renewal projects are expensive both because of the high cost of acquiring already developed sites and demolishing the structures found thereon and because of the ambitious redevelopment programs that follow. The twelve-acre Rockefeller Center project in New York City which was completed during the 1930s at a cost of $150 million was considered a high-cost development at the time. Comparable large urban projects in the 1980s often have price tags of well over $1 billion.

Much of our early road system was supplied at a nominal cost by neighboring property owners who often supplied the horse power and labor on an in-lieu-of-tax basis. The construction of our modern highways cost much more, particularly in urban areas, where new expressways often cost more than $20 million per mile. Similar large outlays, sometimes running into many billions of dollars, are associated with other developments such as surface mining operations, metropolitan subways, urban mass transit systems, and nuclear power facilities.

Social Costs in Land Development

In addition to the cash outlays required for the improvement and processing of land resources, land development frequently results in social costs. These costs can be divided into two classes: social opportunity costs and social diseconomies or negative externalities. *Social opportunity costs* involve the returns and satisfactions forgone by society and its members because of the choices followed in resource developments. *Social diseconomies* in turn involve the external cost and negative spillover effects projects and developments can have on other individuals, communities, and society at large. Externalities can be positive and beneficial to others as well as negative and costly. A private development, for example, may provide open space or free recreation opportunities for the residents of an area or it may attract supporting industries that bring better business, credit, and transportation conditions for the entire community.[8]

[8]For additional comments on externalities, cf. Ezra J. Misham, *The Costs of Economic Growth* (New York: Praeger, 1967), Part II; Willis L. Peterson, *Principles of Economics Micro* (Homewood, Ill.: Richard D. Irwin, 1971), pp. 134–39; and Robert H. Haveman and Julius Margolis, eds. *Public Expenditures and Policy Analysis* (Chicago: Markham, 1970), pp. 81–95.

High social opportunity costs were a common feature of frontier life. The early settlers on the American frontier had to clear, stump, fence, and prepare their lands for farming. They erected their own buildings and joined with neighbors in providing schools, churches, roads, and other community facilities. They worked hard, invested their savings in capital developments, and went without comforts of life they could have enjoyed had they remained in older settled areas. When their developments bore fruit, they usually felt that their efforts were worthwhile and that their children and grandchildren could benefit from their sacrifices. Too often, though, time and labor were wasted on poor and unproductive sites. Deserted cabins and shacks still stand on many abandoned clearings as monuments to some settler's broken hopes and as mute reminders of the social waste that comes with misguided land developments.

Similar social opportunity costs can arise with developments of land resources for urban uses. Family sacrifices are often needed to finance the purchase of new houses and to provide business operators with necessary supplies of working capital. Urban redevelopment projects normally involve the social waste that comes with the scrapping of earlier developments. Promoters of urban developments can experience setbacks and heartaches. Their projects sometimes involve unwarranted expenditures and result in business losses or failures. The social costs and waste associated with these losses can be great, particularly when they imperil the savings and security of innocent investors.

Social diseconomies are a problem when public or private projects have adverse effects on others. A land-clearing operation may disrupt the natural ecology and desecrate scenic and other values that have long been enjoyed by others; a smelter may pollute the air around a city; or an industrial plant may dump untreated wastes in a neighboring stream. Costs in each of these instances are shifted from the operators to others in society. In times past, these costs were often dismissed as parts of the necessary price of progress. With the emphasis now given to environmental concerns, the problem of identifying these diseconomies, minimizing their occurrence, and more closely associating the responsibility for coping with them with the parties who created them has become a key issue in environmental economics.

Time Costs

Land development always involves the passing of time. Weeks and sometimes years elapse before improvements are completed and developed resources are ready for productive use. Throughout this time interval, land developers ordinarily find that their investments are tied up in assets that are not as yet ready to yield an economic return. They also find that they must gauge their present plans and operations to expectations concerning market conditions that will prevail after their properties have been developed. This process involves elements of risk and sometimes unexpected extensions of the time intervals that elapse before developed properties can be sold or put to economic use. The costs associated with the holding of land developments under these conditions may be described as time costs. This cost concept includes two closely related types of costs—waiting costs and ripening costs.

Waiting costs may be defined as those costs that arise because of the waiting period that elapses between the time of an operator's first outlay of capital and labor and the time when he or she can either sell the development or put it to productive use. Allowances for interest on invested capital and taxes paid during the development and normal sales period are two leading types of waiting costs. Of these two items, property tax payments represent a definite cash outlay promoters cannot avoid without endangering their investments. Allowance of interest on the capital needed for acquiring, holding, and preparing a development for its ultimate sale or use frequently represents a cash outlay. Substantial portions of this allowance go for interest payments when promoters operate with borrowed capital, and an opportunity cost return should be credited to those equity funds promoters supply from their own resources.

Significant examples of waiting costs occur with most types of land development and building activity.[9] People who have houses built to order must pay interest, and often taxes as well, on their land and building investments for the weeks or months that elapse before their houses are ready for occupancy. The only real differences between the waiting costs that arise in these cases and those associated with planting an orchard, drilling an oil well, or building a Rockefeller Center stem from differences in the size of the total investments involved and the duration of the time periods over which the waiting costs accrue. Waiting costs arise even in the development of tax-exempt public properties such as schools and reclamation projects. In these examples, as in instances of private ownership, payments of interest are required and should be calculated from the time construction funds are made available until the completed projects are ready for use.

Closely related to the concept of waiting costs is the parallel concept of *ripening costs*. This concept applies to those increases in the cost of holding property that stem from the ripening or imagined ripening of properties from lower to higher uses. Typical examples occur when tax assessors treat cutover land in the same property value class as farmland, farmland in the same value class as residential sites, or residential lots in the same value class as industrial and commercial sites. With each example, the increase in property taxes paid in the period that elapses before the land actually shifts to the higher use may be regarded as a ripening cost.

Ripening costs are usually associated with actual or assumed increases in land values. Farmland, for example, is seldom assessed as potential residential land until the outward growth of urban communities and the residential development of adjacent sites indicate a market demand for the higher use. In cases of this type, ripening costs in the form of higher tax assessments may be used as a lever to force the sale or development of land for a "higher and better" use.

---

[9]The concept of waiting costs is by no means limited to cases of land development. Merchants who buy a stock of Christmas toys in August always have capital tied up until they can turn over their investments. Similarly, farmers who buy seed, fertilizer, and machinery in the spring must allow interest on these and other production costs until they sell their harvest in the fall. Reforestation projects provide another excellent example of waiting costs because interest and taxes must often be carried for fifty years or more before the planted seedings are ready for harvest as commercial timber.

The concept of ripening costs also includes the costs sometimes incurred in holding developed properties for the anticipated higher values associated with their potential uses. A land subdivider, for example, may expect to sell a group of lots within three years but find that some must be held for seven years. The additional carrying charges (interest and taxes) incurred during the extra four years represent ripening costs. They arise because the promoter's development is premature and because additional "ripening" is needed before all of the lots can be sold at the initial asking price. Other examples occur when new developments such as apartment houses or office buildings have additional capacity beyond that which can be immediately absorbed by the market.

The ripening costs that occur when a land development or speculative holding must be held a few months or years can often be absorbed or passed on to the buyers, particularly if both market prices and demand remain high. Real problems arise, however, when investors find themselves caught in the onset of a recession. Under these conditions, some investors may find it possible to wait out the downswing of the business cycle. Others, particularly those operating with borrowed capital, may find their equities seriously reduced or wiped out by mounting ripening costs. With them, the alternative is often a choice between distress sales and bankruptcy, debt foreclosure, or tax reversion. Classic examples of these situations occurred along the frontier during every depression and with several ambitious residential subdivions of the late 1920s and condominium developments of the 1970s and 1980s.

Unfortunately, for their investors, numerous highly touted land resource development projects have been based on wholly unwarranted assumptions concerning economic demand. These projects have often been premature. Once examined in the cold light of reality, it has become obvious that they were "picked too green" to go through a normal ripening process. Instead of ripening to a higher use and acquiring higher values with more earning capacity, they have stagnated and sometimes remained idle and even been abandoned. Their development has often led to tax reversion and land title problems as well as considerable social waste. Prohibitive ripening costs have assured their failure and illustrate the fact that it takes more than high hopes and high taxes to make land resources shift to higher uses.

Costs of Supersession

Much of the world's land area is used with little change over long periods of time. Some sites, however, have experienced a succession of uses. With these sites, changing value and use patterns often make new types of development economically desirable, even though this calls for the writing off of investments in improvements already located on these sites. The costs involved in this process are known as *costs of supersession*.

Typical examples of supersession costs arise when city lots that have been used for residential purposes ripen for commercial use. With this situation, property owners can redevelop their sites for some higher and better use. An owner with a house valued at $50,000 producing a land rent of $400 a month, for example, may see an opportunity to erect a $500,000 office building that will provide a manyfold increase in net rental returns. If the lot were vacant, there would be little question as to what should be done. But the owner has a $50,000 house on the lot. Before he or she can proceed with the redevelopment, a decision must be made to either move or tear down the present building. Either approach involves the writing off of a considerable portion of the owner's present investment.

The problem with supersession costs thus boils down to whether owners are willing to sacrifice all or part of their present investments in improvements to capitalize on opportunities for realizing higher future returns. Should they decide to shift to a higher use, they must write off most of the value of their existing development. On the other hand, if they fail to shift to the higher use, they will suffer an opportunity cost (will forgo income) equal to the difference between the net income they receive and that they could receive with the new development.

Other examples can be cited to illustrate the concept of supersession costs. A fruit farmer may be able to make more money by shifting to dairying. But he or she may logically hesitate if the shift requires the destruction of already developed orchards, which currently provide a good livelihood, plus the cost of supplying a productive dairy herd along with the necessary dairy buildings. Supersession costs are also involved when workers refuse to shift to higher-paying jobs because of their fear of losing seniority and pension rights. They arise again when business operators question whether they should suspend operations and scrap part of the value of their present quarters while they remodel or rebuild with the hope of improving their future competitive and income status. Public and private agencies also incur large supersession costs when they spend millions of dollars for properties in blighted neighborhoods and for relocating displaced families to secure sites desired for new housing developments.

Property owners who are either unwilling or unable to undertake complete redevelopment programs often work out piecemeal compromises. These arrangements usually try to tap part of a property's potentially higher production capacity while minimizing the actual cost of supersession. Familiar examples are provided by the many houses that are converted into shops, restaurants, and office buildings; by the houses that have commercial structures built on in what once were their front yards; and by merchants who remodel their stores while carrying on business as usual. The results of this expedient are often less satisfactory and less sightly than those provided by complete redevelopments, but they do involve lower supersession costs.

BENEFIT-COST ANALYSIS

Few aspects of the land resource development process are more important than the assumptions decision makers have concerning future benefit-cost relationships at the time they approve or reject development proposals. Developers and investors normally proceed with the expectation that benefits will exceed their costs. To act otherwise would be irrational. Simple observations show that a wide range of practices are associated with the weighing of benefits and costs. Individual consumers, private operators, and public officials frequently give little thought to precise evaluation of benefits and costs in their purchase decisions. This is particularly true with habitual transactions and purchases that can be justified on the basis of earlier precedents. More thought is given to benefit-cost considerations and to justifications of purchasing decisions when one buys something new or when the purchase involves a large consideration. Buyers in these cases may or may not consider the appropriate facts. They may be overly receptive to the claims made for a product or underestimate or ignore probable costs. But attention is normally given to some weighing of expected benefits and costs, and a decision to buy can be interpreted as an expectation that the purchase will provide an excess of benefits above costs.

Project evaluation techniques of a more sophisticated nature are used with most public and private land and water resource development projects. The decision-making process in these cases is guided by public investment criteria and by private desires for the maximization of returns. Neither private operators nor public agencies have unlimited resources. They want to make certain that every project has a potential for paying for itself. They seek that project design for any given development that promises the highest surplus of benefits above costs. Moreover, if they can choose between several developments, they may apply an equi-marginal returns approach as a guide in the allocation of their investment resources. They may argue that each last input in a public or private resource development should provide marginal economic and/or social benefits at least as high as the returns, satisfactions, or benefits that could be attained through investments in alternative projects or programs.

Benefit-cost analysis provides a leading example of the techniques used in evaluating the economic prospects of resource development proposals.[10] This approach is used by public agencies in the United States in the formal evaluation of all water-resource development projects proposed for federal funding. Final decisions on the appropriation of public monies for these projects are made in the political arena and are subject to various pressures. Congress requires, however, that every development proposal pass a test of economic feasibility as a prior condition for approval.

Benefit-cost analysis can be used to rate or establish priorities between comparable alternative proposals, but it is ordinarily used simply to determine whether or not single-project proposals are economically justified in the sense that they promise a surplus of benefits above costs. Students of government have recognized the potential for using similar cost-utility (or cost-effectiveness) analysis techniques to

[10]Cf. Mark M. Regan and Elco L. Greenshields, "Benefit-Cost Analysis of Resource Development Programs," *Journal of Farm Economics*, 33 (November 1951), 866–78. A general requirement that project benefits exceed costs is stated in the Reclamation Act of 1902, and a more specific requirement for positive benefit-cost ratios appears in the Flood Control Act of 1936. Since the passage of the latter, Congress has moved steadily in the direction of requiring an excess of economic benefits above cost with all public water resource developments. It has also been argued that this technique should be applied in the evaluation of public investments in highways, urban renewal, outdoor recreation, civil aviation, public health, and government research. [Cf. Robert Dorfman, ed., *Measuring Benefits of Government Investments* (Washington, D.C.: Brookings Institution, 1965)].

compare the productive potential of alternative public projects and programs. Aspects of a cost-effectiveness approach were incorporated in the planning-programming-budgeting system (PPBS) initiated by the United States government in 1965 and are incorporated in most current budget analyses.

Nature of Benefit-Cost Analysis

Benefit-cost analysis is designed "to provide a guide for effective use of the required economic resources, such as land, labor and materials, in producing goods and services to satisfy human wants."[11] It is not the only basis for approving or disapproving resource development projects, as national defense and political considerations often play governing roles. But insofar as economic considerations prevail, benefit-cost evaluations can point the way to efficient use of public funds in land resource developments. It assumes that (1) projects have economic value only to the extent that a need or desire exists for their services; (2) each project should be developed at the scale that provides the maximum excess of benefits above cost; (3) every project or separable segment thereof should be developed at the least practicable cost commensurate with the overall objectives of the project; and (4) the development priorities assigned to various projects should follow the order of their economic desirability.

Concept of benefits and costs. Before one can proceed with a benefit-cost analysis, it is first necessary to define the terms *benefits* and *costs*. Prevailing practice calls for recognition of three types of benefits—primary, intangible, and secondary benefits—and four types of costs—project, associated, external diseconomy, and secondary costs.[12]

[11]Federal Inter-Agency River Basin Committee, *Proposed Practices for Economic Analysis of River Basin Projects*, rev. (Washington, D.C.: Government Printing Office, 1958), p. 5.

[12]Prior to 1950, several different definitions and practices were used by federal agencies. This situation caused the Federal Inter-Agency River Basin Committee to appoint a Subcommittee on Benefits and Costs in 1946 to formulate "mutually acceptable principles and procedures for determining benefits and costs for water-resource projects." These principles and procedures appear in the committee's 1950 report, *Proposed Practices for Economic Analysis of River Basin Projects*. Further standardization of procedures came with the publication by the President's Water Resources Council of *Policies, Standards, and Procedures in the Formulation, Evaluation, and Review of Plans for Use and Development of Water and Related Land Resources*, Senate Document 97, 87th Congress, 2nd session, 1962, the creation of the U.S. Water Resources Council in 1965 with authority to establish appropriate standards and procedures, the issuance of an Executive Order in "Water and Land Resources—Establishment of Principles and Standards for Planning," Part III, *Federal Register*, 38, no. 174 (1973), and the 1979 and 1980 revisions of these standards reported in Vol. 18 *Code of Federal Regulations*, pp. 711 and 713. For other discussions, cf. Otto Eckstein, *Water Resource Development: The Economics of Project Evaluation* (Cambridge, Mass.: Harvard University Press, 1965); U.S. Water Resources Council, *Procedures for Evaluation of Water and Related Land Resource Projects*, Report of Special Task Force, June 1969; Daniel W. Bromley, A. Allan Schmid and William B. Lord, *Public Water Resource Planning and Evaluation: Impacts, Incidence and Institutions* (Madison: University of Wisconsin Center for Resource Policy Studies and Problems, 1971); L. Douglas James and

Primary benefits are associated with the first level of resource values secured with completed projects. With the large reclamation projects in the West, this product involves the water stored in reservoirs. A market value of a certain price per acre-foot could be assigned to these waters; but since no monetary charge is normally made for their use, it has become accepted practice to treat the value of the first level of products secured from use of the stored water as primary benefits. These benefits include the value of electric power generated, crops grown, municipal water supplied, flood protection, recreation amenities, and other intangible benefits provided by a project. Intangible benefits are defined as "those benefits which, although recognized as having real value in satisfying human needs or desires are not fully measurable in monetary terms, or are incapable of such expression in formal analysis."[13]

Secondary benefits involve additional values that result from activities stemming from or induced by a project. These benefits usually involve a second level of resource utilization made possible by completion of a project. They may involve calculations of the value added by using electric power in industrial activities, processing grain into flour, or operating recreation concessions. These benefits have a potential for inducing regional economic growth and could provide a major justification for some resource development projects.[14]

Project costs include the full value of the land, labor, and materials used in developing, maintaining, and operating the project. Associated costs arise with the expenditures of capital and effort needed to secure the primary benefits. They include the cost of generating and marketing electric power, the cost of taking water to fields and growing an irrigated crop, the cost of distributing municipal water supplies, and so on. External diseconomies involve the social costs associated with possible negative externalities such as losses of scenic or environmental values or loss of access to flood waters that may result from a project. Secondary costs involve the value of any expenditures needed to produce secondary benfits. If an

---

Robert R. Lee, *Economics of Water Resource Planning* (New York: McGraw-Hill, 1971); Robert H. Haveman and Julius Margolis, eds., *Public Expenditures and Policy Analysis* (Skokie, Ill.: Rand McNally, 1977); Mark S. Thompson, *Benefit-Cost Analysis for Program Evaluation* (Beverly Hills, Calif.: Sage Publications, Inc., 1980), and Edward M. Gramlich, *Benefit-Cost Analysis of Government Programs* (Englewood Cliffs, N.J.: Prentice-Hall, 1981).

[13] President's Water Resources Council, *op. cit.*, pp. 8–9.

[14] Measurement of secondary benefits involves a complicated process. This factor, together with the possible double counting of benefits in the measurement process, has caused some authorities to recommend that benefit-cost analyses be based only on primary benefits and costs. Cf. S. V. Ciriacy-Wantrup, "The Role of Benefit-Cost Analysis in Public Resource Development," *Water Resources and Economic Development of the West*, Report No. 3, 1954, pp. 17–28; and Maurice M. Kelso, "Evaluation of Secondary Benefits of Water-Use Projects," *Water Resources and Economic Development of the West*, Western Agricultural Economics Research Council, Report No. 1, 1953, pp. 49–62. Senate Document 97 authorizes the counting of secondary benefits, such as positive effects in providing employment or in bringing desired redistributions of income, if they are documented and have value from a national point of view. Secondary benefits and costs that bring local or regional gains at the expense of other areas are not accepted in national account analyses.

excess of value of bread or flour above the value of wheat is claimed as a secondary benefit, the cost of transporting and storing the wheat, milling it into flour, operating the bakery, and distributing the bread to customers must be charged as a secondary cost.

Once these calculations are made, any surplus of primary benefits above project and associated costs are called *net primary benefits*, and any surplus of secondary benefits above secondary costs are called *secondary benefits*. These two types of benefits and costs are treated together in the determination of benefit-cost ratios.

Planning and formulation of projects. To be truly practicable, every resource development project must meet the tests of physical and biological capability, engineering and economic feasibility, and institutional acceptability. Water storage projects call for the presence of both adequate water supplies and suitable reservoir sites. Engineering designs and economic calculations are needed to determine their technological and economic feasibility; and political decisions must be made concerning their political, social, and financial acceptability. Early in the project planning and formulation process, serious consideration should be given to the question of whether proposed projects are actually needed and, if it is decided that they are, to determinations of their optimum scale and to ascertaining the most economic means for their development.

As a first step, care should always be taken to establish the fact that a need or demand exists for the products or services of the proposed project. Unless a real need exists, there is little point in trying to justify a project. If a definite need is found to exist, consideration should be given to the probable benefits and costs associated with projects of varying size. Information of this type is needed for decisions relative to the optimum scale of proposed developments.[15]

Projects are at optimum economic size when they produce more net benefits than can be secured at any larger- or any smaller-scale level. The determination of this scale level may be illustrated by the two diagrams presented in Figure 7-4. Both diagrams show the changing relationship between benefits and costs that occurs as projects increase in size. Point *B* indicates the scale of development that has the highest ratio of benefits to costs. Point *C* represents the scale level that produces the greatest excess of benefits above costs. As the lower diagram indicates, this is the scale level at which the marginal benefits associated with increasing size or scale of project equal the marginal costs—the point at which the ratio between marginal benefits and marginal costs becomes unity. Points *A* and *D* represent the levels at which total benefits equal total costs and the points at which a unity ratio exists between benefits and costs.

Under conditions of perfect competition, the optimum scale for a project is always found at scale level *C*—the level at which the benefits added by the last incre-

---

[15]Speaking on this point, Regan and Greenshields, op. cit., p. 871, observe: "In too many instances, projects are conceived and designed simply on the basis of their physical potentialities for utilizing to the limit available physical resources. Then, after the project is set up, the economist is invited to come in to show whether the plan is economically feasible. The amount of genuine guidance that economic analysis can provide in decisions affecting the public interest at this stage of project development is often quite limited."

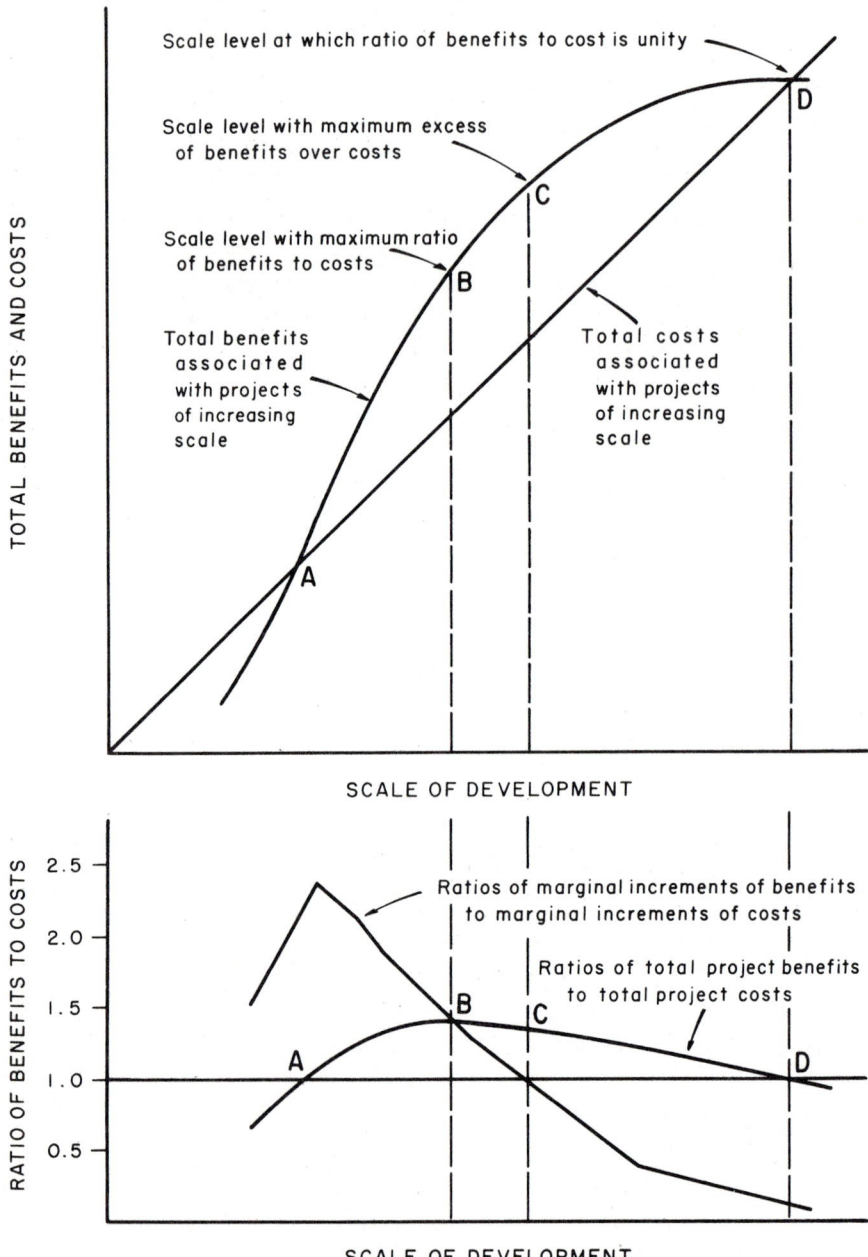

FIGURE 7-4.   Relationship between benefits and costs with projects involving different scales
of development.

ment of increased scale just equal its incremental cost. As is the general case with input-output relationships (see Chapter 5), this situation only holds with the assumption of unlimited resources. With the more realistic assumption of limited development funds, the optimum scale level depends on the location of the point of equi-marginal returns. It thus shifts to some scale level between B and C, to a point at which the marginal benefit-cost relationships of all the operating projects are in balance.

In practice, less than adequate attention has usually been given to determinations of optimum scale. As Regan and Timmons have observed,

Decisions concerning scale are sometimes based upon comparison of only a limited number of projects, which represent variations in scale. More frequently, formulation is based largely on the judgments of those designing the projects. Economic considerations often receive little attention unless those responsible for design are familiar with the economic concepts involved. In any case, the application of proper formulation procedures is extremely difficult because of the absence of sufficiently precise information on incremental relationships.[16]

Once the size of the project has been determined, steps should be taken to make sure that the project and all its separable parts have the lowest practicable cost. A project is poorly formulated when its objectives or the purposes of some of its separable parts can be attained at less cost by other means or when separable parts do not provide more benefits than costs.

Determination of economic feasibility. As benefit-cost analysis involves current evaluations of the expected flows of benefits that projects will produce in response to current and future cost outlays, care must be exercised in determining realistic estimates of the present values of both the expected benefits and costs. Consistent standards have been developed by the U.S. Water Resources Council to guide federal agencies in their analyses. Expected benefits can be counted for periods of up to 100 years. Guidelines have been prepared to standardize the approaches followed in estimating product prices and crop yields, in determining project costs, and in calculating risk allowances and probable salvage values for project items. Discount rates are set by a formula administered by the U.S. Water Resources Council.[17] Agencies are also required to base their evaluations on conditions with and without the proposed projects rather than on before and after project assumptions. This procedure recognizes that aftereffects can occur without as well as with project developments.

Once the appropriate data on benefits and costs have been assembled and ana-

[16]Mark M. Regan and John F. Timmons, "Current Concepts and Practices in Benefit-Cost Analysis of Natural Resource Developments," *Water Resources and Economic Development of the West*, Western Agricultural Economics Research Council Report No. 3, 1954, p. 5.

[17]Federal agencies were free to use discount rates of their own choosing until 1968, when the Water Resources Council posted a standard rate of $4\frac{5}{8}$ percent for all federal agencies. This rate can be raised or lowered by a maximum of $\frac{1}{4}$ percent each year. A rate of $7\frac{7}{8}$ percent applied for the 1983–1984 fiscal year.

lyzed and their estimated totals discounted to provide a measure of their present values, determinations can be made of the economic feasibility of individual projects. Four different approaches can be used to indicate the relative desirability of single or alternate projects.

As the first approach, one could subtract the total cost (C) of each project from its benefits (B) and evaluate projects according to their excess of benefits above costs. This approach ($B - C$) measures the net economic benefit or return but gives no weight to the relative costs incurred in each case. It is unacceptable as a measure of benefit-cost relationships, because it gives the same weight to a $1,000,000 project that costs $999,000 as to an $8,000 project that costs $7,000.

A second approach involves measurement of the rate of net return on the expected total cost outlay. With this procedure, total costs are subtracted from total benefits, and the difference divided by total cost to get a percentage rate of return. This method ($B - C/C$) gives a rate of return on total costs associated with the project. Comparable answers are secured from a third approach, under which the present value of the total expected benefits is divided by the present value of the expected costs (B/C) to provide a benefit-cost ratio. A positive ratio (a ratio of more than 1.0) indicates that a project proposal is economically feasible in the sense that it promises to produce more benefits than costs. This is the approach accepted by federal agencies in their benefit-cost analyses.

A fourth possible approach differentiates between project construction and investment costs and the operation and maintenance costs associated with the productive use of the developed project. With this approach, the present annual value of the expected operating costs is subtracted from the present annual value of the expected benefits, and the difference is divided by the annual value of the project investment costs ($B - OC/IC$) to provide a rate of return on project investment costs.

The (B/C) and ($B - OC/IC$) approaches yield comparable answers in many instances, but they can provide conflicting guidelines for the comparative priorities that might be assigned to alternative project proposals. Of the two, the (B/C) approach provides the better guide when emphasis is focused on allocations of limited construction funds and secondary concern is felt for future operation maintenance costs. The ($B - OC/IC$) approach in turn provides the best measure of rates of return to initial construction investments over time.[18]

Critique of Benefit-Cost Analyses

Several observations both pro and con may be made by way of critique of the benefit-cost approach to project evaluations. On the positive side, it may be argued that some method of project evaluation is definitely needed to guide the allocation of public and private investments, that benefit-cost analysis provides a logical and

---

[18]Cf. Roland N. McKean, *Efficiency in Government through Systems Analysis* (New York: John Wiley, 1958), pp. 108–13; Richard J. Hammond, *Benefit-Cost Analysis and Water Pollution* (Stanford, Calif.: Food Research Institute, 1960), misc. Publ. No. 13, pp. 17–21; Federal Inter-Agency River Basin Committee, *op. cit.*, p. 16; and Eckstein, *op. cit.*, pp. 53–65.

useful technique for this purpose, and that the resulting benefit-cost ratio is easily understood. Furthermore, the benefit-cost approach has been accepted for many years and numerous improvements have been incorporated in the techniques applied.

Critics of benefit-cost analysis have argued that it is at best a system of partial analysis; that the actual decision to proceed with projects is a political rather than an economic one; and that the data used in computations of benefits and costs are often inadequate and incomplete, with the result that benefits are sometimes underestimated and on other occasions inflated. It is also charged that there has been a lack of consistency in the standards used by different agencies, that the discount rates used with large public projects can be unrealistically low, and that significant impacts such as the effects projects may have on the natural environment or on local prospects for economic growth are ignored.

In practice the federal government has used benefit-cost analysis mostly to determine the economic feasibility of individual project proposals. Congress requires that project proposals have a positive benefit-cost ratio as a condition for approval. This practice discourages absurd projects, but it falls short of full maximization of investment returns. Projects are eligible for congressional approval as long as they have a positive ratio and fall in the range between points A and D in Figure 7-4. Projects with positive benefit-cost ratios may or may not be planned at their optimum scale and may include separable features that cannot be justified on a marginal value productivity basis.

Another problem stems from the frequent disassociation of benefits and costs. The assumption that a project proposal is eligible for funding if it has a positive benefit-cost ratio may be wholly justified if the same party pays the costs as receives the benefits. Complications arise when different parties are involved. Property owners in a proposed small watershed project area, for example, may oppose a project that has a positive overall ratio because most of the benefits will be received by downstream residents while they are expected to bear most of the costs. Similar questions may arise when government bears the cost of a harbor-dredging project that will be of primary benefit to a few local industries or when local groups endorse federal "pork barrel" projects that are expected to provide positive benefits for local communities while the tax costs are spread over the entire nation.

One of the most critical issues associated with benefit-cost analysis centers in the selection of the appropriate interest rate for discounting the value of future benefits and costs back to the present.[19] Significantly higher present values are secured with the use of low discount rates than with high rates. Proponents of public investments

---

[19] Economists are far from unanimous concerning the proper criteria that should be used in selecting an appropriate discount rate for use in public project evaluations. Cf. John V. Krutilla and Anthony C. Fisher, *The Economics of Natural Environments* (Baltimore, Md.: Johns Hopkins University Press, 1975), chap. 9; Raymond F. Mikesell, *The Rate of Discount for Evaluating Public Projects* (Washington, D.C.: American Enterprise Institute, 1977); and Gramlich, *op. cit.*, chap. 6. Some argue that discount rates should correspond with the opportunity cost of securing capital in commercial markets. Cf. William J. Baumol, "On the Discount Rate for

are often inclined to favor low rates such as the yield rate on long-term government bonds, while critics of public investment programs argue for acceptance of the higher interest rates associated with the operation of commercial money markets.[20]

In practice, federal agencies have used discount rates ranging from 3 to 12 percent, and in some cases have not discounted future benefits.[21] Questions have been raised about this lack of consistency in the use of discount rates both within and among public agencies. Differences made it easier for projects of doubtful value to secure positive ratios from some reviewers and agencies than from others. A general tightening up of federal agency standards during the 1960s made benefit-cost analysis a more acceptable yardstick for evaluating the economic questions affecting the overall desirability of individual project proposals.

Formal proposals for making benefit-cost analysis a more well-rounded and comprehensive technique for project evaluation were proposed to the U.S. Water Resources Council in 1969 and accepted in 1971.[22] The revised arrangement called for establishing four separate accounts in the evaluation process. These accounts require separate evaluations of proposed projects in terms of their impacts on national economic development, environmental quality, social well-being, and regional development. Most of the refinements in procedures announced after 1971 have involved these accounts. Limited use has been made of the new standards as few new projects were proposed by the Carter and Reagan administrations. Plans for repealing the existing standards were announced by the Water Resources Council in 1982.[23]

Public Projects," in Haveman and Margolis, *op. cit.* Others argue for the acceptance of social time preference rates. Cf. Steven A. Marglin, "The Social Rate of Discount and the Optimal Rate of Investment," *Quarterly Journal of Economics,* 77, (1963), 95–111; and David F. Bradford, "Constraints on Government Opportunities and the Choice of Discount Rate," *American Economic Review,* 65 (1975), 887–99. Still others favor weighting approaches. Cf. Arnold C. Harberger, "On the Use of Distributional Weights in Social Cost-Benefit Analysis," *Journal of Political Economy,* 86 (1978), S87–S120. Joseph A. Maciariello, *Dynamic Benefit-Cost Analysis* (Lexington, Mass.: Lexington Books, 1975), pp. 118–21, suggests that since precise calculations of optimal social discount rates are not possible that a practical course calls for government recognition of a need to allocate funds each year for resource developments and the taking of appropriate steps to make certain that the funds are used for those projects that promise to yield the highest ratios of return on investment costs.

[20]For examples of the effects of the use of alternative interest rates on the benefit-cost ratios associated with selected projects, cf. John V. Krutilla and Otto Eckstein, *Multiple Purpose River Development* (Baltimore, Md.: Johns Hopkins University Press, and Resources for the Future, 1958).

[21]Cf. Report of Comptroller General in *Interest Rate Guidelines for Federal Decisionmaking*, Hearings of Subcommittee on Economy in Government, 90th Cong., 2nd sess., 1968, p. 34.

[22]Cf. *Procedures for Evaluation of Water and Relative Land Resource Projects*, pp. 20–25. This report was revised in 1970 and published under the same title by the Senate Committee on Public Works, as Committee Print Serial No. 92–20, 92nd Cong., 1st sess., 1971. The filing of environmental impact statements for new development proposals, required by the National Environmental Policy Act of 1970, represents an important addition to this broader approach to project evaluation.

[23]Cf. *Federal Register*, 47, no. 55 (March 22, 1982).

Thus far benefit-cost analysis has been used primarily for the single purpose of determining the economic eligibility of proposed projects to qualify for Congressional funding. Little effort has been made to use it for assigning priorities to alternative investment opportunities. The final decision concerning project funding is, and probably will continue to be, a political one. Yet, insofar as emphasis is given to economic feasibility arguments, benefit-cost analysis can supply helpful guides for the rational allocation of limited public funds between alternative projects and programs.

— SELECTED READINGS

Dorau, Herbert B., and **Albert G. Hinman,** *Urban Land Economics* (New York: Macmillan, 1928), chaps. 11, 13, and 14.

Eckstein, Otto, *Water Resource Development* (Cambridge, Mass.: Harvard University Press, 1965).

Ely, Richard T., and **George S. Wehrwein,** *Land Economics* (Madison: University of Wisconsin Press, 1964), chap. 5. Originally published by Macmillan, 1940.

Federal Inter-Agency River Basin Committee, *Proposed Practices for Economic Analysis of River Basin Projects,* rev. (Washington, D.C.: Government Printing Office, 1958).

Gramlich, Edward M., *Benefit-Cost Analysis of Government Programs* (Englewood Cliffs, N.J.: Prentice-Hall, 1981).

James, L. Douglas, and **Robert R. Lee,** *Economics of Water Resource Planning* (New York: McGraw-Hill, 1971), chaps. 8 and 9.

Ratcliff, Richard U., *Urban Land Economics* (New York: McGraw-Hill, 1949), chaps. 12 and 13.

U.S. Water Resources Council, *Procedures for Evaluation of Water and Related Land Resource Projects,* Report of Special Task Force (Washington, D.C.: U.S. Water Resources Council, 1969).

Water Resources Council, "Water and Related Land Resources—Establishment of Principles and Standards for Planning," Part III, *Federal Register,* XXXVIII, No. 174 (1973), 24777–869.

8

LONG-TERM MANAGEMENT AND CONSERVATION OF LAND RESOURCES

Proportionality, attempts to maximize economic returns, and efforts to secure predictable flows of land rents and profits provide useful guides for the successful day-to-day management of land resource investments. They also have important implications for the long-term management and conservation of land resources. But with long-term management, operators often have other managerial goals, goals that frequently vary with the types of resources handled.

With fragile resources such as a wilderness or an endangered species, the principal goal may be that of protecting and perpetuating the resource base. With the development of water or solar power, the key question may involve choice of the optimum scale of operations and timing, that is, when to bring them into operation so that they will yield an optimum return. With farming, forestry, and the use of common properties, the goal may be to maintain productivity and profitable operations over time. With mining, concern may center on whether to start operations now or later and on determination of an optimum rate of exploitation. With man-made resources, efforts may focus on extending the economic use-lives of developments. Decisions in each of these cases are affected by the relative values operators associate with their expected future flows of net returns and by comparisons of these values with those that could be secured from alternative investment opportunities.

Operator attitudes and decisions are usually influenced to some extent by ethical, moral, and environmental considerations. Managers must think in economic and social terms, however, every time they make decisions concerning how their resources can best be utilized over time. At this point, they must decide between practices that promise to maximize immediate returns but that will possibly deplete or undermine their future production potentials and other approaches that emphasize the maintenance or saving of these resources for use over longer time periods.

Which of these approaches should the rational operator follow? What position should society take regarding these policy alternatives? These questions strike at the economic heart of the long-term management and conservation problem and provide the basis for the following discussion of the economics of conservation as it applies to land resources. Emphasis is given in this discussion to the economic meaning of conservation and to some leading factors that affect conservation decisions.

THE ECONOMIC MEANING OF CONSERVATION

Conservation is a concept of many meanings. Environmentalists visualize it as a moral issue tied up with man's responsibility to safeguard certain resources for the use of future generations. Technical workers sometimes identify it with the physical techniques they use to retard soil erosion, plant trees, or manage a deer herd. Sportsmen frequently think of it in terms of better fishing or hunting. Politicians often treat it as a political "sacred cow" closely allied with voter interests. Conservation evangelists regard it as the symbol of a better life, as an almost mystical means for securing "the greatest good to the greatest number—and that for the longest time."[1]

Conservation can be defined in different ways. In a dictionary sense, it involves the preserving, guarding, protecting, or keeping of a thing in a safe or entire state. This strict definition calls for "the preservation in unimpaired efficiency of the resources of the earth, or in a condition so nearly unimpaired as the nature of the case or wise exhaustion will permit."[2]

The idea of preserving land resources intact for future use has never gained much popular acceptance. To be sure, conservationists stress the need for saving certain resources for future use, and some have probably overemphasized this point. Most people, however, react negatively to policies of nonuse. They favor the maintenance and saving of land resources, but only to the extent to which these policies are consistent with programs of effective current use. Because of this rationale, much of the emphasis in conservation discussions is on the need for orderly and efficient resource use, elimination of economic and social waste, and maximization of social net returns over time.

From an economic and social point of view, *conservation* may be defined quite simply as the wise or optimum use of resources over time—wise use being viewed in an economic and social sense as those types of use that will permit operators and/or society to maximize the values they associate with their expected net returns both

---

[1]Charles R. Van Hise, *The Conservation of Natural Resources in the United States* (New York: Macmillan, 1910), p. 379.

[2]Richard T. Ely et al., *Foundations of National Prosperity* (New York: Macmillan, 1917) p. 3.

now and in the future. Viewed in this context, conservation is concerned primarily with choices in the timing of resource use. It deals with public and private decisions regarding the allocation of resources between the present and future and with policies and actions that are designed to increase the future usable supplies of particular resources. It involves the *when* of resource use.[3]

As we explore the economic meaning of conservation, it is important that we emphasize the goal of optimum use of resources over time and its interrelationship with the concepts of orderly and efficient resource use, elimination of waste, and maximization of social net returns over time. These issues are considered in some detail later in this section. Before turning to this discussion, consideration should again be given to two important facets of the conservation problem: (1) the classification of land resources for conservation purposes, and (2) the use of discount and compound interest rates in conservation decisions.

Classification of Land Resources for Conservation Purposes

Although people often speak generally of the conservation of land resources, it is more meaningful to talk of the conservation of particular classes of resources. This situation exists because of wide variations in land resource characteristics. Some have longer use-lives, are more exhaustible, or can be more easily renewed than others. These differences call for a meaningful classification of resources for conservation purposes that distinguishes between fund resources such as metals and mineral fuels that are nonrenewable and relatively fixed in supply; flow resources such as sunlight, precipitation, and changing climate that come in a continuous or predictable flow over time; and certain composite groups of resources that have both fund and flow characteristics.

Biological resources provide the leading example of a composite grouping. They include all forms of plant and animal life and have flow characteristics in that they are replaceable over time, provided care is taken to safeguard the seed stock needed for each new generation. Yet at any given time, they may also be treated as fund resources in a manner that will greatly reduce or even destroy their potential for future growth or reproduction. Unlike fund and flow resources, the productivity of biological resources "may be decreased through exploitation, maintained at the present level, or increased by the actions of man."[4]

Soil resources are a class that represents a combination of fund, flow, and biological resources. Farmers can exploit or destroy the fund of fertility stored over several centuries. They may use land in such a way as to draw only on the annual flow of fertility created by the action of plant roots, soil solutions, and organisms in releasing soil nutrients for possible plant use. They may carry on soil-building programs

[3]Cf. Siegfried V. Ciriacy-Wantrup, *Resource Conservation* (Berkeley: University of California Press, 1952), p. 51. Originally published by the University of California Press; reprinted by permission of the Regents of the University of California.

[4]Arthur C. Bunce, *The Economics of Soil Conservation* (Ames: Iowa State College Press, 1945), p. 4. Reprinted with permission from HETEROSIS, edited by John W. Gowen, © 1952 by The Iowa State University Press, Ames, Iowa.

(use of legumes, manure, and green manure crops) that emphasize the action of plant roots and soil microorganisms in building up the productive capacity of soil. Soils lack the life-cycle characteristics of plants and animals. Except for peat soils, which are better treated as fund resources, they are comparable to biological resources in the sense that their productivity can be decreased, maintained, or increased by human action over time.

Man-made improvements represent a third composite class of land resources whose natural characteristics can be modified through the addition of inputs of capital and labor. Leading examples include houses and buildings, neighborhoods, streets, and multipurpose dams. These improvements usually have a predictable economic life, but they may be treated for conservation purposes in much the same way as soil resources. Their productivity over any given time period can be adversely affected by abuse or destructive action. Yet with good management and the timely application of an appropriate flow of inputs for repairs and improvements, their long-run productivity can be definitely enhanced.

Use of Interest Rates in Conservation Decisions

Conservation decisions call for deliberate choices between the present and the future use of resources. In this decision-making process, operators weigh the benefits expected from the holding of resources during some given planning period against the costs associated with their holding. On the benefit side, operators consider the expected value of their resource at the end of given planning periods together with the value of any expected flows of land rents they may secure during these periods. Their costs include the present investment value of their resources and any operating, holding, or resource-improvement costs that may arise during their planning period. Strong cases can be made for conservation practices when expected future values and benefits exceed the present values and expected holding costs. When expected benefits fall below their costs, conservation can usually be written off as economically impracticable.

This balancing of the value of a sum of expected benefits to be received at some future date (or of an expected flow of benefits to be received over a given planning period) and current investment outlays plus expected operating costs is complicated by interest rate considerations. Operators who make investment commitments must think of interest charges on borrowed capital and of an opportunity cost return on their equity funds. Economic logic requires that they charge compound interest on the value of their fixed investments throughout the weeks, months, or years these investments must be held before the operators receive sufficient returns to recoup their costs. They can also be expected to place a higher value on the present possession or receipt of a given income than on promises of comparable income in the future. Operators accordingly find it logical to consider and compare the discounted present values of expected future returns and the discounted present values of expected costs when they make long-term investment decisions.

Most people are quite familiar with the workings of compound interest and of the increasing values one can expect if a given investment in a savings account is allowed to accumulate at compound interest without withdrawals for periods of several years. They see the need for charging interest on capital borrowed to pay for initial investments (buying land, constructing necessary buildings and improvements, and planting trees) and for operating costs incurred before a project starts to pay off. They also recognize that operators have a justifiable claim for an interest rate return on equity funds that otherwise would be earning a return in some alternative use. The concept of discounting, or of applying a discount rate to determine the present value of an expected future return, is equally valid but less understood.

As Figure 8-1 suggests, discounting can be viewed as compounding in reverse. Instead of starting with a current investment and compounding it at an x rate of interest to secure a given return y years in the future, one asks what is the present value of a given return y years in the future when discounted at an x rate of interest. The concept of discounting is frequently employed in banking and investment operations. Sears and Roebuck Company, for example, marketed zero-coupon ten-year corporate bonds with a maturity value of $1,000 for $290 each in early 1982. This purchase price represented the current market value of the 1992 maturity value of the bonds when discounted at 13.5 percent.

Discount rates can vary over a wide range. If an operator applies a low discount rate of only 2 percent, an expected return of $10,000 fifty years hence would have a current value of $5,521. As Table 8-1 indicates, discounting at 5 percent gives a current value of $823.67; 10 percent, $71.65; and 15 percent, only $6.34. The choice of the interest rate used in these calculations is a matter of strategic consequence in long-term investment and conservation decisions. As Gray has observed: "The primary problem of conservation . . . is the determination of the proper rate of discount on the future with respect to the utilization of our natural resources."[5]

Under the conditions of perfect competition, operators are expected to use the interest rate prevailing in the current money market in discounting their expected future values and in compounding their cost outlays. Operators could thus be expected to use a 7 percent rate if this were the current market rate of interest. They

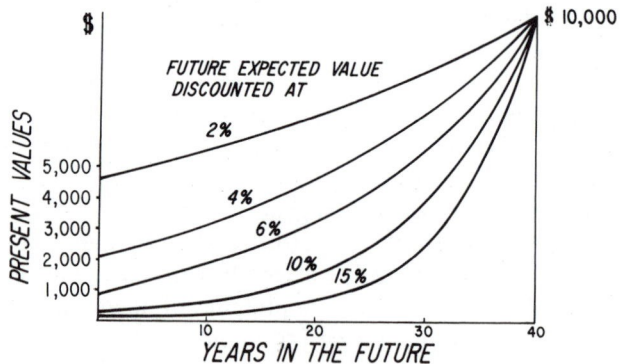

FIGURE 8-1.   Present value of a future income of $10,000 when discounted at selected rates of interest.

[5] Lewis C. Gray, "The Economic Possibilities of Conservation," *Quarterly Journal of Economics*, 27 (May 1913), 515.

TABLE 8-1. Present Value of an Income of $10,000 at Four Future Dates When Discounted at Selected Rates of Interest

DISCOUNT INTEREST RATE	$10,000 OF INCOME AT THE END OF:			
	20 years	30 years	40 years	50 years
2 percent	$6,729.71	$5,520.71	$4,528.90	$3,715.28
4 percent	4,563.87	3,083.19	2,082.89	1,407.13
6 percent	3,118.05	1,741.10	972.22	542.88
8 percent	2,145.48	993.77	460.31	213.21
10 percent	1,486.44	573.09	220.95	85.19
12 percent	1,036.67	333.78	107.47	34.60
15 percent	611.00	151.03	37.33	9.23

would shift to a 5 percent rate (more favorable to conservation) or a 9 percent rate (less favorable to conservation) if the going market rate shifted to either of these levels.

This assumption concerning operator acceptance of market-dictated discount and compound interest rates breaks down in practice. Factors such as imperfect competition; lack of perfect knowledge and foresight, different institutional settings, capital rationing, ethical adherence to conservation objectives; and differences in operator goals have brought the use of a wide range of interest rates in conservation decisions. Some operators tend to accept the going market rates—often with adjustments for the relative certainty or uncertainty of the expected future income—in their calculations. Many others use higher or lower rates. Still others make little or no attempt to identify the specific interest rates they use. Far from making detailed calculations on their probable benefits and costs, these operators often act on the basis of hunches and subconsciously determined interest rates, which may be high or low depending on the operator's inclinations at the particular moment.

The interest rates operators use in their conservation calculations ordinarily depend upon two important factors: (1) the operator's time-preference rate, and (2) adjustments made for uncertainties. Of the two, *time-preference*—the relative weight one gives to the receipt of a given quantity of income or satisfactions at some future date as compared with receipt of the same quantities at the present time—is usually more important.[6] Some people place high emphasis on the current use and exploitation of their resources—on a philosophy of "eat, drink, and be merry, for tomorrow we die." Others may go to an opposite extreme in following a miserly policy of setting aside all of their income and resources above that needed for subsistence living to provide for some future rainy day.

Individual time-preference rates vary widely between these two extremes. They vary from person to person and from day to day for some operators, depending on the operator's alternative opportunities, immediate need for income, desire to put

[6] The "rate of individual time preference" may be defined more precisely "as a ratio between the present marginal utility of . . . money in more distant future intervals and the present marginal utility of the same amount of money in intervals nearer the present and reduced by unity." Ciriacy-Wantrup, *op. cit.*, p. 104.

something aside for old age or one's heirs, zeal for pursuing a conservation or "stewardship of the land" philosophy, and a general feeling of optimism or pessimism at the moment. Individual operators may apply different rates over long planning periods as compared with short periods, and they may use one interest rate in discounting the value of an expected income and quite a different rate in compounding the interest charged on current conservation-cost outlays.

Optimum Resource Use over Time

The question of what constitutes conservation and the optimum or wise use of resources over time differs with each type of resource. Conservation of fund resources calls for spreading the use of the relatively fixed supplies of these resources over extended time periods. It "involves a reduction of the rate of disappearance or consumption, and a corresponding increase in the unused surplus left at the end of a given period."[7]

A very different situation exists with flow resources. Except for storage of resources such as water, there is no practical way to save these resources for future use. Good conservation practices call instead for elimination of the economic and social waste that comes with the nonuse of these resources and for their maximum practicable economic use under existing circumstances. Wise use of biological, soil, and man-made resources in turn calls for practices that yield the highest possible net return throughout one's planning period while maintaining, and if possible improving, their expected productive capacity.

Complications regularly affect decisions concerning the optimum rate and timing of the uses made of the different classes of land resources. Some of these arise because of the limited duration of operator planning periods, some because of the choice of interest rates, and some because of difficulties encountered in estimating expected costs and returns. Major problems also stem from the two-stage nature of conservation decisions: (1) the initial choice between developing a resource now or holding it for future development; and (2) determination now or later of the optimum timing of one's resource use or exploitation activities.

Factors such as expectations of sizable economic and social gains; high time-preference rates; high resource-holding costs; and uncertainties regarding future supply, demand, and price conditions often favor the early development and use of resources. Other factors such as operator inertia, lack of financial backing, insufficient market demand for the product, or an expectation of higher future market prices or technological improvements can have an opposite effect in favoring the postponement of possible developments. Decisions to exploit or utilize a resource now or to postpone its use until a future date involve an economic weighing of the values associated with immediate use as compared with the anticipated values associated with postponed use. As Hotelling has indicated, postponement can be justi-

---

[7]Erich W. Zimmermann, *World Resources and Industries* (New York: Harper & Row Publishers, Inc., 1933), p. 790.

fied in an economic sense only if the expected appreciation in recoverable resource value equals or exceeds the amount by which one discounts the value of the expected future value of the resource.[8]

Fund resources are often saved (and flow resources are lost or wasted) simply because resource owners choose a policy of nonuse. Some owners follow nonuse policies because of their social outlook, their desire to hold certain resources for future use. Some maintain reserves for use in future business operations. Some speculate by holding resources in the hope that they can realize significantly higher returns by postponing their development to a later date. Others hold back because they doubt that their contemplated developments will pay off.

A second-stage decision is made when operators decide to go ahead with the development of a resource. At this point, planning decisions must be made concerning the optimum rate and timing of resource use. These decisions are geared to expectations concerning future costs and returns and the possible impacts interest rate considerations can have in dictating the optimum time periods over which operators should plan their activities if they are to maximize their economic returns and satisfactions. The resulting rationale can best be illustrated with examples involving the various classes of land resources.

Flow resources. Operators who visualize current opportunities for the successful and profitable use of flow resources have a definite incentive to proceed with the early development of their plans. Examples include the possible use of oceans and streams for commercial navigation, construction of hydroelectric and solar power facilities, and recreation and resort developments at sites that boast climatic attractions.

The decisions associated with these situations parallel the resource development decisions described in Chapter 7. As long as a market demand exists for the product or service visualized, and the cost of providing this product or service is below its expected selling price, the development is economically feasible. Projects may be postponed for various reasons; but unless the delay is prompted by valid economic expectations of lower development costs in the near future or an emerging surge in market demand that will justify projects of larger scale than are now feasible, postponement ordinarily involves a loss of land rents and profits that could otherwise be realized.

Figure 8-2 illustrates the nature of the conservation decision with flow resources. As long as one envisages a continuing flow of annual returns that exceeds expected annual costs, early development of the resource should be encouraged. Postponement of a promising development results only in a loss of net benefits (land rent and profits) that could otherwise be received. Conservation and the minimization of waste accordingly call for early development and utilization of the flow resource.

---

[8]Harold Hotelling, "The Economics of Exhaustible Resources," *Journal of Political Economy*, 39 (1931), 137–75. Cf. also R. L. Gordon, "A Reinterpretation of the Pure Theory of Exhaustion," *Journal of Political Economy*, 75 (1967), 274–86; and Talbot Page, *Conservation and Economic Efficiency* (Baltimore, Md.: Johns Hopkins University Press, 1977), chap. 7.

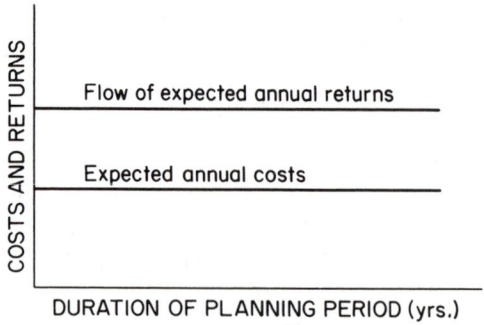

FIGURE 8-2. Relationship of expected annual returns and costs over an investor's planning period with a contemplated development of flow resources.

Fund resources. Developments that call for the exploitation, extraction, or mining of fund resources require a different rationale. Operators who utilize deposits of oil, coal, or iron ore, for example, must recognize the fixed and nonreplaceable nature of these resources. They seldom have specific information concerning the exact quantities of their deposits, and they usually lack technical ability to extract and utilize all of the fund resources they control. They know, however, that once the resource is removed from their land, it is no longer there and the supply will not be replenished.

Once they decide to proceed with a mining operation, operators could wish for the immediate recovery and sale of the entire supply of their mineral deposits. This, of course, is not possible. Drilling and mining operations require installations and equipment, and time is needed for capturing and removing the resource. Plans for the early capture of a fund resource can easily call for drilling numerous oil wells or opening several mine shafts or pits and for other high investment costs of questionable economic feasibility. Operators who want to maximize the present value of possible future returns must plan for both the optimum scale and timing of their mining operations.

Much of the problem of optimum timing with mining operations centers in the choice of an optimum scale. Operators should seek that scale that allows them to recover the greatest amount of fund resource that can be removed at a profit and also the scale that permits exploitation of the resource in an optimum time period. Their range of choices in selecting the scale that permits maximum recovery of a typical buried deposit of fund resource is illustrated in Figure 8-3. An operator with a deposit of oil or mineral ore can drill a single oil well or sink a single mine shaft and plan to use this facility to advantage for a substantial time period. Provision of additional oil wells or mine shafts will facilitate the capture of portions of the resource deposit that cannot be tapped effectively by a single well or shaft. An optimum scale measured in the number of oil wells or mining shafts provided will soon be reached, however. Beyond this point, additional wells or shafts will add little if anything to total volume of resource recovered. The addition of more wells, shafts, or other mining exploitation units beyond this optimum can speed the rate of total recovery but will involve less output per input-unit together with fewer net returns.

From a conservation point of view, operators should seek the scale level that per-

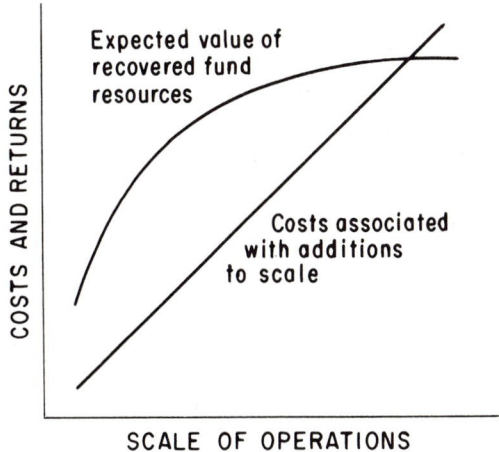

FIGURE 8-3. Relationship of expected value of recovered fund resources to costs associated with scale of operations needed to effect their recovery.

mits the highest rate of economic recovery and the least wastage of their fund resource. An operator's ability to attain this goal in practice is usually clouded by lack of knowledge concerning the precise extent of the resource deposit. Decisions concerning optimum scale are also affected by factors such as government regulations concerning the spacing of oil wells and the operator's ability to assemble needed installations, equipment, and work crews.

Choices concerning scale of operations have strong conservation implications, because they determine the timing of the operator's mining and resource-exploitation practices. Operators with several oil wells, mining shafts, or units of mining equipment can exploit a given deposit of fund resources in far less time than if they operate with fewer mining units. The scale of operations thus bears an inverse relationship to conservation. Large-scale operations facilitate the early and rapid exploitation of fund resource deposits, whereas smaller-scale operations are more conservation-oriented in the sense that they spread the resource-exploitation process over a longer time period. Optimum timing from an economic point of view requires choice of the scale of operations that can provide the highest present value of expected excess of future returns above operating costs.

The rationale associated with the optimum timing of the exploitation of a deposit of a fund resource can be illustrated with an example of a large surface deposit of some fund resource such as sand, gravel, limestone, or a mineral ore (see Table 8-2). Surveys show that the deposit contains 3 million tons of the resource and that it can be mined at an average cost of $4 per ton. The mining operation can be handled by blasting-mining-loading-and-trucking units, which are capable of handling 100 tons per day or 30,000 tons annually at an initial investment cost of $240,000 each. These units have an assumed economic life of 15 years and limited salvage value if used for shorter periods. The operation also involves a general overhead investment outlay of $1 million for an office building and office equipment.

With these assumptions, the operator can choose between a variety of rates and scales. Fifty blasting-mining-loading-and-trucking units could be used to mine the entire deposit in a 2-year period. Twenty units could be used for 5 years of opera-

TABLE 8-2. Illustration of Planning Model Used in Determining Optimum Timing Period for Exploitation of a Known Deposit of a Fund Resource*

OPERATION PLANNING PERIODS (years of expected operations)	QUANTITY OF RESOURCE TO BE MINED EACH YEAR	NUMBER OF OPERATION UNITS REQUIRED	DISCOUNTED PRESENT VALUE OF BUILDINGS, EQUIPMENT, AND OPERATIONS UNITS†	DISCOUNTED PRESENT SALVAGE VALUE OF OPERATIONS UNITS†	PRESENT EXPECTED VALUE OF TOTAL OPERATING COSTS ($12 million + column 4 − column 5)†	AUC_1 (present value of average unit cost of resource recovery)	AUC_2 (present value of average unit costs with capital recovery charge at 8 percent)	$AUR - AUC_1$ (present value of average net return per ton with no charge for investment cost recovery)	$AUR - AUC_2$ (present value of average net return per ton with investment cost recovery charge of 8 percent)
1	2	3	4	5	6	7	8	9	10
2.0	1,500	50	$13,000	$5,144	$19,856	$6.62	$6.92	$1.38	$1.06
2.94	1,020	34	9,160	2,159	19,001	6.33	6.68	1.67	1.32
4.0	750	25	7,000	1,021	17,079	5.99	6.41	2.01	1.59
5.0	600	20	5,800	517	17,283	5.76	6.22	2.24	1.78
6.25	480	16	4,840	200	16,640	5.55	6.02	2.39	1.98
8.33	360	12	3,880		15,880	5.29	5.82	2.71	2.18
10.0	300	10	3,400		15,400	5.13	5.69	2.87	2.31
11.11	270	9	3,160		15,160	5.05	5.63	2.95	2.37
12.5	240	8	2,920		14,920	4.97	5.57	3.03	2.43
14.29	210	7	2,680		14,680	4.89	5.52	3.11	2.48
16.67	180	6 + 6	2,894	248	14,646	4.88	5.62	3.12	2.38
20.0	150	5 + 5	2,578	76	14,502	4.83	5.70	3.17	2.30
25.0	120	4 + 4	2,263		14,263	4.75	5.77	3.25	2.23
33.3	90	3 + 3 + 3	2,019	53	13,966	4.66	5.89	3.34	2.11
45.0	90	3 + 3 + 3	2,019		14,019	4.67	6.50	3.33	1.50

*Example assumes a known surface deposit of 3 million tons of a fund resource that can be mined at an average variable cost of $4 per ton. Overhead costs include an investment of $1 million in building and equipment plus outlays of $240,000 for each blasting-mining-loading-and-trucking unit. Each operations unit can mine 30,000 tons of ore annually, has an economic life of fifteen years, and has limited salvage value if used for shorter periods. Investments in buildings and equipment are assumed to have no salvage value at termination of operations. Future outlays for operations units and salvage values are discounted at 8 percent in determining present costs (columns 4-6). Columns 8 and 10 assume product sales at a uniform price of $8 per ton. Compound interest at 8 percent is charged on the operator's investment outlay for buildings, equipment, and operating units by multiplying average unit costs for these items by years of operations by an 8 percent capital recovery factor. Discounted present values assume flows of annual net returns over the operation planning periods and are computed by multiplying the expected total net returns associated with each alternative operation planning period by the present worth of annuity factors for the three discount rates.

†Figures are given in thousands of dollars.

					ASSUMPTION OF SLIDING SCALE MARKET PRICE				
ASSUMED PERIOD OF EXPECTED MINING OPERATIONS	DISCOUNTED TOTAL VALUE OF EXPECTED NET RETURNS WITH ASSUMED UNIFORM PRICE OF $8 A TON WHEN DISCOUNTED AT†			AUR (expected average unit return)	AUR − AUC₁ (present value of average net return with no charge for investment cost recovery)	AUR − AUC₂ (present value of average net return with charge for investment and recovery)	Discounted total value of expected net returns when discounted at †		
	5%	10%	15%				5%	10%	15%
11	12	13	14	15	16	17	18	19	20
2.0	$2,965	$2,767	$2,592	$2.285					
2.94	3,651	3,291	3,016	3.632					
4.0	4,237	3,787	3,411	4.904					
5.0	4,621	4,047	3,578	5.607					
6.25	5,016	4,295	3,709	6.312	$0.76	$0.29	$ 733	$ 628	$ 542
8.33	5,318	4,336	3,544	7.129	1.84	1.31	3,200	2,609	2,132
10.0	5,356	4,262	3,481	7.576	2.44	1.89	4,374	3,480	2,843
11.11	5,371	4,178	3,361	7.810	2.76	2.18	4,940	3,843	3,091
12.5	5,352	4,054	3,212	8.052	3.08	2.48	5,466	4,141	3,281
14.29	5,228	3,870	2,991	8.301	3.41	2.78	5,863	4,341	3,355
16.67	4,770	3,409	2,578	8.558	3.68	2.94	5,888	4,209	3,182
20.0	4,301	2,938	2,160	8.823	3.99	3.12	5,840	3,990	2,933
25.0	3,778	2,433	1,733	9.906	4.34	3.33	5,632	3,627	2,583
33.3	3,151	1,827	1,253	9.377	4.72	3.49	5,043	3,019	2,072
45.0	2,398	1,478	898	9.377	4.71	2.88	4,601	2,553	1,722

tions, ten for 10 years, five for 20 years, or three for 33.3 years. The expected cost of operations at the various alternative scales and operation planning periods is itemized in Table 8-2. These calculations show that when costs are limited to recoupment of investment outlays plus variable operating costs, the lowest average unit cost per ton (AUC_1 in Figure 8-4) comes with a scale that spreads the mining operation over 33.3 years (column 9 in Table 8-2). Most operators expect com-

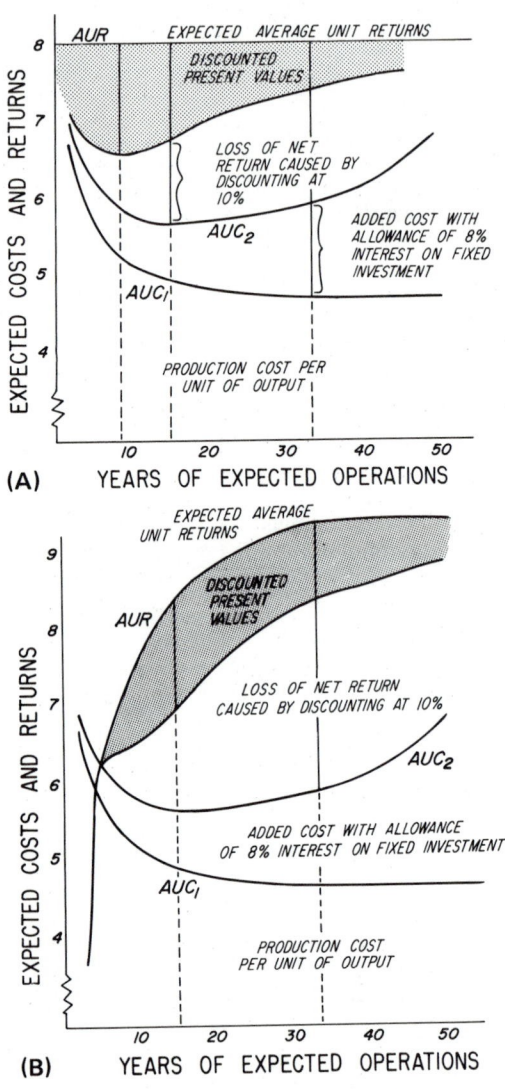

FIGURE 8-4. Use of planning models that assume expected average rates of returns, average costs, and the discounting to their present values of expected average net returns per unit of output for operations scheduled over alternative time periods to indicate optimum duration of extraction production periods.

pound interest on the amount of their investment outlay for the simple reason that payments are made for these funds if they operate with borrowed money and payments could be received if their funds were invested in alternative enterprises. Allowance of an interest charge of 8 percent compounded annually raises the average unit costs (AUC_2 in Figure 8-4) associated with each longer planning period. The lowest average unit costs now come with operations scheduled over a 14.3-year planning period (column 10).

The operator in Table 8-2 has a natural desire to maximize net returns. This objective is attained with the use of seven operations units over a 14.3-year period, at which point the operator receives an average net return of $2.48 per ton, when there is no discounting of the value of expected future returns. If the value of these returns is discounted at 5 percent, the optimum planning period drops to 11.1 years with nine operating units (see column 12 in Table 8-2). Acceptance of a 10 percent discount favors the use of twelve operations units over a 8.3-year period (column 13), and use of a 15 percent discount favors use of sixteen operations units over a 6.25-year period (column 14).

The operator may feel that the assumption of a uniform average unit return is unrealistic, that dumping all or most of the resource on the market within a short time period would depress the average price, and that higher market prices will be received if the resource is mined and sold gradually over a number of years. With an assumption of a sliding scale of market prices starting at $10 a ton if only 30,000 tons are marketed annually and dropping 3 percent for each additional 30,000 tons offered for sale (Figure 8-4B and column 15 in Table 8-2), the operator will find that the highest net return per unit comes when operations are planned over a 33.3-year period. Application of a 5 percent discount rate favors shortening the planning period to 16.7 years. Use of the 10 and 15 percent discount rates favors shortening the planning period to 14.3 years.[9]

Operators seldom know as much concerning the extent and value of their fund resources as is assumed in this example. Higher compound interest and discount

[9]The points of optimum scale and timing shown in Table 8-2 and Figure 8-4 correspond with the periods of operations at which one can expect the largest discounted present values of net returns per average unit of expected output. In marginal analysis terms, the ratio of the operator's expected added costs to expected returns per output unit with no discounting becomes unity at the operating time period that provides the maximum spread between AUR and AUC_2 (see Figure A); but the discounting process favors selection of a shorter operations time period to the left of this point at which one can expect a maximum present discounted value of net returns. As an alternative, one can also determine this optimum period of operations by visualizing the present discounted values of the expected costs and of returns per output unit for the various possible operations periods (see Figure B) and plan to operate to the point at which the ratio of the discounted expected costs to discounted expected returns per output unit becomes unity.

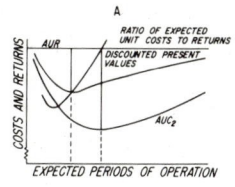

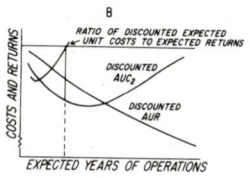

rates are often used to compensate for uncertainties associated with lack of precise information concerning the extent and quality of deposits, ease or difficulties of capture, and future costs and prices. Regardless of the interest rates used, a rationale similar to that used in the example above should govern economic decisions concerning the optimum timing of fund resource mining operations. As in the example, the charging of compound interest on investment outlays and the discounting of expected future returns to obtain measures of the present values of these returns tend to favor shorter operation planning periods than would be the case if no interest rate assumptions were used. High operator time-preference rates prompt the use of high compound interest and discount rates and lead logically to a shortening of the time periods over which mining operations can be expected to occur.

Planning models such as those suggested by Figure 8-4 can provide helpful guides for operator decisions. They are never more accurate, however, than the assumptions on which they are based. Adjustments must be made in the models whenever new or better planning data become available. Successful operators must always be ready to adjust to changing conditions. If market prices increase or operating costs decrease, one may extend the production period and attempt to recover oil, coal, or ore deposits that would appear uneconomic under less favorable conditions. On the other hand, when prices drop or costs increase, it may be necessary to cut back or even abandon production plans.

Biological resources. The conservation and optimum use of biological resources call for managerial practices that maximize the current values of operator net returns over time while maintaining or improving their future productive capacities. The practices used to attain these ends vary with different resources. Some operators plant crops that mature in the space of a few months. Others deal with resources (grass, livestock, fish, and wildlife) having life cycles running into several months or years; and some work with resources such as forests or people who have productive life spans covering several decades.

Some operators are concerned with products and services secured from resources such as honey bees, draft animals, orchards, or scenery. Others deal with products (crops, forests, or meat animals) that involve the eventual taking of the resource itself. Some use managerial practices that call for complete harvesting of the resources found in given areas (field crops and rotation cutting of forests). Others maintain herds and forests with animals and trees of mixed ages from which selected animals are sold or trees are cut while other young stock is always coming along.

A major concern in the management and conservation of biological resources centers in the optimum timing of harvest operations. Operators have little choice concerning the best time to harvest some resources. A wheat crop, for example, must be harvested when it is ripe. As Figure 8-5 indicates, the crop has little value before the grain is ready for harvest; it can be harvested to advantage during a limited time period only; and, if it is not harvested during this period, most of its commercial value is lost. Fruit crops such as peaches and strawberries must also be harvested during the brief period when they can be picked and marketed before they are overripe.

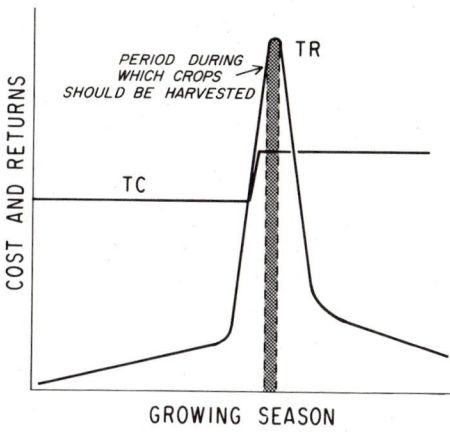

FIGURE 8-5. Typical relationship between expected costs and returns at different periods in a growing season with a biological resource such as wheat.

A wider range of choices is possible with meat animals and forests. These resources can be harvested early, or they can be stored "on the hoof" or "on the stump" for later use. Thus a rancher may choose between the sale of cattle as veal, baby beef, or mature beef, and the forester between holding trees for sale as Christmas trees, posts, pulp logs, or sawlogs. Both operators will find it to their economic advantage to harvest their products before they reach a point of maximum growth and before they suffer from decadence or decay.

Optimum timing of harvest operations is a matter of economic arithmetic and may be illustrated with a planning model for a forestry enterprise such as that depicted in Figure 8-6. This example assumes that an operator starts with a tract of essentially bare land that has been acquired and afforested at an initial investment cost of $100,000. Annual cost outlays for taxes and management are $2,000. The forest has little commercial value for the first twenty years. Thereafter, its value increases rapidly until it reaches its highest economic value of $855,000 in its seventieth year. This expected increase in total value product is shown by the *TVP* curve in Figure 8-6.

If the operator's cost calculations are limited to actual cash outlays for initial investment plus annual management and taxes, the costs can be represented by the total factor cost curve *TFC* in Figure 8-6A. Maximization of net returns then calls for harvesting the forest in the sixty-fifth year, the year in which the operator can expect the highest possible spread between the *TVP* and *TFC* curves.[10]

A decision to discount the value of the expected net return in Figure 8-6 has a greater impact in lowering the current value of a return expected sixty-five years hence than a return expected after a shorter waiting period. Adjustments according-

[10]Marginal analysis can be used, as in the lower diagrams in Figure 8-6A and B, to indicate the optimum time period for operations before adjustments are made for discounting. With the diagrams shown, the points of maximum spread between *TVP* and *TFC* correspond with the points in time at which the ratio of marginal increments of returns to marginal increments of cost becomes unity. The net returns that occur at this point and in shorter time periods must still be discounted to determine the economically optimum period for operations. Similar marginal analysis could be used to identify the time periods that provide operators with their highest discounted net returns if one worked with discounted present value *TVP* and *TFC* curves.

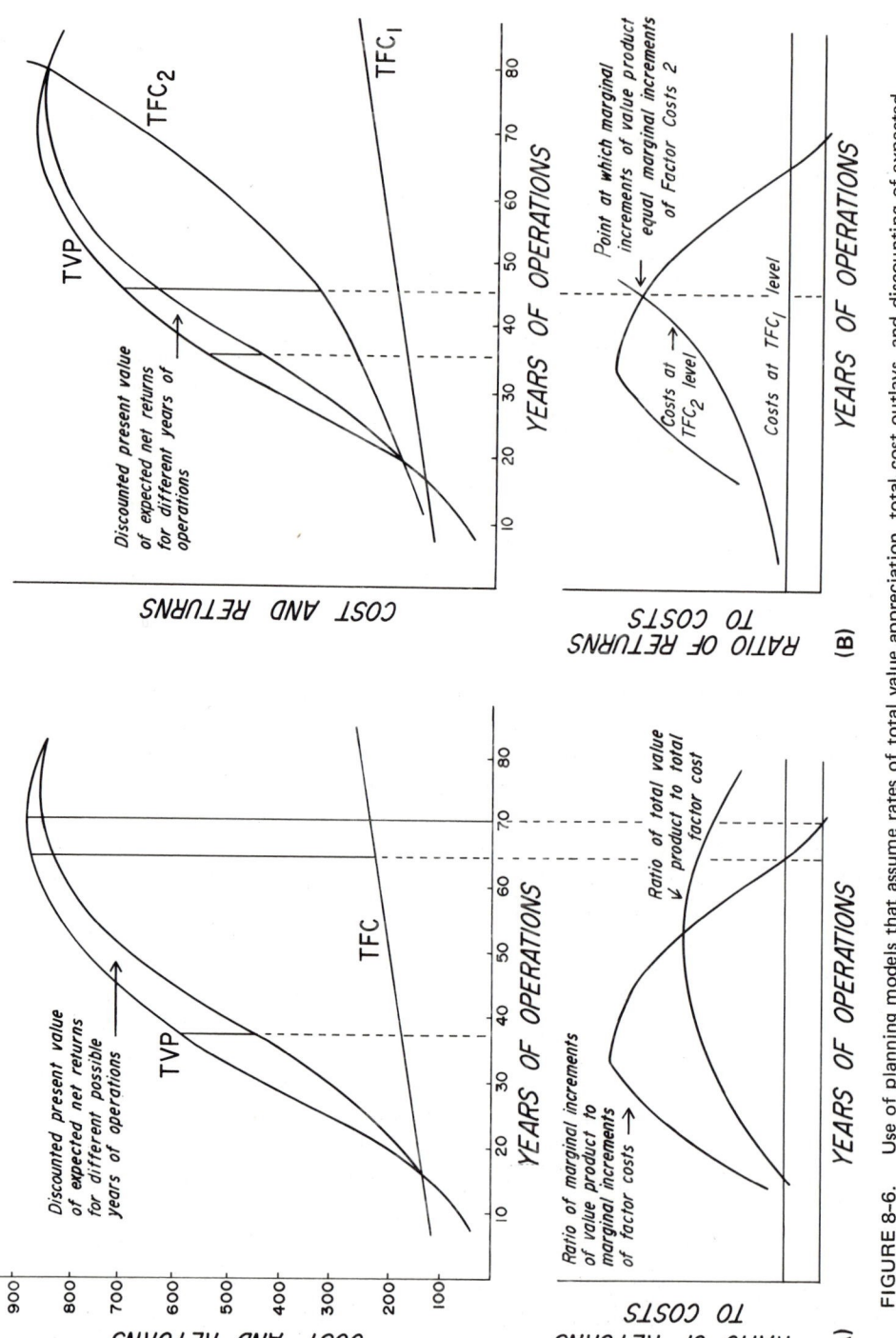

FIGURE 8-6. Use of planning models that assume rates of total value appreciation, total cost outlays, and discounting of expected future net returns to their current values to determine the optimum timing of harvest of a forest of uniform age.

ly are needed in the calculation. The curved upper portion of Figure 8–6A shows the current values of the expected net returns for every possible year of operations when discounted at 3 percent. Lower discount rates would produce higher present values, and higher discount rates would produce lower values.[11] With a 3 percent discount rate, an operator can maximize the present value of the net returns by limiting the planned production period to thirty-seven years.

The operator in this example can logically insist that compound interest be charged on the initial investment and on the series of annual charges that must be paid until the forest resource is ready for harvesting. *TFC$_2$* in Figure 8–6B shows the effect of charging a 2 percent compound interest rate on these costs over the life of the project. With the addition of these costs, the optimum operating planning is forty-five years when there is no discounting of net returns. When the net returns are discounted at 3 percent, as shown by the shaded portion of Figure 8–6B, the optimum planning period drops back to thirty-six years.

With situations such as that assumed in Figure 8–6, questions can be raised as to why people invest in long-term forestry when other more promising alternative investment opportunities are available. The simple truth is that few commercial operators start with isolated investments in raw land that they plant to trees and hold over long periods for eventual harvest. Many of those who have operated in this way have benefited from monetary inflation—their major investments were made when costs and interest rates were lower, and their harvested timber has had a higher value than they could have expected earlier. Others have received much of their compensation in the form of recreation or in the pleasure of working with nature and seeing otherwise underutilized land shift into production.

A somewhat different rationale applies with those commercial operators who manage and harvest their forests on either a selective-cutting or a rotation-cutting basis. Operators who periodically harvest mature, malformed, and diseased trees in their forests, while leaving young stock for continued growth, follow a pattern such as that depicted in Figure 8–7A. With each cutting, the market value of the remaining forest declines and then gradually rises again as the time for the next selective cutting approaches. The owners' holding costs also rise; but by paying off all accumulated costs to date out of the proceeds of the timber harvested in each cutting, operators can realize a profit and not be as much concerned with the discounting of expected net returns as they would be if their calculations covered the entire life span of each tree.

A similar situation applies with operators who work on a long-term cutting cycle. These operators normally own or control large forest acreages, which they manage as single-age stands. To keep their crews and facilities operating, they must plan to harvest one or more tracts each year and then move on to other tracts in succeeding years and eventually repeat the cycle. Meanwhile, the cutover areas are reseeded and managed so that they will be ready for harvest when the next cutting cycle

---

[11]Low discount and compound interest rates are used in Figure 8–6 for the obvious reason that higher rates would so reduce the discounted value of the net returns that they would hardly be visible in the two diagrams.

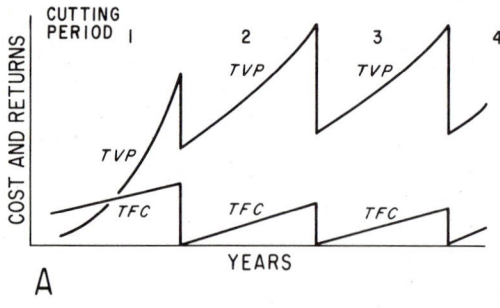

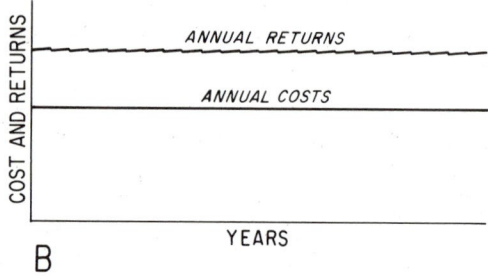

FIGURE 8-7. Interaction of total value product and total factor cost considerations with typical selective forest-cutting and rotation-cutting operations.

begins. These operators follow the pattern suggested by Figure 8-7B. Firms pay each year's management and holding costs out of that year's proceeds. This practice, like those of operators who grow crops for annual harvest, accepts a planning model similar to that for flow resources shown in Figure 8-2. The important distinction, however, is that foresters and farmers must nourish and protect their seed stock from year to year if they are to benefit from its continued bounty.

Common property resources. Common properties pose a unique problem for long-term resource management and conservation. They include such varied examples as air, water in streams and lakes, oceans along with their fisheries and mineral deposits, outer space, public grazing areas, fish and wildlife resources, wilderness areas, public parks and recreation resources, and the radio spectrum. Some of these are flow resources; others are fund or biological resources. The characteristic that binds them together as a group is the fact that they are, or within recent time periods have been, treated as free goods that individuals could use at will.

With pure examples of common resources, every operator enjoys a right of access and use; no one can exercise proprietary ownership rights by excluding others from the right to use; no operator can lose rights through nonuse; and no one can benefit from transfers of previously exercised rights to others.[12] In an ultimate sense, all environmental resources and all of the world's land resources can be viewed as resources that were given to mankind as common property. Rules governing the use

[12]Cf. A. Allan Schmid, *Property, Power and Public Choice* (New York: Praeger Publishers, 1978), p. 53.

of these resources and permitting the exercise of private property rights have been adopted to facilitate their effective use; but increasing population pressures and possible overusage can destroy the value of the resource base for all users.[13]

With operators exercising equal rights of access and use, conservation problems arise with common properties because of a basic inability of the resources to support unlimited amounts of usage. Fragile resources such as a wilderness can support only limited use efforts before they lose their wilderness characteristics. Overuse can easily bring the depletion of ocean fishery, game hunting, and grazing resources. By exceeding carrying capacities, overuse and inappropriate uses can also destroy important values we associate with the radio spectrum, park and recreation areas, and air and water resources.

The conservation problem with common properties can be illustrated with an example involving the free grazing of public lands, a situation that existed with 150 million acres of public domain lands in the western United States prior to 1934. Volume of grazing effort is shown on the base axis in Figure 8-8, and costs and returns are shown on the vertical axis. Operator costs rise steadily as larger numbers of animals are grazed.[14] The value of the sustainable yield of the range rises steadily until a point of maximum carrying capacity is reached. Usage beyond this point results in reduced carrying capacity and a reduction in total value product. Wise administration accordingly calls for limiting the grazing effort to the number of units

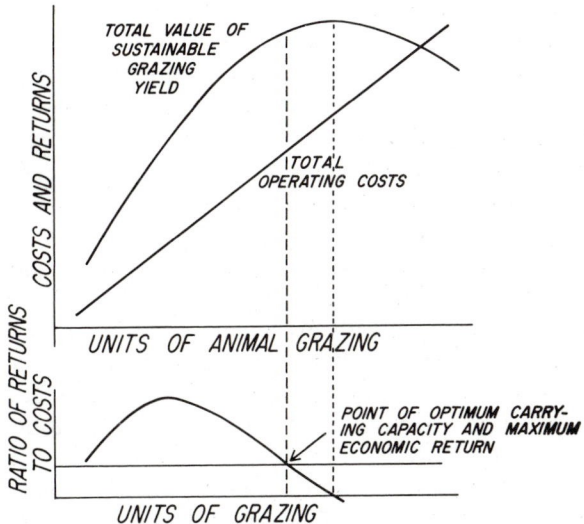

FIGURE 8-8.   Determination of the optimum level of grazing on a common property grazing area.

[13] Cf. Garrett Hardin, "The Tragedy of the Commons," *Science*, 162 (December 13, 1968), 1243–48.

[14] Cf. Francis T. Christy, Jr., and Anthony Scott, *The Commonwealth in Ocean Fisheries* (Baltimore, Md.: Johns Hopkins University Press, 1965), chap. 2, for a similar analysis involving fishery resources.

that bring the maximum spread between total value of sustainable yield and total cost or to the point at which the declining ratio of marginal benefits to marginal costs drops to unity.

Without regulations that provide economic order by limiting the units of the grazing effort to the carrying capacity of the range, several problems involving disassociations of benefits and costs develop. Individual ranchers find it personally advantageous to increase the size of their herds to gain a larger share of the available grazing resource. Those who operate from nearby headquarters or who benefit from better transportation facilities, superior livestock, or other technological advantages can enjoy higher benefits than others. Yet no one enjoys true security of their grazing expectations. No one has an economic incentive to improve or conserve the range. New graziers can invade the range used by others at any time. Feuding is common between graziers. Overgrazing leads to depletion of the range and a consequent reduction of its carrying capacity, and a dry summer or severe winter can mean death for many animals.

These conditions prompted demands for grazing regulations and the creation of a more rational system of grazing land administration. Similar types of regulations occur with our acceptance of fishing and hunting licenses and seasons; rules that ration uses of recreation areas; and international agreements that limit the takings of seals, whales, and various commercial fish species.

Soil resources. With proper management, most soils can be used and still retain their productive capacities over long periods of time. Their conservation does not require the saving of every particle of soil. New soil resources are formed every year; and although wide differences exist in soil loss tolerance levels, soil conservation specialists generally agree that soils can retain their productive capacity if steps are taken to prevent their depletion and to limit average soil losses from erosion to maximum rates of 4 tons per acre per year.[15]

Different opinions are held concerning the precise meaning of soil conservation. In a strict economic sense, one might distinguish between those practices that maintain the productive capacity of a soil resource and those that go further to develop, build up, or improve its productivity. When emphasis is given to the maintenance concept, *soil conservation* may be defined as "prevention of diminution in future production on a given area of soil from a given input of labor or capital with the techniques of production otherwise constant" or as "the retention of a given production function over time."[16]

Most soil conservationists include provisions for the development and improvement of soils in their definitions of soil conservation. For them, soil conservation is "a system of using and managing land based on the capabilities of the land itself,

---

[15]Cf. *Soil Degradation: Effects on Agricultural Productivity*, National Agricultural Lands Study Interim Paper No. 4, 1980; and J. V. Mannering, "The Use of Soil Loss Tolerances as a Strategy for Soil Conservation" in R. P. C. Morgan, ed., *Soil Conservation: Problems and Perspectives* (New York: John Wiley, 1981).

[16]Earl O. Heady, *Economics of Agricultural Production and Resource Use* (Englewood Cliffs, N.J.: Prentice-Hall, 1952), pp. 781–82.

involving the application of the best measures or practices known, and designed to result in the greatest production without damage to the land."[17]

With this broad definition, soil conservation is mostly a matter of good land use and management. Operators can usually choose from a variety of managerial practices. In so doing, they ordinarily try to maximize their returns and satisfactions both now and throughout their planning periods. Insofar as they understand the consequences of their actions, they consider the costs and returns expected with different managerial practices, the probable distribution of these costs and returns throughout their expected operating periods, and the effects of these practices on the market value of their soil resource base.

Whether operators accept and use soil conservation practices depends on their understanding of the soil conservation problem; the urgency of their conservation needs, their calculations regarding the effects of proposed conservation programs on their income expectations both now and in the foreseeable future, their capital positions, their time-preference rates, and their general willingness to accept a conservation philosophy. Some of the major problems that arise in this regard are illustrated by the four problem situations depicted in Figure 8-9.

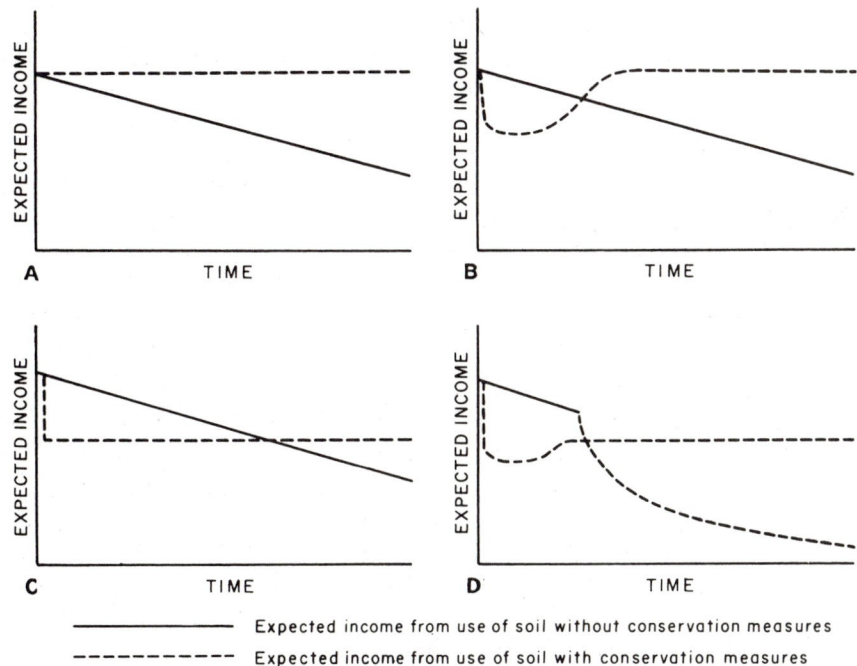

Expected income from use of soil without conservation measures

Expected income from use of soil with conservation measures

FIGURE 8-9. . Use of projection curves showing incomes expected from the use of soil resources over a period of years with and without conservation measures to illustrate four type-situations frequently encountered in conservation decisions.

[17]William R. Van Dersal, "What Do You Mean: 'Soil Conservation,'" *Journal of Soil and Water Conservation*, 8 (September 1953), 227.

Figure 8-9A involves a situation in which operators can expect a gradual but steady decrease in the income and production received from their soil resources. They can remedy this situation and stabilize expected crop yields and income by adopting conservation practices. Possible examples include application of lime and fertilizer and the adoption of strip-cropping or summer-fallowing practices. With these conditions, operators who understand their opportunities will shift to conservation practices with little prompting, while educational programs may be needed to acquaint the uninformed of their opportunities. Both groups can realize returns from their conservation investments almost immediately, and reluctance to adopt these practices may be regarded as a mark of poor or uninformed management.

A more perplexing situation arises with Figure 8-9B. In this example, operators who would use conservation practices to stabilize the income-producing capacity of their soil must first accept a period of reduced income while they invest in conservation practices or shift to a cropping system that emphasizes use of soil-building rather than soil-depleting crops. They may sacrifice income from cash crops while they use fields to grow crops that will be plowed under as green manure. They may give up income that would be available for other purposes by spending money to build terraces and check dams or to provide a better drainage system. They may also shift from primary dependence on row crops to the use of a forage crop and pasture program—a shift that often brings a period of reduced income while they build up livestock enterprises capable of replacing income that could have been secured from sales of cash crops.

A big question with Figure 8-9B centers on the willingness of operators to forgo income in the immediate future in order to maximize expected returns over a longer future time period. The case for adopting conservation practices is not as clear cut as in Figure 8-9A. The line of action operators will choose to follow reflects the duration of their planning periods, their current need for income, and their ability to secure credit to tide them over until the expected period of higher returns.

An additional complication is introduced in Figure 8-9C. This example assumes a situation in which operators have small prospects for restoring the productive capacity of their soil to a level that will maintain present income. Long-run sustained use of the soil now calls for a permanent shift from soil-depleting crops to forage crops, grass, or trees. By delaying a shift to these uses, operators can expect a higher annual return from continuation of their present use pattern up to the year when the alternative production curve would intersect their currently decreasing production curve. But by waiting, they would also suffer a continued loss of top soil from sheet erosion, which would further reduce the productive capacity of their soil resources for lower alternative uses.

Operators in Figure 8-9C may be reluctant to shift to the lower income-producing alternative for understandable reasons. Their willingness to shift, however, might be heightened if they face a situation such as that pictured in Figure 8-9D. The operators in this example are aware of the declining productivity of their soils and also of the fact that sheet erosion has now taken all but a few inches of topsoil or that gullies are threatening to ruin their most productive fields. They recognize that they are fast approaching a *critical danger point*, after which their soil resources will be almost worthless for their present use. With this prospect, they may be quite willing to employ conservation measures (terraces, check dams, or sodded waterways) and shift their fields to a lower use, because this may be the only practicable way of keeping them in use.

These four examples illustrate some of the circumstances that complicate soil conservation decisions. No emphasis has been given thus far in this discussion to the effect discounting of net returns may have on operator willingness to accept soil conservation practices. This is not an important problem with the example presented in Figure 8-9A because the operators in that example can expect an early return from their conservation decisions. It becomes a problem, however, whenever operators must forgo immediate expected income while they invest in conservation measures that promise to increase total income expectations during later time periods.

This problem is illustrated by the example pictured in Figure 8-10, which assumes a farm with a cropping program that now produces a net return of $20,000 but is associated with soil losses that will reduce the expected annual net return (ERC_1) by a predicted $500 each year over the next twenty years. The operator understands that the present level of income (ERC) can be stabilized but that this will require conservation investments (foregone income) of $20,000 over the next five years. If no attempt is made to discount the expected future net returns, the prospect of shifting to the conservation program will seem economically feasible if the operator's planning horizon extends five or more years beyond the fifth year or for a long enough time period to recoup the $20,000 investment.

If the operator reacts as most investors do, the $20,000 of income to be forgone in the next five years will seem to have a higher current value than the $20,000 of added income expected in the second five years. Time-preference considerations will cause the operator to discount the expected net returns with conservation and with no conservation $(ERC_2$ and $ERN_2)$ and may result in the use of a higher dis-

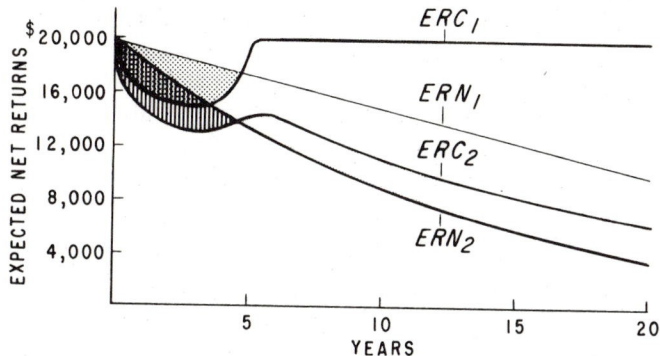

FIGURE 8-10.   Illustration of the effect the discounting of the expected net returns to be secured from the use of a soil conservation program as compared with continued use of an exploitive soil use program may have on an operator's decisions concerning the possible adoption of soil conservation practices.

count rate with the conservation investment than with the alternative of no conservation.

As ERC_2 and ERN_2 indicate, discounting the expected returns with the conservation investment at 6 percent and the net returns associated with the no conservation option at 5 percent complicates the conservation planning decision. A planning horizon that extends at least seven years beyond the fifth year is now needed if the operator is to visualize a present value of future net returns adequate to compensate for the present value of the investment outlay needed for the conservation program. If money is borrowed to finance the conservation investment or to replace the income foregone during the first five years, the charge for carrying the loan could be treated as an additional cost that would further prolong the breakeven period.

Man-made resources.   Most man-made real estate resources such as houses, office buildings, shopping centers, highways, and multipurpose dams have economic lives of predictable length. Under some circumstances, operators find it good economic policy to exploit these resources. This may be the situation when the site occupied by a building has ripened for some better use. In this event, an owner may try to get all of the use possible from a property during a relatively short time period with a minimum outlay for maintenance and operating costs, so that the site can be shifted to a higher use with a minimum write-off of present investment. Cases of this type are more the exception than the rule. Far from wanting to exploit their properties, most owners find it advisable to carry on property-use programs designed to prolong the economic life of their structures.

Conservation practices with man-made resources ordinarily involve the use of techniques that can extend the economic usefulness and life of these developments. The rationale involved in the conservation of this type of resource usually parallels that pictured in Figure 8–10A, B, and D for soil conservation projects. Changes in current use practices can sometimes stabilize development values. More often, some current income must be forgone as funds are invested in renovation, remodeling, or improvement programs.

A common problem in the conservation of man-made resources arises with the conservation of urban neighborhoods. The desirability of a neighborhood as a place to live usually depends on the separate actions of its various residents. By exploiting or abusing their properties, a few families living in a neighborhood can often lessen its desirability and create the cancerous conditions that lead to area blight. Group action is often needed to prevent such developments. This action takes two principal forms: (1) joint efforts by property owners to maintain and improve the appearance, utility, and value of their properties, and (2) use of area renewal programs to improve the physical layout and environment of neighborhood areas.

SOME CONSERVATION ISSUES

Consideration has been given in the discussion above to some of the more important economic and social factors that affect conservation decisions. Supplemental emphasis is given here to three additional aspects of the land resource conservation problem: (1) the question of whether conservation pays, (2) the nature of society's interests in conservation, and (3) the problem of overcoming obstacles to conservation.

Does Conservation Pay?

One of the first questions practical business operators ask concerning conservation is, Does it pay? As citizens, they approve of conservation in principle and endorse its general objectives. As business operators they are profit-conscious and have little interest in forgoing present income for conservation reasons unless there is reason to believe that these investments will pay off.

Experience shows that conservation measures can and frequently do pay off, particularly when they represent economically sound uses of resources over time. There are also instances in which the expected payoffs are small or nonexistent. Whether a conservation program will prove profitable depends primarily on the costs of the program, the volume of expected benefits, the time period that will elapse before these benefits can be realized, and the interest rates used in its present valuation. Beyond these items, the question of whether conservation really pays depends on a miscellany of factors. Important among these are (1) choice of a discount rate, (2) the duration of the operator's planning period, (3) investment and disinvestment considerations, (4) operator choices between alternative conservation measures, and (5) the effects of conservation programs on other resources.

Choice of discount rates. How well conservation pays depends in large measure on one's choice of a discount rate. Individuals with low time-preference rates and low discount rates often derive major satisfactions from their conservation activities and feel that these activities pay off in an economic sense as well. Operators who discount the value of their expected flows of future returns at higher rates, on the other hand, can easily find reasons for regarding conservation practices as economically impracticable.

No absolute criteria exist for selecting a correct discount rate. It is often assumed that operators should look to their costs of borrowing capital for an indication of the rates at which they should discount future returns. In support of this view, it is argued that economic efficiency requires the use of discount rates that reflect the rates of return operators can expect from alternative investment opportunities. Unfortunately, market rates of interest fluctuate over a wide range. During recent decades, they have ranged from a 4 to 5 percent level during the 1940s to rates of from 15 to 22 percent during the early 1980s.

Fluctuations of this order have significant impacts on the investment decisions of individual operators. But from the overall position of society, it is questionable whether they provide a sound basis for determining the discount rates that should apply with conservation practices. The need for conserving minerals, forests, and soils was certainly as great, if not greater, in 1982 than in 1947. Yet acceptance of the market cost of money as a guide would have required operators who were following conservation programs in 1947 to shift their emphasis to the acceptance of rapid and sometimes irreversible resource-exploitation practices by the latter date.

Operators who accept and use conservation practices ordinarily do so because of the weight they give to noneconomic considerations. They have low time-preference rates at least in part because they accept a stewardship concept of resource use, are interested in saving resources for future generations, and see conservation as the right thing to do. Fortunately for society, this willingness of large numbers of operators to accept ethical as well as economic goals provides a promising alternative to the anticonservational bias that strict adherence to short-run economic efficiency criteria can have on longer-term resource management practices.

Duration of operator's planning period. Conservation decisions must be forward-looking.[18] They are made in advance or at the time of resource use and almost always assume some expected planning period. When operators decide to carry on conservation programs, they commit themselves, at least temporarily, to given lines of action. These commitments are not necessarily binding for all time and can be adjusted for changing conditions. Forest owners who plan to hold their forests for twenty-five years, for example, can amend their plans to harvest their resources in fifteen years or in thirty-five years. But plans are usually geared to some time period over which they are able and willing to plan. If their planning horizons are short, operators will use the short-term nature of their calculations to justify the early cutting of forests, mining of the soil, or other types of resource exploitation.

The relationship between operator planning horizons and their conservation decisions may be illustrated by an example of a soil conservation program that promises to pay off in eight years. An operator in this instance may be willing to adopt conservation measures if his or her planning period covers eight or more years of continuous operations or if it is felt that the land can be sold within this time period for a sum sufficient to more than pay for the conservation investments. One's attitude may be much different, however, if the operator's planning horizon is limited by a one-year lease, the leasing arrangement contains no provision for compensation for unexhausted improvements, or other factors make the operator unable or unwilling to plan ahead for as much as eight years of continuing operations.

As this example indicates, the question of whether conservation pays depends in many instances on the operator's ability or willingness to plan operations over a long enough time period to permit recoupment of conservation costs. Wise resource use over time calls for planning horizons of sufficient duration to permit resource harvesting and use programs that will return a maximum of economic and social net returns. Operations that involve shorter planning periods can be both exploitive and wasteful, in the sense that they favor either the underuse or overuse of resources.

Planning periods are frequently influenced by factors other than the operator's tenure status or willingness to plan ahead. Mine owners, for example, may find that they could maximize their returns over time from given coal or ore deposits by operating on a smaller scale. Once they sink mine shafts and purchase mining equipment, however, they find it best to continue with their present scale of operations. Forest owners with substantial investments in large sawmills may find it unprofitable

---

[18] Cf. Ciriacy-Wantrup, *op. cit.*, pp. 20–33, 54.

to shift to otherwise desirable sustained-yield production programs if this adjustment makes it impossible to operate their present mills on an efficient basis. Farmers with large inventories of relatively new equipment may decide to continue exploitive soil-use programs rather than write off part of the value of their machinery investments.

Investment and disinvestment considerations. Most operators who practice conservation are not interested in simply saving or storing resources for some vague future use. They expect their investments to pay off within definable planning periods. Thus if they postpone mining operations or pump oil at a conservative rate, they expect a positive economic return for their conservation efforts. This return may come in the form of higher prices from holding the resource, the prospect of recovering larger quantities of product, or both. Forest owners, fishers, or hunters who postpone the harvesting or taking of resources expect larger trees or more fish and game in the future. Farmers who invest in soil-building practices expect to build up reserves of fertility they can use over time or draw on when they need additional capital. Building owners invest in structures and improvements with the idea that they can gradually reap the full benefit of their investments.

Wise use with most land resources—particularly those with characteristics of both fund and flow resources—involves a continual process of investments and disinvestments. During periods of peace and plenty, we often use the soil-bank principle to invest in soil improvements, enlarge and improve our forests, and increase our inventory of buildings and improvements. Under wartime and emergency conditions, this process can be reversed as emphasis is placed on increasing food production, larger timber harvests, and restricting private construction activities.

Important conservation considerations arise in the timing of operator investments and disinvestments. Ordinarily, it is assumed that operators should make a series of investments in conservation practices before they disinvest. In practice operators may find it both economically and socially expedient to capitalize on the investments made by nature or by previous operators. Individuals thus can use disinvestments at times to secure needed capital for desired improvements. Nations, such as the United States during the 1800s, can also use widespread resource disinvestment policies to provide the capital and resources needed for economic growth.

The American frontier settlers started with the accumulated natural resource investments of many centuries. In developing and exploiting these resources, they followed a disinvestment policy. This policy had its regrettable aspects. All things considered, however, it was desirable from the standpoint of both the average operator and society at that time. A disinvestment policy was favored because it provided the quickest means operators could use to maximize their individual returns and because it facilitated the economic development of the frontier and the nation.

Conservation investment policies were scorned on the frontier largely because of the plentiful nature and low monetary value of the available forest, wildlife, and soil resources. Once these conditions changed—as soon as it became obvious that the supply of these resources was definitely limited relative to potential future needs—

the American people became conservation-conscious. Conservation policies were then developed to protect and safeguard oil, mineral, forest, wildlife, and soil resources. But exclusive emphasis was not given to the idea of saving these resources for future generations. Reasonable amounts of disinvestment are expected every year, and disinvestments in excess of the amounts justified by current investments are expected during wartime and emergency periods.

This concept also applies to private conservation investments. Some operators carry on forestry or soil-improvement practices with the idea of maintaining a continuous level of future productivity. Some lay aside a reserve of productivity they can draw on in periods of unusual income need. Others—particularly those operators with a high debt load and pressing need for additional operating capital—borrow from their resources with the intent of repaying their loans with future conservation investments. This disinvestment practice can result in exploitation in the short run. Over time, it can represent a good management practice if operators follow through with a planned reinvestment program.

Choices among alternative conservation programs. Another important factor affecting operator attitudes and decisions concerning the profitability of conservation involves their ability to choose conservation programs that meet their particular needs or inclinations. In this respect, operators can usually choose from a variety of practices, some of which are more exploitive than others. Forest owners, for example, may follow inactive management programs under which forest growth is left entirely to nature. They may employ intensive management practices such as tree planting, spraying, or selective thinning. They may follow sustained-yield harvesting approaches by cutting selected trees while leaving seed trees and young growth for future harvest; or they may clear-cut their forests and then either reseed or regenerate them by transplanting young seedlings.

Farmers can choose between programs that will build up their soil, merely maintain it at its present productive level, or permit some acceptable amount of soil depletion or erosion. Once this decision is made, they can usually decide between alternative means for achieving the goal they have in mind. This range of choices is illustrated by the indifference curves as shown in Figure 8-11.[19] Three goals are assumed: (1) a farming program that will build up the soil and result in an annual loss of only two tons of top soil per acre, (2) a program that will result in an annual loss of four tons of soil per acre—the maximum loss consistent with the goal of maintaining yields at their current level with present technology, and (3) a program that will lead to annual losses of six tons of soil per acre. Each of these goals can be attained by a crop rotation program that will put part of the operator's land in grass or forage crops and require some combination of conservation practices on the remaining cropland.

If the operator in this example identifies the soil conservation goals to be pursued and has specific knowledge concerning the effects of alternative practices on soil

---

[19]This example is suggested by data assembled by William H. Heneberry and Elmer L. Sauer in a study of net returns from alternative soil conservation practices on two types of soil found in Illinois.

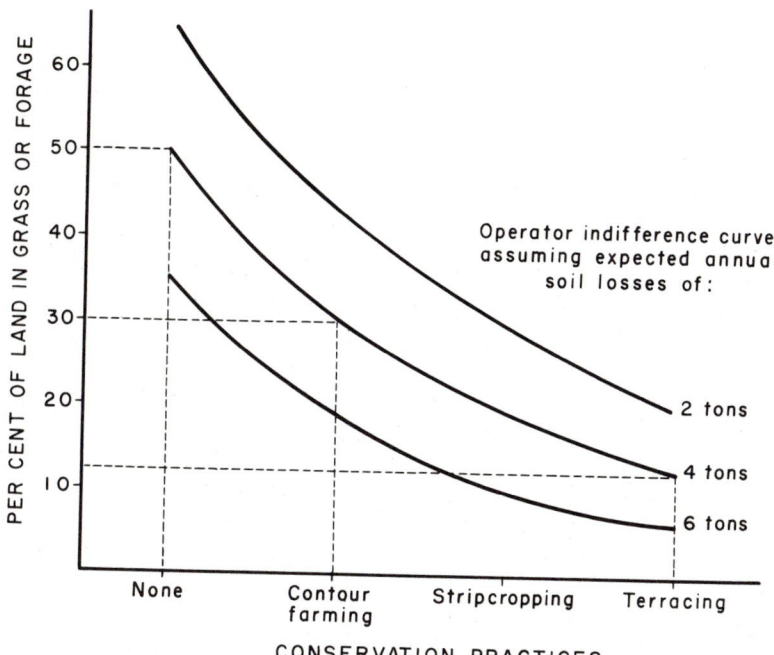

FIGURE 8–11. Use of indifference curves to illustrate the range of choices between alternative management programs available to individual operators.

productivity, he or she can then choose between the possible combinations found along the two-ton, four-ton, or six-ton indifference curves. The four-ton goal, for example, can be attained by using a crop rotation that puts 50 percent of the land in grass or forage crops without other conservation practices, or by using a rotation that puts 30 percent of the land in grass or forage crops and requires contour-farming, stripcropping, and terracing practices on the remaining land.

As this example suggests, operators with several conservation alternatives can often find programs that appeal to them and that have prospects of paying off under their operating conditions. The option of using different combinations of practices greatly improves the prospect of operators accepting and carrying out conservation programs.

Effects of conservation programs on other resources. The question of whether conservation practices really pay cannot be answered in an ultimate sense until consideration is given to the impact of these practices on other resources. Operators frequently find it practicable to substitute one type of resource for another in their production programs. These adjustments normally result in reduced use of the displaced factor. Substitution of one scarce resource for another may save the operator money or lead to higher production or a better product and thus make substitution economically desirable. It may also contribute, however, to depletion of the supply of the new factors and thus have a negative or neutral overall effect.

Whether resource substitution complicates or simplifies the conservation problem often depends on whether conservation is viewed in physical or economic terms. Physical conservation calls for limiting the use of exhaustible and nonrenewable resources, for substituting flow resources for fund resources. Substitution practices that work to this end may appear economically infeasible because they do not lead to monetary savings or higher profits. Quite the opposite, operators are often eager to substitute fund resources for flow resources, even though this process leads to physical depletion, if it saves them time and money.

Numerous examples may be cited to illustrate the frequent tendency of accepted practices to cause physical resource depletion. Workers on the American frontier substituted land resources for capital and labor in their production combinations because the supply of land resources was plentiful while capital and labor were scarce. Natural gas and oil have been widely substituted for coal as a source of heat and industrial power, even though they involve basic resources that are more limited in supply. Commercial fertilizers are used on farms as a partial substitute for fertility that might otherwise be gained or retained by soil-building practices. Gasoline-powered automobiles and tractors have replaced the hay-and-oats-fed horse as our principal source of transportation power.

Substitutions of fund resources for biological and flow resources will continue as long as they provide a cheaper or economically superior product. Once the more readily available supplies of these resources are used, however, the prices of the more critical of these resources can be expected to rise. This will favor more intensive mining practices and the treatment of these resources as a limiting factor in production. It will also prompt a search for additional substitutes—a process that must eventually lead to the increased use of renewable and reusable resources.

Society's Interest in Conservation

Mention is frequently made in conservation discussions of differences between the interests individuals and society have in conservation. Individuals are often assumed to have high time-preference rates and short planning periods. Society in turn supposedly has a longer planning horizon and uses lower discount rates because of its interest in the welfare of future generations and its ability to borrow money at a lower rate of interest.

A realistic view of this supposed dichotomy of interests shows that the interests of society are not necessarily contrary to those of individual operators. Society is made up of individuals, and its interests necessarily reflect those of its members. The real division comes between each individual's desire to maximize personal satisfactions and his or her desire to stress social and community interests. Business operators emphasize personal and firm goals and often find that they must pit their interests against those of other operators. The responsibility for attaining social goals, in contrast, is frequently delegated by individuals to society and is fulfilled through collective action—through the marshaling of public opinion, the joint action of individuals in groups and organizations, and action by the state.

Rational individuals are always concerned with survival and the returns and satisfactions they can secure for themselves and their families. They are also interested to a greater or lesser degree in the future of the race, the welfare of their heirs, and the well-being of others. Every person has some combination of these sometimes complementary, sometimes conflicting, interests. These combinations make for a wide range of attitudes regarding conservation, varying from extreme conservation-mindedness to almost exclusive emphasis on policies of resource depletion or exploitation.

This range of interests in conservation also applies to business organizations and public agencies. Corporations, which are ordinarily assumed to have longer planning periods and lower interest rates than individuals, sometimes stress the conservation and sometimes the rapid depletion of particular resources. Governments also vary in their conservation practices. They frequently live up to Pigou's admonition:

> It is the clear duty of Government, which is the trustee for unborn generations as well as for its present citizens, to watch over, and if need be, by legislative enactment, to defend the exhaustible natural resources of the country from rash and reckless exploitation.[20]

On other occasions, as during a war when the continued life of the nation is at stake, they may engage in resource-disinvestment policies that are every bit as exploitive as those of the self-seeking business operator.

Important questions may be raised concerning the positions governments, and the societies they represent, should take respecting the long-term management and conservation of land resources. Should they emphasize current economic efficiency goals that call for maximization of current resource values, or do they have a moral responsibility to work for intergenerational equity in the use of land resources? If they stress economic efficiency, they may logically base their social rates of discount on the government's cost of borrowing money. But if this course is followed, should the social discount rate reflect the low cost of money of around 2 percent in the 1940s, the higher cost of 11 to 13 percent of the early 1980s, some compromise between these extremes, or simply fluctuate from year to year? Differences of opinion exist on the answers to these questions.[21]

Most conservationists feel that current short-run market interest rates provide a poor criteria for long-term resource management and conservation decisions that will affect the welfare and well-being of generations decades and centuries hence. As Randall has observed: "Present-value and expected-value concepts derived from

[20] A. C. Pigou, *The Economics of Welfare*, 4th ed. (New York: St. Martin's Press, 1962), pp. 29–30.

[21] Cf. footnote 19 in Chapter 17; Talbot Page, *op. cit.*, chaps. 7–10; and Alan Randall, *Resource Economics: An Economic Approach to Natural Resources and Environmental Policy* (New York: John Wiley, 1981), chaps. 10 and 11. The fact that most individual operators delegate much of the responsibility they have for ensuring adequate supplies of resources for future generations to society justifies an assumption of lower social time-preference and discount rates than those used by business operators who are concerned primarily with short- and intermediate-run market operations.

traditional economics have proven to be of little use for the solution of decision problems involving very long time horizons, massive uncertainty, and/or irreversibility."[22] With a continuing prospect of resource exploitation and with persistent uncertainties concerning the extent to which new technology can maintain and increase the productive capacity of the earth's limited land resource base, society has good reasons for insisting that governments sponsor and carry out resource conservation policies as a necessary form of social insurance against possible unwanted eventualities.

Important questions can be raised concerning the proper criteria for social action to secure conservation goals. Prevailing political philosophy in the United States favors limited social intervention and broad individual free agency in the management of one's personal affairs. But no landowner or land user lives in isolation. Land-use practices frequently affect one's neighbors and the whole community. Individual practices become a matter of group and public concern anytime they have an adverse effect on the productivity, value, or cost of maintaining an area's resource base.

A clear case can be made for social action to promote conservation anytime an operator's practices are regarded as detrimental to the nation's security and anytime public programs are needed to facilitate desired resource developments. Social controls are justified when they are used to prevent individual property-use practices that contribute to neighborhood blight or that cause drainage, erosion, fire, siltation, or soil-drifting problems on other properties. Comparable action may also be needed at times to help individuals help themselves. As Bunce has indicated:

> ... social action to achieve conservation is desirable: (1) when it would be economic for the individual entrepreneur to conserve but he does not; (2) when conservation is not economic for the individual but is economic for society; and (3) when intangible ends desired by the majority of individuals in a democracy can be attained only by collective action.[23]

Public agencies can use the same types of tools in promoting society's interests in conservation as in other programs for social direction of land use. (See the discussion in Chapters 17 and 18.) Educational measures and subsidies for the acceptance of conservation practices can be used when lack of knowledge is the principal obstacle. Credit facilities and technical assistance can be provided with soil and neighborhood conservation programs to help operators finance and adjust to their conservation practices. Tax incentives can be used to foster forest and other types of resource conservation. Police power measures can be used with forest-cutting restrictions and oil well spacing regulations for conservation purposes; and the public spending, public ownership, and eminent domain powers can also be used to attain conservation goals.

[22] Randall, *op. cit.*, p. 242.
[23] Bunce, *op. cit.*, p. 105. Reprinted by permission from HETEROSIS, edited by John W. Gowen, © 1952 by The Iowa State University Press, Ames, Iowa.

Overcoming Obstacles to Conservation

Obstacles frequently prevent the acceptance of conservation practices. Some of the more important of these involve the contrary attitudes and goals of individual decision makers. Others, such as the hidden nature of most mineral resources and the tendency of some soils to erode faster than others, are primarily physical in nature. Still others involve economic, institutional, and technological barriers.

Many operators refuse to accept conservation practices because they are interested primarily in maximization of current returns and satisfactions. Their attitudes can frequently be credited to shortsighted thinking and reliance on faulty information. They may not realize that they have opportunities to enjoy high economic returns while using conservation practices. They may also have little understanding of the adverse effects their practices can have on others. These problems can be handled at least in part by educational programs that make operators both more conservation-conscious and conservation-wise. Beyond its effect on conservation attitudes, knowledge is needed for sound, long-term resource management and conservation. Rational operating decisions call for considerable information about the characteristics and extent of one's resource base, one's marketing and cost expectations, and the possible benefits and costs associated with alternative management practices.

With physical obstacles, operators must usually accept the resource base as they find it and use resources when and where they occur. In this process they often sink mine shafts, build dams, plant trees, use terraces and sodded waterways, and employ other techniques to modify natural situations and overcome physical obstacles to optimum resource use. Social action of quite a different type is often needed to overcome economic, institutional, and technological obstacles.[24]

Economic obstacles. Lack of capital frequently poses a problem for the adoption of conservation practices. Most operators have some capital; but few have all of the capital they need to operate in the manner they wish. This constraint may cause them to disinvest their resources or accept high time-preference rates simply because they feel that they must increase the current incomes they have available for living and operating purposes.

Special credit facilities are needed in many instances to help operators finance conservation practices. Credit and outright income grants can be used to help tide operators over the periods of reduced income that sometimes elapse before they can capitalize on the longer-run benefits of their conservation programs. Compensation arrangements can be worked out to encourage individuals to carry on conservation practices that are desired by society but that might not be profitable to individual operators.

[24] For other discussions on these obstacles, cf. Ciriacy-Wantrup, *op. cit.*, pp. 111–219; Anthony Scott, *Natural Resources: The Economics of Conservation* (Toronto, Canada: University of Toronto Press, 1955), pp. 99–152; John C. Frey, *Some Obstacles to Soil Erosion Control in Western Iowa*, Iowa Agricultural Experiment Station Research Bulletin 391, 1952; and John F. Timmons, "Institutional Obstacles to Land Improvement," *Journal of Land and Public Utility Economics*, 22 (May 1946), 140–50.

Economic instability provides another economic barrier to conservation practices. Many operators use short planning periods and high discount rates because they find themselves unable to predict future cost, price, and market conditions. This situation can be improved by measures to reduce uncertainties, stabilize the economic system, and minimize the fluctuations in net returns that come with periods of inflation and depression.

Institutional obstacles. Like other types of human behavior, conservation decisions are often influenced by institutional factors. Many people practice conservation as a matter of habit or custom or because they are imbued with "a stewardship of the land" philosophy. Others find that their conservation plans are favored by the relative stability of governments, the clear titles they have to lands, their eligibility for public conservation payments, their expectations regarding public price supports, or guarantees they have against exploitive taxation.

Institutional arrangements may also operate to discourage conservation. Customary practices, inertia, and ignorance of the principles of wise land use can lead to resource exploitation. Operators with limited tenure rights, a mortgage that may soon be foreclosed, or a vacillating government have little incentive for using low discount rates or long planning periods. High property or severance taxes, operating units of inadequate size, and a lack of suitable credit facilities are other important causes of resource exploitation.

Society can play an important role in overcoming these obstacles. Educational, demonstration, and subsidy measures can be used to acquaint people with their conservation opportunities and to persuade them to try conservation practices. Positive action can be taken to stabilize political institutions and to clarify titles and the various use rights people have in property. Programs can be developed to promote leasing and tenure arrangements that encourage investments in conservation practices and to devise tax systems or provide subsidies that favor private acceptance of conservation measures.

Technological obstacles. How people use resources is always conditioned by the existing state of the arts. Our primitive forebearers often practiced conservation unknowingly because they lacked the motivation and knowhow for exploiting land resources. As they acquired motivation and knowledge, they often wrecked their resource base, not so much because they wanted to, but rather because of their inability to discipline their actions in the use of new technology.

Many people still endorse an undisciplined policy of resource exploitation on the assumption that technology will solve our resource-scarcity problems. They argue that "necessity is the mother of invention" and that science will come to the rescue with improved production processes and with new substitutes as the prospect of resource exhaustion approaches. Whether technology can supply all the needed answers to our future resourse-use problems is a question beyond immediate answer.

The task of technology in this instance is simple but challenging. Technology can help us discover more fund resources, facilitate the easier and more complete extraction and use of these resources, and extend their useful life over time. It can promote the economic development and more extensive productive use of flow resources. It can bring the development of improved breeds and species of biological resources that will yield better products at lower cost. It can point the way for improved soil conservation techniques and for construction and repair practices that will extend the economic life of man-made improvements.

Society has definite responsibilities for promoting the development and use of improved techniques for each of these purposes. At the same time, precautions must be taken not to assume the unassumable. Policies that promote the saving and conservation of resources for future use will always have their place as long as doubts and uncertainties exist concerning the economic and technological feasibility of programs for securing needed future resource supplies.

— SELECTED READINGS

Bunce, Arthur F., *The Economics of Soil Conservation* (Ames: Iowa State College Press, 1945).

Ciriacy-Wantrup, Siegfried V., *Resource Conservation*, rev. ed. (Berkeley: University of California Press, 1963).

Fisher, Anthony C., *Resource and Environmental Economics* (Cambridge, U.K.: Cambridge University Press, 1981), chaps. 3 and 4.

Gray, Lewis C., "The Economic Possibilities of Conservation," *Quarterly Journal of Economics*, 27 (May 1913), 497–519.

Heady, Earl O., *Economics of Agricultural Production and Resource Use* (Englewood Cliffs, N.J.: Prentice-Hall, 1952), chap. 26.

Howe, Charles W., *Natural Resource Economics* (New York: John Wiley, 1979), chaps. 8–13.

Johnson, V. Webster, and **Raleigh Barlowe,** *Land Problems and Policies* (New York: McGraw-Hill, 1954), chap. 7.

Randall, Alan, *Resource Economics: An Economic Approach to Natural Resource and Environmental Policy* (New York: John Wiley, 1981), chaps. 9–11.

Scott, Anthony, *Natural Resources: The Economics of Conservation* (Toronto, Canada: University of Toronto Press, 1955).

Timmons, John F., *et al.* (Committee on Soil and Water Conservation), *Principles of Resource Conservation Policy*, National Academy of Sciences, National Research Council Publication No. 885, 1961.

9
LOCATION FACTORS AFFECTING LAND USE

In the world of economic theory, it is common practice to ignore differences in spatial location. With the frequently used concept of perfect competition, for example, it is ordinarily assumed that all buyers, sellers, and products in the market have perfect mobility—that they are either located at the market or can be moved instantaneously and without cost to that point.[1] This assumption has its place and value in theoretical analysis, but it does not meet the conditions of reality. Land resources in particular are fixed in their location. They are always found at varying distances from the centers of economic activity, and costs are involved in moving land products to market and in bringing capital and labor to the land. Locational differences play a highly significant role both in determining the economic uses that can be made of land and in affecting the rent and value levels associated with its use.

Most people prefer to live in areas that boast a pleasant climate, low living costs, and opportunities for the satisfaction of their various wants and desires. In deciding where they and their families will live, individuals are often torn between the counterpull of their wants as consumers and their wants as producers. As consumers, they "seek to settle where living is secure, cheap, and agreeable. As producers, they seek to locate where earnings will be large and assured and the working conditions pleasant."[2]

Despite frequent complaints about the weather, most people are reasonably content with their present locations. This is particularly true if they have lived there all

[1] Cf. Walter Isard, *Location and Space-Economy* (New York: John Wiley, 1956), pp. 42 and 53.

[2] Edgar M. Hoover, *The Location of Economic Activity* (New York: McGraw-Hill, 1948), pp. 4–5.

their lives or if they have come to regard these locations as home. Yet conflicts do exist between consumer and producer wants and goals. Ambitious, energetic, and productive individuals often move to areas of greater economic opportunity. This willingness to move encourages successful operators to seek out locations at which they can maximize their economic returns without forfeiting valued noneconomic benefits. Their activities and the jobs they create attract others to the same locations. In this way the lure of higher returns causes many people to live and work in areas that may appear less than ideal from a consumer's viewpoint.

All things considered, operators and areas ordinarily concentrate on the production of those goods or services for which they have the highest comparative advantage—the greatest opportunity for realizing surpluses in their trade with others. The following discussion logically begins with a brief examination of the concepts of economic specialization and comparative advantage. Consideration is then given to the impact of location factors on the allocation of land areas between particular uses; to the factors that influence the location of cities and the determination of their internal land-use structures; and to factors affecting the location of industrial, commercial, and residential areas.

ECONOMIC SPECIALIZATION AND COMPARATIVE ADVANTAGE

Economic specialization is a common phenomenon in the present world. Individual workers tend to specialize and hopefully find employment doing those types of work they can do best. A similar type of specialization affects the uses made of land areas. Every area could attempt to provide the products needed by its residents. The Midwest could try to provide its needs for cotton, coffee, and bananas. But even if it were possible to produce a sufficient supply of these products, this process would prove both expensive and wasteful. It makes much more sense for areas to concentrate on the production of products for which they have a natural or economic advantage and to trade their surplus of these products for goods they need that can be better produced in other areas. By encouraging areas to specialize in the types of production for which they have high comparative advantage, we have been able to produce larger supplies of products and enjoy higher average levels of life.

Principle of Comparative Advantage

Generally speaking, each area tends to produce those products for which it has the greatest ratio of advantage or the least ratio of disadvantage as compared with other areas. This concept is known as the *principle of comparative advantage*. The operation of this principle can best be illustrated by a few simple examples involving two areas and two products. Cases 1 through 4 compare the abilities of areas A and B to produce physical units of two products. To keep the comparison simple, it is assumed that both areas are able to produce the minimum needed supplies of either product for both areas, that each would prefer to concentrate on the production of one product and trade its surplus to the other for supplies of the second product, and that considerations of market prices, market structures, transportation costs, and relative production costs can be ignored. With the first example (see Case 1), it is assumed that areas A and B each produce all of the rice and beans they need and that each receives the net outputs of product indicated in Case 1. Under these circumstances, neither area has a production advantage for either product. This same situation would hold true if the production in area B dropped to thirty and forty-five units or increased to fifty and seventy-five units of rice and beans, respectively. In each case, both areas would have identical ratios between the units of rice and beans they could produce, and neither would find it to its advantage to specialize.

Case 1

Land use	Area A	Area B
Rice	40	40
Beans	60	60

If the production situation is changed in area B as in Case 2, it immediately becomes profitable for each area to specialize. Area A finds its ratio of advantage highest when it concentrates on beans, whereas area B finds it most profitable to concentrate on rice. In this case, each area has an absolute advantage in the production of one product.

Case 2

Land use	Area A	Area B
Rice	40	<u>50</u>
Beans	<u>60</u>	40

Under real-life conditions, some areas occasionally have an absolute advantage for more than one use, while most areas fail to enjoy an absolute advantage for any use. The disadvantaged areas in these instances do not go unused. Instead, they are ordinarily used for those purposes for which they have the least comparative disadvantage. In Case 3, area B has an absolute advantage for the production of both rice and beans. Yet since it lacks sufficient productive capacity to supply the needs of both areas for both products, it will concentrate mostly on rice production—the use for which it has the highest comparative advantage, and area A will concentrate on beans—the use for which it has the least comparative disadvantage.

Case 3

Land use	Area A	Area B
Rice	40	<u>70</u>
Beans	<u>60</u>	65

To push this analysis further, it should be recognized that areas sometimes find it advantageous to concentrate on their second or third rather than their most productive use. In Case 4, for example, area B again has an absolute advantage in the production of both rice and beans, but it concentrates upon beans because this use has the highest comparative advantage. Area A would find its least comparative disadvantage with the production of rice. Area B in this instance can produce two units of beans for every unit of rice, and area A can produce only three units of beans for every two units of rice. In terms of trade, B would be willing to give up two units of beans for one unit of rice as compared with three units of beans for every two units of rice in the case of A. At the same time, A would be willing to give up two units of rice for every three units of beans, and B would give up only one unit of rice for every two units of beans. These conditions make it advantageous for B to concentrate on the production of beans while area A concentrates on the production of the less productive of its two products, because B has a definite competitive advantage in the production of beans while the rice market is left to A more or less by default.

Case 4

Land use	Area A	Area B
Rice	40	45
Beans	60	90

The joint operation of these principles may be illustrated by the hypothetical situation outlined in Case 5. This example assumes four separate producing areas and predictable estimates of the average amounts of land rent associated with four alternative uses. For illustrative purposes, area A may be considered as representative of parts of the Northeastern or Lake States region, area B of parts of the Midwest, area C of some irrigated sections of the West, and area D of the nonirrigated dry farming areas of the western Great Plains.

Case 5

Land use	Area A	Area B	Area C	Area D
Wheat	$10	$14	$11	$8
Corn-hogs	19	30	20	2
Potatoes	18	16	17	—
Dairying	25	25	10	1

Examination of the data in Case 5 shows that area B has an absolute advantage for the production of wheat and corn and that it can earn as high a net return in dairying as any other area. Its highest comparative advantage lies in corn-hog production, and a high proportion of its resources accordingly are used for this purpose. Area A has an absolute advantage in potato production, but its highest comparative advantage lies in dairying. Areas C and D do not enjoy an absolute advantage in any of the four enterprises. Area C could diversify and engage in any of the enterprises but would probably find its least comparative disadvantage in potato production. Area D has the lowest wheat yields of any area but would concentrate on wheat production because of the limited nature of its alternatives.

Scope of Comparative Advantage

Comparative advantage is often associated with natural advantages such as favorable climate, soils, and topography. Viewed in this manner, it is easy to assume static situations in which the successful use of certain areas for nature-favored purposes is more or less guaranteed. In practice, the concept of comparative advantage is both more dynamic and more all-inclusive. Some comparative advantages stem from natural endowment factors. But others involve favorable combinations of production inputs, favorable location and transportation costs, favorable institutional arrangements, and desired amenity factors.

Natural endowment. Numerous examples may be used to illustrate the contribution of the natural endowment to comparative advantage. Minerals must be available in economically attractive concentrations if commercial mining is to take place. Favorable climatic conditions and specific natural resources such as sand beaches, good fishing waters, or ski slopes are a "must" for many types of recreation developments.

The relatively frost-free climates of southern Florida and southern California favor the use of these areas for the production of citrus fruits and other semitropical crops. The long growing season enjoyed by the South gives it an advantage for cotton production. Rich soils favor corn production in the Midwest. Level and rolling fields provide distinct advantages for mechanized farming. Mountain valleys frequently provide excellent sites for reservoirs and power dams.

Favorable production combinations. Comparative advantage implies ability to realize an economic return for one's fixed inputs in the production of goods or services. It calls for favorable combinations of the inputs needed for production and also for satisfactory markets. A shortage of capital or skilled management or a lack of adequate marketing or credit facilities may easily outweigh an area's natural advantages. Skilled labor may be essential and low labor costs may give an area a distinct production advantage. Operators must consider the availability and cost of raw materials, water, power, and other utilities and also the services offered by a community's developed infrastructure.

Operators who find it possible to work out low-cost combinations of their factors of production often enjoy comparative advantages over other producers. Their low-cost combinations may result from favorable raw material, climatic, and other natural advantages or from superior management or utilization of the agglomeration economies associated with developed infrastructure and locations near established industries. Comparative advantage may also be created. With dynamic leadership, areas of limited natural advantage can develop their own supplies of skilled labor and management, capital, water and utilities, and even build up market areas that give them high comparative advantage.

Transportation considerations. Location and transportation considerations are a third group of factors that significantly affect comparative advantage. Business operators are always concerned about the distances they must ship raw materials and finished products. Local producers benefit from ability to move products to market at lower cost, in less time, and in fresher condition than more distant competitors. Savings in transportation costs frequently make it possible for local producers to compete favorably with producers who live in areas that boast strong natural advantages for producing particular products.

Transportation improvements have greatly facilitated the economic development of many areas during the past two centuries. As late as 1816, the market price of flour in the United States did not justify its transportation for distances of over 150 miles overland, and bulky and heavy articles could be shipped 3,000 miles across the Atlantic Ocean at about the same cost as 30 miles overland.[3] The cost of shipping wheat from Buffalo to New York City was approximately $100 per ton in 1817, or roughly three times its delivered value in New York City. With the opening of the Erie Canal in 1825, this shipping cost dropped to $8.81 a ton, and it suddenly became economically feasible for farmers in western New York and along the Great Lakes to ship produce to the eastern market.[4]

As late as 1850, limited transport facilities and high shipping costs favored concentrations of land settlements along navigable streams. Distant overland transportation was limited primarily to objects with high value-to-weight ratios, and cities depended on immediate hinterland areas for much of their food as well as other products with low value-to-weight ratios. The building of canals, railroads, highways, airports, and pipelines and the introduction of improved modes of transportation have greatly relaxed the transportation constraints of the past and have made it possible for producers to look to far-off places for raw materials and potential market areas. But as long as transportation involves cost and time, it will continue to influence location and production decisions and affect the comparative advantages of different sites.

Institutional advantages. Comparative advantages can also be created or enhanced by institutional arrangements. Nations with histories of political stability offer greater attractions for investors than nations threatened with frequent revolutions. Tariff barriers and trade restrictions have long been used to shut off outside compe-

---

[3] See letter from Robert Fulton to Secretary of the Treasury Albert Gallatin, December 8, 1807, *American State Papers—Miscellaneous, I, 919*; and Caroline E. MacGill et al., *History of Transportation in the United States before 1860* (Washington, D.C.: Carnegie Institution, 1917), p. 78.

[4] Cf. MacGill *et al., op. cit.,* p. 84.

tition and augment the production advantages of domestic producers. Protective tariffs can provide definite market advantages for agricultural and industrial enterprises that might otherwise be hard pressed for survival. Other institutional controls such as quarantine restrictions, city milk market inspection requirements, and zoning ordinances can have similar impacts in favoring or discouraging particular land uses.

At an institutional level, economic, military, or political dominance can produce powerful advantages for favored nations. The OPEC nations enjoyed tremendous economic power during the 1970s because of their near-monopoly position in controlling world export prices for petroleum. Colonial powers in the past often realized important economic benefits from the marketing and trading arrangements they were able to impose on dominated areas.

Amenity factors. The cultural and aesthetic attractions associated with local amenities provide a fifth facet of comparative advantage. Amenity considerations are often ignored when significant economic advantages are associated with particular sites. But producers, workers, and consumers are more conscious of amenity factors now than in the past. Moreover, choices of prospective operating sites can often be narrowed to several sites that offer quite comparable economic opportunities. When this situation exists, final decisions are frequently influenced by the general attractiveness of a community and the climate, cultural, educational, recreational, and other opportunities it offers for potential residents.

Interrelation of comparative advantage factors. Operators must consider all of the above factors in their calculations of comparative advantage. Marked advantages associated with any one set of factors can be and often are neutralized by others. The interrelationship of the various factors may be illustrated by a hypothetical example of a Detroit industrialist who seeks a plentiful supply of a particular item that she plans to market in combination with her product. Suppliers at five different locations can produce and deliver the needed item at the prices reported in Table 9-1.

Examination of these costs shows that a prospective supplier in Brussels enjoys a natural endowment advantage with the lowest unit cost for raw materials and that a prospective supplier in Seoul has the lowest production costs, while a supplier from

TABLE 9-1. **Hypothetical Example of Costs Associated with Delivered Prices Quoted for a Product Needed by a Detroit Industrialist**

SITE OF PRODUCTION	COSTS PER UNIT OF PRODUCT FOR:				DELIVERED PRICE PER UNIT
	Raw materials	Production costs	Shipment to Detroit	Customs duty	
Windsor	$1.00	$2.35	$ .15	$ .50	$4.00
Seoul	.95	1.55	.90	.50	3.90
Los Angeles	.85	2.20	.65		3.70
Brussels	.80	2.10	.40	.50	3.80
Cleveland	1.00	2.45	.25		3.70

across the Detroit River in Windsor, Ontario, has the lowest transportation costs. The two domestic suppliers enjoy an institutional advantage with their freedom from the need to pay customs duties. When all costs are considered, it appears that suppliers in Cleveland and Los Angeles can deliver the needed item at the same lowest price. At this point, personal considerations and amenity factors may be the deciding factors in determining which supplier gets the contract. If our industrialist wants a supplier close at hand, she may choose the Cleveland producer. Should she seek opportunities to visit friends in southern California or spend winter weekends at Palm Springs, she may favor the Los Angeles supplier.

Comparative advantage involves ability to compete on favorable terms with alternative sites in the production of goods or services for a given market. High comparative advantage for particular uses does not necessarily mean that sites will be used for these purposes. Sites near metropolitan centers often enjoy high comparative advantages for many competing uses. Selection of the highest and best use involves the counterbidding of the marketplace. Unless prevented from doing so by institutional constraints, individual sites tend to move to their highest and best economic uses. Uses that cannot pay top price move to less favorable locations where they can become the highest and best uses of those sites even though they may represent uses of least comparative disadvantage for the sites in question.

SPATIAL RELATIONSHIPS AFFECTING LAND USE

Land-utilization patterns frequently reflect geographic differences in location with respect to market. This is particularly true when one deals with land areas of like productive capacity located at different distances from market. Transportation cost is the key factor in these cases. Since these costs ordinarily increase with distance, sites near a market usually enjoy an element of comparative advantage over sites located farther away. Areas close to market accordingly receive higher net prices for their products, yield more land rent, and have higher capitalized values than areas located at greater distances.

Von Thunen's Model

Most of our theory regarding the effect of spatial location on land-utilization patterns stems from a model presented by Johann Heinrich von Thunen in his book *Der isolierte Staat*, written in 1826.[5] Von Thunen assumed the case of an isolated state (which freed his example from the impact of other economies and markets) with one central city located in the midst of a level productive plain surrounded by

[5]Cf. Peter Hall, ed., *Von Thunen's Isolated State* (Elmsford, N.Y.: Pergamon Press, 1966). For other discussions of this topic, cf. Richard T. Ely and George S. Wehrwein, *Land Economics* (New York: Macmillan, 1940); August Lösch, *The Economics of Location*, trans. William H. Woglom and Wolfgang F. Stolper (New Haven, Conn.: Yale University Press, 1954); Edgar S. Dunn, Jr., *The Location of Agricultural Production* (Gainesville: University of Florida Press, 1954); and Isard, *op. cit.* Von Thunen's approach has been reformulated in a linear programming model. Cf. Benjamin H. Stevens, "Location Theory and Programming Models: The von Thunen Case," *Regional Science Association Papers*, 21 (1968), 19–34.

a wildnerness area. He also assumed a village type settlement (with farm families living in the central city rather than the open country), uniform climate and soils, uniform topography, and relatively uniform transportation facilities. Since railroads and superhighways were not as yet known, he assumed that farm produce would be hauled to the central market in horse- or oxen-drawn wagons, carried by man, or driven in the case of livestock.

Except for location and distance to market, von Thunen's analysis held constant all of the natural factors affecting land use. Differences in land use could be attributed directly to variations in transportation costs. These in turn were dependent on such factors as distance to market; ease of transportation; and the bulk, weight, and perishability of the products sent to market. With these assumptions, von Thunen visualized a central city surrounded by a series of concentric land-use zones (see Figure 9-1A). The zones closest to the city were utilized for intensive purposes and uses that involved highly perishable products or those that were heavy and hard to transport. The direct relationship between effort and time required for transportation favored utilization of the outlying zones by enterprises with lower transportation costs.

Von Thunen assumed that the first concentric zone around the city was used for gardens, truck crops, and the facilities needed for stall-fed milk cows and laying hens. This use was logical as this area was subject to intensive use, was visited frequently, and most of its products were hand-carried to the city. The second zone was used to produce forest products. This use may seem unusual today, but it must be re-

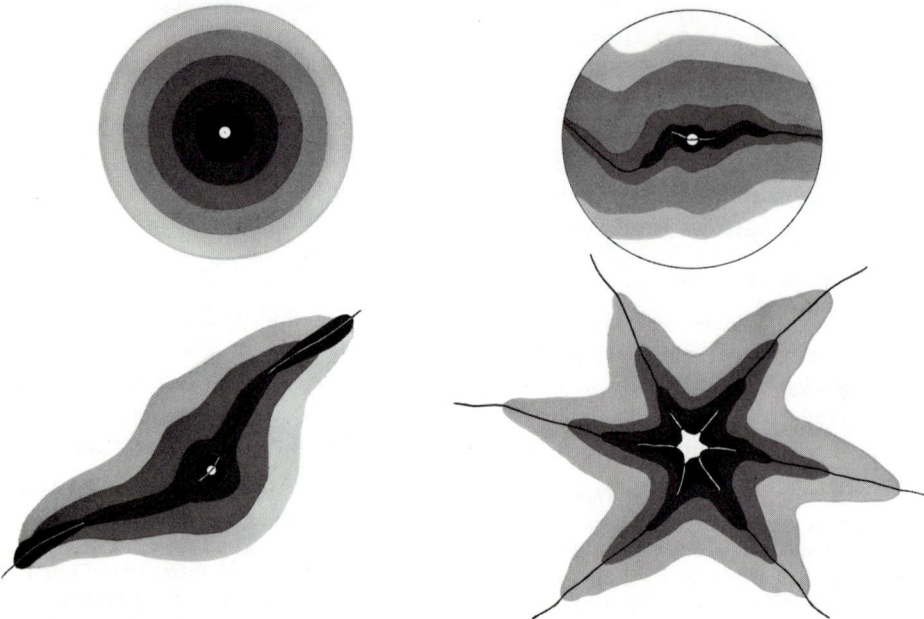

FIGURE 9-1.   Modified presentation of von Thunen's theory of the relationship between resource location and land utilization.

membered that forest products provided both fuel and a source of building materials. And since this product is both bulky and heavy to haul, it seemed important that it be produced near the city.

Immediately beyond the forest zone the land was used for the more intensively cultivated field crops—for bulky and heavy crops such as potatoes, root crops, and hay, and for grain grown in rotation with these crops. The fourth zone was planted to cereal crops, which call for less intensive operations. The fifth zone was used for grazing purposes, with the cattle and sheep produced and fed in this area being driven to market. The surrounding wilderness area was a sixth zone used for hunting purposes.

Von Thunun's simple model can be modified by adjustments in its many assumptions. For example, if one assumes that a navigable stream flows through the "isolated state," opportunities for water transportation may warrant changes in land utilization. With some series of uses, each zone could be expected to take on the elongated pattern suggested in Figure 9-1B. With the example described above, however, the market garden area associated with the first zone would probably remain pretty much unchanged, while it would become practicable to shift the areas used for forest production to sites along the navigable stream at greater distance from the city (Figure 9-1C). The introduction of additional improved transportation routes, as in Figure 9-1D, would lead to star-shaped land-utilization patterns.

Importance of Transportation Costs

Von Thunen was primarily concerned with the role transportation costs play in allocating the land resources found at varying distances from market among competing agricultural uses. His villagers used land as they did because of their rational desire to minimize the effort, inconvenience, and loss of time associated with their use of various sites and the movement of their products to market. Their location decisions may be quantified in economic terms as in Table 9-2. Basic land rent

TABLE 9-2. Illustration of the Joint Effect Transportation Costs and the Rent-Producing Capacity Associated with Different Land Uses Have in Allocating the Land Areas Around a Central Market between Alternative Uses

TYPES OF LAND USE	LAND RENT TYPE OF USE COULD EARN IF CARRIED ON AT MARKET	TRANSPORTATION COST PER MILE OF DISTANCE FROM MARKET	DISTANCE FROM MARKET TO EXTENSIVE OR NO-RENT MARGIN	RANGE OF DISTANCES FROM THE MARKET WITHIN WHICH THIS USE HAS FIRST CHOICE
A	$10.00	$2.50	4 miles	0.0–1.7 miles
B	7.00	0.70	10 miles	1.7–5.0 miles
C	4.50	0.18	25 miles	5.0–19 miles
D	2.00	0.05	40 miles	19.0–40 miles

levels may be assumed for each land use at the market; average transportation costs per mile can be calculated for the products associated with each use; and distance from the central market to the extensive or no-rent margin for each use can be computed by dividing the level of land rent at the market by the transportation cost per mile. Zones of highest and best use occur between the transference margins of the more profitable uses.

Land rent triangles can be drawn to depict the relationships reported with the four types of land use identified in Table 9-2. In Figure 9-2, the land rent triangle for use *A* starts with $10.00 of land rent on its vertical axis and stretches horizontally for 4 miles. Use *B* is depicted by a land triangle that has a vertical apex at $7.00 and stretches out for 10 miles. The rent triangle for uses *C* and *D* start with land rent values of $4.50 and $2.00 and extend horizontally for 25 and 40 miles, respectively.

When the four rent triangles are brought together in a margin-of-transference diagram as in Figure 9-2, it appears that use *D* alone will be carried on to its extensive margin. Use *A* promises the highest land rent for the sites immediately adjacent to the city. It accordingly is the highest and best use for the concentric zone that extends out to transference margin *ab*, which occurs 1.7 miles from the city. Use *B* is most profitable in the zone that extends from *ab* to *bc*, or from 1.7 to 5 miles from the central market. Use *C* is the highest and best use between *bc* and transference margin *cd*, which is located 19 miles from the city. The margin-of-transference points for the overlapping rent triangles correspond with the respective boundaries of the concentric zones in the von Thunen model.

Figure 9-2 illustrates the relationship between the land rent triangles employed in margin-of-transference diagrams and von Thunen's concentric zones of land use.

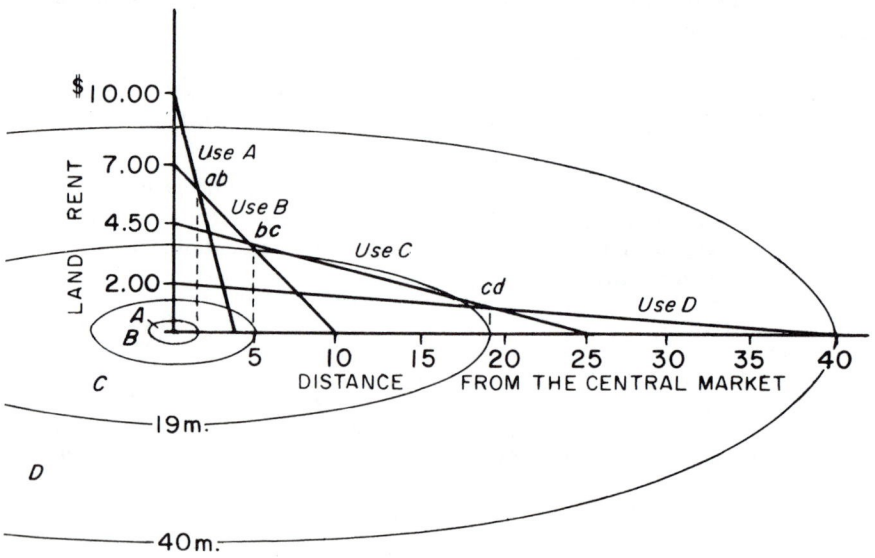

FIGURE 9-2.   Example illustrating allocation of land around a central city market between four competing types of land use.

A third dimension can be added to von Thunen's concentric zones as in Figure 9-3. Land rent cones now rise above the concentric land-use zones and find their highest points at the central market. The surface of each overlapping cone depicts both the amount of land rent and the slope of the land rent function associated with a particular use at increasing distances from the central market. The height of the highest cone surface for each site around the central market indicates the amount of land rent it can produce when utilized for its highest and best use. Viewed in this way, it is obvious that the overlapping land rent triangles used in the margin-of-transference diagram represent a cross section of the portion of the land rent cones found on a single side of the central city.

Von Thunen's basic concept, as depicted in Table 9-2 and Figure 9-2, illustrates the relationship of land rents to the cost of overcoming "the friction of space." As Haig observed,

> Site rents and transportation costs are vitally connected through their relationship to the friction of space. Transportation is a means of reducing that friction, at the cost of time and money. Site rentals are charges which can be made for sites where accessibilty may be had with comparatively low transportation costs. While transportation overcomes friction, site rentals plus transportation costs represent the cost of what friction remains. . . . The two elements, transportation costs and site rentals, are thus seen to be complementary. Together they may be termed the "cost of friction."[6]

Transportation costs rise as products must be shipped greater distances to market. These costs explain the higher land rents associated with sites near the

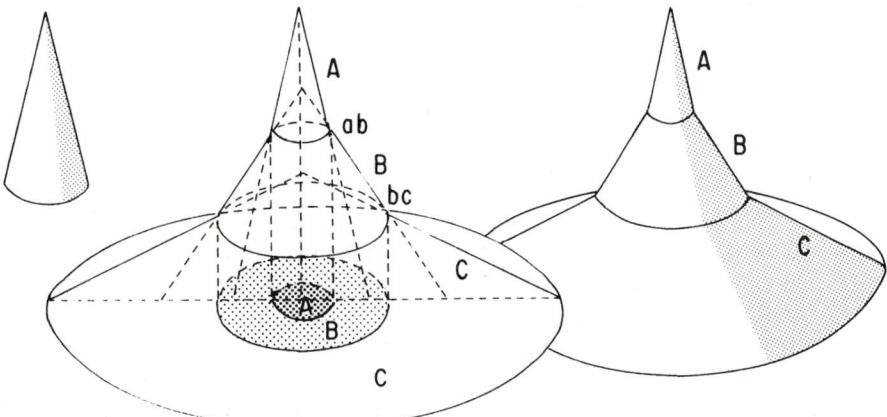

FIGURE 9-3. Use of land rent cones to illustrate relative levels of land rents or land values associated with highest and best uses of sites located at different distances from central markets.

[6]Cf. Robert Murray Haig, "Towards an Understanding of the Metropolis," *Quarterly Journal of Economics*, 40 (May 1926), 421–22.

market; and, as offsets to land rent, they set the area limits within which specific uses can be carried on to advantage. Location phenomena can often be explained simply in terms of the relation of land rent to nearness of the market.[7] This formulation, however, usually oversimplifies the situations found in real life.

Differences in Land Quality

Relaxation of von Thunen's assumption of uniform climate, fertility, and topography can also have significant effects on land-use patterns. If the land west of a city is fertile, level, and easy to work, while the areas to the east are handicapped by rough terrain, one can logically expect more expansion of the concentric use zones to the west than to the east (see Figure 9-4A). This situation results because the higher productivity and lower unit production costs associated with the better lands provide a larger economic surplus that can be used in paying shipping costs.

The effect of differences in fertility and topography may also be illustrated as in Figure 9-4B and C. In the first of these two models, it is assumed that a central city is located at the apex of four evenly divided areas of different productive capacities. The first area is well suited for the uses carried on in zones 1 through 4. Area II, however, is primarily suited for the intensive cropping practices carried on in zone 1 but poorly suited for uses 2, 3, and 4. Area III is best suited for uses 3 and 4 but less well suited for uses 1 and 2, and area IV represents an area of low productive value for all uses. When these areas lie side by side, sharp breaks may be visualized in the land-use pattern as one moves from one area to the next. When the areas are scattered throughout the region, the resulting land-use pattern may assume a complex crazy-quilt form such as that suggested by Figure 9-4C. Another variation of the von Thunen model occurs when climatic and other natural productive advantages enjoyed by areas located at some distance from the city permit them to produce and ship products at a lower cost than they could be supplied near the city (see Figure 9-4D).

The overall impact of differences in land quality and improved transportation facilities can be visualized with the example of an isolated state and a single city located in a valley at the foot of a mountain. With urban growth, the city will sprawl outward primarily on lands of level terrain. New commercial and industrial enterprises will seek flatlands and highly accessible sites that offer good transportation facilities. Residential developments will be built on land adjacent to the city, mostly on sites of gentle terrain. Bottomlands near the city will provide logical sites for intensive agricultural uses. The remaining bottomlands together with some of the

---

[7]Haig, *ibid.*, p. 423, advanced a hypothesis that land-use patterns in metropolitan centers should "be determined by a principle which may be termed the minimizing of the costs of friction." See also Richard M. Hurd, *Principles of City Land Values* (New York: The Record and Guide, 1903), p. 78; Lowdon Wingo, Jr., *Transportation and Urban Land* (Washington, D.C.: Resources for the Future, 1961), pp. 63–80; and William Alonso, *Location and Land Use* (Cambridge, Mass.: Harvard University Press, 1968).

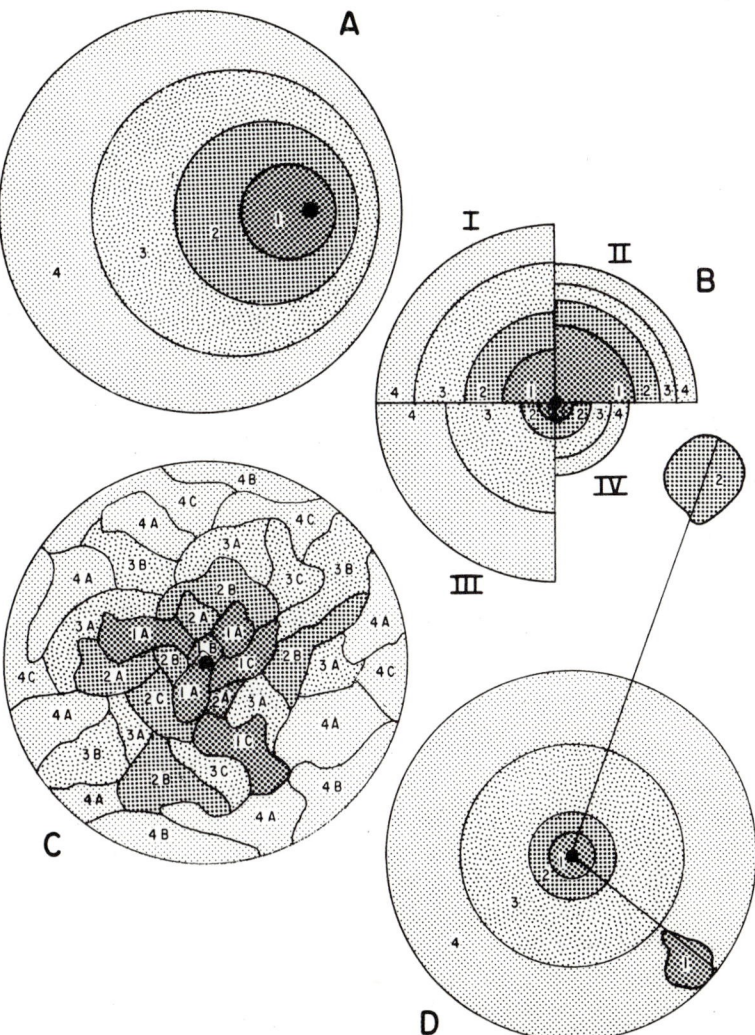

FIGURE 9-4. Examples of the effects of varying fertility and topographic conditions on land utilization.

more accessible uplands will be used for less intensive crops. Grain and hay crops may be produced on the uplands, but the higher and rougher lands will likely be used for grazing and forestry purposes. The surrounding mountainous areas will be used mostly for forest recreation and wildlife uses. If there is a need for additional production and the city is connected by a pass to a second mountain valley, it may be economic to use the fertile bottomlands of this valley for crop or grazing purposes rather than attempt to reclaim additional lands for these uses in the first valley.

Impact of Satellites and Other Markets

Central cities are ordinarily surrounded by natural supply areas upon which they have often depended for supplies of agricultural and other materials. This pattern is complicated when two or more cities are located near each other. Several market centers then compete for the products of a single supply area, and the resulting land-utilization patterns reflect the pull of the markets they serve. The impact of this pull of additional markets depends on the size and needs of the markets together with their location and transportation ties to the lands in question. When two cities of comparable size and function are located next to each other, it is logical for them to divide their outlying areas, with each city drawing on and servicing the areas closest to it. Complications may arise because of differences in transportation facilities and urban functions. A highway or railroad connecting one city with the natural hinterland of a second city will often claim much of the area served by these facilities for the first city. Similarly, if one has the only furniture factory while the other has the only flour mill, considerable overlapping of supply and market areas can be expected.

When one or more smaller cities are located within the natural hinterland of a central city, the central city must compete with its satellites for the use of certain areas. The central city under these circumstances is still surrounded by a series of generalized land-use zones—zones that take the form of irregular bands rather than concentric circles because of differences in transportation facilities, topography, and land productivity. Each of the smaller cities also has need for surrounding land areas and can ordinarily outbid the central city for particular sites as long as its uses have a higher economic or social priority than those of the central city (see Figure 9-5). When the central city's uses have highest priority, the satellite cities must seek alternative use sites—usually at greater distances and away from the central city—where they can better compete with the prices offered by the central city.

Product prices in satellite cities often reflect the cost of transporting goods to the larger market of the central city. As long as a surplus supply of a product is produced in the immediate area, its local price floor will represent the price offered in the larger city less the cost of transportation. For example, if milk is priced at $8 a cwt. (hundredweight) in the larger city and can be shipped from the satellite city area for 60 cents per cwt., the minimum local cost will be $7.40 per cwt. If local producers are offered a lower price, they have the option of shipping to the larger market. In actual practice, satellite cities often pay more than this minimum. This situation may exist because of less desirable or less stable local market conditions or because of a need to attract supplies from areas lying between the satellite cities and the central market.

Prices in satellite city markets are sometimes higher than in the central market. This is particularly true when a satellite area does not produce all of its own supplies or when the products grown in the area must go to the central market for processing. In meat-producing areas, for example, local livestock and meat prices may reflect the major packinghouse center price less freight. If an area is dependent on

FIGURE 9–5.   Illustration of the effect of satellite cities and variations in highway facilities and topographic conditions on land-utilization patterns around a central city.

Omaha packers for part of its meat supply, however, local prices are more apt to represent Omaha prices plus shipping costs.

Figure 9–5 illustrates the effect of satellite cities and variations in transportation facilities along with differences in land quality on the generalized land-use zones found around a central city. The impact of these land-use patterns on land rents and property values is shown with rent value profiles in Figure 9–6. They can also be depicted on three-dimensional topographic maps of rent and value levels on

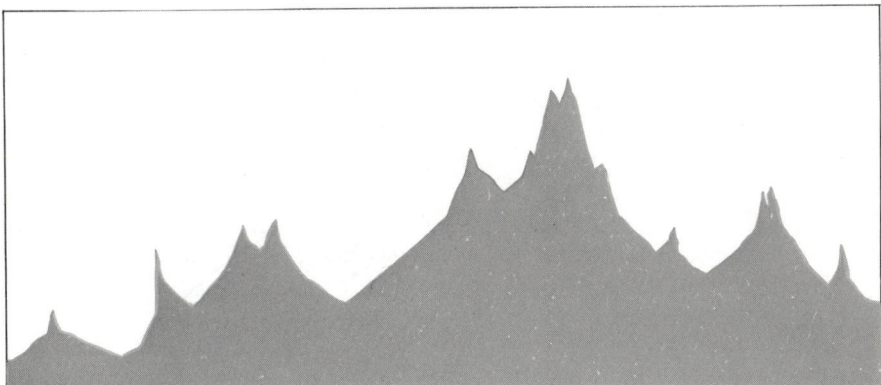

FIGURE 9–6.   Profiles of land rents (or land values) associated with the example depicted in Figure 9–5.

which the land rents associated with the higher and better uses rise like mountain peaks, ridges, and hilltops above the surrounding plains and valleys.

Competition for market areas. Cities compete not only for the raw materials and supplies they secure from their hinterland areas but also for markets for the goods they produce. When two or more producers compete for the same market, price competition and even price wars may develop. But cutthroat competition is not likely to occur as long as producers refuse to sell at less than their actual costs of production plus transportation.

When producers residing in different areas quote standard f.o.b. prices to their customers, their actions can have an automatic effect in dividing and allocating market areas. This situation is illustrated in Figure 9–7, which assumes three producers of comparable products who are located in different cities and who market their goods at f.o.b. prices of $45, $35, and $50, respectively. As long as the producers hold to these prices and as long as one assumes uniform transportation costs, it is easy to determine the areas within which each producer can undersell the other two. Adjustments must be made in this concentric circle model when consideration is given to the impacts of different types of transportation facilities and the fact that sites equidistant from a given city seldom enjoy equal transportation advantages. Complications also arise when some producers enjoy more favorable transportation rates than others, absorb their freight charges, or adhere to a basing point system. Manufacturers of many much-advertised products absorb transportation costs so that their products may be sold at a uniform price all over the country.

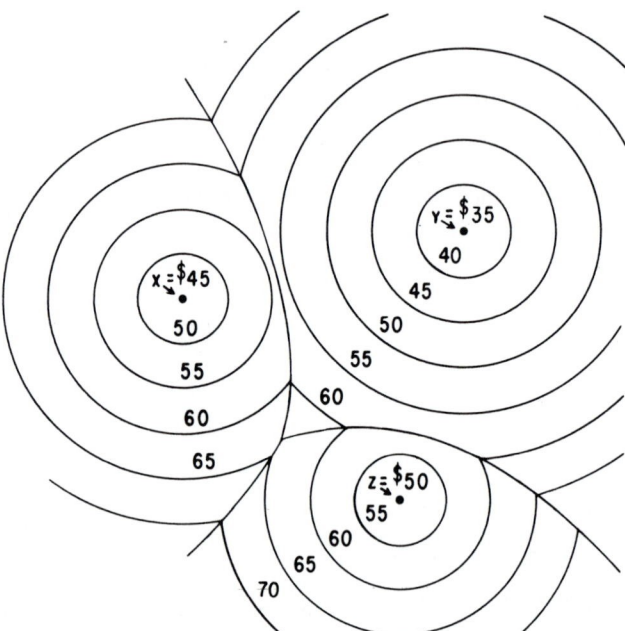

FIGURE 9–7.   Division of market areas between three producers with different f.o.b. prices (assuming uniform transportation facilities and costs).

This practice naturally favors buyers in distant locations in comparison with those who reside near the points of production.

Other Modifications of von Thunen's Model

Two other important modifications of von Thunen's model involve his assumptions of a village-type settlement and an isolated state. Rural people tend to live in central villages in many European farming communities. Farmland holdings often involve several small separated tracts that are frequently used for the same purposes as the adjoining tracts owned by other villagers. Families may thus own parcels of land in each land-use zone, and individual tracts may be used in accordance with the von Thunen model.

A different pattern applies in most parts of the United States and Canada. Farm families in these areas ordinarily live on their farms rather than in villages; they usually farm solid blocks of land; and their farms often involve holdings in a single land-use zone rather than a range of zones. With these circumstances, operators ordinarily let location and comparative advantage determine their major enterprises.

Complementarity in farm organization usually favors a mixture and often a rotation of land uses. Operators in a cash-crop farming area may use a high proportion of their land for cotton or corn but also some of it for pasture and grazing purposes. Those living in wheat or ranching areas will use most of their land for these purposes but still have gardens and possibly a small dairy herd.

Von Thunen used the concept of an isolated state to reduce the number of variable factors that affected his model. In so doing, he found it possible to disregard the effects that interarea trade, distant markets, tariff barriers, and public price and production programs have on the land-use practices of local producers. In practice, local land-use decisions are often affected by regional and national economic developments. Also with each new development in transportation facilities, areas come into closer contact with one another, and local producers find themselves competing more and more with producers from other areas. With these factors at work, producers must gear their operating decisions to national and international conditions. Instead of worrying about local market demand, they may look to distant cities for markets. Instead of calculating their prospects for profit in terms of local competition, they may find that the demand for their products is affected mostly by industrial employment levels or public price policies.

Von Thunen's model provides a meaningful basis for explaining the principal relationships between spatial location and land utilization.[8] Its focus on the land-use patterns associated with a single central market, however, ignores the effects diverse locations of sources of materials and markets can have on location decisions. An important contribution dealing with this facet of location theory stems from the work of Alfred Weber.[9]

[8]Cf. Michael Chisholm, *Rural Settlement and Land Use* (London, U.K.: Hutchinson University Library, 1962), for a discussion of current applications of von Thunen's concept in different parts of the world.

[9]Cf. Alfred Weber, *Über den Standort der Industrien* (Tübingen, West Germany: 1909); also Carl J. Friedrich, *Alfred Weber's Theory of the Location of Industries* (Chicago: University of Chicago Press, 1929).

Weber's approach. Like von Thunen, Weber started his analysis with basic assumptions about climate, topography, and the location of basic resources. He visualized several cities scattered over a region and noted that (1) some inputs in the manufacturing process are "ubiquitous"—available almost anywhere—whereas others are found at particular sites, and (2) cities can have both agglomerating attractions that draw industries to them and deglomerating features that have the opposite effect. With these factors in mind, he asked where an industry should locate if deposits of its chief raw material are found at a single site (point B in Figure 9-8A) and the principal market for the product is at point C. If all other inputs in the manufacturing process are ubiquitous and no loss of product bulk or weight takes place as the raw materials are processed, the processing can logically take place at B, C, or some site in between. If the raw material from B is such that considerable weight is lost in the manufacturing process, transportation cost savings would favor location of this phase of the production process at B.

Weber went on to ask how the decision concerning industrial site location might be affected if a second necessary ingredient is found only at point A. If no loss of bulk or weight is associated with the use of this input, it would be expedient to ship it to point B for use in the manufacturing process. If the ingredient is a necessary supply of waterpower, as in the days before electricity, the manufacturing process would have moved to the fixed location of the power site at A. If it is a bulky or heavy input that loses all or part of its bulk or weight in the manufacturing process, transportation costs could be minimized by locating the processing plant at some intermediate site such as at point D (see Figure 9-8B).

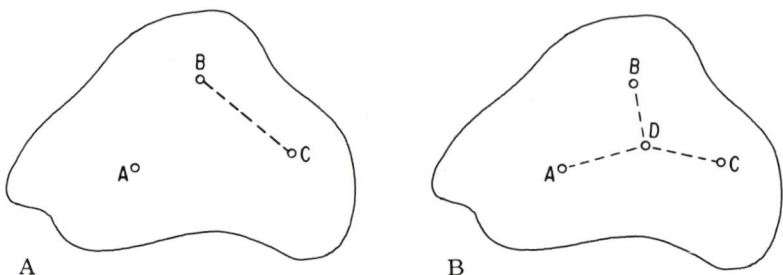

FIGURE 9-8. Illustration of Weber's location of industry model.

Precise calculations concerning the optimum location for a processing plant call for detailed information on transportation and other costs. If transportation cost is the only variable, operators can draw a series of isotim curves around each of three sites—concentric circles with the width of each indicating the distance units of the raw material or final product can be shipped for $1.00—to determine the least-cost transportation site. With Figure 9-9A and the assumption that all other costs are equal, the optimum site will be somewhere in the ABC triangle. Its location can be determined by adding the costs indicated by the isotims around A, B, and C for different sites and then selecting the site with the lowest sum of the three costs. Point

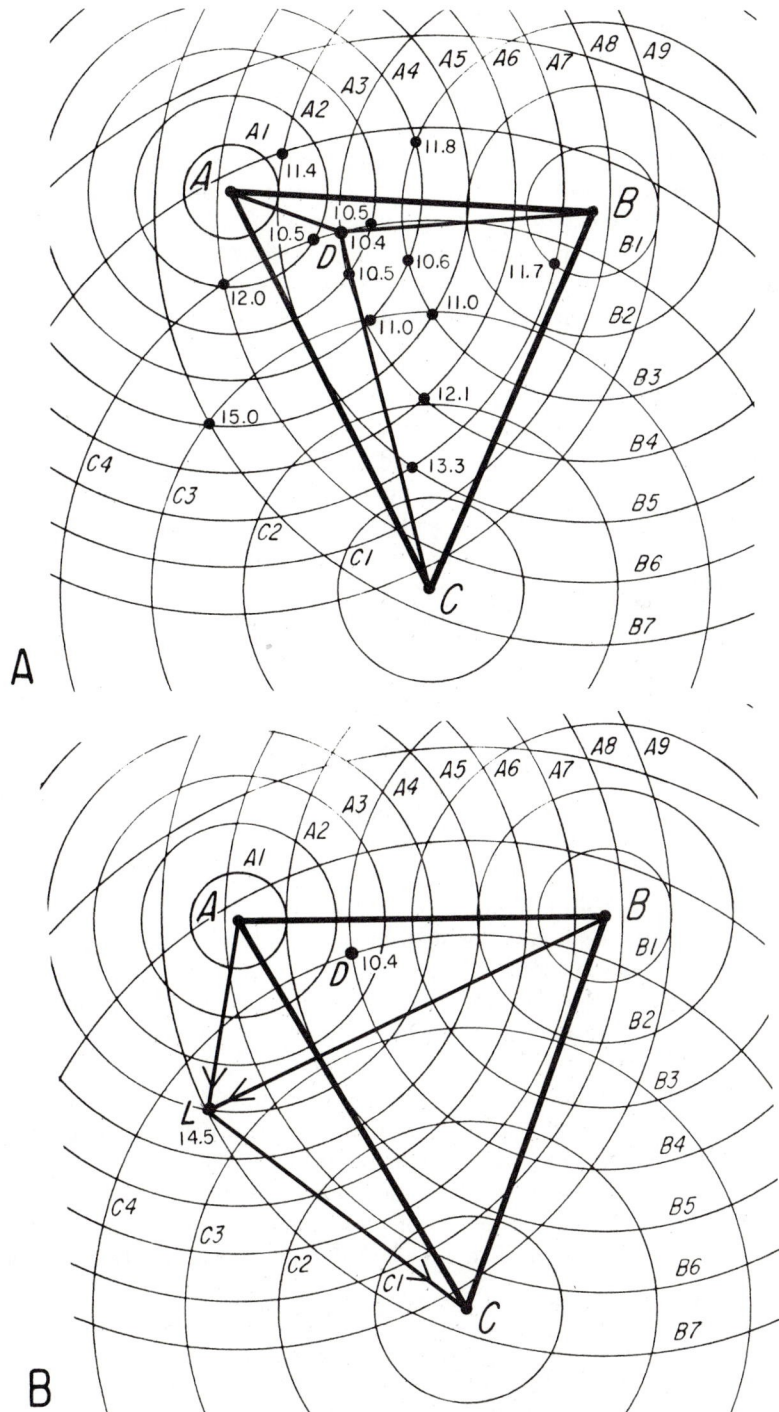

FIGURE 9-9.   Use of isotims to illustrate Weber's assumptions on optimum location of industries.

D, with a transportation cost of $10.40 per unit, represents the site with the lowest transportation cost.[10]

Weber also considered the impact of other variations in manufacturing costs on industrial location decisions. Figure 9-9A assumes uniform labor costs. If one assumes that the labor cost for producing the product is $25.00 per unit for all sites within the ABC triangle but is $20.00 per unit at site L, it would be advisable to move the industry to L (see Figure 9-9B) because the operator's combined costs for transportation and labor at that point would come to $34.50 a unit, compared with $35.40 at point D.

Changing impact of location factors over time. A dynamic dimension is added to the von Thunen and Weberian models when consideration is given to the effects changing supply and market conditions and new technology have on the location of economic activities over time. History provides numerous examples of cities that existed initially as agricultural trade centers, expanded as their rulers used military power to command trade and tributes from others, and then lapsed back to their earlier status. The development of new trade routes, new transport facilities, and the rise of new industries have also caused thriving cities to spring up at sites where few people lived before.

Excellent examples of the changes that come with new trade and industrial arrangements are provided by the cities of East Asia. Prior to 1600 the major cities of this world region were almost always located at interior sites.[11] Cities and villages were largely dependent on agricultural hinterlands, although some enjoyed additional attractions as military, government, or trade centers. Port cities were virtually nonexistent, until the arrival of European traders brought the establishment of bases at sites such as Bombay, Calcutta, Rangoon, Jakarta, Manila, Singapore, Hong Kong, and Shanghai. With international trade, the new cities grew, industrialization processes were introduced, and new metropolitan centers emerged, while many of the older cities remained as they were.

New technological developments have also brought notable changes. Most early settlers along the American frontier tried to minimize transportation costs by locating along navigable streams. Where this was not possible, they often treated easily transportable commodities such as furs, livestock, and whiskey as their primary cash products. The development of canals and railroads between 1840 and 1890 opened up vast new empires. Railroads made it feasible for farmers in the South and West to sell their products in the industrial centers of the East and for wheat growers on the Great Plains to produce for world markets. Land-utilization practices that were not profitable more than a few miles from cities in von Thunen's day are

---

[10] For a more detailed discussion of this and other aspects of Weber's analysis, cf. Richard S. Thoman, Edgar C. Conkling, and Maurice H. Yeates, *The Geography of Economic Activity*, 2nd ed. (New York: McGraw-Hill, 1968), pp. 187–94. Weber's model has numerous applications in modern spatial economics. Cf. Louis LeFeber, *Allocation in Space: Production, Transport, and Industrial Location* (Amsterdam: North Holland Publishers, 1958).

[11] Cf. Rhoads Murphey, "Colonialism in Asia and the Role of Port Cities," *East Lakes Geographer*, 5 (1969), 24–29.

now carried on thousands of miles away. Lumber from the Pacific Northwest is used in far away building operations; truck crops are shipped from Mexico and the Imperial Valley to New England kitchens; Australian grain and New Zealand butter are standard commodities in the British market; and far-off attractions have become the playgrounds of world tourists.

LOCATION OF PARTICULAR LAND USES

Some of the most relevant and most significant issues in location economics involve decisions concerning the optimum siting of particular land uses and the land-use patterns that accompany them. Five of these issues are discussed here. They include the location of cities, urban land-use patterns, industrial locations, location of commercial establishments, and location of residential developments.

Location of Cities

Cities have existed almost since the dawn of civilization. Much of their basis is found in the gregarious nature of mankind and also in the cultural, economic, and political advantages that stem from the agglomeration or clustering together of people. Many early cities started as religious centers, the home of a royal court, or as fortified areas; but they also benefited from the opportunities they provided for trade and labor specialization. Throughout the modern era, the presence or potential development of a strong economic base has always been a prime requisite for urban growth. Viewed from another angle, it may be noted that the rise of cities has had a marked impact on the economic development of the areas around them. Experience shows that there is a distinct tendency for economic development to center at particular locations and for the developments taking place at these locations to push forward at a faster rate than in other areas. These growth centers or "localized matrices" are increasingly of industrial-urban composition. They provide the centers from which economic development spreads into surrounding areas. Because of this situation, the highest levels of economic development are usually found at or near industrial urban centers, whereas lower levels of economic development and less satisfactory types of economic organization are usually found (as von Thunen's concentric zone hypothesis suggests) toward the periphery of the areas that surround these centers.[12]

The urban growth prospects of today's urban sites depend largely on their location with respect to (1) tributary or hinterland areas, (2) transportation facilities and trade routes, and (3) supplies of resources that can be used to advantage by local industries. In this sense, cities can be classified into four functional groups:

---

[12] Cf. T. W. Schultz, *The Economic Organization of Agriculture* (New York: McGraw-Hill, 1953), chap. 9; John R. P. Friedmann, "Locational Aspects of Economic Development," *Land Economics,* 32 (August 1956), 231–37; *L'Economie du XXe Siecle* (PUF, 1961) and other works by Francois Perroux; and J-R. Boudeville, *Problems of Regional Economic Planning* (Edinburgh, Scotland: Edinburgh University Press, 1966).

trade centers, transportation centers, specialized function centers, and cities representing combinations of these types.[13]

Most cities exist primarily as trade and commercial centers. They provide goods and services for surrounding hinterland areas and in return draw food supplies, raw materials, and workers from tributary areas. The spacing of these centers reflects the nature of the population and land resource base. In an ideal model, with even distribution of population and land resources of uniform quality, local trade centers would be evenly distributed. Every local trade center would be surrounded by a service area, which in the days of horse and buggy travel would have been small enough to permit easy commuting from the outskirts of the area to the trade center.

If these trade centers were some distance apart and enjoyed uniform ease of transportation, every center would be surrounded by a circular service area. When they are located alongside each other in a crowded country, the boundaries of their service areas press against each other, and the idealized trade areas take the form of hexagons, the form that most nearly approaches that of a circle and still permits division of the entire territory into trade areas of comparable shape and size. Viewed together, these hexagonal areas suggest a huge honeycomb, each cell representing a separate neighborhood or community center with its surrounding area (see Figure 9-10A).

This honeycomb pattern supports several hierarchical levels of trade and service centers. A cluster of six hexagons (*A*-level regions) around a seventh hexagonal area might logically look to the *B*-level trade center of the enclosed hexagonal area for services not normally provided in the other six. Groupings of six *B*-level regions located around a seventh cluster, in turn, may look to the *C*-level city of the central cluster for higher levels of services. The regions represented by these groupings in turn may look as in Figure 9-10B to still larger metropolitan centers (*D*- and *E*-level centers) for more specialized levels of service.[14]

This model is more suggestive of the spatial relationships that should exist under idealized conditions than of those found in practice. Even so, examples approximating this model have been observed in parts of Europe and the United States. Residents of rural areas generally look to a local village or town for selected commercial, educational, postal, and social services. For banking, medical, hospital, legal, department store, and supermarket services, they frequently look to some other center that has grown faster than its neighbors and taken on the function of providing

[13] Cf. Chauncey D. Harris and Edward L. Ullman, "The Nature of Cities," *Building the Future City, Annals of the American Academy of Political and Social Science*, No. 242, November 1945, pp. 7-17. For other, more detailed functional classifications, cf. Chauncey D. Harris, "A Functional Classification of Cities in the United States," *Geographical Review*, 33 (January 1943), 86; and Otis Dudley Duncan *et al., Metropolis and Region* (Baltimore, Md.: Johns Hopkins University Press, 1960).

[14] For other discussions of this theory, cf. Walter Christaller, *Die Zentralen Orte in Suddeutschland* (Jena, East Germany: Gustav Fischer, 1933), translated by Carlisle W. Baskin as *Central Places in Southern Germany* (Englewood Cliffs, N.J.: Prentice-Hall, 1966); Ely and Wehrwein, *op. cit.*, pp. 432-33; Edward L. Ullman, "A Theory of Location for Cities," *American Journal of Sociology*, 46 (May 1941), 853-64; and Losch, *op. cit.*, pp. 109-37.

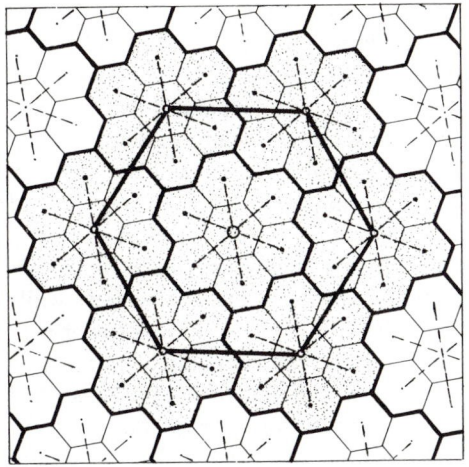

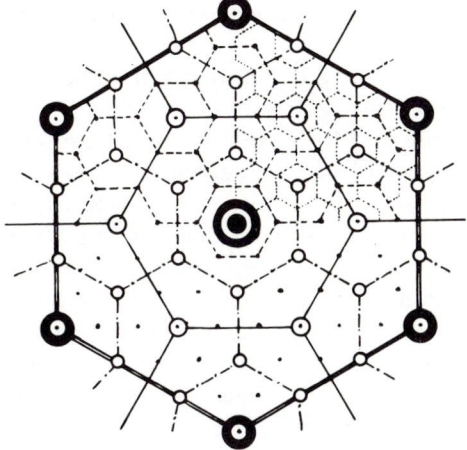

FIGURE 9-10. (A) Idealized distribution of cities (assuming uniform distribution of population and land resources, tendencies for small trading centers to develop at fairly equal distances throughout the countryside, and a tendency for these centers to form hexagonal patterns in their patronage of larger service centers, which in turn fit into hexagonal patterns around still larger commercial and service centers). (B) Illustration (after Christaller, *Central Places in Southern Germany,* p. 66) of hierarchies of service centers with idealized distribution of cities in which

· represents the center and

the boundary of the *A*-level region;

○ and - - - - the *B*-level region,

⊙ and — · — the *C*-level region,

◉ and _____ the *D*-level region,

◎ and _____ the *E*-level region.

specialized trade and service functions for the surrounding communities. Groups of counties often look to state capitals and regional metropolitan centers for higher levels of services such as wholesaling, regional office headquarters, larger department and specialty stores, and opportunities for particular types of entertainment. Residents of the areas served by these centers may look in turn to large metropolitan centers for special services such as face-to-face contacts with corporation and financial leaders, convention sites, better selections of specialty goods, and opportunities to attend major league sports events. At each of these levels, the central city plays a hierarchical service role for its hinterland area and also supplies all the services provided by lower-level centers for its immediate area.

Location with respect to hinterland areas and other cities is only one of the factors that affect urban growth. Complications arise because of the uneven distribution of population, land resources, and local trade centers. Plus factors such as location along favored transportation routes, development of local industries, and far-sighted local leadership have often caused villages to become cities when they might have remained as hamlets under the hexagonal approach. The growing competitive power of these urban centers has often discouraged the parallel rise of neighboring trade centers that may have boasted an initial advantage in location.

Since the beginnings of recorded history, locations along ocean and lake harbors, near the fall lines of navigable streams, at intersections of land trade routes, and at transshipment or break-in-bulk points along water and land trade routes have usually favored urban growth. The development of railroads, highways, and air travel has brought the advantages of good transportation facilities to many new areas while enhancing the advantages already enjoyed by cities with good locations. The growth of port cities such as Montreal, Boston, New York, New Orleans, and San Francisco can be attributed both to their world trade advantages and to the industries and commercial establishments that have benefited from location at these transshipment points. Inland cities such as Chicago, St. Paul, Kansas City, and Dallas enjoy comparable advantages because of their location at railroad and highway centers. In contrast, many once-thriving villages bypassed by railroads and early highways have virtually disappeared.

Prospects for urban developments are also affected by location with respect to particular types of resources. The presence of valuable forests, mineral deposits, or special recreation attractions can favor the rise of cities in out-of-the-way locations. Many industrial cities are located where they are because they specialize in the production or processing of goods that require local supplies of raw materials. Other specialized function cities such as political capitals and educational centers often owe their locations to historical accident or design.

Most large cities function as trade centers and also as transportation and specialized manufacturing or service centers. The joint impact of these functions on the growth of urban centers is suggested by Figures 9-10 and 9-11. Figure 9-10 depicts an idealized hexagonal arrangement of trade centers, while Figure 9-11A shows the effects transportation facilities and an uneven distribution of resources can have on the rise of cities. When these patterns are combined, one gets a complicated distribution pattern such as that shown in Figure 9-11B. At first view, this figure suggests that cities are located more or less at random without respect to any set of principles. Considered in light of the factors discussed above, however, logical explanations involving applications of comparative advantage usually explain why cities are found where they are.

Urban Land-Use Patterns

Very few cities start as planned developments with neat allocations of given areas for commercial, industrial, residential, and other uses. Instead the average city begins as a village and gradually expands. This growth process is often haphazard, poorly planned, and frequently expensive. As cities grow, they usually sprawl outward. Business districts spill over into surrounding residential areas. Sometimes this

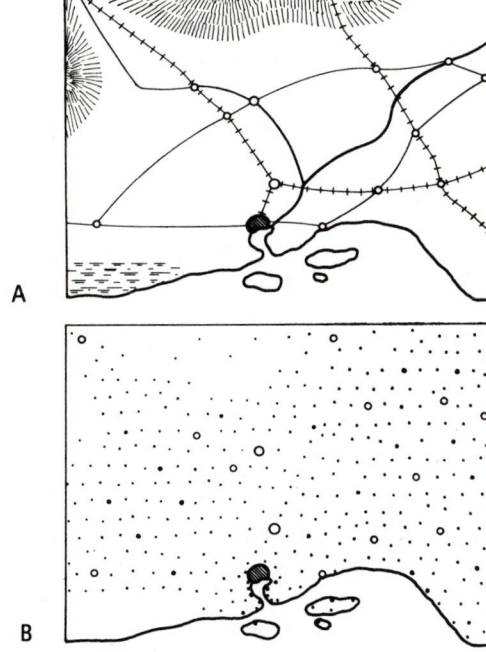

FIGURE 9-11. (A) Illustration of the impact differences in land resource patterns and in transportation facilities have on location and the distribution of cities. (B) Composite effect of the hexagonal approach, differences in resource patterns, and differences in transportation facilities upon urban locations.

expansion has a relatively uniform effect on all the blocks surrounding the original 100 percent spot. Sometimes it is all in one direction or may follow a single street; and in some instances, business districts migrate with their 100 percent spots to new locations. Industrial areas are also affected by this growth process. The original industrial sites—ordinarily located near the outskirts of small cities—are soon engulfed by the growing city and frequently cut off from contiguous areas that could be used for plant expansion purposes.

Of the various land uses affected by the squeeze of urban growth, the residential area located around the commercial core of the original city is usually the first to give way. With the encroachment of commercial establishments and light industries on this area, the higher valued residential districts usually shift in the direction of the city's outskirts. This movement brings a succession of lower-valued residential uses in the transitional zone surrounding the heart of the city and frequently results in blighted neighborhoods. Timely redevelopment or redesigning of the transitional areas can contribute to the vigor and vitality of the urban economy. But when the succession process brings lower uses faster than properties are redeveloped, blight often heralds the emergence of slums.

The variety of growth and changing land-use patterns found in different cities complicates the task of identifying simple principles that govern the allocation of urban land uses. Several explanations of urban land-use patterns have been advanced. Three of the most important of these are the *concentric-zone, sector*, and *multiple-nuclei* theories.

Concentric zones in urban land use. One of the first explanations of the internal land-use structure of cities was presented by Ernest W. Burgess in 1925.[15] His approach employs a concentric-zone concept, which in many ways parallels von Thunen's explanation of rural land use.

As Figure 9-12 indicates, Burgess designated the central zone as the loop area. This zone encompasses the 100 percent spot and includes the principal stores, office buildings, banks, theaters, and hotels. It is the business center of the city—the focal point of its commercial, social, and civic life. The concentric zone surrounding the loop area is a transitional zone. It is made up, for the most part, of older homes and tenement houses. Factories and business establishments are encroaching on the inner portion of this zone, and most of the remaining area is blighted. Many single-family homes have been converted to rooming houses or small apartments; some properties are boarded up; and much of the area is ill-kept and speaks of poverty and neglect. A working-class housing area lies beyond the transitional zone. Its residents live in modest single-family houses, rowhouses, and two- or three-decker dwellings. They prefer to live here because of the lower rents and values and because they are within easy commuting distance of the central business district and their places of work. Higher-cost and more sumptuous residential districts are located near the city's outskirts in zone IV, while a suburban or commuters' zone is found still farther out in zone V.

Burgess's model suffers from the same weaknesses as the von Thunen approach. Allowances and modifications are needed to explain the roles played by important streets and transportation routes, physical barriers such as lakes and rivers, changing social preferences in land use, the impact of satellite cities and shopping centers, and by changing land use-capacities.

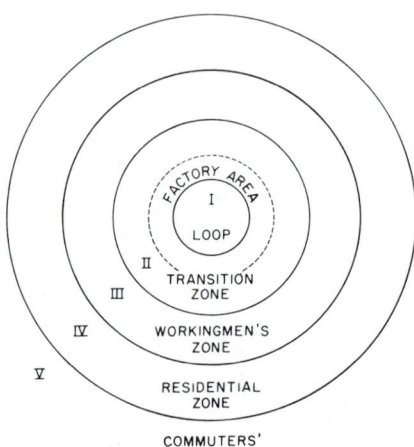

FIGURE 9-12. Burgess's concentric zones in urban land use.

[15] Cf. Ernest W. Burgess, "The Growth of the City," in Robert E. Park *et. al., The City* (Chicago: University of Chicago Press, 1925), chap. 2. Burgess's use of the concentric-circle approach in connection with urban land uses was by no means new. Variations of this approach had been used by several earlier writers, including Plato and Aristotle.

Some of the principal weaknesses of the concentric-zone formulation can be remedied by shifting to a radial zone concept such as that depicted in Figure 9-13 in which the major land-use zones are aligned along the leading transportation routes. This approach is closer to reality, but allowances must still be made for the not infrequent failure of some classes of people and land uses to gravitate to their predestined zones. In other words, allowances must be made for the failure of the self-regulating aspects of this theory to operate as envisaged.

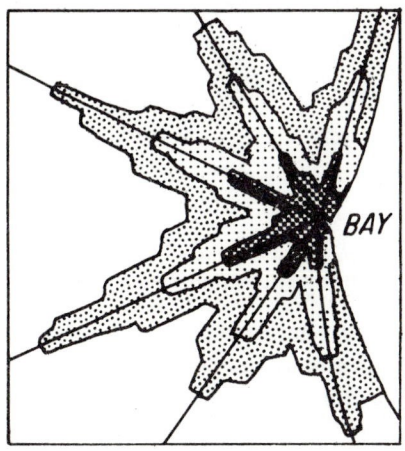

FIGURE 9-13. Illustration of a radial pattern of urban development.

Sector theory. An important alternative to the concentric-zone hypothesis is provided by the sector theory of urban growth. This theory was developed by Homer Hoyt during the late 1930s and resulted from his analysis of residential neighborhood trends in a study involving more than 200,000 blocks in approximately seventy American cities.[16]

Hoyt assumed a pie-shaped city with a central business district and numerous sectors or slices extending out from the central district to the city's outskirts. He then argues a theory of axial development in which the land uses found in the various sectors tend to expand outward usually along the same axis, along principal transportation routes, and along the lines of least resistance (see Figure 9-14). This theory provides a logical explanation for string-street developments and for the tendency of commercial districts to expand along important streets and to sometimes jump several blocks and then reappear along the same streets. Where possible, factory and industrial districts also tend to continue their expansion along railroads, waterways, and sometimes principal streets.

The sector theory assumes urban growth with succession in land uses in already developed areas and in new developments around the fringe of the city. Commercial areas are usually contained by surrounding areas devoted to other uses and can be

---

[16] Cf. Homer Hoyt, *The Structure and Growth of Residential Neighborhoods in American Cities* (Washington, D.C.: Government Printing Office, 1939); and Hoyt, "Recent Distortions of the Classical Models of Urban Structure," *Land Economics*, 40 (May 1964), 199–212.

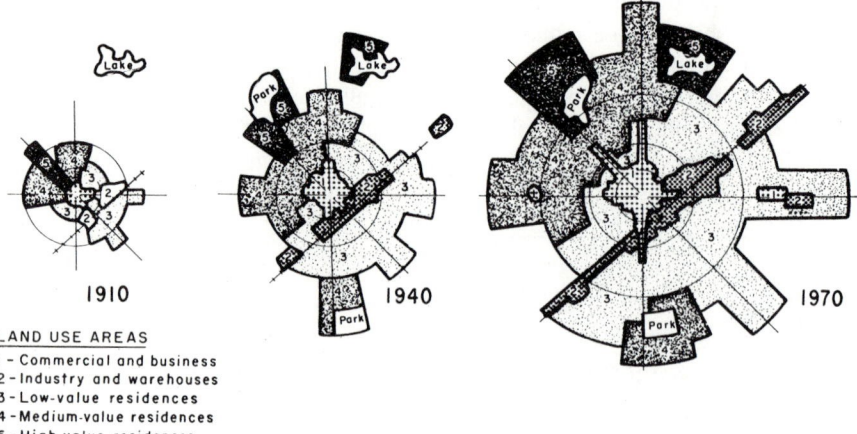

LAND USE AREAS
1 - Commercial and business
2 - Industry and warehouses
3 - Low-value residences
4 - Medium-value residences
5 - High-value residences

FIGURE 9-14.   Illustration of Hoyt's sector theory showing the generalized urban land-use patterns of an expanding city at three different time periods.

expanded only through the acquisition and redevelopment of neighboring uses. Properties in high-value residential areas filter down to lower-cost residential uses as their occupants shift to newer, high-prestige locations. Some intermediate- and low-cost housing results from the filtering-down process, but a high proportion of the housing occupied by low- and intermediate-income groups is built on new ground as urban growth causes the sectors used for these purposes to expand outward toward and beyond the city's outskirts.

The trend toward outward growth is particularly apparent with the high-cost and high-value neighborhoods. As Hoyt observed:

> The wealthy seldom reverse their steps and move backward into the obsolete houses which they are giving up. On each side of them is usually an intermediate rental area, so they cannot move sideways. As they represent the highest income group, there are no houses above them abandoned by another group. They must build new houses on vacant land. Usually this vacant land lies available just ahead of the line of march of the area because, anticipating the trend of fashionable growth, land promoters have either restricted it to high-grade use or speculators have placed a value on the land that is too high for the low-rent or intermediate-rental group. Hence the natural trend of the high-rent area is outward, toward the periphery of the city in the very sector in which the high-rent area started.[17]

The sector theory provides a reasonably realistic explanation of the basic land-use structures of many North American cities of fifty years ago.[18] It still explains

[17] Hoyt, *The Structure and Growth of Residential Neighborhoods in American Cities* (Washington, D.C.: Government Printing Office, 1939), p. 116.

[18] The sector theory has been most operative in modern industrial cities where sites have gone to the highest bidder and where the process of urban development has not been constrained by

many new urban land-use developments, but changing conditions have made it a less meaningful explanation than it once was. In this respect, we must note that the urban growth process is not mechanistic. Land areas are frequently developed for different uses from what the theoretical models specify, and theoretical explanations must often be modified to match the facts.

Critics of the sector theory argue that urban land-development patterns are "too variable to be conceived in terms of two-dimensional cartographic generalizations."[19] They point out that urban growth decisions are affected by economic, social, and cultural factors. Historical accidents, changes in family incomes, aspirations for better housing, and the cultural associations of particular neighborhoods may have important effects on urban land uses. In similar fashion, the direction and nature of neighborhood growth may be affected by street layouts, changes in transportation facilities, the location of parks and educational institutions, deed restrictions, zoning ordinances and city plans, and public housing and redevelopment programs.

Adjustments for multiple nuclei. Hoyt's sector pattern is definitely a product of an accumulation of individual decisions in a laissez faire society that was affected to a greater extent by transportation constraints than are most present cities. An updating of his theory is provided by Harris and Ullman's concept of *multiple nuclei.*[20] This concept envisages cities and metropolitan areas with more than one business district (see Figure 9-15). These urban areas have a downtown district that provides a central core, but they also have additional business districts located along major

---

cultural and institutional controls. In contrast to the cities that fit this pattern, most of the older cities of Latin America have developed according to a "plaza plan." An open square or plaza provides the civic and social center of these cities. The cathedral, city hall, and state government buildings are located around this plaza, while the municipal market and the business and commercial district are usually concentrated in an adjacent area. Upper-class dwellings occupy most of the blocks immediately surrounding the central plaza, while the homes of the lower classes tend to be farther out toward the periphery of the community. This urban development pattern represents the reverse of the "gradients of status" ordinarily found in North American cities and stems in part from regulations issued by Spain's Council of the Indies during the 1500s, which limited the subdivision of residential lots near the urban centers and thus prevented a filtering down of the higher-cost residential sites. Cf. Peter W. Amato, "Population Densities, Land Values, and Socioeconomic Class in Bogota, Colombia," *Land Economics,* 45 (February 1969), 66–73; and Leo F. Schnore, "On the Spatial Structure of Cities in the Two Americas" in Philip M. Hayser and Leo F. Schnore, *The Study of Urbanization* (New York: John Wiley, 1967).

Gideon Sjoberg's research in *The Preindustrial City* (New York: Free Press, 1960), pp. 95–103, indicates that the "plaza plan" is typical of the land-use patterns found in most preindustrial cities and is directly related to the class system that existed in these cities. As these cities have become more industrial society–oriented and as they have experienced growth and urban redevelopment, they have tended to follow growth patterns more in keeping with those suggested by the sector theory.

[19] Walter Firey, *Land Use in Central Boston* (Cambridge, Mass.: Harvard University Press, 1947), pp. 84–85. Also cf. Lloyd Rodwin, "The Theory of Residential Growth and Structure," *Appraisal Journal,* 18 (July 1950), 295–317.

[20] Cf. Harris and Ullman, "The Nature of Cities," *loc. cit.,* pp. 13–15.

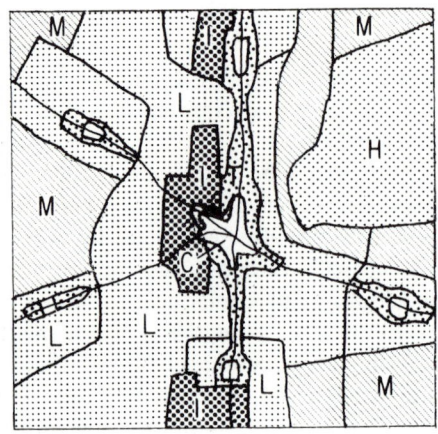

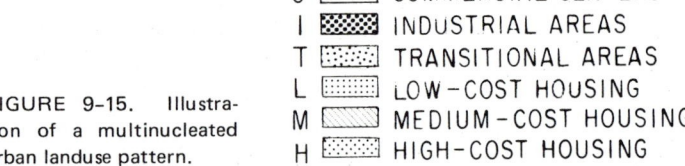

FIGURE 9-15. Illustration of a multinucleated urban landuse pattern.

C ☐ COMMERCIAL CENTERS
I ▨ INDUSTRIAL AREAS
T ▨ TRANSITIONAL AREAS
L ▨ LOW-COST HOUSING
M ☐ MEDIUM-COST HOUSING
H ▨ HIGH-COST HOUSING

streets at some distance from downtown. Each of these districts becomes a nucleus for a competing hierarchy of land uses comparable to those shown for all land uses in Figures 9-5 and 9-6.

This multiple-nuclei concept provides a realistic description of the land-use patterns that are appearing in more and more cities. New nuclei have appeared in some cases because rapidly expanding cities have annexed or engulfed already existing commercial centers, which have continued to operate as small commercial nuclei within larger metropolitan areas. Additional nuclei have appeared with the development of new industrial sites, new neighborhoods, and regional shopping centers that serve the needs of new urban residents. The creation of new nuclei has also been affected by transportation developments that have relaxed the mobility constraint that once made the central business district the transportation hub of the entire city, and by a noticeable flight of residents and commercial and industrial activities from the inner city to suburbia.[21]

Variations in urban land-use patterns. Changing conditions have had important impacts on the land-use structure of most cities. Radial patterns of growth and land use are still observable in small towns and cities. They also provided good descriptions of the uses made of urban lands as long as workers either walked to work or rode streetcars or buses to work and to market. Sector or axial patterns of land-use

---

[21]Cf. Robert G. Healy and James L. Short, *The Market for Rural Land: Trends, Issues, Policies* (Washington, D.C.: The Conservation Foundation, 1981), pp. 6-7.

structure were common with hundreds of cities during the early decades of this century. Population and transportation changes favored multinucleated patterns of development in most cities after the 1940s.

Thousands of rural residents moved to cities during and after World War II. This movement along with a high birth rate brought an urban population explosion. Cities grew both in numbers of residents and in areas occupied. At the same time, automobile ownership became more common and thousands of miles of streets, highways, and expressways were constructed or upgraded to speed the flow of traffic. Public planning programs also emphasized the provision of new housing and community facilities in the newly urbanized and suburbanized areas around cities. A major suburbanization movement resulted. With their improved mobility, thousands of city residents moved to the outskirts, to satellite suburban communities, or to the open country. Banks, department stores, and other commercial establishments set up branches to serve their clientele in neighborhood and regional shopping and commercial centers. Factories joined in the flight from older city locations. The jobs they offered went with them; and in many instances, the outward centrifugal pull of the suburbs drained downtown areas of much of the attraction they once had along with the economic base and viability they needed to operate as truly 100 percent locations.[22]

New patterns of urban land use are still evolving. Outlying residential, commercial, and industrial areas are still expanding and thriving at the expense of inner city locations. But progress is being realized in the replanning, rebuilding, and renewal of numerous downtown areas. Construction of new multistoried office buildings, and high-rise apartments and condominiums in downtown areas indicates that some of these areas are still very much alive. The spread of blight in other urban areas unfortunately suggests that many inner city areas are also dying and that huge capital investments are being abandoned and left with residual residents who frequently lack the necessary economic and financial ability to undertake needed programs of renewal and rejuvenation.

Aerial views of metropolitan areas often provide excellent examples of mixed patterns of urban land use. Central business districts frequently stand out as great hubs of business and commercial activity. Farther out, hierarchies of uses surround regional and local shopping and commercial centers. Still farther out, two types of development frequently appear. Suburban bedroom communities often exist primarily as residential communities with little land utilized for commercial purposes and frequently none for industrial uses. Somewhat in contrast, recently urbanized rural areas frequently display gridiron patterns of development. With these areas, the principal avenues and cross streets follow the old section line roads and are lined with commercial and industrial uses or sites zoned for these uses while the enclosed blocks are reserved for residential uses. The residential streets found in these

---

[22]The declining significance of the 100 percent spots of these cities can be contrasted with Richard M. Hurd's description of their importance around 1900. His study of property values at that time indicated that the land rent triangles and profiles of urban property values rose like sharp needles above the surrounding areas. Cf. Hurd, *op. cit.*, chap. 10.

superblocks may intersect with the commercial avenues or may connect with feeder streets that insulate the residential areas from commercial traffic.

Industrial Locations

Experience shows that most industrial enterprises start with small plants and that location at good sites often makes the difference between success and failure. These plants frequently owe their location more to historical accident than to economic design. When operators deliberately seek sites at which they can optimize their profits, they often find that their choices are complicated by factors such as availability and cost of raw materials, labor costs and problems, product demand, and marketing issues.[23] Their principal objective, however, should be that of selecting sites that meet the physical requirements of the industry and at the same time facilitate high productivity, low costs, and a large volume of sales.

Two principal types of costs—processing costs and transfer costs—affect the optimum location of industrial plants. Processing costs include the many expenses that arise in the industrial production process as labor and other factors are used to transform raw and semifinished materials into manufactured goods. Transfer costs, in turn, deal with the expense of moving materials to processing plants and finished goods to their points of sale or use. Industrialists naturally attempt to minimize both types of costs. In their search for optimum industrial locations, they recognize that transportation cost considerations cause some industries to be material-oriented and some to be market-oriented. Others are less affected by transportation costs and may be attracted by the agglomeration economies of particular cities or regions where the presence of other industries provides pools of skilled labor and management, capital availability, services of complementary industries, public utilities, and public services.

Material-oriented industries. Material-oriented industries may be divided into four principal groups. The first of these involves extractive and primary industries such as agriculture, fishing, lumbering, mining, and outdoor recreation, which are bound to the location of the basic natural resources on which they depend. A second group includes those with processing activities that involve elimination of waste materials and excess weight. Copper and iron ores are usually found in combination with large quantities of rock and slag. These ores are smelted or subjected to flotation, beneficiation, or other weight-reduction processes at sites near the mines. Most of the waste material is thus eliminated before the metal is shipped to other processing points. Except in those instances in which timber can be floated to a mill, sawmill operations also tend to be material-oriented. This situation is favored by the reduction of bulk and weight that comes with the processing of logs into

---

[23] Much of the basic thinking in industrial location theory is rooted in the pioneer work of Alfred Weber (cf. Friedrich, *op. cit.*). For other discussions of this subject, cf. Hoover, *op. cit.*; Losch, *op. cit.*; Isard, *op. cit.*; M. Greenhut, *Microeconomics and the Space Economy* (Glenview, Ill.: Scott, Foresman, 1963); and Glen E. McLaughlin, "Criteria in the Selection of Cities for Industrial Location," *Appraisal Journal*, 17 (April 1949), 168-72.

rough or finished lumber. Numerous agricultural-processing operations also fall into this class. Farmers who sell butterfat usually separate their cream from the skim milk and thereby reduce the bulk of their marketable product. The high transport costs associated with bulky products such as sugar beets and sugar cane favor the location of sugar factories within reasonably short distances of their supplies. Other products such as cheese, dried and condensed milk, vegetable oils, maple and cane syrup, turpentine, and rosin are also processed near their sources of supply.

A third group of material-oriented industries involves processes that require large quantities of fuels, power, or water that do not appear in the final product. The large quantities of coal and coke required for iron and steel production in times past often favored the location of steel mills near sources of coal supply. High electrical power requirements favor the location of synthetic nitrate, aluminum, and electrometallurgical plants near hydroelectric power sites. Similarly, the high water requirements of some industries together with the economies of water transportation favor industrial locations along navigable waterways.

A final group of material-oriented industries benefit from processing changes that make their product less bulky, easier to handle, less perishable, or more susceptible to bulk handling. These changes lead to transportation economies and thus favor material orientation. Cotton gins and compresses reduce the bulk of the raw cotton by forcing it into compact bales. Metals are often processed into ingots or sheets to facilitate their handling and reuse. Canning and preserving operations reduce the perishability of fruits and vegetables and thus lower their transportation and storage costs. Another type of material-oriented service is provided by local grain elevators, commission agents, junk dealers, and others who assemble carload lots of materials for shipment to other points.

Market-oriented industries. While some products benefit from weight-and-bulk-reduction operations in the early stages of their production, many others experience a reversal of this process as they approach the final stages of their production-distribution process. As these products "approach by stages the form in which they will be delivered to the final consumer, they become progressively more fragile, more cumbersome to pack and handle, more valuable in relation to their weight, and differentiated into more separate types and sizes."[24] These developments favor market-orientation.

House building, construction activity, home food preparation, and ice making provide classic examples of market-orientation, as do the functions performed by electricians, plumbers, and local service workers. Some other leading market-oriented industries of recent decades, such as bakeries (which must provide a fresh and bulky product), beverage-bottling plants (which add large quantities of carbonated water to sugar and concentrated syrup), dairy product distribution plants, and breweries, have tended to move to local regional operations for economy-of-scale reasons.

Most large industrial and mercantile concerns seriously consider the advantages of market orientation when they locate new plants and stores. Several mail-order

[24] Hoover, *op. cit.*, p. 36.

houses operate with both regional and local outlets. The American automobile industry centers in the populous North Central region but operates branch production and assembly plants in other regions and countries, just as some foreign producers have established branch plants in the United States.

Footloose and other attraction-oriented industries. Many industries are neither material- nor market-oriented. From a transport cost standpoint, these industries are frequently relatively footloose and free to locate where they will. Transportation cost considerations often cause businesses to locate at points between their principal sources of raw materials and their major markets. Locations at other sites have the disadvantage of double transfer costs, that is, shipment of raw materials to the factory and backshipment of goods to market. Within the zone of optimum transfer costs, these industries usually find it best to locate at transshipment points or at locations that favor low processing costs.

Labor is the most important and most expensive ingredient in the manufacturing process for many industries. These industries naturally seek factory sites in areas that offer adequate supplies of labor. Frequently they share in the agglomeration economies of other firms by locating in populous centers where they can draw on the supplies of skilled and semiskilled workers used by other industries. Location in these areas makes it possible for them to increase or decrease their labor forces if such a need arises with a minimum of social repercussion. Industrialists may hesitate to relocate or open branch establishments in areas where they will be the leading industry until they have had years of management experience, are competent to develop their own skilled labor force, and feel confident that they can offer employment stability in the new plant area.

Labor-oriented industries often seek particular types of employees. Textile mills have sometimes located in heavy industry towns where they could draw on the large potential supplies of women workers. Plants requiring highly skilled workers may avoid expensive recruiting costs by locating in cities with comparable plants. Some industrialists have tried to hold down their operating costs by locating new plants in low-wage areas, areas where labor has not as yet been unionized, and areas that offer facilities that contribute to low living costs.

Industries not normally classified as material-oriented, market-oriented, or labor-oriented are sometimes attracted to particular locations by the availability of special attractions. "Clean industries," for example, may be attracted to industrial parks located near large universities because of desired contacts with academic personnel and the cultural advantages associated with college communities. Other enterprises find it logical to locate alongside similar industries in metropolitan regions where they can share in the local pools of managerial and consultant talent and enjoy the availability of banking and other needed industrial services and the prospect of providing goods and services for complementary enterprises.

Importance of the land factor. Final decisions on new industrial locations often hinge on secondary issues. Operators may be primarily interested in finding favorable raw-material, market, transportation, and labor situations. But they are also in-

terested in locating at sites that offer good living conditions for employees; adequate space for parking and plant expansion; adequate supplies of water, power, and other utilities; and moderately low land values and tax levies. In this respect, land-factor considerations can easily tip the balance in their choices between alternative sites.

Throughout the horse and buggy period, sites that provided water power, favorable water or railroad transportation facilities, and a good labor supply ordinarily had first choice for industrial use. The shift to major uses of electrical power and motorized transportation has freed most industries from this earlier dependence on a narrow choice of available sites and has contributed to industrial decentralization. But even though many industries have chosen to locate around the periphery of cities where land is less expensive, taxes are lower, and large areas are available for use, considerable emphasis is still given to sites that provide access to water, railroad, or highway transportation routes.

Experience shows that many industries have reserved too little space for future expansion. Modern technology frequently favors shifts from multistory to single-floor factories. At the same time, good worker relations call for the provision of large parking areas. Both of these situations require more space. Yet many industries originally located at the outskirts of cities now find themselves hemmed in with limited opportunities for expansion in contiguous areas. Urban redevelopment offers a high-cost answer to this problem. Plant relocation suggests another somewhat costly answer. Knowledge of these situations has caused many industrialists to emphasize space considerations when they select new plant locations.

Industries often occupy high-value sites and occasionally invest large sums in structure- and site-improvement programs. As a rule, their land costs are small compared with cash outlays for raw materials, marketing, and labor. This situation can cause people to dismiss land-cost factors as relatively unimportant. But the importance of the land factor should not be downgraded. An industrialist who pays $1,000 a month in rent or ownership costs benefits from lower costs per output unit, other things being equal, than a competitor who pays $5,000 a month.

Personal and noneconomic considerations. Industrial location decisions often involve noneconomic considerations. Most industrial plants start as small businesses in the operator's home community at sites that the operator finds available, acceptable, affordable, and convenient. Some of these sites enjoy considerable comparative advantage. With good management and the smile of fortune, businesses located at these locations may prosper and expand, whereas businesses founded in less favorable locations may frequently fail.

One cannot assume, however, that all industries started at unfavorable locations are doomed to mediocrity or failure. An investor with a new idea, a new industry with exceptional business management, an ambitious city with strong leadership, or some historical accident such as a large government investment in a defense industry can easily compensate for an initial lack of locational comparative advantage.

Following this line of thought, one can argue that some of our leading industries could have found more profitable sites than those at which they have developed.

The automobile industry of Detroit and the rubber industry of Akron, for example, might have found it more profitable to locate at sites closer to their sources of material supply. Once these industries were established, however, they attracted skilled labor forces and developed industrial economies that gave these cities a relative advantage over potential competitive sites. Much of the success enjoyed by these industries can be attributed to these advantages and to their favorable location with respect to their major markets.

Institutional arrangements such as favorable public regulations and tax policies, far-sighted planning for the provision of needed utilities and local services, and programs that contribute to civic pride and enhancement of local amenity values can also add to comparative advantage.

Shifts in industrial locations.   History shows that industries frequently migrate to new locations and that this shift can have important impacts on property values and the economic life of both the communities they leave and the communities they enter. Industrial migration sometimes results in stranded communities and ghost towns. This has often been the situation with mining, lumbering, and other one-industry towns. On other occasions, potential adverse effects are cushioned by the growth of replacement industries.

Decisions to move are ordinarily prompted by factors such as (1) technological developments that favor processing economies, permit new products, and require new facilities; (2) exhaustion of a raw material base such as occurs when an ore deposit is mined out; (3) changes in material requirements such as the location of power-using plants away from water-power sites once alternative sources of power become available; (4) changes in transportation costs made possible by the provision of new facilities; (5) adjustments in individual processing costs, which cause particular sites to either gain or lose low-cost production advantages; (6) increasing tax loads; (7) site restrictions, which prevent desired plant expansions at existing locations; and (8) changing market tastes, such as the substitution of automobiles for buggies, which create new markets for some products while heralding the phasing out of others.

Location of Commercial Establishments[25]

Compared with other types of land use, the total area used for business and commercial sites is usually small. From the standpoint of intensity of use, rent-paying capacity, and land values, however, the areas occupied by commercial business districts represent some of our most valuable lands.

Most business operators recognize that their success or failure depends to a considerable extent on their choice of a suitable site for operations. Accordingly they

---

[25]For more detailed discussions of this subject, cf. Richard U. Ratcliff, "The Problem of Retail Site Selection," *Michigan Business Studies*, 9, no. 1 (1939). Ratcliff, *Urban Land Economics* (New York: McGraw-Hill, 1949), chap. 13; and Richard Lawrence Nelson, *The Selection of Retail Locations* (New York: F. W. Dodge, 1958).

have a definite incentive to seek locations that promise the greatest opportunities for profit. Their actual decisions in this regard are affected by a variety of factors. Naturally, most business operators prefer sites that promise a high volume of business activity. Until recent years in the United States, these locations were usually found in central business districts at or near the spot most accessible to the greatest number of potential customers. But in making their site selection decisions, operators must weigh the costs associated with the use of each site against its business advantages. If the added volume of business expected at the 100 percent site does not more than match the additional cost, the operator will usually find it advisable to locate in an outlying shopping center, at the outskirts of the central business district, on an upper floor of a downtown office building, or perhaps along a commercial street.

Other factors also favor locations away from the downtown area. Neighborhood grocers and druggists choose decentralized locations so that they may better serve the needs of the people living in particular neighborhoods. Furniture stores and drycleaning establishments frequently locate on the outskirts of the business districts or in the suburbs because of their space requirements and their need for lower rents and more adequate parking facilities than are available in central shopping districts. Other establishments such as lumber yards, warehouses, and freight depots have large space requirements and often locate along railroads or waterways.

In the idealized urban land-use pattern accepted in the United States in the early 1900s, most business, professional, and commercial activity was expected to take place in central business districts around the 100 percent spot. This general situation still holds in many small cities. Significant concentrations of commercial activity are also found in the central business districts of larger cities, but these cities typically have experienced considerable decentralization of their retail activities. Several factors including the flight of large numbers of upper- and middle-income families to the suburbs, increasing urban area size, shopper convenience, the acceptance of standard brands that can be purchased just as easily at outlying locations as at 100 percent sites, widespread ownership and use of automobiles, concern over parking problems in congested downtown areas, and decreasing reliance on the urban masstransportation facilities that radiate from central business districts have favored local shopping center developments that cater to neighborhood and multineighborhood needs.

Central business districts grow and expand or stagnate and wither in response to demands for the services they provide. In cities where the central business district has retained its attractiveness and economic strength and viability, the central business district (CBD) is almost always found near the hub of the city's traffic and transportation system and at sites both accessible and convenient to large numbers of people. Considerable concentrations of people are attracted to the district during business hours. This creates a potential for high volumes of retail and other business activity, which in turn justifies intensive land-use practices, high rents, and high land values.

Street sites around the 100 percent spot are ordinarily used for retailing purposes. In the past, large office buildings, banks, hotels, and first-run theaters congregated around this point. Among retailing establishments, the CBD has long been a focal point for large department stores, apparel shops, variety stores, restaurants, drugstores, and specialty shops that serve the many shoppers who flock to this district. Surrounding the area of most intensive retail activity—and often interpenetrating it—are a number of less intensive retail uses. Music, video, television, sporting goods, and hardware stores often appear in this class. Still farther out along arterial streets, one finds a lower grade of business uses such as auto supply shops, beer gardens, taverns, and pawn shops and also businesses such as automobile showrooms, used car lots, grocery stores, and lumberyards, which require considerable parking facilities.

Sites near the 100 percent spot supposedly offer the greatest opportunities for profitable use. They have the highest site values and command the highest rents. The use-capacity and profit opportunities associated with the surrounding areas often decline rather rapidly. Sites located a few blocks away on a main street, a block away on a back street, or only a few floors above the street may have only a fraction of the income-producing value of a ground-floor location near a strategic business corner.

The scarcity factor in this situation causes considerable bidding and counterbidding between firms and operators for the choice locations. This process often results in land-use patterns in which retail space is allocated in accordance with the rent-paying capacities of the various operators. This pattern is seldom stable. New adjustments are always taking place. Operators are often tempted by the opportunities suggested by site vacancies in the 100 percent district. Few of them, however, can estimate the exact effect a move may have on their volume of business. As a result, most site bids involve an element of trial and error. Some blind bids turn out very favorably. Others sometimes involve higher rental commitments than operators can pay and eventually lead to bankruptcy, closing-out, and removal sales, and in the vacating of sites for use by new operators.

Successful location factors. Many commercial firms—particularly those with chain operations—make a science of their selection of successful commercial locations. Before they choose a site, they analyze the advantages, disadvantages, and income-producing prospects of several alternative locations. They consider factors such as floor space, parking, and location relative to other stores. Questions are raised concerning the size and characteristics of the potential pool of customers, their levels of income, and their buying habits and tastes. Pedestrian traffic counts are made of the number of people who pass various store sites during shopping hours. Consideration is also given to the problem of competition with other commercial establishments.

In their search for good commercial locations, individual business operators must consider the characteristics and buying habits of potential customers together with the nature of the goods or services they provide. Men and women frequently vary in their buying habits. Except for their purchase of articles such as automobiles and hobby goods, men tend to be more hurried and impatient in their shopping than women. They are often "prone to buy the first article that approaches their requirements or taste. Convenience is more important in their minds, and the opportunity for comparison is less important."[26] Most women, on the other hand, seem to enjoy shopping and attack their shopping problems with an enthusiastic thoroughness that often leads them to compare numerous articles both within and between stores before they make a purchase. "They are more observant, more susceptible to display, and hence indulge more generally in impulse buying."[27]

Differences in buying habits, tastes, and levels of consumer incomes often have important effects on retail locations, the types and volume of goods sold, and the manner in which goods are displayed. Stores in low-income neighborhoods seldom stock luxury items. Some commercial establishments cater to men and thrive because of the convenience of their locations. Ladies' apparel shops, on the other hand, usually find it profitable to prepare attractive eye-catching displays, give their customers plentiful opportunities to compare products, and facilitate milady's comparison process by locating near clusters of similar shops that deal in comparable and complementary products.

Another important consideration involves the type of goods or services supplied and the relative frequency with which they are purchased. Frequently purchased products may be described as convenience or shopper goods. With relatively inexpensive convenience goods such as chewing gum or newspapers customers tend to patronize the closest and most conveniently located vendor. Food stores depend on a wider range of patronage. Their storage and display space requirements, need for parking space, and the frequency with which most buyers purchase grocery products favor location at points convenient to customers. Accordingly, while they are often found in the central shopping districts of small cities, they usually locate at strategic outlying locations and in the neighborhood shopping centers of larger cities.

In contrast to convenience goods, articles such as pianos, TV sets, automobiles, and diamond rings represent sizable purchases that can easily be postponed until the buyer has made comparisons in other stores and has "thought it over." Because of the bulk of their products and the tendency of their buyers to shop around for their types of goods, automobile showrooms, furniture stores, and other comparable establishments usually locate outside the central retail districts. Stores that feature specialty products such as china or jewelry, on the other hand, usually find it to their advantage to locate in central shopping districts where they can use glittering window displays to attract additional customers.

As this discussion suggests, stores that feature convenience and shopper goods ordinarily find it advantageous to locate along principal paths of pedestrian traffic. Similar locations can be advantageous for service establishments such as barber, beauty, and shoe shops. Locations near the 100 percent spots are less important in the sale of postponable, bulky specialty products such as automobiles, furniture, or

---

[26] Ratcliff, *Urban Land Economics* (New York: McGraw-Hill, 1949), p. 378.
[27] *Ibid.*

home appliances. Dealers in these goods ordinarily locate outside the high-rent districts and often substitute larger advertising expenditures for the sums they could have paid as higher rents. Electricians, plumbers, and other service workers who depend on telephone contacts find it just as well to locate outside the central business district.

Rise of shopping centers. Almost every city has neighborhood shops that exist and sometimes thrive because of their ability to fulfill the convenience needs of nearby residents. As with the hexagonal concept of urban locations, these shops provide a low hierarchical level of services while the customers served look to higher hierarchical levels (central business districts and shopping centers) for the filling of more specialized needs.

Prior to the 1960s most cities offered two general levels of commercial services— neighborhood stores and a central business district. Public transportation patterns typically required the movement of large numbers of people to and through central business districts as they went to work or visited friends or businesses in other parts of the city. This situation, with all major arteries of travel radiating out from the CBD like spokes from the hub of a wheel, supported the downtown area's claim as the commercial and business center of the city. This picture has changed. Improved individual mobility and the development of travel routes that bypass the CBD have made it possible for more and more urban residents to live and work in cities and still experience little need to go downtown. This change, in combination with rapid urban growth, outward expansion of cities and their suburbs, and the acceptance of standard brands, has favored the emergence of shopping centers as intermediary hierarchical commercial service centers.[28] These centers customarily provide a clustering of retail shops that specialize in the provision of convenience and shopper goods and services. Ordinarily, they are located at sites convenient to large numbers of shoppers and offer abundant parking facilities.

Several factors including ease of accessibility, the attraction of new facilities, wide variety of convenience goods offered, and special services such as free parking and opportunities to shop under one roof in air-conditioned comfort have contributed to the prosperity of shopping centers. These advantages have made it possible for these centers to siphon off much of the trade advantage once enjoyed by CBD establishments and by neighborhood shops. Not all shopping centers, however, are a commercial success. Some are poorly designed, lack a desirable mix of shops, have inadequate parking facilities, or soon lose their luster of newness. Some also are overbuilt or suffer from competition with other centers better located to serve the same market area.

From the standpoint of overall successful location, shopping centers should be located at strategic sites that enable them to handle the convenience and shopper

---

[28] Cf. Paul E. Smith, *Shopping Centers: Planning and Management* (New York: National Retail Dry Goods Association, 1956). Nelson, *op. cit.*, pp. 26–34, suggests a concept of interceptor rings, which places shopping centers at predictable distances from the downtown center along the major streets that radiate from the urban core.

goods needs of large numbers of potential customers. An idealized location model for commercial establishments in the typical American metropolitan region calls for three (and sometimes four) hierarchical levels of service centers. Neighborhood shops, where they exist, provide the lowest level of service. Shopping centers provide the next level of service and should be located to service the shopping needs of several contiguous neighborhoods. Two levels of shopping centers can be envisaged in some areas, with small centers serving several neighborhoods and larger regional shopping centers often duplicating these services but offering additional attractions for larger areas. Central business districts should provide the highest level of services. In so doing, they duplicate the lower orders of services for nearby residents and for those who choose to use their facilities while at the same time offering many specialty goods and services not provided at the shopping centers. As is the case with the idealized location of villages and cities, this pattern of hierarchical levels of commercial establishments is more indicative of what ought to be than of what is. The particular talents, interests, and clientele of operators at even the neighborhood store level can cause them to provide specialized services normally available only at higher hierarchical levels. Similarly, in a number of cities where the CBDs have suffered from declining attractiveness and losses of customer patronage, department stores and other commercial establishments have closed their downtown headquarters and moved their operations to the more profitable and more popular regional shopping center sites. Branches of large department stores found in regional shopping centers frequently rival or even surpass their downtown headquarter stores in attractiveness.

In addition to selecting a site of adequate size, which is readily accessible to large numbers of potential customers, successful operation of a shopping center calls for careful selection of the shops and services found at the center. The final mix of shops and services should provide for a wide gamut of shopper needs and at the same time have definite drawing power. Success almost invariably calls for the presence of one or more *generative* businesses such as branches of department stores or large food markets, which use active advertising programs to attract customers to their sites of business.[29] Several *sharing* businesses (e.g., clothing, shoe, book, bakery, and liquor stores), which are usually smaller and do less advertising than the self-generators but which benefit from location near them, should be included in the mix. Some *suscipient* establishments such as barber, magazine, and tobacco shops, which benefit from locations near concentrations of people, may also be included.

Location of Residential Developments

Location decisions affecting where one lives can be influenced by economic considerations in much the same way as decisions involving commercial and industrial establishments. Typical householders want convenience as well as space and a pleas-

---

[29] Cf. Nelson, *op. cit.*, pp. 52-53.

ing environment. They prefer living near their places of employment, near market and service centers, and at the same time minimize their housing costs. Although the problem of residential location is similar in many ways to that of commercial and industrial locations, it is also different. Consumer satisfactions and personal preferences play a bigger role and economic considerations a lesser role with residential location decisions. The commercial operator or industrialist who fails to heed the economic implications of location decisions courts business failure. Householders who choose inconveniently located or overly expensive residential properties, in contrast, can often justify their choices on noneconomic grounds even though they may have to subsidize them with considerable outlays of time and money.

What people want and what they get in residential locations is a product of time and circumstances. Farm families in the United States tend to live on the farm tracts they operate. In contrast, most urban workers live at sites separated from their place of work. Before the advent of modern transportation facilities, these sites were almost always found near the worker's place of employment. Today they can be located anywhere within a large commuting zone depending on the worker's personal choices and willingness and ability to pay the costs associated with his or her location decision. Convenience of access is still an important factor for many people. The relative emphasis given to it, however, involves other considerations, several of which are associated with the workings of the succession process.

Succession in residential developments. Single-family homes often line the streets of towns and small cities to sites within a block or two of the 100 percent spot. The more pretentious and higher-cost homes are usually located on one or two streets near the center of town. Lower-cost housing is ordinarily found at less favored locations, often on back streets, near local industries, and sometimes across the railroad tracks.

With urban growth, this pattern gradually changes. Expansion of the commercial district calls for the acquisition and redevelopment of adjacent residential sites and automatically brings nearby sites into a transitional-use zone. New higher-cost housing is built in open areas, usually as an extension of the existing higher-cost housing district. New medium- and low-cost housing is also built around the fringe of the city. Meanwhile, occupants of the higher-cost older houses consider the prospect of modernizing and remodeling their houses but frequently decide instead to sell and move to more prestigious new locations. In selling their houses, they start a "filtering-down" process in which the higher-cost and medium-cost houses of yesteryear gradually decline in relative market value, are occupied by a succession of lower-income owners and tenants, and eventually are converted into apartment or rooming houses or are acquired for other purposes.

With urban growth, land areas farther and farther from downtown centers are developed for residential purposes, and properties located nearer downtown are allowed to deteriorate both in function and in appearance. These properties could be refurbished or replaced with new housing developments, but their basic site values for possible commercial use often appear too high to justify new long-term commitments to residential developments. Meanwhile, new housing is constructed along the principal streets and highways that radiate from the city and in the back-street areas that lie between these streets. High-cost, medium-cost, and low-cost residential units are normally constructed in neighborhoods with properties of similar values. Frequently new housing in each of these value classes is built farther out in a continuation of the same sectors in which similar developments already exist. Hoyt has observed that when no restraints are present, high-cost residential neighborhoods "do not skip about at random" but rather "follow a definite path in one or more sectors of the city." They gravitate to sites easily accessible to fast transportation lines, "grow toward the homes of the leaders of the community," "progress toward high ground which is free from floods," "spread along lake, bay, river and ocean fronts, where such waterfronts are not used for industry," and "grow towards the section of the city which has free, open country beyond the edges."[30]

Higher-cost residential neighborhoods frequently have first choice of the available sites for new residential developments. It must be noted, however, that real estate developers and promoters play strategic roles in determining the value-class levels of new sites. They may not be able to make poor sites attractive for higher-cost developments, but through their decisions concerning lot sizes and prices, building restrictions, preservation and enhancement of amenity features, and construction of model houses, they can bend the direction of residential growth and make sites attractive for their intended markets.

Exceptions to the outward pattern of higher-cost residential growth occur when factors such as sentiment, the presence of particular amenities, or private or public redevelopment programs are involved. Deluxe high-rent, high-rise residential properties are sometimes built in redeveloped, highly accessible, once downtrodden areas. The location of these developments at sites adjacent to or near low-income housing suggests some income-level integration in housing. This integration is largely illusory, however, because the residents of the near high-rise apartments tend to live apart from their lower-income neighbors. Their numbers are made up mostly of professional people who have decided to trade the privacy and open space offered by suburban living for the convenience and easy accessibility of downtown locations.

Explanations of residential locations. Personal and family choices play a major role in residential site location decisions. A husband may want to live close to work. The wife wants to be reasonably close to commercial and social service centers. Both want to live in a "good" and prestigious neighborhood and, yet, secure housing at what they consider a reasonable price. They want indoor living space plus outdoor play areas for their children and easy access to schools and parks. They may want to live near particular friends or relatives or be near public transportation facilities. They may desire a spot away from the city, close to nature, and perhaps where they can build their dream house. These objectives are seldom filled by any

---

[30]Hoyt, *op. cit.*, pp. 114–19.

one location. Some goals must be emphasized at the expense of others in the decision-making process, and final decisions are often hard to explain.

In an early discussion of the economics of residential site locations, Robert Murray Haig observed:

> In choosing a residence purely as a consumption proposition, one buys accessibility precisely as one buys clothes or food. He considers how much he wants the contacts furnished by the central location, weighing the "costs of friction" involved—the various possible combinations of site rent, time value, and transportation costs; he compares this want with his other desires and his resources, and he fits it into his scale of consumption, and buys.[31]

Haig emphasized two factors—the effect of nearness to central locations on site rents and personal choice considerations.

If one assumes a single central market and work area, standardized units of residential space, and exclusive concern with Haig's "costs of friction," it would make little difference where in a city a person lived. At all locations, one's "costs of friction"—the sum of one's site rent and costs of transportation—would be the same. These assumptions do not hold true in the real world. Residents are oriented to many different employment, shopping, social, and other service nodes; some residential sites are larger, more prestigious, of higher quality, and involve access to more amenities than others; and individual assumptions concerning transportation costs vary with choices in transportation methods and time-cost considerations.

Haig's prospective resident obviously exercises personal choices in the selection of a residential site. Lack of income could cause one to sacrifice space, privacy, and housing services by locating in a congested, low-rent ghetto area. With more income or stronger desire for good housing, one might logically seek higher quality housing at a downtown location or pay the additional transportation costs associated with the more spacious and better quality housing that is available at outlying locations.

Decisions concerning where individuals and families will live involve a weighing of spatial and environmental, rental and land cost, and accessibility and travel cost considerations. People seek as much enjoyable living space with the safest, most congenial, and most pleasing environment they can get per dollar of housing expenditure. They need to provide for their basic housing needs at a price they can afford and must accept the fact that this constraint may force them to accept housing conditions and facilities they otherwise would reject. They also have a natural interest in minimizing the time, bother, and monetary cost of travel to work, markets, and other activities. This combination of interests results in various weights because the costs people associate with commuting vary over a wide range and because some have high preferences for living within walking distance of work and shopping facilities, whereas others may enjoy or make constructive use of their travel time.

Housing expenditures represent a consumptive expenditure for most families, and family goals are often weighted heavier in residential location decisions than are

[31] Haig, *loc. cit.*, p. 423.

more strictly economic considerations. The family's choice of the site where it will live involves its self-image and the home and the residential neighborhood environment it desires. High-income families tend to place more emphasis and value on site amenities than lower-income families. They tend to locate in neighborhoods where they can enjoy above-average homes, more open space, and more privacy, because they can afford to do so. Many medium- and lower-income families also aspire to these housing amenities and move to suburban and rural surroundings where they can enjoy these advantages, even though their "costs of friction" require that they support their housing expenditures with family income that could well be used for other purposes.

The typical high-income family that selects a suburban residential location may only give secondary emphasis to the fact that the husband must travel a considerable distance to his office four or five times a week. A house in the suburbs may represent an inseparable part of their picture of the successful executive. The entire family may associate a wide spectrum of values with their opportunity to live in a suburban environment. The wife's activities, if she does not work outside the home, may call for frequent visits to stores, friends, and social functions in the area. The children will be close to "their" school and "their" friends. The husband may see professional, social, and recreational advantages in living near the country club or near business associates. Similarities of interests as well as incomes also tie large numbers of middle- and low-income families to neighborhoods composed mostly of individuals and families with comparable incomes.

— SELECTED READINGS

Alonso, **William**, *Location and Land Use* (Cambridge, Mass.: Harvard University Press, 1968).

DeSousa, **Anthony R.**, and **J. Brady Foust**, *World Space-Economy* (Columbus, Ohio: Chas. E. Merrill, 1979), chaps. 5–9.

Hall, **Peter**, ed., *Von Thunen's Isolated State* (Elmsford, N.Y.: Pergamon Press, 1966).

Hoover, **Edgard M.**, *The Location of Economic Activity* (New York: McGraw-Hill, 1948).

____, *An Introduction to Regional Economics*, 2nd ed. (New York: Knopf, 1971).

Hoyt, **Homer**, *The Structure and Growth of Residential Neighborhoods in American Cities* (Washington, D.C.: Government Printing Office, 1939).

Isard, **Walter**, *Introduction to Regional Science* (Englewood Cliffs, N.J.: Prentice-Hall, 1975).

____, *Location and Space-Economy* (New York: John Wiley, 1956).

Nelson, **Richard L.**, *The Selection of Retail Locations* (New York: F. W. Dodge, 1958).

Nourse, Hugh O., *Regional Economics* (New York: McGraw-Hill, 1968).

Ratcliff, Richard U., *Urban Land Economics* (New York: McGraw-Hill, 1949), chaps. 2 and 13.

Richardson, Harry W., *Regional Economics* (Urbana: University of Illinois Press, 1979).

10

LAND RESOURCE VALUES AND THE REAL ESTATE MARKET

Land resources are frequently viewed primarily as economic commodities that can be bought and sold. Their possession involves rights people want and for which they often are willing to pay substantial sums of money. The sale, mortgaging, and taxation of these resources require special value appraisal techniques. The fact that they are wanted and have economic values gives them an attractive investment potential, and their frequent purchase and sale have brought the rise of real estate markets.

THE NATURE OF PROPERTY VALUE

Justice Brandeis once observed that "value is a word of many meanings."[1] The truth of this dictum is demonstrated by the dozens of different meanings we apply to this term in popular usage. In a broad sense, "value implies capacity to satisfy wants", and "there are as many kinds of value as there are classes of wants."[2] Thus we may deal with aesthetic values, political values, psychic values, social values, spirtual values, and the like. Economists and appraisers are primarily concerned with economic and market values, but they too use the term *value* in many different contexts and with different adjectives to mean different things.[3]

[1] *Southernwestern Bell Telephone Company* v. *Public Service Commission,* 262 U.S. 276, 310 (1923).

[2] Edwin R. A. Seligman, *Principles of Economics* (New York: Longmans, Green, 1905), p. 174.

[3] McMichael lists some seventy-three types of value that appraisers deal with in their work. Cf. Stanley L. McMichael, *How to Operate a Real Estate Business*, rev. ed. (Englewood Cliffs, N.J.: Prentice-Hall, 1967), p. 132.

Some of the more important concepts of economic value as it applies to real estate may be illustrated by the example of a building site that is acquired for $150,000 and on which $600,000 is spent to construct an office building. At this point, $750,000 is invested in the property, a sum that may be taken as a measure of its investment cost. When appraised for a mortgage loan, it is found to have a total loan value of $650,000. The tax assessor assesses it for property taxation purposes at $350,000. The owner decides to sell the property, and after discussing the matter with a real estate broker, decides to list it for $800,000. But before it is actually listed, notice is received that the site is needed for a public project and that it has a condemnation value of $600,000.

Each of these five figures represents a measure of economic value, and each has its justification. Taken together, they indicate the aura of confusion often associated with the term *value* and the need for a more explicit understanding of what is meant by the "economic value of property."

Economic value has three important components.[4] The property in question must have use-value or *utility* to its owner or user. Otherwise, no one would want it. Coupled with the idea of utility are the assumptions that properties such as land resources promise expected future flows of returns and satisfactions and that effective demand exists for these flows of products or services. A second necessary component is *scarcity* in supply. Regardless of its utility, a product must be scarce if it is to command a price. Otherwise, it would be a free good. To have economic value, an object must also be *appropriable*. It must be something that can be possessed and be transferable from one owner to another.

Most economists identify the economic value of property with the market price of property objects. In this sense, economic value depends on the interaction of the forces of supply and demand. It represents the worth of given properties in given markets at a given time and place. Far from being something that is fixed for each type of property, economic value is a subjective concept that is dependent on the desires of people to possess and use property objects and on their willingness and ability to offer money or other considerations in exchange for the privilege of ownership or possession.

Various writers in the past have confused the meaning of economic value by identifying it with other concepts such as utility, production cost, and "fair price."[5] Care should be taken to distinguish between economic value and each of these concepts. It may be noted with the concept of utility, for example, that property objects must have some type of usefulness if they are to have economic value. But mere possession of utility does not give an object economic value. As Adam Smith pointed out in his classic distinction between value and utility:

> The things which have the greatest value in use have frequently little or no value in exchange; and on the contrary, those which have the greatest value in

[4]Cf. Alfred A. Ring, *The Valuation of Real Estate*, 2nd ed. (Englewood Cliffs, N.J.: Prentice-Hall, 1970), p. 10.

[5]Cf. James C. Bonbright, *The Valuation of Property* (New York: McGraw-Hill, 1937), chap. 2; and Edmund Whittaker, *A History of Economic Ideas* (New York: Longman, 1940), chap. 9.

exchange have frequently little or no value in use. Nothing is more useful than water: but it will purchase scarce anything; scarce anything can be had in exchange for it. A diamond, on the contrary, has scarce any value in use; but a very great quantity of other goods may frequently be had in exchange for it.[6]

Smith and several other early economists associated economic value with the cost of producing economic objects. This view has some validity, as the market price of a product must equal or exceed its production cost over time if operators are to have an incentive to continue production. During the typical short-run period, however, supply and demand conditions often cause the market prices of products to rise above or drop below their production cost. Land resource developments provide frequent examples of this distinction between economic values and costs. Desirable building lots may command market prices far in excess of their development costs. A ten-story hotel built on a site where little demand exists for its services, on the other hand, would have a market value far below its production costs.

The association of economic value with the concept of fair price can be traced back to the writings of the Greek philosophers and several medieval clergy. An assumption that economic value corresponds with "some notion of a fair or ethical price higher or lower than the price at which the commodity or service in question is being sold or can be sold" is also used by the courts at times in their determination of reasonable values.[7] Under active market conditions, this fair-price concept of value can be rejected as unrealistic—as a measure of someone's idea of what ought to be rather than the actual exchange prices found in the market. It may have considerable judicial or political significance, however, when there are no established guidelines for determining market values, value figures are not available for comparable properties, a court must determine reasonable values, or public action is needed to fix prices.

Appraisers often distingiush between the concepts of economic value and market price.[8] They see determination of market value as the end product of the appraisal process. Yet they speak of "justified" or "warranted" price as the price at which a property should sell as compared with possible sales at higher or lower price levels. The American Institute of Real Estate Appraisers has defined its concept of market value as follows:

Market Value—The most probable price in terms of money which a property should bring in competitive and open market under all conditions requisite

[6]Adam Smith, *The Wealth of Nations* (London, 1776: Modern Library edition; New York: Random House, 1937), p. 28.

[7]Bonbright, *op. cit.*, p. 23.

[8]Cf. Frederick M. Babcock, *The Valuation of Real Estate* (New York: McGraw-Hill, 1932), pp. 12–16; Earl F. Crouse and Charles H. Everett, *Farm Appraisals* (Englewood Cliffs, N.J.: Prentice-Hall, 1956), pp. 19–20; and Paul F. Wendt, *Real Estate Appraisal* (New York: Henry Holt, 1956), pp. 4–11.

to a fair sale, the buyer and seller, each acting prudently, knowledgeably and assuming the price is not affected by undue stimulus.[9]

This definition assumes well-informed buyers and sellers who are motivated to buy and sell, a reasonable sales period, sales for cash or financing arrangements generally available to other buyers, and market prices not affected by special considerations or arrangements.

Market values seldom remain constant for long periods. They fluctuate with changing supply and demand conditions and, more important, with shifts in business and group psychology. Theoretically, buyers should calculate the worth of objects in terms of their expected flows of utilities and satisfactions. In practice, most of us judge the future in terms of our knowledge of the present and recent past. As Lord Keynes has observed:

1. We assume that the present is a much more serviceable guide to the future than a candid examination of past experience would show it to have been hitherto. In other words we largely ignore the prospect of future changes about the actual character of which we know nothing.

2. We assume that the *existing* state of opinion as expressed in prices and the character of existing output is based on a *correct* summing up of future prospects, so that we can accept it as such unless and until something new and relevant comes into the picture.

3. Knowing that our own individual judgment is worthless, we endeavor to fall back on the judgment of the rest of the world which is perhaps better informed. That is, we endeavor to conform with the behavior of the majority or the average. The psychology of a society of individuals each of whom is endeavoring to copy the others leads to what we may strictly term a *conventional* judgment.[10]

Our acceptance of conventional judgment causes us to adjust our attitudes about the economic value of property to the shifting tides of the business cycle. During periods of business prosperity, we are ordinarily optimistic about the future and frequently add fire to the business boom by expanding our activities and bidding property values up to new heights. Let something happen to shake this mass faith in the soundness of the economy, and we make an abrupt about-face. "The practice of calmness and immobility, of certainty and security, suddenly breaks down. New fears and hopes will, without warning, take charge of human conduct. The forces of disillusion may suddenly impose a new conventional basis for valuation."[11] Panic and economic fear become the order of the day with some operators. Others are

---

[9] Bryl N. Boyce, ed., *Real Estate Appraisal Terminology* (Cambridge, Mass.: Ballinger Publishing Company, 1981), p. 160. Copyright 1975, 1981 by The American Institute of Real Estate Appraisers and The Society of Real Estate Appraisers. Reprinted with permission from Ballinger Publishing Company.

[10] John M. Keynes, "The General Theory of Employment," *Quarterly Journal of Economics*, 51 (February 1937), 214.

[11] *Ibid.*, p. 215.

slower to change. But they too are frequently engulfed in a tide of conventional pessimism, which leads to sometimes short, sometimes extended, periods of depressed business conditions and reduced property values.

VALUATION AND APPRAISAL OF REAL ESTATE

Property owners, prospective buyers, real estate credit agencies, tax assessors, and others have frequent need for value figures indicating the worth of individual properties for various purposes. The determination of these value figures is usually accomplished through a property appraisal process. Thousands of these appraisals are made each day. Many are made by owners who seek the highest sale price at which their properties can be expected to clear the market and by individual buyers who try to determine the maximum prices they should consider paying for properties. Many others are made by loan appraisers, tax assessors, and professional appraisers who make a regular business of appraising properties.

Conscious efforts have been made in recent decades to improve and formalize the techniques used in appraisal work. As a result, most professional appraisers proceed on a more scientific basis than in times past. But property appraisal is still more an art than a science. Appraisers start with a given property and try to determine a value figure that reflects "the current attitude of typically informed users and investors as to the probable future utility of that property."[12] They try to discover the market price at which properties would sell in a willing buyer–willing seller market.

The leading principles and techniques of sound appraisal can often be learned from textbooks. Success in appraisal work, however, calls for more than booklearning. Appraisers' success is highly dependent on their ability to assemble pertinent information concerning individual properties and to provide realistic interpretations of these data in light of present market conditions and probable future productivity trends. According to the American Institute of Real Estate Appraisers (AIREA), "The appraiser is not a fortune teller and . . . does not attempt to foretell the future." Yet he or she must be a "keen observer of known facts and of the current attitude of informed persons toward future probabilities as reflected by current market action," should be thoroughly familiar with current local market conditions, and display both skill and judgment in assembling and interpreting the value data for individual properties.[13]

Several steps are involved in the property appraisal process.[14] The appraiser first

[12] American Institute of Real Estate Appraisers, *The Appraisal of Real Estate*, 2nd ed. (Chicago: AIREA, 1951), p. 28.

[13] *Ibid.*

[14] For more detailed discussions of the steps involved in the appraisal process, cf. American Institute of Real Estate Appraisers, *The Appraisal of Real Estate*, 7th ed. (Chicago: AIREA, 1978), chap. 4; Sanders A. Kahn, Frederick E. Case, and Alfred Schimmel, *Real Estate Appraisal and Investment* (New York: Ronald Press, 1963), pp. 14–19; and Arthur A. May, *The Valuation of Residential Real Estate*, 2nd ed. (Englewood Cliffs, N.J.: Prentice-Hall, 1953), chaps. 3 and 19.

determines the precise location and legal description of the property; the nature of the property rights (ownership, lease, mineral, and other rights) being appraised; the purpose of the appraisal; the type of value (sales, loan, insurable, or other value) desired; and the specific date for which the appraisal applies.

After determining the nature and purpose of the appraisal, the appraiser examines the property and its surroundings and assembles relevant information concerning the various physical, economic, and institutional factors that affect its value. These data are interpreted, and one or more of the standard valuation approaches are used in computing the value of the property. Any differences that may result from use of alternative valuation techniques must then be reconciled, and an appraisal report is submitted that indicates the appraiser's opinion of the value of the property.

Three principal methods of determining real property values are now in use in the United States: (1) the market-comparison or market approach, (2) the net income-capitalization or income approach, and (3) the replacement-cost or cost approach.

Market-Comparison Approach

The market-comparison approach (also described as the sales-comparison or market approach) provides a basic and highly realistic method for appraising the market value of land resources. With this approach, an appraiser determines the expected price a property should command in the current market by comparing its value characteristics and sales circumstances with those of similar properties that have been sold in the recent past. This approach has particular merit when one is seeking the current market value of a property, because it relates appraised values to current supply and demand conditions.[15] It recognizes that market prices frequently rise above or drop below the averages suggested by longer-run trends and that market price levels may vary from one community to the next.

The market approach finds its rationale in the economic principle of substitution. Informed buyers will not pay more for given properties than it costs them to buy comparable substitute properties. With this reasoning, appraisers can look to the current real estate market for indications of the actual going market values of the properties they appraise. In their analyses of comparable sales, however, appraisers often encounter wide variations in market circumstances. They find that buyers and sellers frequently operate with incomplete knowledge of their market opportunities, that properties have a wide variety of characteristics, and that some properties sell for more than they probably should while others sell for less. This lack of standardization among properties and fluctuations in actual market prices for comparable properties complicate the appraisal process. But as long as one deals with cases of comparable properties that have sold recently at prices agreeable to willing

[15]Cf. Wendt, *op. cit.*, pp. 253 and 255–260.

buyers and willing sellers, market comparisons indicate the range of prices accepted by typical investors under actual market conditions.

A major problem with the market-comparison approach stems from the appraiser's need for bona fide sales data for comparable properties sold under comparable market conditions. Sales data for comparable properties may be scarce or nonexistent when local real estate markets are relatively inactive, the local market is small, or the property being appraised is of a type that is seldom sold. Comparability questions also arise because of differences in property characteristics and sales conditions.

No appraiser expects to find properties that are exactly alike. Even when two houses have identical floor plans or when two farms are of the same size and general layout, they are at different locations and often in different neighborhoods. The most appraisers can hope for is general comparability. In their search for comparable examples, they usually find it necessary to study the characteristics of numerous properties listed or sold in the market along with the conditions and circumstances of their sale.

Market comparability requires that careful consideration be given to the circumstances, conditions, location, and time of the benchmark property sales.[16] Information is needed on current and recent sales in the local market. Changing market conditions can easily strip sales made even a few months earlier of their relevance for market-comparison purposes. Care must also be taken to check the circumstances of the benchmark sales to make certain that they involved bona fide willing buyer–willing seller transactions and that they did not involve unduly long sales periods, heavy advertising costs, or special sales-financing arrangements.

Once they have assembled the relevant sales data on comparable properties, appraisers look for similarities and dissimilarities among properties. They note the plus and minus features associated with the properties they are appraising. They consider them from the standpoint of market trends and their knowledge of buyer preferences and desires regarding architectural styling, floor plans, building materials, site locations, neighborhood amenities, and other features. With these facts in mind, they estimate the probable market values of the properties at the date of appraisal.

The market comparison approach works well when one seeks an appraisal of the current market value of properties such as urban and suburban houses and apartments or farms that are comparable to other properties that are sold in sufficient numbers to provide a suitable basis for current market comparisons. It is less applicable with commercial, industrial and other properties that are seldom sold. Problems also arise with its applications to residences and farms when the properties sold are

---

[16]May, *op. cit.*, p. 166. Also cf. Wendt, *op. cit.*, pp. 267–90; AIREA, *op. cit.*, 7th ed., chap. 15; and William G. Murray, *Farm Appraisal and Valuation*, 5th ed. (Ames: Iowa State University Press, 1969), chaps. 5–10; and Robert G. Parvin, "Market Approach to Value," in Edith J. Friedman, *Encyclopedia of Real Estate Appraising*, 3rd ed. (Englewood Cliffs, N.J.: Prentice-Hall, 1978), pp. 21–35.

dissimilar or involve small samples of sales. The approach is poorly suited for determinations of the long-term justified values of properties.

Income-Capitalization Approach

Theoretically, the market value of a property should equal the present worth of all its future incomes. It should equal the discounted present value of the expected future flow of its land rents. The logic of this reasoning has prompted widespread use of an income-capitalization approach in property appraisal work. As indicated in Chapter 6, this valuation approach involves the simple formula $V = a/r$, in which V represents the value of the property, a represents the estimated average annual land rent or net return to land expected in the future, and r represents the rate of interest used in the capitalization process. With this formula, a property with an expected average annual net return of $1,000 is worth $20,000 when this income is capitalized at 5 percent ($1,000 ÷ .05 = $20,000).

This formula is designed for use with productive properties that are expected to yield an even flow of land rents year after year into the future. Where this situation exists, appraisers are concerned with only two variables. They must determine the average annual land rent attributable to the property, and then choose an appropriate capitalization rate. Complications arise when the property being appraised will not provide an even flow of land rent in the future. Typical examples of this situation occur with the appraisal of apartment buildings, mines, and other properties with limited economic lives. With these properties the standard capitilization formula may be used to compute the market value of the site, but adjustments are needed to correct the total value estimate for the limited duration of the returns associated with the building improvements.[17]

Modifications are needed when land rents are expected to increase or decrease in the future and when they are expected to continue for only a limited number of years. With the first of these situations, the appraiser may either adjust the estimate of average annual land rent to take the expected changes into account or use a modified capitalization formula:[18]

$$V = \frac{a}{r} \pm \frac{i}{r^2}$$

In this modified formula, a represents the average land rent received by the prop-

[17]Cf. AIREA, *op. cit.*, 7th ed., chaps. 20–22. Appraisers use two general approaches in dealing with this problem. Some appraisers use compound interest tables (either the Hoskold annuity or Inwood tables) to determine the present discounted value of the expected returns attributable to building improvements. Others use the present net return attributable to the building as their *a* factor in the capitalization formula and add a straight-line or a sinking-fund depreciation correction factor to their capitalization rate. This correction factor results in a market-value estimate for the building that equals the present discounted value of its expected flow of net returns less a depreciation allowance sufficient to cover its investment cost.

[18]Cf. Clyde R. Chambers, *Relation of Land Income to Land Value*, U.S. Department of Agriculture Bulletin No. 1224, 1924, pp. 28–29. When this increase or decrease is expected to

erty, and i represents the average increment of increased or decreased return expected to result from more intensive use of sites now utilized at less than their highest and best use, from a proposed land improvement, from continued soil erosion, or from some other factor. With the second situation, as when one appraises a mine that is being liquidated at a rate that will reduce its value to zero in x years, the following formula[19] should be used:

$$V = \frac{a}{r} \left[1 - \frac{1}{(1 + r)^x} \right]$$

Additional adjustments are sometimes needed when properties have characteristics such as a prestige address or a desirable view that provide greater buyer satisfactions and that are not adequately measured by the monetary values of the a factor. Final determination of value in these instances may call for the addition of allowances for locational advantages, amenity factors, and other items that add to property value or possible subtraction of varying sums for detriments that may reduce the value of the property.

Estimating future land rents. Computation of the a or average annual land rent factor poses problems with use of the income-capitalization approach. In theory, appraisers should try to determine the average annual flow of land rent that a property can be reasonably expected to earn in the future. Most appraisers give serious attention to this problem. In practice, however, their estimates of future net returns are usually weighted quite heavily by knowledge of the returns received in the recent past. It is easy to estimate an appropriate level of future land rents when a property is leased on a long-term basis at a mutually satisfactory rental rate to a tenant who is using it for its current highest and best use. The appraiser in these instances can use the landlord's current net rent (contract rent less outlays for taxes, insurance, and operating costs) as a measure of expected future land rents. Situations of this order are more the exception than the rule. Appraisers usually find that they must compute land rental rates for properties that are owner-operated or that are leased at rates that may be out of line with their present or future income-producing potential.

---

continue for a specific number of years and then be followed by a period of relatively constant returns, the appraiser should use the formula

$$V = \frac{a}{r} \pm \frac{i}{r^2} \left[1 - \frac{1}{(1 + r)^n} \right]$$

For variations of this formula, cf. James O. Wise, "Modifying the Income Approach to Farm Appraisal," *Appraisal Journal*, 40 (October 1977), 505–10; and Robert G. Healy and James L. Short, *The Market for Rural Land: Trends, Issues, Policies* (Washington, D.C.: The Conservation Foundation, 1981), pp. 112–13.

[19]With appraisals of mineral deposits, appraisers usually employ a sinking fund approach which uses Hoskold coefficients along with information on expected annual productivity to determine the annual payments to a sinking fund that are needed to recoup the value of the mining investment by the end of its assumed working life. Cf. AIREA, 7th ed., *op. cit.*, pp. 416–23; and Friedman, *op. cit.*, pp. 617–30.

In determining the land rents associated with urban properties, appraisers usually give first attention to the character of the neighborhood, the economic base of the city, and various trends evident in the national economy. They then use operating income, expense statements, and other available information to determine expected average gross income, and to calculate expected operating costs. These costs may then be subtracted from the expected gross income to provide the estimated land rent of the property before depreciation.[20]

Farm appraisers usually start by considering the physical resource base of the farm, the productivity of the soil, and the average crop yields for the preceding five- to ten-year period. These data are used along with information concerning the typical or most likely cropping system to provide a picture of the overall productivity of the farm. Average farm-product price levels are determined and used to compute the average expected gross income. The appraiser then subtracts the estimated operating expenses (including an allowance for the operator's labor and management) from the estimated gross income to determine the land rent attributable to land and buildings.[21]

Fairly well defined procedures are now used by both urban and rural appraisers in estimating the future income flows of properties. Skill and judgment are needed if the appraiser is to arrive at an approximately correct answer. Even under ideal conditions, the process of estimating future production, business conditions, price levels, and net returns is fraught with hazards. Small errors can have magnified effects on final value determinations. Minor errors in the assumptions regarding future contract rental rates, apartment vacancy rates, or the effective economic life of a building can inflate or deflate the values of urban properties. Similarly with farm appraisals, a mistake of two bushels per acre in an appraiser's crop-yield calculations or 5 cents a bushel in long-run price estimates may make a $40 to $60 difference in the acreage value of farm land.

Choice of a capitalization rate. The capitalization rate is the second important variable in the capitalization formula. Appraisers must use discretion and judgment in their determinations of the appropriate capitalization rates that should apply to individual properties.[22] If all properties were held without income risk or uncer-

---

[20]Cf. AIREA, 7th ed. *op. cit.*, chaps. 17 and 18. Many urban appraisers split the net rental return between the site and the building improvements so that the value of these two items may be computed separately. Land and building residual techniques have been developed to guide appraisers in this income-splitting process. Cf. Wendt, *op cit.*, pp. 172–85. Urban appraisers are not unanimous in their acceptance of this practice, and most rural appraisers regard it as non-applicable to farm properties. Cf. Richard U. Ratcliff, "Net Income Can't Be Split," *Appraisal Journal*, 18 (April 1950), 168–72; Kahn, Case, and Schimmel, *op. cit.* pp. 151–53; Murray, *op. cit.*, chap. 15; and Crouse and Everett, *op. cit.*, p. 17.

[21]Cf. Murray, *op. cit.*, chaps. 11 and 12; and Crouse and Everett, *op. cit.*, chaps. 7 and 8.

[22]It should be noted that the capitalization rate is the reciprocal of the number of years of expected income it takes to equal the property's present value. A property with an annual land rent of $1,000 capitalized at 5 percent to give a $20,000 value has a reciprocal of 20 and may thus be described as a twenty years' purchase property. This years' purchase concept is widely used in Great Britain and some other European countries. Farmland values in England are often

tainty under the conditions of perfect competition, owners could capitalize their property incomes at a low safe rate of interest. But this situation does not exist in practice. Investors take risks when they invest in property; they often suffer from illiquidity when they tie their capital up in particular investments; and they frequently assume considerable burdens of management. Each of these factors justifies the use of capiatlization rates that are higher than the relatively safe, nonrisk rate used with government bonds in periods of relatively plentiful money supplies.

A simple approach has often applied in past determinations of the capitalization rates used with farms and other properties that have production potentials with infinite economic lives. Appraisals were frequently made by mortgage credit agents who accepted prevailing farm mortgage interest rates of around 5 or 6 percent as an appropriate capitalization rate.[23] Rural property appraisers still tend to apply uniform capitalization rates with different properties. This situation does not apply, however, with urban appraisals. Differences in the risk, capital liquidity, and burden of management associated with different types of properties and sometimes with properties of the same type have brought the use of a wide range of urban capitalization rates with urban properties. Four principal approaches are used in the determination of these rates: (1) the summation method, (2) the band-of-investment approach, (3) an equity-yield method, and (4) a comparative-market method.[24]

With the *summation method*, the appraiser constructs a capitalization rate by assuming a series of independent rates for the various components considered in the capitalization process. One may start with a safe, nonrisk rate of 5.5 percent, add 1.5 percent as a rate for the income risks associated with the property, add 1.5 percent as a penalty for nonliquidity of capital, and add 1.0 percent as an allowance for burden of management. Addition of these separate rates gives a total capitalization rate of 9.5 percent.

Different combinations can be used in this summation process. In one prominent example (Table 10-1), the Federal Housing Administration recommended in 1947 that appraisers rate properties according to five risk features: safety of principle, certainty of return, regularity of return, liquidity, and burden of management.[25]

---

expressed in terms of twenty, twenty-five, or thirty years' purchase of their annual rents. The value of residential and other urban properties ranges from as little as three years' purchase if the buildings are old, in a poor state of repair, or prevented by rent control regulations from charging more than a low nominal rent, to many times this number of years' purchase with more desirable properties. This approach to urban and rural land resource valuation is particularly applicable in areas where large numbers of properties are held under leaseholds.

[23] Some rural appraisal authorities have criticized this identity of farm capitalization rates with mortgage interest rates and argued "that farm ownership carries more risk than mortgage lending and therefore is entitled to a higher rate of return." Cf. Crouse and Everett, *op. cit.*, p. 35; also Roland R. Renne, *Land Economics*, 2nd ed. (New York: Harper & Row Publishers, Inc., 1958), p. 236.

[24] For a more detailed discussion of the last three of these approaches, cf. Robert G. Cox, "Income Approach to Value" in Friedman, *op. cit.*, pp. 39–61.

[25] Cf. Federal Housing Administration, *Underwriting Manual* (Washington, D.C.: Government Printing Office, 1947), sections 1239–41.

TABLE 10-1. Summation Method Used by the Federal Housing Administration in Determining Capitalization Rates for Residential Income Properties

RISK FEATURES	RATES FOR 5 GRADES OF PROPERTIES RANGING FROM POOR TO EXCELLENT RISKS					RATING FOR PROPERTY
	1	2	3	4	5	
Safety of principal	3.50	3.25	3.00	2.75	2.50	—
Certainty of return	2.00	1.75	1.50	1.25	1.00	—
Regularity of return	1.75	1.50	1.25	1.00	0.75	—
Liquidity	1.50	1.25	1.00	0.75	0.50	—
Burden of management	1.25	1.00	0.75	0.50	0.25	—
Total capitalization rate						—

Source: Federal Housing Administration, *Capitalization Rates and Rent Multipliers.*

With the *band-of-investment* theory approach, the capitalization rate is computed as the weighted average of the market interest rates that apply to various portions or "bands" on one's investment. As an example, one may assume a case in which 50 percent of the value of a property can be covered by a first mortgage bearing 9 percent interest and 25 percent by a second mortgage that pays 11 percent interest, while the final 25 percent involves the buyer's down payment or equity on which a 12 percent return is allowed. As the following tabulation shows, addition of the fractional rates secured by multiplying each interest rate by the percentage share it represents of the total investment gives a capitalization rate of 10.25 percent.

	PERCENT OF VALUE	INTEREST RATE	FRACTIONAL RATE
First mortgage	50	9	4.5
Second mortgage	25	11	2.75
Owner's equity	25	12	3.0
Total capitalization rate			10.25

The *equity-yield* or *Ellwood method* is a relatively new approach that has particular value for the appraisal of mortgage loans and that has gained acceptance as a means of determining capitalization rates for other types of property appraisal.[26] It involves the use of sophisticated techniques for determining the internal rate of return needed to attract capital to specific types of real estate investment. As Cox indicates, it calls for capitalization rates that reflect

the ratio of annual net income before debt service to value, including that portion of the total investment which must satisfy all debt service, provide the desired equity yield, and compensate for any changes in property value.[27]

[26]Cf. L. W. Ellwood, "Appraisal for Mortgage Loan Purposes" in Friedman, *op. cit.*, pp. 1095–1116.

[27]Cox, *loc. cit.*, p. 57.

Property depreciation factors are frequently added to the capitalization rates derived by use of the summation, band-of-investment, and equity-yield methods. The size of these correction factors depends on the method of their calculation and the estimated remaining economic life of the building improvements. They should be large enough to enable owners to recoup the investment cost of their improvements but not so large as to rob properties of their fair present value. With a typical example involving a building that has an estimated remaining economic life of twenty-five years, a 4 percent depreciation rate may be added to a 9 percent capitalization rate to provide an overall rate of 13 percent.

Most urban and rural appraisers use a *comparative-market method* in determining their capitalization rates. With this approach, they base their capitalization rates on the percentage relationship that exists between the annual net returns and going market values of comparable properties.[28] This approach is less precise than the others, but it may be noted that they all call for subjective choices. After his review of three of these approaches, Wendt concludes that none of them "provides a scientific and objective method for establishing the capitalization rate for a given property" and that the choice of a capitalization rate "is essentially one of subjective estimation."[29] The reliance of the comparative market method on the percentage relationship between estimated annual flows of land rent and the market values of comparative properties involves considerable dependence on the market comparison approach to property valuation. In this comparison process, consideration may also be given to other factors such as the income quality of the property, the risk associated with the receipt of future incomes, the likelihood of serious competition arising from the development of comparable enterprises, the marketability of the property, the stability of its value, and the burden of management.

Capitalization rates can vary over a considerable range depending on the method of determination and the appraiser's inclinations. Sample appraisals of several different types of property are reported by Friedman.[30] Examples of the capitalization rates used include rural lands, 7 percent; apartments, 9.5 percent; office buildings, 9.7 percent; cooperative apartments, 9.9 percent; commercial properties, 10.3 percent; condominiums, 11.7 percent; and motels, 14 percent. In practice, many appraisers employ the comparative market method and frequently use customary capitalization rates, rates such as 5 or 6 percent with farm properties and 11 to 16 percent with urban residences that they and their colleagues have used with success over extended time periods. In accepting this approach, they treat their capitalization rates more as the reciprocal of an accepted years' purchase approach to valuation than as rates influenced by commercial market interest rates. The commonsense of following this approach was well illustrated by property value and market interest rate trends during the 1970s and early 1980s. Without the separation of capitalization rates from market interest rates, property values should have declined as

---

[28]Cf. Cox, *loc. cit.*, pp. 44–45; AIREA, 7th ed., *op. cit.*, pp. 373–85; Wendt, *op. cit.*, pp. 155–62; and May, *op. cit.*, p. 177.

[29]Wendt, *op. cit.*, pp. 300–301.

[30]Friedman, *op. cit.*

market interest rates more than doubled. In practice, property values also increased at a rapid rate.

Replacement-Cost Approach

A third important method of determining property values is provided by the *replacement-cost approach*.[31] This approach is rooted in the early classical assumption of a close relationship between production costs and value. It assumes that properties should be worth their present replacement cost (or the cost of providing an acceptable substitute property) less an allowance for accrued depreciation and possible obsolescence.

Like the market-comparison and income-capitalization approaches, this method has advantages and disadvantages. As a rule, replacement costs set an upper limit on property values. Just as well-informed rational buyers refuse to pay more for properties than for comparable substitutes, so also do they refuse to pay more than it would cost to replace them with other properties capable of providing comparable utilities and satisfactions.

True as this statement appears, there are occasions when imperfect competition and inadequate or faulty knowledge result in sales prices above this level. Time considerations and a demand for the immediate use of a property may also cause buyers to bid up prices rather than wait the usual number of months it takes to construct new buildings.

Replacement-cost valuations can also result in property values somewhat in excess of those justified by current market conditions. This is particularly true of the appraisal of overdeveloped properties such as outdated mansions and other "white elephant" structures. The valuation of these properties is complicated by the fact that they are often outmoded or were developed to fulfill particular goals that frequently have limited resale value. Large portions of their original production costs and their current replacement costs must often be written off in the valuation process by deducting an allowance for economic obsolescence along with other depreciation from the estimated replacement cost of the property. Even then the prospect of continued high operating costs may reduce the number of potential buyers.

The replacement-cost approach is best used with properties that fit between these two extremes. As Babcock noted, "A building is worth its cost of replacement provided it is new, represents the highest and best use of the site, and provided its construction is justified by the expected returns which it will produce."[32] With the passing of time, the related concept of replacement cost less an allowance for accrued depreciation can be used as a continuing measure of the value of most build-

---

[31]This method is also described at times as the *reproduction-cost approach*. Some appraisers think in terms of the cost of reproducing an improvement with the same or similar materials. Others prefer to think of the cost of replacing an improvement with one of similar utility.

[32]Babcock, *op. cit.*, p. 477.

ings. This is especially true "when the appraiser is justified in thinking that, if the property were lost or destroyed, the owner would rationally replace it."[33]

Widescale use is made of this valuation method in appraisal work, particularly in the appraisal of residential and other urban-oriented properties. This broad acceptance can be attributed to five principal factors: (1) need for a standardized technique that can be applied to mass appraisals. (2) the general acceptance of replacement costs as a ceiling on value estimates, (3) the relative simplicity of this approach and ease with which it may be applied, (4) the tendency of many appraisers to reject market sales as a measure of justified value,[34] and (5) difficulties encountered in use of the income-capitalization approach.

Three major problems arise in the use of this approach. Appraisers must determine the cost of providing sites of comparable value. They must determine the cost of replacing present improvements or of providing a suitable substitute. And they must make appropriate allowances for accrued depreciation and obsolescence of present improvements.

The first problem—that of placing a value on the site—ordinarily calls for use of a market-comparison approach.[35] Lots and building sites are normally valued in terms of the going prices of other sites of comparable size, location, and use-capacity. More difficult problems usually arise with determinations of the replacement costs and depreciation allowances associated with building improvements.

Estimating replacement costs. When appraisers compute the cost of replacing buildings, they are seldom concerned with the cost of constructing exact replicas. Changes in materials and building techniques make this impractical. But they do try to determine the cost of providing replacements of comparable size, design, and use-capacity.

Important decisions must be made at this point concerning the quality of the construction and the level of building costs appraisers are to assume in their calculations. They must decide whether they will think in terms of the costs associated with the highest standards of workmanship, average quality standards, or those that may apply with more slipshod construction. They must choose whether they will "seek the costs of the most efficient builder, the marginal builder, or some imaginary 'typical' builder."[36]

Once these decisions are made, the next appraisal problem calls for choosing the specific method of calculating the estimate of building replacement costs. Three

---

[33]Wendt, *op. cit.*, pp. 219–20.

[34]Cf. *ibid.*, pp. 213–18.

[35]Three other possible methods may be used: (1) a distribution, abstraction, or allocation method, which can be used to allocate value between land and improvements when a price that corresponds to value is known; (2) an anticipated use or development method, which assumes a value for a developed property and subtracts the probable development costs to secure a measure of the value of the undeveloped land; and (3) a land residual method, which capitalizes the imputed rent of land with a hypothetical improvement to secure its value. Cf. AIREA, 7th ed. *op. cit.*, pp. 145–51.

[36]Cf. Wendt, *op. cit.*, p. 225.

principal methods are used for this purpose: (1) the quantity survey method, (2) the inplace unit-cost method, and (3) the square-foot and cubic-foot methods.

With the *quantity survey* method, appraisers duplicate the contractor's original procedure in determining the actual construction cost of building improvements. They may work with blueprints and floor plans of the present building to determine the quantity and quality of materials needed, the probable labor costs required, the contractor's profit and overhead costs, the architect's fee, and any other costs one might encounter in replacing the structure at the appraisal date. This method provides the most accurate measure of building replacement costs. It is time-consuming and costly, however, and accordingly seldom used. A substitute approach known as the *repeat case* method is sometimes used with residential appraisals. With this approach, the cost of constructing comparable dwellings along with lump-sum additions or subtractions for variations between properties is used as a basis for estimating building costs.

The *inplace unit-cost* method eliminates some of the detail required with quantity surveys. With this approach, the appraiser divides the building into component parts and applies appropriate unit prices in calculating the cost of putting each part in place. Calculated costs can thus be applied for each linear or square foot of exterior and interior wall space, each square foot of roof area, plumbing installations, the heating system, electric wiring and fixtures, tile work, concrete driveways, and other component parts of the structure.

The *square-foot* and *cubic-foot* methods provide another commonly used shortcut for computing replacement costs. The appraiser calculates the total square feet of floor space or cubic feet of interior space in the building and applies the current square- and cubic-foot cost figures quoted by local builders for comparable structures to determine its replacement cost. At $40 a square foot, a house with 1,500 square feet of floor space would thus have a replacement cost of $60,000. In similar fashion, a structure that cubes at 20,000 cubic feet has a replacement cost of $54,000 when one assumes a building cost of $2.70 per cubic foot.

Allowances for depreciation. After the appraiser has estimated the cost of replacing the building and other land improvements, deductions are made from the calculated replacement cost for accrued depreciation and obsolescence. Three general methods are used in calculating depreciation.[37] Of the three, the breakdown-observed-depreciation method is ordinarily favored with real estate appraisals. On some occasions, particularly in determinations of book value, use is made of one of the theoretical approaches (straight line, years digit, equal percentage, sinking fund, annuity, or liability to replace methods) to calculate depreciation. Use also may be made of an engineering-observed-depreciation approach, under which percentage deductions from value are estimated.

With the breakdown-observed-depreciation approach, the appraiser starts by considering the loss in value the property has suffered because of physical deteriora-

---

[37]Cf. AIREA, *op. cit.*, 7th ed., pp. 238–52; and Ring, *op. cit.*, pp. 151–61.

tion and possible functional obsolescence. A distinction is usually made between these two concepts. *Physical deterioration* comes as a result of wear and tear, cracks, decay, and the like, whereas *functional obsolescence* is attributed to outmoded floor plans and functional inadequacy owing to the size, style, or age of the structure.[38] The items that contribute to these two types of depreciation are usually classified as curable or incurable. *Curable* items such as a faulty roof, a worn-out heating system, or inadequate electric wiring can be replaced. *Incurable* items such as the advanced age of a structure or a generally outmoded layout, on the other hand, must be accepted for appraisal purposes as they are.

For depreciation-cost computing purposes, the cost of curable items is ordinarily measured in terms of their cost of replacement or repair. Three alternatives are used in measuring the cost of incurable items. Appraisers may use the observed-condition method by observing the property and expressing opinions concerning the monetary or percentage loss of value it has suffered in comparison with a possible replacement. They may use age-life tables to compute the depreciation that should be allowed for structures of varying types, ages, and economic life expectancies. They can also capitalize the difference between the assumed rental value of the property with all the curable items replaced or repaired and the rental value of a new replacement property to determine the amount of depreciation that should be allowed.

As a final step, appraisers frequently make allowances for possible economic obsolescence. These allowances apply when building improvements are overdeveloped for their use, when their value suffers because they are located in areas of declining property values, or when they do not represent the highest and best use of the site. *Economic obsolescence* represents the difference between the capitalized rental value of the building with all its physical and functional deficiencies corrected at its present site and the capitalized rental value of the same building at some ideal location where it would represent the highest and best use of the site. In calculating economic obsolescence, the appraiser looks for the difference between the current market value of a property and its depreciated cost-of-replacement value. As Kahn, Case and Schimmel note, at this point the "significant measure of depreciation can only be found in the marketplace, which leads the appraiser back to the sales-comparison approach.[39]

Appraisers add up their allowances for depreciation to determine total accrued depreciation. This figure is then subtracted from the estimated replacement cost (site plus building improvements) to provide an estimate of value. For example, a property may have an estimated replacement cost of $10,500 for the site and $76,500 for the building, giving a total of $87,000. Its depreciation allowances may include a charge of $5,400 for curable deterioration and funcational obsolescence items, $15,500 for incurable items, and $13,500 for economic obsolescence—a total of $34,400. Subtraction of this total from the estimated total replacement cost leaves $52,600 as the estimated value of the property.

[38] Cf. AIREA, 7th ed., *op. cit.*, pp. 251–58.
[39] Kahn, Case, and Schimmel, *op. cit.*, p. 180.

Choice of a Valuation Method

Most appraisers find it desirable to use two and sometimes all three property valuation methods in their appraisal work.[40] This practice often results in more than one answer. Questions then arise as to which value estimate appraisers should accept in their final determination of value. When appraisers find themselves in this position, they should recheck their calculations and try to narrow the differences between their high and low estimates. They should reexamine the purpose of the appraisal, the adequacy of the data used, and the applicability of their appraisal assumptions. In this correlation process, they ordinarily give maximum weight to the appraisal approach they consider as most reliable while treating the other methods as checks to guide them in the determination of reasonable value figures.[41]

Individual appraisers naturally differ in the emphasis they give to different appraisal methods. Some appraisers feel that the market-comparison approach provides the only practicable means for estimating market values. Others reject this approach because of its acceptance of fluctuating values. Some argue that income capitalization provides the only sound basis for estimating value; others reject this approach because of difficulties that complicate the calculation of average expected net returns and the selection of a capitalization rate. Many appraisers like the sense of certainty they get with the replacement-cost approach; others argue that this approach has little practical use until its depreciation allowances are adjusted to bring its value determinations in line with those found by market comparisons.

As the old adage "The worth of a thing is the price it will bring" suggests, the market-comparison method provides a logical and direct approach to the determination of property values. This method provides a definite bridge between the theory of economic value and the actual exchange values of the market. It is accordingly given considerable weight in most appraisals made to determine sales values. It is also widely used as a check on other methods with other kinds of appraisals.

The income-capitalization approach is usually emphasized with the appraisal of income-producing properties such as industrial plants, commercial establishments, rental housing, and commercial farms. Loan agencies give major weight to this approach in their appraisals for mortgage lending purposes because of its emphasis on the future income-producing capacities of individual properties. Emphasis is also given to this approach in the appraisal of such items as the value of a tenant's interests under a long-term lease and the value of the severance damages that result from condemnation of some portion of an owner's property holdings.

The replacement-cost approach receives considerable use along with the market-comparison method in the valuation of residential and certain other urban properties. Its ease of application and frequent dependence on standardized cost and

---

[40]The American Society of Farm Managers and Rural Appraisers recommends that its members use the income-capitalization and market-comparison methods in farm appraisals. Several government agencies require their appraisers to use all three approaches. The use of all three approaches is also recommended by the American Institute of Real Estate Appraisers.

[41]Cf. AIREA, 7th ed., *op. cit.*, pp. 506–9.

depreciation allowance tables have brought its widespread acceptance in tax assessment work. It is also favored by many courts as an appropriate method for use with condemnation appraisals.

LAND AS AN AREA OF INVESTMENT

Real estate has long been regarded as a prime area of investment. Ownership and investment conditions vary considerably, however, among different cultures and areas. Most landed properties are held by the crown, the state, or a few families in some societies, whereas widespread opportunities for private ownership are available to most individuals in others. Custom and family status considerations discourage sales of landholdings in some countries, while nearly every tract of land is viewed as potentially for sale in areas such as the United States.

Opportunities for land ownership have always rated high among the average American's goals. The prospect of easy access to ownership provided a major incentive for the migration of thousands of Europeans to the United States and Canada during the 1800s. The relative freedom with which landed properties can be bought and sold has made the average citizen of these two nations more willing to move and less land-hungry than their ancestors but has done little to lessen their desire for land ownership.

Several factors help explain the high regard usually associated with investments in land. These include (1) the traditional tendency of people to rate property ownership as desirable, (2) the durability and long life of land investments, (3) the feeling of most investors that they understand land and real estate, (4) their belief that they can manage their investments, and (5) the expectation that ownership of real estate provides an excellent hedge against inflation.

Much of the prestige associated with real estate investments may be attributed to the advantages and privileges associated with land ownership in the past. Under feudalism, those individuals who held rights in land almost invariably enjoyed special economic, social, and political status. Most of these advantages have been greatly diluted by various reforms, but their vestiges still clothe landowners with special status in many communities.

Real estate is favored by many investors because of its durability and relative immobility. Most land resource developments have the advantage of long life. They can be used now but will still have considerable use-value many years hence. Their values may deteriorate because of exploitation, depletion, or depreciation, but losses of this type can be minimized.

Compared with other types of investments, land resources enjoy several distinctive advantages. Individual investors usually find it possible to acquire complete ownership control rather than just a share of the total ownership rights. Investments in real estate involve physical assets investors can see and easily inspect. Furthermore, their management involves familiar types of operations that many investors feel they can take over and handle without the help of hired management.

Real estate resource investments usually compare favorably with other investment alternatives during periods of stable or rising prices, although they may show
up somewhat less advantageously during deflationary periods. As the comparison of
increases in market values of alternative investments reported in Table 10-2 indicates, investments in farm real estate and in urban residential building sites provided
good hedges against inflation between 1950 and 1980. Average farm properties in
the United States increased 10.3 times and new residential lots 11.4 times in value
during this thirty-year period. Measured in constant 1950 dollars, average farm real
estate market prices increased 200 percent and average new residential lot values
233 percent, as compared with increases of 23 percent for investments held in typical savings accounts and 89 percent for investments in average common stocks.
Typical single-family urban residences increased approximately 394 percent in value
between 1950 and 1980 but increased only 44 percent when evaluated in 1950
constant value terms. Of the six investment alternatives listed, it may be noted
that investments in housing, farm real estate, and common stocks provided their
owners with continuing flows of economic returns and services while their market
values were increasing.

Goals in land investment. Investments in land, along with holdings of stocks
and bonds, savings, and insurance, are often cited as basic components of a rational
investment program. This does not mean that all investments in land resources are

**TABLE 10-2. Comparison of Average Market Values of Selected Alternative
Investments, 1950 to 1960, 1970, and 1980**

| TYPE OF INVESTMENT | VALUES OF AN INVESTMENT OF $1,000 MADE IN 1950 | | | | | |
| | Current dollar values | | | Value in constant 1950 dollars* | | |
	1960	*1970*	*1980*	*1960*	*1970*	*1980*
Cash held in a safety deposit box	$1,000	$1,000	$ 1,000	$ 813	$ 620	$ 292
Savings account[†]	1,489	2,447	4,226	1,211	1,517	1,234
Average common stock[‡]	3,038	4,522	6,457	2,470	2,804	1,885
Average farm[§]	1,716	2,943	10,277	1,395	1,825	3,001
Average urban residential building lot[‖]	2,386	4,785	11,410	1,939	2,966	3,333
Well-kept urban single-family residence[#]	1,431	1,992	4,944	1,163	1,235	1,444

*Deflation factor used in calculation of constant 1950 dollar values is based on Bureau of Labor
Statistics' average consumer price indices for the three years.

[†]Interest rates of 4 percent during the 1950s, 5 percent during the 1960s, and 5.5 percent during the 1970s, compounded quarterly, assumed with savings account.

[‡]Standard and Poor's index of average market prices for 500 common stocks in March of each
year used as basis for average common stock prices.

[§]Farmland values are based on farm real estate market price indices reported by U.S. Department of Agriculture for March of each year.

[‖]Values for residential building lots are based on average appraised values of lots for single-family
houses covered by Federal Housing Administration–insured loans each year.

[#]Value trends for single-family houses are based on appraised values of used single-family houses
covered by Federal Housing Administration–insured loans each year.

equally good. Some offer higher prospects of profit or capital gain than others, and some offer special advantages that appeal to particular investors. Others involve risks and undesirable features that make them opportunities of questionable value.

Potential investors should ask themselves what they want of their investments. Their economic goals may well include the following:

1. Safety of investment—assurance that investment will produce sufficient yield or increase enough in value to permit resale at a price in excess of acquisition cost
2. Certainty of yield—ability of properties to provide periodic returns or increments of additional capital value that will give investors an acceptable investment return
3. Liquidity of investment—relative ability of owners to sell or liquidate investments at their full market value within short periods of time
4. Capital appreciation—the hope that investments will increase in capital value, that there will be little or no depreciation in values, and that investments will provide a good hedge against inflation
5. Taxation advantages—expectations that investments will qualify for special tax treatment advantages that will reduce the taxes paid by investors
6. Managerial responsibilities—opportunities, as investors wish, to participate in or to be free from managerial responsibilities and the possible liabilities associated therewith
7. Reinvestment of net returns—opportunities to reinvest flows of land rents and profits in the investment
8. Opportunities for leverage—ability to use one's equity funds along with borrowed capital to spread one's realm of economic control over larger amounts of property[42]

In addition to the economic goals listed above, investors may also have goals associated with personal satisfactions and desired amenities. An investor may acquire a tract of forestland, for example, as much for its open space and recreation advantages as for its economic potential for timber production. Others may acquire land resources because of the satisfactions they attribute to ownership.

Alternatives for real estate investments. Most investments in land resources involve either the purchase of residential lots and houses for family use or the acquisition of properties for business purposes. Real estate investments of this order are recommended for people who need land for residential and business reasons. A variety of alternatives are available for potential investors who plan additional investments in real estate ventures.

One possibility involves the sometimes questionable activities of real estate development promoters who specialize in the development and sale of home sites for recreation and retirement purposes. During past boom periods, promoters have used effective selling tactics involving complimentary dinners and free trips to view the sites, along with arguments that stress the upward trend in land values, opportunities for speculative profits, and the obvious clincher that "God doesn't make

[42]For further discussion see Kahn, Case, and Schimmel, *op. cit.*, pp. 307–16.

land anymore" to sell their lots.[43] Congress enacted the Interstate Land Sales Act of 1968 and several states have passed land sales laws to protect buyers against high-pressure and fraudulent land sales tactics. These laws provide a measure of consumer protection; but buyers must still investigate numerous details that affect the true value of potential investments. As a general rule, one should *never* buy landed property on impulse. One should always allow periods of a day or more to consider the merits and demerits of proposed purchases. Many land sales laws specify "cooling off" periods of several days during which buyers can change their minds and request cancellation of signed contracts to buy land. Fortunately most investment arrangements leave more of the responsibility for initiating interest in particular properties and for deciding what is or is not a good investment to the potential investor.

Operators who have money to invest and a willingness to assume risks can acquire real estate that they can lease to others or turn over to management agencies to handle. They may also acquire prospective building sites that can be held for resale or later development. With speculative investments of this sort, operators can realize capital gains if they pick the right sites. Their expectations may go unfulfilled, however, if their properties are zoned for lower uses than they planned, if undesirable developments are located in the areas, if cities do not grow in their direction, of if the expected ripening of a site fails to materialize.

Investors frequently find that they can operate on a bigger scale by combining their investment funds with those of others. Operating in this way, they may pool their funds with those of friends to acquire an apartment house, an office building, or a shopping center, which they manage for their mutual benefit. Partnerships of this order involve joint responsibilities for the actions of others. Investors can avoid these responsibilities along with the direct responsiblities of management by investing in the shares offered by incorporated real estate syndicates.

Investments in syndicate shares provide many of the advantages of direct investments in land resources.[44] These organizations are frequently large enough to spread their investments over a wide spectrum of properties, a situation that reduces the risks associated with investments in single properties. They also assume responsibility for management. As corporations, they are subject to corporation income taxes, but they can be exempt from this tax if they are incorporated as real estate investment trusts (REITs) that pay out at least 90 percent of their current net earnings to stockholders. Some 210 real estate investment trusts, most of them listed on stock exchanges, provided approximately one-fifth of the funds used for construction and development loans during the early 1970s. Unlike the investment syndicates, the REITs used most of their funds along with additional borrowings

---

[43]Cf. Morton C. Paulson, *The Great Land Hussle* (Chicago: H. Ragnery Co., 1972).

[44]Cf. Robert Kevin Brown, *Real Estate Economics* (Boston: Houghton Mifflin, 1965), pp. 231–34; Kahn, Case, and Schimmel, *op. cit.*, pp. 319–21; Hugo Rothschild (revised by Daniel S. Berman), *How to Invest and Protect Your Profits in Real Estate Syndicates* (New York: Doubleday, 1964); and Bertram Lewis, *Profits in Real Estate Syndication* (New York: Harper & Row Publishers, Inc., 1962).

to provide mortgage construction financing for new apartment and condominium complexes, shopping centers, and other real estate developments. Overcommitments during the building boom of the early 1970s led to serious financial setbacks for most REITs during the 1974–1977 period and to subsequent filings for reorganization under the nation's bankruptcy laws.

Credit leverage and tax incentives. Investors enjoyed unusual opportunities for realizing substantial capital gains from investments in income properties between 1940 and 1980. This situation was fostered by a steady appreciation in real estate values, mortgage financing arrangements that permitted considerable credit leverage, and favorable income and capital gains taxation policies. Property values appreciated at a slower rate in some instances and actually declined in others after 1980; but it appears that opportunities for credit leverage will probably continue, and investors can expect future benefits from favorable tax treatment, even though these benefits may be less than those enjoyed prior to 1984.

A simple example may be used to illustrate the investment opportunities available to higher-income operators who have surplus funds they can use to buy income properties. Table 10–3 assumes a potential investor who has a taxable income after deductions in the 50 percent federal income tax bracket. The investor can buy a ten-unit apartment development for $480,000 with a down payment of $96,000 plus $4,000 in closing costs. An 80 percent mortgage for $384,000 at 11 percent interest for thirty years can be arranged to cover the balance of the investment cost. The credit leverage permitted with this mortgage makes it possible for the operator to acquire control of five times as much property as one could with full payment in cash. Use of credit calls for amortized monthly payments of $3,659.92 for principal and interest, or total payments of $43,883 a year for this purpose.

In calculating expected costs and returns, the operator assumes average contract rental payments of $550 a month for each of the ten apartment units. This assumption gives a gross return of $66,000 in the first year, which, when reduced by a 5 percent allowance for vacancies, yields an expected rental income of $62,700. Total costs are listed in the following first-year cost and returns statement:

Expected rental income (deducting 5 percent as a vacancy allowance)	$62,700
Expected annual expenses	
Mortgage payments	43,883
Property taxes	9,500
Property insurance	1,520
Maintenance allowance (assumed at 12 percent of rental income)	7,524
Total expenses	62,427
Net cash flow	273
Equity buildup (portion of mortgage payments credited to principal)	1,759
Net return on operator's investment	$ 2,032

TABLE 10–3. Expected Costs and Returns on Investment in a Ten-Unit Apartment Development During First Ten Years of Ownership

YEAR OF INVESTMENT	INCOME FROM RENTALS*	ESTIMATED COST[†]	NET CASH FLOW	EQUITY BUILDUP	INVESTMENT RETURN[‡]	RATE OF RETURN ON INVESTMENT (percent)
1	$62,700	$62,427	$ 273	$1,759	$ 2,032	2.03
2	63,954	62,798	1,156	1,966	3,122	3.12
3	65,233	63,176	2,057	2,193	4,250	4.25
4	66,538	63,562	2,976	2,442	5,418	5.42
5	67,868	63,956	3,912	2,726	6,638	6.64
6	69,226	64,357	4,869	3,045	7,914	7.91
7	70,610	64,767	5,834	3,395	9,229	9.23
8	72,023	65,184	6,839	3,790	10,629	10.63
9	73,463	65,610	7,853	4,428	12,281	12.28
10	74,932	66,045	8,887	4,723	13,610	13.61

*Rental income assumes average rentals at $550 a month for each of the ten apartments with increases of 2 percent annually, with a 5 percent allowance for vacancies.

[†]Mortgage payments are fixed at $43,883 each year. Taxes and insurance are expected to rise at an average rate of 2 percent annually. Maintenance costs are budgeted at 12 percent of total rental income.

[‡]Initial investment includes a downpayment of $96,000 plus $4,000 in closing costs.

An additional allowance for anticipated appreciation of the market value of the investment is sometimes listed with this balance sheet. Expectations of value appreciation definitely influence investment decisions. The fact that these expectations cannot be realized until the investment is sold, however, suggests that they should not be counted as more than expectations until that point.

If the operator assumes that rental returns will increase approximately 2 percent each year and that property taxes and insurance costs will rise by a similar 2 percent annually while the mortgage payments remain constant, he or she will be able to look forward to the projection of costs and returns reported in Table 10-3. With these assumptions, the investor's expected rate of return on the initial cash investment of $100,000 will rise from 2.03 percent in the first year to 13.61 percent in the tenth year.

Additional calculations are necessary if the potential investor utilizes his or her opportunities for favorable federal income tax treatment. With the federal tax regulations in effect in 1984, operators could charge off depreciation with income property investments as operating costs for income tax calculation purposes. A fifteen-year depreciation period and a 175 percent declining balance shifting to a straight-line accelerated cost recovery system (ACRS) depreciation schedule was authorized by the Economic Recovery Act of 1981. This law liberalized the provisions of the Tax Reform Act of 1976, which had favored the use of straight-line depreciation over the economic life of investments. Accelerated depreciation at a double declining balance rate was permitted prior to 1976 with properties held for a minimum of 200 months. A new law in August 1984 required the use of an eighteen-year ACRS depreciation schedule on properties acquired after March 1984.

Using the eighteen-year ACRS depreciation schedule approved by the Internal Revenue Service in 1984, annual tax savings ranging from $17,280 in the first year to $9,600 in the tenth year could be realized. Addition of these to the investment returns raise the annual rate of return to 19.3 percent in the first year and 23.2 percent in the tenth year (see Table 10-4).

If the property is sold at the end of the tenth year, its book value will have declined by the $253,440 claimed as depreciation. Should the property sell at its initial investment cost of $480,000, the operator would have to pay a capital gains tax equal to a maximum of 20 percent of the capital gain of $253,440 or $50,688. This tax would represent 40 percent of the operator's federal income tax savings of $126,720 over the ten-year investment period.

Residual nature of investment in land resources. Investments in land are far from infallible. While an operator's initial prospects may be bright, there is no guarantee that property values will continue to appreciate, that future rental market conditions will permit the leasing of properties at reasonable rates, or that taxes and other operating costs will not rise. Past experience shows that many investors have suffered from either lower incomes or higher costs than they expected.

The actual returns that can be realized on land resource investments vary considerably. Sometimes they are high, sometimes low, depending both on individual

TABLE 10-4. Effects of Depreciation Allowances on Returns from Investment Assumed in Table 10-3

YEAR OF INVESTMENT	INVESTMENT RETURN	DEPRECIATION CLAIMED*	TAXABLE INCOME FROM INVESTMENT	VALUE OF TAX SAVINGS[+]	INVESTMENT RETURN WITH TAX EFFECT	RATE OF RETURN ON ORIGINAL INVESTMENT
1	$ 2,032	$34,560	($32,528)	$17,280	$19,312	19.3
2	3,122	34,560	(31,438)	17,280	20,402	20.4
3	4,250	30,720	(26,470)	15,360	19,610	19.6
4	5,418	26,880	(21,462)	13,440	18,858	18.9
5	6,638	26,880	(20,242)	13,440	20,078	20.1
6	7,914	23,040	(15,126)	11,520	19,434	19.4
7	9,229	19,200	(9,971)	9,600	18,829	18.8
8	10,629	19,200	(8,571)	9,600	20,229	20.2
9	12,281	19,200	(6,919)	9,600	21,881	21.9
10	13,610	19,200	(5,590)	9,600	23,210	23.2

*Depreciation is allowed only on the building and improvement portion (assumed here to be 80 percent) of the total property investment value of $480,000. Depreciation is calculated at the 18-year 175% declining balance ACRS rate approved by the Internal Revenue Service for real properties acquired after March 15, 1984. Cf. *P-H Federal Taxes 1985* (Englewood Cliffs, N.J.: Prentice-Hall, 1985), Section 54,604.

[+]Calculation of annual tax savings assumes an investor who is in the 50 percent federal income tax bracket.

cases and the relationships between costs and prices. Under typical operating conditions, land resources have no more than a residual claim on the gross returns received from their use in combination with other productive factors. In allocating the receipts received from the operating of an apartment house or office building, first consideration must be given to the payments made for wages, supplies, and utilities. Without these payments, properties would soon lose their earning power. Office managers give top priority to operating outlays because they realize that they must provide elevator and janitorial service, heat, and electric power if they are to retain their present tenants and attract new occupants for their vacant office suites.

Second priority is usually claimed by taxes and insurance. These costs can go unpaid when business incomes are low; but the risk of losses from insurable mishaps, penalties for late payment of taxes, and the threat of property tax reversion ordinarily favor their prompt payment. More leeway is possible with payments covering long-term capital costs. Operators can often postpone maintenance and repair operations for several months, forgo the setting aside of depreciation reserves, and even default on mortgage and bond interest payments. When the choice is between the payment of these costs during short-run periods, operators ordinarily prefer to keep up their interest payments while they postpone outlays for repairs and depreciation. When this problem continues for longer periods, these priorities are often reversed. Mortgagees and bondholders may then have to wait for their interest because it is necessary to replace worn-out equipment and carry on basic maintenance and repair operations if plant deterioration and losses in production capacity are to be avoided.

Any income that remains after payment is made for the operating costs noted above can be treated as land rent or as interest on the operator's equity. The size of this residual return is affected by all the factors that bear on the relationship between costs, prices, and volume of business. Between 1940 and 1980, commodity market prices tended to increase faster than costs, and substantial profits were realized by most productive enterprises. This net return was often capitalized into higher property values and thus contributed to the upward trend in real estate and stock values reported in Table 10-2. When prices falter and cost-price squeezes develop, the residual return attributable to land resources often declines and may disappear entirely.

One of the most important factors affecting returns to land resource developments is the problem of "sunk costs." While projects are being planned and developed, careful consideration is usually given to the relationship between costs of development and potential receipts. At this stage, every project must be justified by its favorable prospects for producing a surplus of returns above costs. Once a project is completed, however, the actual cost of development loses significance as far as its further effect on operations in concerned. Production and investment outlays already made in the development process become "sunk costs" because they can no longer be withdrawn or recovered. Instead of enjoying guaranteed incomes, owners must now accept whatever return they can secure in the market. Naturally, they still want to break even on their overall investments. But they will take higher returns if they can get them and accept less if they must.

As the discussion above suggests, property owners can never plan to price their products in terms of production costs. In each successive short-run period, they must accept whatever return the market offers. Sometimes this means profits; sometimes it involves operating at a loss. Yet over the long run, most well-planned and well-managed properties can count on a surplus of returns above production and operating costs. This situation prevails because of the recurring need for new resource developments that comes with the wearing out of old properties and with changes in supply and demand conditions. The entrance of new developments into the market is always prompted and guided by prospects of profits above production costs. Before new capital is invested, profits must usually be realized on the developments already in existence.

OPERATION OF THE REAL ESTATE MARKET

What is often referred to as the "real estate market" is really a conglomerate concept made up of thousands of smaller markets that operate in different areas and deal with different types of properties. Separate markets exist for every type of property that involves different groups of buyers and sellers. Distinctions are ordinarily made between the markets for farm, residential, commercial, industrial, mining, forest, and recreation properties and at times for various subtypes of properties.

Emphasis in this discussion is centered primarily on the characteristics and general operation of two types of aggregate markets—the farm real estate market and the market for urban and suburban residential properties. These are the largest and best organized land resource markets. Neither market involves enough sales of closely comparable properties to permit day-to-day market quotations, but both have sufficient sales in most communities to permit the use of market-comparison valuation techniques and to facilitate a certain continuity in market-price relationships.

Characteristics of the Real Estate Market

The real estate market deals for the most part with the buying and selling of land resources that have already been developed and brought into use. New developments can result when market transactions lead to new construction or the shifting of land areas to higher uses. But most of the productive properties offered for sale have already benefited from development work on the part of previous owners and have been used for crop production, grazing, forest, mining, industrial, or commercial purposes. Similarly, the average residential property placed on the market also represents a used second-, third-, or fourth-hand structure that has been occupied and used by earlier owners. Unlike many commodity markets, new products are the exception rather than the rule in the real estate market.

Real estate markets are characterized by (1) concern with only that portion of the total supply of properties offered for sale, (2) fixed location of product, (3) nonstandardized products, (4) special legal requirements that affect transfers of real estate, (5) dependence on local supply and demand conditions, (6) large considerations with most transactions, (7) customary use of credit arrangement to supplement the limited equity interests held by many sellers and most buyers, (8) infrequent participation in the market by the average buyer and seller, and (9) common use of a broker's services.

While all properties may be considered as available for purchase if the price is right, real estate markets deal only with the fractional portion of the total supply of different types of land resources (not more than 5 or 6 percent in most years) that is actively offered for sale. Moreover, like the legal descriptions of property with which it deals, the real estate market is definitely site- or location-oriented. Real estate brokers frequently find it possible to sell comparable lots, houses, or farms, but they can never standardize their product in the same sense that they could provide duplicate models of an automobile, typewriter, or can opener. They cannot quote real estate prices f.o.b. Detroit. Landed property must always be sold where it is.

The fixed-location factor makes the real estate market primarily a local market. Florida real estate may find a market among northern investors, but this is not the usual case. Once prospective buyers decide where they want to acquire property, they must look to local property owners to supply the market listings. Similarly, local property owners who wish to sell ordinarily expect to find buyers among the people who live or who expect to live in the area. Local market conditions often buoy up property values and the level of real estate market activity in the immediate area without having an appreciable effect on property values outside the local commuting zone.

Sales and other transactions that involve the holding of rights in real estate are subject to a number of legal requirements not ordinarily applied to other sales commodities. Every tract of land has its legal description. All claims against land titles must be registered to have legal standing. Titles are carefully checked for flaws and possible claims against them. These procedures, which are designed to protect ownership rights, highlight the fact that real estate is legally different from other market-commodities.

Unlike most markets in which buyers and sellers participate on a day-to-day basis, real estate market transactions usually involve substantial considerations. Property sale agreements ordinarily involve cash outlays of thousands of dollars. Buyers quite often spend their life savings plus all the money they can borrow on the properties they buy in one market transaction.

Most sellers are private owners with only a single property to sell. They ordinarily offer used properties that are frequently encumbered by mortgages that have not as yet been paid off. This means that they can sell only their equity interest and that they must plan to either transfer their mortgages to buyers or use part of the proceeds from their sales to satisfy this claim against their properties. They are not in the business of selling and trading real estate, and their experience in the marketing of real properties is definitely limited. Their decisions to sell are usually associated with outside circumstances such as desire to move to a larger or smaller house or farm, plans to retire or transfer to a job in another area, or need to liquidate an investment.

In addition to their other characteristics, sellers frequently have somewhat erroneous notions regarding the value of their properties. During inflationary periods, they sometimes fail to realize how much the market value of their properties has climbed. More often their failing lies in the direction of overvaluing the product they expect to sell.

Like average sellers, average buyers may come from any walk of life. In most cases they too have had little experience in the buying and selling of real estate. Their supplies of capital are usually limited, and they ordinarily plan to mortgage the properties they buy to secure substantial portions of their purchase outlay. If they are in the market to buy a house or a farm, they have probably shopped around for a while and may have good ideas concerning the size and quality of the property they can expect to get for the prices they are able to pay. If they have not checked market prices and listings, their first reaction will often be one of shock or surprise when they are informed of the prices asked for real estate.

The inexperience and infrequent participation of average buyers and sellers in the real estate market are often compensated through use of a broker's service. Most real estate brokers have considerable knowledge about the operation of the market and the transfer of properties. Through office contacts, advertising, and joint listing arrangements with other brokers, they provide facilities for bringing buyers and sellers together; and their knowledge of market conditions and real estate price trends often make them good judges of property values.[45]

These factors make it possible for brokers to provide a valuable service to average buyers and sellers. They can advise prospective buyers regarding the types of properties available and the prices they must expect to pay. At the same time, as supposedly neutral parties to the transaction, they can often do a better job in pointing out the advantages of and actually selling a property than can the seller.

There is probably no market in the American economy where as much haggling and "horse trading" over price terms take place as in the real estate market. Offers and counteroffers are common, even expected in most cases. Far from the usual approach with many commodities in which the buyer expects to pay the amount

---

[45] Real estate brokerage started in the United States around 1800 and became an established business during the 1840s and 1850s. Cf. Pearl J. Davies, *Real Estate in American History* (Washington: Public Affairs Press, 1958), pp. 18–27. Brokers' offices range from small one-person operations to large firms with several departments and numerous personnel. Some brokers limit their activities to the listing and selling of properties. Others act as buying and leasing agents, manage properties for investors, appraise real estate, sell property insurance, act as real estate mortgage credit agents, and plan and supervise real estate developments. Brokers are expected to follow a professional code of ethics and are licensed to operate by the state. The term *realtor* is a coined title used to describe those brokers who are members of local real estate boards associated with the National Association of Real Estate Boards.

listed on the price tag, most buyers of real estate expect to start their bargaining with an initial offer somewhat below the price asked by the seller. In consequence, many sellers list their properties at higher prices than they expect to get. Brokers often play important intermediary roles in this bargaining process. They not only carry offers and counteroffers back and forth between buyers and sellers but also frequently find themselves advising both parties concerning the adequacy or fairness of an offer, the buyer's prospects for getting the property at less than the list price, and the seller's prospects for securing a higher price than offered.

Farm Real Estate Market

Factors such as size of holdings, extent of improvements, quality of production resources, and location cause farm property values to vary considerably. Properties in the same community frequently differ in average acreage values. Wider variations come with regional differences as farming areas vary greatly in their climates, soil and water resources, and locations with respect to major markets. Farms located near metropolitan centers usually sell for higher prices than farms in more rural areas, and farms in the Corn Belt for more than ranch land. Average farmland values by states in 1982 ranged from highs of $3,118 per acre in New Jersey and $2,804 in Rhode Island to lows of $211 in New Mexico and $170 in Wyoming.

Thanks to the services of the U.S. Department of Agriculture, considerable data have been collected and reported concerning the situation and trends in the farm real estate market.[46] Since 1912, the first year for which annual trend data are reported, this market has witnessed two important periods of rising prices plus an intervening period of falling and depressed farm property values.

Wartime demands caused farmland values to almost double between 1912 and 1920 when a sharp drop in farm product prices brought an abrupt end to a long period of rising farmland prices. Average farmland values then dropped steadily from a current dollar index value of 14.1 in 1920 (1977 = 100) to 5.7 in 1933. Prices rebounded slowly after 1933 to 6.7 in 1941. World War II brought higher farm prices and an increasing demand for farm properties. Vivid memories of the disastrous effects of the 1920 land boom and the wave of property foreclosures that followed throughout the 1920s and 1930s helped keep prices down. Even so, farmland values increased at a rate of approximately 1 percent per month throughout most of the war and early postwar period. By 1949, the index of farmland values reached a high of 14.4. The upward trend stalled briefly with the recession of 1949 and then started a steady upward climb during the Korean War that brought average price index levels of 24.2 in 1960 and 41.5 in 1970. A farmland boom of major proportions then caused the national index to soar to 75.2 in 1975, 144.9 in 1980, and a peak of 158.3 in 1981. High interest rates and a generally depressed

---

[46] The U.S. Department of Agriculture publishes a situation report entitled *Current Developments in the Farm Real Estate Market* twice each year. These reports provide current and up-to-date information on farm real estate values and market trends by states, regions, and for the nation.

farm economy prompted a reversal of this trend after 1981 that brought the national index down to 128 in early 1985.

The upward spiral in average dollar prices per acre between 1915 and 1920 seemed large at the time but was not sufficient to keep farmland values on par with average changes in the consumers' price level (see Figure 10-1). Farmland values actually declined in constant-dollar value terms between 1916 and 1942. A very different situation prevailed between 1962 and 1980 during which average acreage values increased more than sixfold in current value terms and doubled in constant 1977 dollar terms. Farmland values increased at more than twice the general price level rate during this period. This upward trend ended in 1980, and by April 1985 average prices for farm real estate were 31.7 percent lower when measured in 1977 dollars than in 1980.[47] Average prices rose in all sections of the nation during the 1960s and 1970s. A mixed trend occurred between 1980 and 1985 as average prices gained slightly in the New England states and Texas while major declines were experienced (57.1 percent in Iowa) in the agriculturally productive Corn Belt and Northern Plains states.

Number of voluntary transfers. Market listings and the number of voluntary transfers provide a second important facet of the real estate market. The number of farms offered for sale in any given community is always affected by ownership patterns and trends toward parceling units or combining fields into larger farms. It can

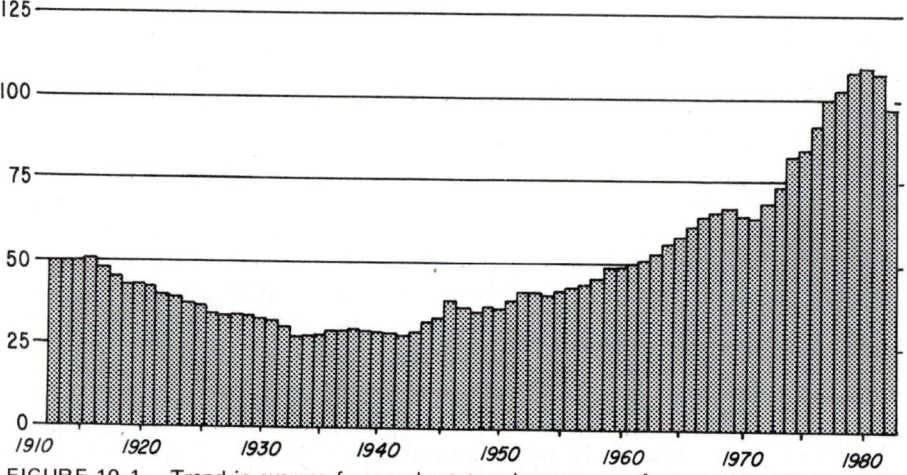

FIGURE 10-1. Trend in average farm real estate values per acre for the United States, 1910 to 1983, measured in constant 1977 dollars.

[47]For more detailed information on the year-by-year indices of average farm real estate values per acre by states, cf. U.S. Department of Agriculture, *Farm Real Estate Market Developments*, CD–64, August 1963, p. 40, and subsequent issues.

also be affected by cultural attitudes. In many parts of the world, the ties between people and land are such that owners are reluctant to sell their land. Farm properties in these areas usually pass by inheritance, and it is an unusual event when farms are offered for sale in an open market. This is not the situation in the United States. Since colonial times, most American farmers have been willing to sell their holdings and move on to new opportunities. Almost every farm may be considered as being potentially for sale, although only a fraction of the total number are listed at one time.

Analysis of the available data on voluntary farm transfers (Figure 10-2) shows that the volume of transfers reached peaks in the early months of the 1916-1920 and post-1941 land booms. The prospects of high selling prices—high in comparison with those of the recent past—brought a considerable flow of properties into the market after World War I and World War II and again in the 1972-1974 period. More than 5 percent of the farms (57.7 farms per 1,000 in 1945 and 1946) were sold in voluntary transfers every year between 1944 and 1947. Once these properties (many of which would have been offered earlier had prices been higher) were sold, the number of voluntary sales dropped to the 28 to 32 farms per 1,000 level until 1972-1974 when the rate rose to 41.2 before dropping back to the 26-28 level in 1976-1980.

High land prices have provided sales opportunities for those farm owners who were approaching retirement or who planned to sell anyway. They have also encouraged operators located in the path of urban growth to sell and relocate in more rural areas. Generally, though, they have had little direct impact, other than raising the costs of farm enlargement, for those operators who planned to stay in farming.

Market conditions in 1983. Farmland prices presented a perplexing picture in 1983. Throughout most of the preceding century, trends in land values had been generally correlated with trends in net farm incomes. But net incomes from farming

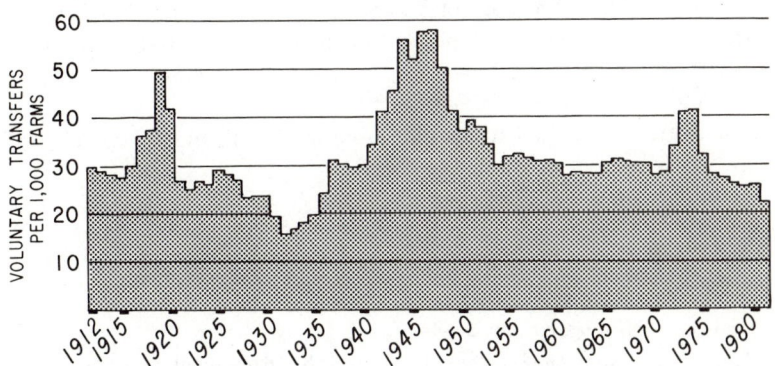

FIGURE 10-2.   Trend in voluntary sales and transfers of farms, United States, 1912-1981.

were not high in the 1950-1980 period. They exceeded the $16 billion level in 1967 dollar terms in only two years (1973, $25.1 billion; and 1974, $17.6 billion).[48] Net returns to farm real estate land dropped from annual averages of $4.5 billion in 1945-1949 and $3.9 billion in 1950-1954 to $2.7 billion in 1955-1959, before rising to $4.4 billion in 1960-1964, $6.9 billion in 1965-1969, and $7.7 billion in 1970-1972. Stated as a return on current farmland values, these returns dropped from 8.0 percent in 1945-1949 to 5.2 percent in 1950-1954, 2.8 percent in 1955-1959, 3.6 percent in 1960-1964, 4.2 percent in 1965-1969, and 3.9 percent in 1970-1972.[49] Yet even though these rates of return were low, most farm owners controlled land and other production assets that were increasing rapidly in value. Gertel and Lewis found that the internal rates of return on farmland investments (which include allowances for increases in asset values together with annual net productivity values) were higher on the average between 1940 and 1970, and much higher during the 1970s, than with holdings of average common stock.[50]

In the absence of high net farm incomes, other factors obviously were responsible for the more than doubling of average farmland values between 1962 and 1980. The more important of these factors included: (1) land acquisitions for farm enlargement purposes; (2) expectations of a more profitable future for farming; (3) the availability of credit for farmland purchases; (4) capitalization of public program benefits into higher land values; (5) acquisitions of farmlands for rural homesites, subdivisions, and other urban-oriented uses; (6) the changing mix of agricultural production inputs, and (7) purchases as a hedge against inflation.[51]

American agriculture changed a great deal during the three decades between 1950 and 1980. The total number of farms dropped from 5.39 million in 1950 to 2.24 million units in 1982, and the average size of farms increased from 216 to 439 acres. The proportion of farm units involving 500 or more acres rose from 5.6 to 16.3 percent, and the proportion involving units of less than 10 acres rose from 6.6 percent in 1959 to 8.4 percent in 1982. With the labor savings associated with mechanization, many operators found it to their advantage to expand the scale of their operations by acquiring additional tillable land. In this farm enlargement process, they frequently felt that the added efficiency associated with larger operations justified payment of above average prices for the additional land. Transfers for farm enlargement purposes rose from 26 percent of the total in 1950-1954 to 37 percent in 1955-1959, 49 percent in 1960-1964, 55 percent in 1965-1969, 57 percent in 1970-1974, and 61 percent in 1975-1979.[52]

Operators were encouraged to acquire more land both by widely felt optimism

---

[48] Cf. *Agricultural Statistics*, 1982, p. 430.

[49] *Agricultural Finance Statistics*, U.S. Department of Agriculture, Economics Research Service AFS-2, 1974, p. 70.

[50] Karl Gertel and James A. Lewis, "Returns from Absentee-Owned Farmland and Common Stock, 1940-79," *Agricultural Finance Review*, 40 (1980), 1-11.

[51] Cf. Healy and Short, *op. cit.*, pp. 33-44.

[52] Cf. *Farm Real Estate Market Developments*, U.S. Department of Agriculture CD-85, 1980, and annual reports for earlier years.

about the future of farming and the ready availability of credit for the purchase of farmland and farm equipment. Farmers enjoyed more demand for their products, higher prices, and higher net incomes in the early 1970s and again in 1978–1979 than they had for several years. New world markets opened up, and farmers were told that additional production would be needed to help feed the hungry people of the world and to pay for the nation's imports of oil. With these prospects, buyers and sellers visualized higher net returns to land.[53] They also made liberal use of mortgage credit as is indicated by the rise in the amount of outstanding farm mortgage debt from $6.1 billion at the end of 1950 and $12.8 billion in 1960 to $30.3 billion in 1970 and $95.5 billion in 1980.

Farm real estate values were pushed upward in some areas by the capitalization of benefits from tobacco allotments, favorable public grazing land leases, and other public agricultural benefits into higher land values.[54] Land values near growing urban centers also were buoyed up by the purchase and prospect for future acquisitions of several million acres of farmland for residential, industrial, commercial, highway construction, and other urban-oriented uses.

Two of the most important factors that have affected farmland values concern the changing mix of agricultural production inputs and the impact of inflation. With the substitution of land and capital for labor that has taken place in recent decades, the share of the total returns from farming attributable to production assets rose from 25 to 69 percent in the twenty-five years before 1979. This means that the share of net returns attributable to land and land improvements has increased substantially. Capitalization of the value of this larger share leads logically to higher farmland values even with some reduction of total net farm incomes.[55]

Approximately 61.5 percent of the increase in farmland values between 1970 and 1981 can be attributed to keeping up with the nation's overall inflation rate. Much of the remaining increase reflects demand among farmers and nonfarmer investors for expected increases in farm real estate asset values. The attraction of farmland investments to nonfarmer and absentee buyers is illustrated by the fact that they accounted for one-third of the farm purchases between 1957 and 1967, between 34 and 39 percent of the purchases between 1968 and 1975, and then 33 percent of the farm purchases in 1980, 30 percent in 1981, and 25 percent in 1982.

[53]This situation adds a dynamic feature to the valuation of farmland. As Robert D. Reinsel and Edward I. Reinsel, "The Economics of Asset Values and Current Income in Farming," *American Journal of Agricultural Economics*, 61 (December 1979), 1093–97, have observed, much of the increase in farmland values during the 1970s reflects the impact of expectations of larger net land rents, which in turn were capitalized into higher land values.

[54]Cf. C. Lowell Harriss, ed., *Government Spending and Land Values* (Madison: University of Wisconsin Press, 1973), chaps. 5, 6, 8, 10; and Robert D. Reinsel and Ronald D. Krenz, *Capitalization of Farm Program Benefits into Land Values*, U.S. Department of Agriculture, Economic Research Service, ERS–507, 1972. Reinsel and Krenz found that capitalization of the program benefits for cotton, peanuts, rice, tobacco, wheat, and feed grains had raised farmland values by $16.5 billion in 1970.

[55]Cf. Emanuel Melichar, "Capital Gains versus Current Incomes in the Farming Sector," *American Journal of Agricultural Economics*, 61 (December 1979), 1085–92.

Seven percent of the purchases in 1981 were for rural homesites, and 3 percent for expected subdivision and commercial or industrial developments.[56]

Market for Residential Properties

Data on residential real estate transfers have been reported by federal agencies for a shorter period than for farm properties. The median sales price for new single-family houses sold in the United States was $18,000 in 1963, the first year for which this information was reported. This average price gradually increased to $27,600 in 1972 and then almost tripled in the dozen years that followed (see Table 10-5). Meanwhile the U.S. Department of Commerce's composite index on single-family house prices and the Boeckh index of residential construction costs followed parallel paths by increasing almost threefold between 1970 and 1983. Deflation of these indices to a 1965 constant-dollar basis shows that the overall cost of new construction rose approximately 20 percent and the cost of new housing rose around 26 percent in real value terms between 1965 and 1980.

Trends with new construction. The nonfarm residential real estate market is concerned primarily with two types of properties—used dwellings and new construction. Of the two, new construction often has a dominant effect on market price trends both in providing new additions to the supply and in setting the replacement cost of existing dwellings. The trends in new residential construction (including shipment of new mobile home units) and in residential building costs for the 1910-1982 period are summarized in Figure 10-3.

A pre-World War I peak in new residential construction came in 1909 when 573,000 new nonfarm dwelling units were started. Construction continued at a reduced level during and immediately following World War I. An urban housing boom followed during the 1920s and peaked with the construction of 937,000 units in 1925. Building activity dropped off later in the decade and fell to a low of 93,000 units in 1933. The annual volume of new construction then gradually increased to 706,000 new units in 1941.

Residential construction was limited throughout World War II; but a major building boom started after the war. New construction rose to 1.02 million units in 1947 and then to a peak of 1.95 million units in 1950.[57] Mortgage credit restrictions were imposed in late 1950 to dampen the inflationary pressures associated with the Korean War. New construction then dropped to 1.49 million units in 1951 and remained at the 1.2 to 1.6 million level until 1971. A building boom in the

---

[56]Cf. *Farm Real Estate Market Developments*, U.S. Department of Agriculture, CD-87, 1982, Tables 15 and 18.

[57]A new method of reporting new housing starts was initiated by the U.S. Bureau of the Census in 1959. This approach, which has been extended back to 1945, covers both farm and nonfarm housing starts, includes Alaska and Hawaii in the national totals, and counts temporary as well as permanent structures. Additions to the stock of mobile homes are counted separately.

TABLE 10-5. Selected Measures of Residential Real Estate Cost Trends in the United States, 1950–1983

YEAR	MEDIAN SALES PRICE OF NEW SINGLE-FAMILY HOUSES SOLD	BOECKH INDEX OF RESIDENTIAL CONSTRUCTION COSTS	COMPOSITE INDEX OF COST OF NEW HOUSES	CONSUMER PRICE INDICES FOR: Home purchase costs	Home ownership costs	Residential rents
1950	n.a.	29.1	n.a.	n.a.	n.a.	45.9
1955	n.a.	33.5	n.a.	48.6	37.6	54.9
1960	n.a.	37.8	n.a.	51.1	42.1	59.7
1965	$20,000	41.8	44.4	54.0	45.2	62.9
1970	23,400	56.5	55.3	69.5	62.7	71.7
1975	39,300	84.7	81.7	89.3	88.7	89.4
1978	55,700	109.0	114.5	109.5	110.9	106.8
1979	62,900	119.0	130.8	124.3	128.1	114.7
1980	64,600	128.9	145.2	141.7	153.2	124.8
1981	68,900	136.0	157.4	149.1	172.1	135.6
1982	69,300	147.5	161.5	156.9	183.9	145.9
1983	75,500	156.3	165.5	161.0	192.2	157.6

Source: U.S. Bureau of the Census, *Statistical Abstract of the United States, 1984*, (Washington: GPO) and earlier years. Indices assume that 1977 = 100.

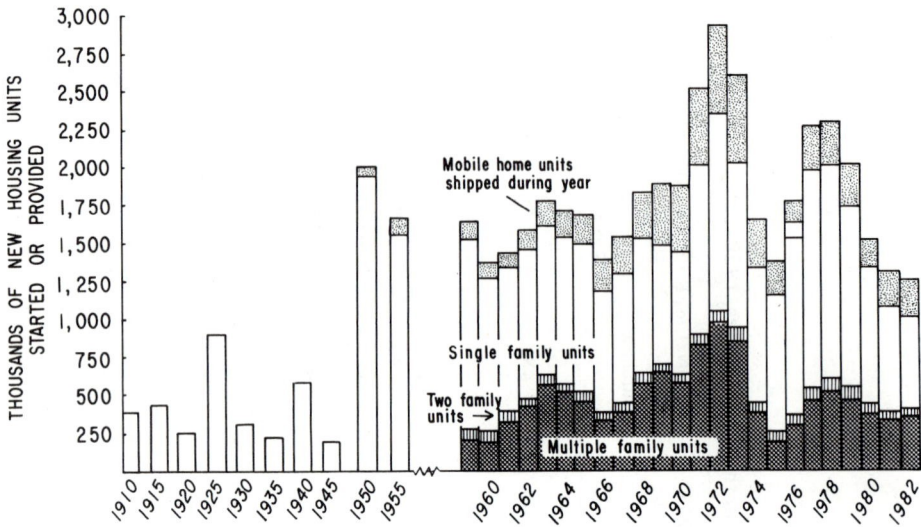

FIGURE 10-3.   Trends in starts of new residential unit construction and shipments of new mobile homes, United States, 1910-1982.

early 1970s brought peaks of 2.08, 2.38, and 2.06 million new units in 1971, 1972, and 1973, respectively, followed by totals of only 1.35 and 1.17 million units in 1974 and 1975, respectively. Another surge of building activity brought totals of 1.99 and 2.02 million new units in 1977 and 1978, respectively, followed by lows of 1.10 and 1.07 million new units in 1981 and 1982, respectively. Most of the new residential units provided since World War II have been single-family units. Duplexes and multifamily units accounted for 40.5 percent of the new construction during the 1971-1973 building boom but for less than 30 percent of the new units later in the decade.

Despite wide fluctuations in residential construction activity, building costs have been quite sticky when they have not been rising to new levels. Between 1910 and 1915, the Boeckh index of residential construction costs (1977 = 100) stood around 6.9. This index rose to 10.7 in 1918 and 16.0 in 1920. It dropped following the business recession of 1920 and then rallied and remained fairly stable at around 12.9 until 1930. It dropped to a depression low of 10.2 in 1932 and 1933 and then gradually climbed back to 12.9 by 1938. With the outbreak of World War II, building costs started a steady upward trend, which reached 28.3 in 1948, 37.8 in 1960, 56.5 in 1970, and 128.9 in 1980. Meanwhile, the construction cost index for apartments, hotels, and office buildings rose from 17.2 in 1945 and 57.5 in 1970 to 125.1 in 1980.

Trends in urban lot values. Higher residential site values have contributed significantly to the upward trend in housing costs since 1950. Reports published by the National Association of Home Builders based on surveys of its membership indicate that building sites accounted for 11 percent of the cost of the average new $10,000 house in 1949. By 1969, land accounted for 21 percent of the cost of the average house which now was priced at $25,600; and by 1977, land costs accounted for 25 percent ($12,150) of the average investment in a new house, which now had a price tag of $48,600.

A semiofficial measure of the upward trend in urban residential lot values is provided by the reports of the Federal Housing Administration on the estimated site values associated with new and existing houses covered by new FHA-insured mortgages (see Table 10-6). These reports, which involve properties of less than average value for the nation, show that the average market value of the building sites with new housing units rose from $1,035 in 1950 to $2,470 in 1960, $4,952 in 1970, and $11,809 in 1980. With this almost twelve-fold increase in lot prices, the ratio of site values to total values rose from 12.0 percent in 1950 to 20.5 percent in 1980.

Some of the increase in average building site costs reflects the higher prices associated with the acquisition and development of raw land for building site purposes. Some also stems from the fact that average lots are now larger than they were in 1940 and 1950 and that lots are often more developed than they once were. But most of the increase can be attributed to inflation and the impact of demand on the supply of available lots.

Residential market trends since World War I. The conglomerate effect of the many factors that affect residential market values can best be described through a recounting of recent market history. World War I gave rise to a housing shortage that was followed by a major urban housing boom during the 1920s. This boom

TABLE 10-6. Appraised Values of New and Existing Single-Family Houses and Their Building Sites, FHA Mortgage-Insured Properties, 1950–1982

YEAR	AVERAGE APPRAISED VALUE OF HOUSES WITH FHA-INSURED LOANS		AVERAGE APPRAISED VALUE OF BUILDING LOTS, FHA LOANS		RATIO OF AVERAGE SITE VALUES TO TOTAL VALUES	
	New units	Existing units	New units	Existing units	New units	Existing units
1950	$ 8,594	$ 9,298	$ 1,035	$ 1,150	12.0%	12.4%
1955	12,118	12,047	1,626	1,707	13.4	14.2
1960	14,899	13,304	2,470	2,356	16.6	17.7
1965	17,190	15,394	3,427	3,219	19.9	20.9
1970	23,559	18,517	4,952	3,973	21.0	21.5
1975	33,172	27,029	6,382	5,468	18.8	19.9
1978	42,091	34,323	7,764	6,985	18.5	19.8
1980	56,583	45,970	11,809	10,105	20.5	21.1
1981	59,870	46,604	12,482	11,410	19.9	21.8
1982	62,425	50,586	12,127	10,055	20.0	21.1

Source: U.S. Department of Housing and Urban Development, *1979 Statistical Yearbook* (Washington, D.C.: Government Printing Office, 1980), pp. 142–43; and Federal Housing Administration, *FHA Trends of Home Mortgage Characteristics*, Mortgages Insured under Section 203b, 1981–83.

tapered off after 1925, and with the onset of the Great Depression in 1930, contract rental rates declined; new building activity dropped to a low level; and numerous owners, pressed by reduced incomes and high mortgage indebtedness, offered their properties for sale at distressed prices.

A low point in the housing market was reached in 1932 and 1933. Property sales, market prices, new construction, building costs, and rental rates were all at their lowest ebb while thousands of owners were losing their properties through mortgage foreclosures or tax forfeitures. Conditions improved gradually after 1933. Property values remained low, but market activity increased. Rental rates climbed slightly, and building costs rose as the volume of residential construction increased with each new year. By 1939, a new building boom was in the making. Employment was up; average incomes were increasing; and real estate market activity was getting back to normal.

World War II brought numerous complications for the housing market. Rental rates and property values started to climb in many defense areas in 1940 and 1941. Rent controls were applied to these areas in early 1942. Private construction was restricted and placed under a priority system, which channeled most new private and public housing construction to military areas and cities with defense industries. Insufficient housing, increased employment, and higher incomes favored property price increases in many cities. Nationwide, however, the war prompted a feeling of uncertainty, which led to a decline in sales activity until late 1943, when an increasing volume of sales and rising property values heralded the beginning of another market boom.

With demobilization of the armed forces following the end of the war in 1945, it soon became apparent that the nation's cities suffered from a critical housing shortage. Rental quarters were difficult to find and often commanded black market prices. Thousands of people were moving to urban job opportunities; new families were being created at a rapid rate; and with full employment, higher incomes, a backlog of wartime savings, veterans' benefits, and public credit programs, strong effective demand existed for all available housing. Housing sales reached a peak in 1946, and average prices rose and kept on rising as building costs increased.

By 1950, the market value of old as well as new homes in many areas stood at double and sometimes triple their prewar level. This upward price trend stablized for a while in 1949 and 1950 and then rose again with the inflationary spiral and building restrictions that came with the outbreak of war in Korea in 1950. The cost of new single-family houses then rose steadily at a rate of roughly 2 percent annually until 1966, after which the rate of increase accelerated. Older house values also increased but at a slower rate until 1961, dropped slightly until 1966, and then experienced an upward surge.

Several factors contributed to the rising price trends of the 1950s and 1960s. Family incomes were rising along with the general cost of living. Average house rental rates increased gradually after the dropping of rent controls in most areas in the early 1950s. Higher rents and new construction permitted vacancy rates on rental units to rise to a high of 8.3 percent in 1965. Housing credit was readily available throughout most of this period, even though average interest rates rose from 4.5 to 5 percent in the late 1940s to 6 percent in 1960.

Housing prices, which had been rising at rates of between 5 and 8 percent annually during the 1960s, started to increase at double-digit rates during the 1970s, with the result that it cost about twice as much to buy an average single-family house in 1980 as in 1973. General price inflation was a principal cause of the upward swing. But rising prices were also caused by new demands in housing. Heavy local demands prompted above-average increases in California and Hawaii and in many metropolitan and resort communities. But a general upswing in demand also occurred nationwide as young people born during the population explosion of the 1950s moved into their household formation years. The upward trend in housing prices was also fueled by a ready availability of mortgage credit in most years. Shortages of mortgage money relative to demands, however, led to a boosting of interest rates to the 8.5 to 9 percent levels in 1970 and again in 1973.

Some 17.6 million housing units were built during the 1971–1980 decade. These units satisfied most of the market demand for new housing. Some communities even found themselves temporarily oversupplied with new apartments and condominiums. But an inordinate proportion of the new units constructed after 1970 were designed for upper-income families. With the rapid rise in housing costs after 1973 and mortgage interest rates rising to the 12 to 15 percent level by the end of the decade, thousands of low- and middle-income families found themselves priced out of the home ownership market. According to the Department of Commerce composite index, the cost of acquiring a new single-family house rose by 185 percent between 1970 and 1981. During this same period, average disposable incomes per capita increased by only 124 percent. Accepting the guideline that families should not spend more than 2.5 times their annual incomes in buying houses, a comparison of the nation's income distribution statistics for 1970 and 1981 shows that 84.2 percent of the families and 76.6 percent of the households had incomes of $10,000 or more in 1970. This was adequate income to purchase a $25,000 house at a time when the average new single-family house cost $23,500. By 1981 only 43.0 percent of the families and 36.5 percent of the households had incomes of $25,000 or more, the amount then needed for acquiring a $62,500 house. Meanwhile the price of the average new single-family house had gone up to $68,900.

Many fortunate families, of course, had the necessary savings and earning capacities to acquire residential properties. Home ownership became possible for many when wives joined their husbands in the labor force. But large numbers lacked the necessary savings and incomes needed to acquire properties and carry the higher costs associated with paying off mortgages. These people often found it less expensive, though frequently less satisfying, to rent rather than buy.[58] With widespread

---

[58] Average residential rental rates increased less rapidly than either home purchase or home ownership costs (taxes, insurance, mortgage interest payments, energy costs, and utilities) during this period. The Bureau of Labor Statistics' indices for home purchase costs, home ownership costs, and residential rents increased 126, 174, and 89 percent, respectively, between 1970 and 1981 (see Table 10–5).

unemployment in the early 1980s, many families doubled with parents or friends. Lacking low-cost housing opportunities, many others purchased mobile homes. The number of these shipped to dealers increased from 103,700 in 1960 to 401,110 in 1970, to a high of 575,940 in 1972, and then dropped to 220,000 in 1980.

Other Real Estate Markets

Little attempt has been made to systematically record the trends with other real estate markets. Their price trends have generally followed those of the farm and urban residential real estate markets.

Sites with recreation and second-home development potential increased substantially in demand and in price between 1950 and 1974 and again between 1977 and 1981. Similar trends affected the market for urban commercial and industrial sites. Office rents rose to new highs in most cities during the 1960s and early 1970s and again in the late 1970s. Demands for more office space, more commercial developments, and more new industrial properties triggered considerable public and private urban renewal in cities as well as development of new sites in suburban areas. A significant measure of the investments associated with these developments is indicated by the construction trend data reported in Table 10-7. These data show that nonresidential building construction provided approximately two-thirds as much new floor space in 1960, 1970, and 1980 as residential construction and that it involved investment costs of $12.2 billion in 1960, $24.5 billion in 1970, and $52.5 billion in 1980.

TABLE 10-7. Value of New Construction and New Floor Space Provided, United States, 1960 to 1980

CLASS OF CONSTRUCTION	VALUE OF CONSTRUCTION			FLOOR SPACE OF BUILDINGS		
	1960	1970	1980	1960	1970	1980
	(billions of dollars)			*(millions of square feet)*		
Nonresidential buildings						
Commercial	3.7	9.1	24.8	283	530	688
Industrial	2.1	3.7	8.4	178	212	216
Education and science	3.0	5.3	6.8	196	195	95
Hospitals	0.8	2.8	5.3	36	75	54
Public buildings	0.7	1.0	1.6	33	29	18
Religious	0.8	0.6	1.2	53	27	28
Social and recreational	0.6	1.1	4.4	44	47	101
Miscellaneous	0.5	1.0		31	42	
Total	12.2	24.5*	52.5	854	1,157	1,200
Residential	15.1	24.8	63.7	1,300	1,781	1,892
Nonbuilding construction	15.1	19.0	32.2	—	—	—

*Totals do not add to $24.5 billion due to rounding.
Source: U.S. Bureau of the Census, *Statistical Abstract of the United States, 1982–83* (Washington, D.C.: Government Printing Office, 1982), p.745.

Land Booms and Their Control

Market prices for most types of real estate increased much more rapidly than the general price level during the 1950–1982 period. (see Figure 10-4). These rapid increases and the large volume of sales brought the rise of land-boom conditions to some markets. In evaluating this situation and the potential threat it posed for the national economy, it should be recognized that land booms have been a familiar feature in the past history of the real estate market. Many booms such as the Chicago canal boom of the 1830s, the Southern California land boom of the 1880s, and the Florida land boom of the 1920s have been localized.[59] Others such as the land booms of 1815–1818, 1832–1836, 1854–1857, and the rural and urban market booms after World War I were nationwide in scope.

Most of these booms have been associated with periods of high demand for local properties, rapid population growth, public improvements such as canals and railroads, discoveries of oil or valuable minerals, and inflationary movements. Most of

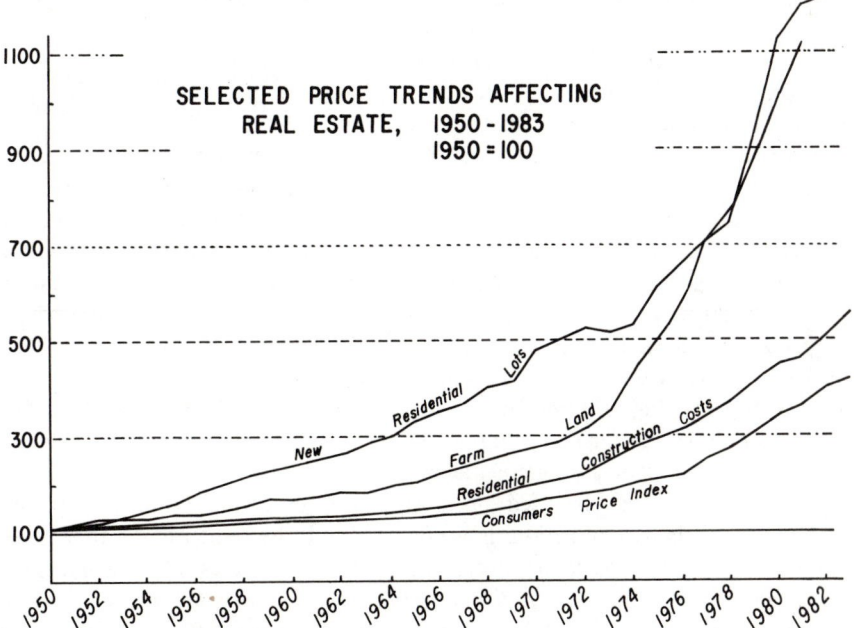

FIGURE 10-4. Selected price trends affecting real estate in the United States, 1950–1982 (1950 = 100).

[59]Cf. Homer Hoyt, *100 Years of Land Values in Chicago* (Chicago: University of Chicago Press, 1933); Glen M. Dumke, *The Boom of the Eighties in Southern California* (San Marino, Calif.: Huntington Library, 1944); Homer B. Vanderblue, "The Florida Land Boom," *Journal of Land and Public Utility Economics*, 3 (May and August 1927), 111–31 and 252–69; and Philip H. Cornick, *Premature Subdivision and Its Consequences* (New York: Institute of Public Administration, Columbia University, 1938).

them have been carried along on waves of optimism and overconfidence, conditions that favored considerable speculation with loose financing and an upward spiraling of market prices to levels that were unreasonably high in light of the immediate income-producing potentialities of the properties involved. Some of these booms were fostered and fed by the promotional activities of particular groups; some gradually petered out; and some blossomed into full-sized bubbles, only to bring substantial losses to many investors at the time they burst.

The collapse of some of the leading land booms of the past led to depressed business conditions for the nation as a whole. Examples include the panics of 1819, 1837, 1857, and 1893. The agricultural depression of the 1920s was seriously aggravated by the high debt loads many farmers carried over from the 1918–1920 land boom. Widescale use of loose financing arrangements during the housing and subdivision boom of the 1920s also brought serious debt problems for numerous urban property owners during the 1930s.

The land boom of the 1970s was fed by factors that bore a marked similarity to those that fueled earlier booms. Recognition of this situation and awareness of the distressing aftereffects of the earlier land booms has aroused interest in ways and means of keeping real estate markets from getting out of hand. Among the more important control measures, mention should be made of the possible use of market education, credit restrictions, taxation of capital gains, price ceilings, sales restrictions, and inflationary control measures.

Land-boom psychologies can often be dampened through the judicious and effective use of educational measures. Newspaper and radio publicity, speeches by public officials and others, and editorial comment may be used to advise prospective buyers of market facts and of the unfortunate consequences of unfettered boom conditions. Public credit restrictions and upward adjustments in interest rates can provide effective control measures to limit the number of potential buyers. Higher taxes on capital gains can be used to discourage speculation and transfers of properties during boom periods. This approach is naturally unpopular with sellers; but a tax approaching 100 percent of the seller's capital gain can provide an effective means for discouraging speculation. Price ceilings might also be used to freeze real estate market values during emergency periods.

Control measures can be used alone or in various combinations to discourage land booms. In their application, however, it should be remembered that there is nothing wrong with rising land values, as long as the increases reflect genuine prospects of additional productivity. A need for controls arises only when increases in land values appear to be unwarranted and when they pose a threat to the nation's well-being.

—SELECTED READINGS

American Institute of Real Estate Appraisers, *The Appraisal of Real Estate*, 7th ed. (Chicago: American Institute of Real Estate Appraisers, 1979).

Bloom, George F., Arthur M. Weimer, and Jeffrey D. Fisher, *Real Estate*, 8th ed. (New York: John Wiley, 1982).

Friedman, **Edith J.**, ed. *Encyclopedia of Real Estate Appraising*, 3rd ed. (Englewood Cliffs, N.J.: Prentice-Hall, 1978).

Kahn, Sanders A., Frederick E. Case, and **Alfred Schimmel,** *Real Estate Appraisal and Investment* (New York: Ronald Press, 1977).

Kinnard, William N., *Income Property Valuation* (Lexington, Mass.: Heath Levington Books, 1971).

Murray, William G., *Farm Valuation and Appraisal*, 5th ed. (Ames: Iowa State University Press, 1969).

Ratcliff, Richard J., *Urban Land Economics* (New York: McGraw-Hill, 1949), chaps. 10–12.

Ring. Alfred A., *The Valuation of Real Estate*, 2nd ed. (Englewood Cliffs, N.J.: Prentice-Hall, 1970).

Seldin, Maury, *Land Investment* (Homewood, Ill.: Dow Jones-Irwin, 1975).

Shenkel, William M., *Modern Real Estate Principles* (Dallas Tex.: Business Publications, Inc., 1977), chaps. 8–11, 14, and 15.

Suter, Robert C., *The Appraisal of Farm Real Estate* (Danville, Ill.: Interstate Printers & Publishers, 1974).

Ventolo, William L., *Fundamentals of Real Estate Appraisal* (Chicago: Real Estate Education Co., 1975).

Wendt, Paul F., *Real Estate Appraisal: Review and Outlook* (Athens: University of Georgia Press, 1974).

11

IMPACT
OF INSTITUTIONAL FACTORS
ON LAND USE

How we use real estate resources is determined in large measure by the institutional acceptability of our activities and practices. Our behavior as individuals and as members of families, groups, communities, and society is affected more than most of us realize by cultural attitudes, custom and tradition, habitual ways of thinking and doing things, legal arrangements, government programs, religious beliefs, household considerations, and other manifestations of the workings of our cultural background and environment.

The factors that contribute to the institutional framework can be classified as cultural, economic, political, religious, and social phenomena. Some of them have considerably more economic significance than others. But economic or not, each factor usually plays an important role—sometimes strategic, though more often routine—in directing the course of our behavior. In this respect, each factor represents a part of the social organization into which individuals are born, which they are always modifying, and which in turn directs and controls them in their various activities.

Institutional factors exert a continuing influence on economic behavior. They help to make it more stable and predictable but at the same time adjustable and dynamic. Their significance is often overlooked or taken for granted in economic discussions. Yet their overall impact is such that they can never be assumed away once economic theory is taken out of its academic vacuum and applied in the arena of real life.

NATURE OF INSTITUTIONAL FACTORS

To understand economic behavior, one must start with the individual. As Ernest S. Griffith has observed, anyone "who would really understand the changes in government or economics must start a long way from his ultimate goal by asking himself what sort of a creature man really is; for changes in man's group or institutional behavior are generated ultimately by his individual nature and must not do too great violence to this nature."[1]

No attempt will be made here to analyze the basic appetites, drives, desires, and other forces that provide the basis for human motivation; nor to explain the effects the biochemical composition of our bodies, our nervous systems, or our social nature have on our outlook and behavior.[2] Suffice it to say that as individuals, we are all products of nature and nurture. We are what we are and we think and act as we do because of our inherited genes and because of the conditioning of the cultures into which we are born. As human beings, we have wide potentials for development, but we differ in our capacities, personalities, and attitudes. Almost everyone has a strong desire for survival. And up to a point, most people tend to maximize their economic self-interests; but their behavior is also strongly affected by family considerations and in many cases by desires for power and prestige; saving of face; the promotion of hobby interests; and cultural, psychic, religious, and altruistic values.

While each individual stands at the center of his or her world, it is important to note that most individual activity involves operations in a larger universe. Men and women are social animals. They are born into the stream of society, and from the moment of birth they share in a social heritage that conditions and controls their attitudes and reactions. Unlike Robinson Crusoe, they are seldom in a position to make their own rules. Instead they tend to accept and conform to the rules, social organization, and institutions that surround them during their formative years. They usually accept the system of "discipline and obedience" into which they are born, because experience tells them that "conformity to repeated and duplicated practices . . . is the only way to obtain life, liberty, and property with ease, safety, and consent."[3]

The rules incorporated in our social heritage often change with time. At all stages, however, they tend to dictate what is and what is not considered acceptable behavior. In this respect, it is well to remember that

> human behavior is made up of two separate and distinct elements, the one biological, the other cultural. . . . On the one hand is the organism, composed

[1] Ernest S. Griffith, *The Impasse of Democracy* (New York: Harrison-Hilton Books, 1939), p. 344.

[2] For a detailed discussion of this approach to economic behavior, see C. Reinold Noyes, *Economic Man in Relation to His Natural Environment* (New York: Columbia University Press, 1948).

[3] John R. Commons, *Institutional Economics: Its Place in Political Economy* (Madison: University of Wisconsin Press, 1959), p. 45. Originally published by Macmillan, 1934.

of bones, muscles, nerves, glands, and sense organs. This organism is a single coherent unit, a system, with definite properties of its own. On the other hand is the cultural tradition into which the organism is born. It could have been born into one cultural tradition as well as another, into Tibetan as well as American or Eskimoan culture. But from the standpoint of subsequent behavior, everything depends upon the type of culture into which the baby is introduced by birth. If he is born into one culture he will think, feel, and act in one way; if he is born into another, his behavior will be correspondingly different. Human behavior is, therefore, always and everywhere, made up of these two ingredients: the dynamic organization of nerves, glands, muscles, and sense organs, that is *man*, and the extrasomatic cultural tradition.[4]

The various aspects of group, collective, or social action that influence and control individual behavior may be described as *institutions* or as institutional factors. This concept is conveyed by Common's well-known definition of an institution as "collective action in control, liberation, and expansion of individual action."[5] It is also suggested by Griffith's definition of an institution as "purposeful cooperative behavior,"[6] and by Barnes's treatment of institutions as "the social structure and machinery through which human society organizes, directs, and executes the multifarious activities required to satisfy human needs."[7]

Institutions can be classified for our purposes as *primary* and *secondary* institutions. *Primary institutions* are the more elemental institutions such as government, property, industry, education, religion, and the family. Each of these primary institutions is associated with a subordinate group of secondary institutions. Government, for example, involves a legion of subordinate institutions such as constitutions, legislatures, political parties, civil service systems, tariffs, police forces, and local zoning ordinances. In similar fashion, religion involves secondary institutions such as beliefs, creeds, rituals, sacraments, symbols, and taboos.

As these examples suggest, widely different factors in our society may be classified as institutions or institutional factors. The broad scope of this coverage was emphasized by Walton H. Hamilton when he wrote:

Institution is a verbal symbol which, for want of a better, describes a cluster of social usages. It connotes a way of thought or action of some prevalence and permanence, which is embedded in the habits of a group or the customs of a people. . . . Institutions fix the confines of and impose form upon the activities of human beings. The world of use and wont, to which we imper-

---

[4] From *The Science of Culture* by Leslie A. White. Copyright 1949 by Leslie White. Used by permission of the publishers, Farrar, Straus and Giroux, Inc., pp. 121–23.

[5] Commons, *op. cit.*, p. 5.

[6] Griffith, *op. cit.*, p. 20.

[7] Harry Elmer Barnes, *Social Institutions* (Englewood Cliffs, N.J.: Prentice-Hall, 1942), p. 29.

fectly accommodate our lives, is a tangled and unbroken web of institutions. The range of institutions is as wide as the interests of mankind. . . . Arrangements as diverse as the money economy, classical education, the chain store, fundamentalism and democracy are institutions. They may be rigid or flexible in their structures, exacting or lenient in their demands; but alike they constitute standards of conformity from which an individual may depart only at his peril.[8]

Institutions represent established arrangements in society and established ways of doing things. They involve the working rules of society. And in many instances— as with such primary institutions as our economic, educational, family, legal, and political systems—they provide systems of control that point the way to what is considered acceptable individual and group behavior. As Steiner observes,

> . . . these systems are neither contiguous nor mutually exclusive. Rather, they are inextricably interwoven and interdependent. Clusters of institutional arrangements and methods of doing things exist in each system. These set the pattern of operation of the system of control. Some of these institutions and methods of doing things are merely formalized codes of human conduct. As such, they tie directly into cultural values. At this point, systems of control and cultural values get rather well tied together and become almost indistinguishable.[9]

Economic institutions. Economic activity is often governed by the joint functioning of economic and institutional factors. In this process, institutional factors frequently take on an economic cast and become forces of economic significance. Our concepts of private property and the sanctity of contracts, for example, are institutions of legal origin. Yet they are also regarded as economic institutions because of the controlling impact they have on the operation of the economic system. Other institutions such as public fiscal policy, taxes, and production regulations also have tremendous impacts on economic life, even though they may be classed as governmental controls.

Those institutions that are closely identified with economic activity may be regarded as economic institutions. As one writer has observed: "Economic institutions are social arrangements . . . by means of which business life is organized, directed, conducted, and regulated."[10] Hundreds of economic institutions operate in modern society.

[8] Walton H. Hamilton, "Institution," *Encyclopedia of the Social Sciences* (New York: Macmillan, 1932), 8, 84.

[9] George A. Steiner, *Government's Role in Economic Life* (New York: McGraw-Hill, 1953), p. 23.

[10] Vernon A. Mund, *Government and Business*, 4th ed. (New York: Harper & Row Publishers, Inc., 1965), p. 4.

IMPORTANCE OF SPECIFIC INSTITUTIONS

The institutional factors that affect the ownership and use of real estate resources may be described as *landed institutions*. Of these factors, the concept of property has the most significance. (see Chapter 12, for a more detailed discussion). Several others—some of them economic and some primarily noneconomic—also have significant impacts on the ownership and use of land resources. Consideration is given here to the effects family, educational systems, government, law, custom, and religion have on the ownership and use of real estate resources.

Family and Educational Systems

Some of our most basic actions and attitudes concerning land resources and their use reflect our conditioning by family and educational systems. The nature of these two institutions has changed over time. It is still normal, however, for young couples to marry, have and raise children, and consider education as an appropriate means for preparing themselves and their children for participation in and enjoyment of modern life.

Families involve social systems of privileges, joys, and responsibilities. They have an extremely important impact on economic behavior, because in most cases it is families rather than single individuals that operate as the planning and resource-using units. Individual operators take the responsibility for making economic decisions; but the decisions they make are tempered by their concern for the support, comfort, well-being, and future welfare of their families. Family goals provide basic incentives for the development and use of land resources. Husbands and wives look to direct and indirect uses of land for their livelihood. They seek homes and often a spot of earth they can call their own. They plan for the future happiness and security of the family and its members. They look to land for a continuing flow of raw materials and products; and once their basic needs for these are supplied, they continue to look to land for many of the amenities, recreation values, and satisfactions they associate with quality living. With adverse conditions, support and protection of the family may necessitate resource disinvestment and exploitation. But ordinarily, family considerations favor resource planning and development practices that extend their use-potential beyond current needs.

Education has become a steppingstone to participation in the opportunities of modern life. People go to school to enhance their income-producing abilities and to develop understandings and appreciation of how the physical world and human society operate. Education teaches us how to make a living by preparing us to function as skilled workers. It can also teach us how to live and appreciate life. It has affected land use by raising individual aspirations, facilitating the development of improved technologies, and pointing the way for better resource management. Unfortunately, acquired skills and knowhow can also be used to exploit and destroy valuable resources if care is not taken to protect their long-run production potential.

Government and Political Institutions

Most of the rugged individualists who settled and developed the American frontier placed high personal values on their economic and political freedom. Many felt that "that government governs best which governs least." This emphasis on individual freedom from governmental restraint has permeated much of our political thought and has caused many people to view extensions of governmental influence with alarm.

Despite this point of view, recent decades have brought a gradual expansion of the scope of government in the United States and in most other countries. Much of this expansion has come with the increased exercise of public power in the resolution of conflicts of interest, in the advancement of the public welfare, and in the pursuit of new social goals. Far from leading to losses of individual freedom, these actions have often enhanced the economic opportunities and freedoms enjoyed by average citizens. They have also contributed to the rise of a "great leviathan," which now plays a dominant role in our lives. As Steiner observes:

> The influence of government . . . is felt today in every home, every manufacturing plant, and every farm. It circumscribes, channels, directs, and controls actions of every description. Economic institutions operate on the basis of government action or the conscious lack of government interference. No corner of economic life escapes the hand of government. The touch is sometimes light and sometimes heavy, at times helpful and at other times restraining; it may be agreeable or arbitrary, and beneficent or greedy. But whatever may be its character at any one point, the power of government affects our economic lives intimately and often irrevocably. Government regulation is a silent partner in all economic activity and an active partner in most economic activity.[11]

Government and political institutions have long exerted an important impact on the ownership and use of real estate resources. The effects of these institutions are manifold. For our purposes, it is sufficient that we emphasize two aspects of these impacts: (1) the overall effect of government policies and restrictions on public and private decisions regarding land resources, and (2) the impact of the organization and framework of government on the development and administration of public land resource policies.

Effects of government policies and restrictions. Almost every decision regarding the ownership or use of land resources is in some way affected by public policies or restrictions. Real property taxes represent an annual levy on real estate ownership and can be used to force lands into more intensive uses. Inheritance taxes—or "death duties" as they are known in Great Britain—can force the breaking up of landed estates. The power of eminent domain can be used for public acquisition of properties from owners who are unwilling to sell. The government's police

---

[11] Steiner, *op. cit.*, p. 2. Cf. also George A. Steiner and John F. Steiner, *Business, Government and Society* (New York: Random House, 1980), chaps. 2 and 10.

power, its sovereign power to make and enforce rules and regulations may be used to protect property rights; prevent fraud; and force citizen compliance with public health standards, building codes, or local land-use ordinances.

Numerous examples illustrate the impact of public policies on the development and use of real estate resources.[12] During past decades, the federal government has acquired substantial additions to the public domain and prescribed a liberal public land sale and disposal program, which facilitated rapid settlement of the Western frontier. More recently it has reserved large areas of public lands as public park, forest, mineral, grazing, and wildlife reserves. It has used public funds to reacquire certain lands from private owners for forestry, military, and other uses. It has undertaken large-scale multipurpose resource-development programs such as the construction of the Hoover, Grand Coulee, and Shasta dams in the West and the development of the Tennessee Valley Authority in the East. It has engaged in numerous public works such as building highways, digging canals, and providing navigation improvements.

The federal government has established agencies to increase the credit facilities available to farmers, homeowners, and business operators. It has used tariffs to protect and favor domestic industries. It has authorized special arrangements and exemptions under its tax laws that provide inducements for home ownership, residential construction, and commercial and industrial investments. It has used subsidies to encourage conservation practices, dispose of farm surpluses, construct factories for defense industries, and promote the construction of low-rent public housing. It has used price supports to bring stability to the agricultural sector of the economy and rent controls to prevent tenant gouging during periods of severe housing shortages. Cost-sharing arrangements have been used with state and local governments to promote new highway construction, the improvement of small watershed areas, metropolitan planning, and redevelopment and renewal of blighted urban communities. Environmental protection regulations affecting air and water quality also have important impacts on land use.

In addition to the federal policies, state and local governments also have policies that affect the uses made of land resources. Some of the most significant of these involve state authorization of land-use zoning ordinances, subdivision regulations, building codes, forest-cutting restrictions, and similar measures that direct private land-use practices in the public interest. Other important programs include the provision and administration of parks and recreation areas, the location and building of streets and highways, and the provision of public parking areas in urban areas. State and local action is needed with area planning, urban renewal, and public housing projects. State legislation also governs the rights we hold in land and water resources and the ways in which rights are allocated between landlords and tenants and between mortgagors and mortgagees.

---

[12] Cf. V. Webster Johnson and Raleigh Barlowe, *Land Problems and Policies* (New York: McGraw-Hill, 1954), chap. 4; and John F. Timmons and William G. Murray, *Land Problems and Policies* (Ames: Iowa State College Press, 1950); Land Economics Institute, *Modern Land Policy* (Urbana: University of Illinois Press, 1960); and Howard W. Ottoson, ed., *Land Use Policy and Problems in the United States* (Lincoln: University of Nebraska Press, 1963).

Framework of government. Governments vary considerably in their organization and in the scope of the powers assigned to their various levels. They range from the absolute monarchies of the past, under which all political power stemmed from the pharaoh, emperor, or king to theoretical, anarchistic societies under which no individual is subject to the political control of others. Most nations now operate under constitutional forms of government. This means that their organization and the scope and distribution of their powers are spelled out either in a written constitution or in legislation and recognized precedents.

Some constitutional governments, such as the Netherlands, are highly centralized with most of the sovereign power concentrated in a central government. Others such as the American States under the Articles of Confederation (1778–1789) may be loosely organized and vest virtually all of their political authority in the various state governments. Still others such as the United States and Canada operate as federal systems with the political power divided between a national government and several states or provinces.

Patterns of governmental organization often have direct effects on land resource use policies because they determine the powers of various levels of government. With constitutional governments, policy proposals must fall within the legitimate powers assigned to government and must not conflict with rights and privileges guaranteed to individuals. Furthermore, policies may be developed and administered only by those branches and levels of government duly authorized to exercise such powers. In the United States, this means that some policies can be instituted on a national basis while others require state or local action.

The federal government of the United States operates with delegated powers. Its *express* powers are limited to those enumerated and conferred by the federal Constitution. Those powers not delegated to the federal government are known as *residual* powers and are reserved to the states or to the people. Strictly construed, this framework of government leaves the states with the powers not delegated to the federal government and not prohibited to them by either the federal Constitution or the constitutions of the several states. Counties, cities, townships, and other local units of government operate with powers delegated to them by the states.

In theory, this organization pattern makes the federal government a government with limited powers. The scope of its powers, however, has been definitely broadened by liberal interpretation. The courts have accepted a doctrine of *implied* powers, which makes it possible for the federal government to do many things not specifically authorized by the Constitution.[13] In its exercise of its implied powers, the federal government must respect the prohibitions of the Constitution, the guarantees

---

[13] The doctrine of implied powers was suggested by Alexander Hamilton in 1791. Twenty-eight years later in the famous case of *McCulloch* v. *Maryland*, 4 Wheaton 315 (1819) this doctrine was enunciated by the United States Supreme Court. In the reasoning of Chief Justice Marshall, the federal government has an implied power to use such measures as are necessary in the exercise of the powers bestowed upon it. "Let the end be legitimate, let it be within the scope of the constitution, and all means which are appropriate, which are plainly adapted to that end, which are not prohibited, but consist with the letter and spirit of the constitution, are constitutional."

of individual liberties and privileges covered by the Bill of Rights, and the general reservation of residual powers to the states and to the people. Infringement on any of these areas involves unconstitutional action. In the *Hoosac Mills* case, for example, the majority of the Supreme Court held the Agricultural Adjustment Act of 1933 invalid because its program for regulating agricultural production invaded "the reserved rights of the states" and involved statutory action "beyond the powers delegated to the federal government."[14]

Division of powers between the federal and state governments requires the involvement of different units of government in the development and administration of policies affecting land resources. The federal government can operate and administer its own lands, set up land disposal policies, and acquire private properties for public use. It can provide funds for land-use research, housing and reclamation projects, agricultural conservation payments, price supports, highway and canal construction, and the administration of government-insured mortgage credit programs. It can use its commerce power and its civil rights and environmental protection programs to influence the uses made of real estate resources. But in practice, except for its jurisdiction in federal areas such as the District of Columbia, it often lacks many of the sovereign powers involved in familiar day-to-day cases of land resource control. It acts for the states, however, in all matters that require treaty negotiations with other nations.

Most public powers affecting the ownership and use of land resources are vested in the several states. Virtually all legislation dealing with problems such as police protection, landlord-tenant relations, water rights, property taxation procedures, plat restrictions, and state land administration must be drafted on a state basis. Local units of government in turn operate as agencies of the states. It is essential that the states establish them as political entities, clothe them with powers and responsibilities, and provide them with appropriate enabling legislation. Local governments must be vested with this authority before they can proceed with the enactment of zoning ordinances or building codes; the creation of planning commissions; and the establishment of conservation, grazing, irrigation, drainage, or levee districts.

Counties, cities, villages, townships, and other local districts occupy a bottom rung in the hierarchy of political power. Despite this position, they often have a far greater impact on land-use practices than the states or the federal government. These are the units of government that levy and collect property taxes; enact and enforce zoning ordinances; and provide the officials who protect individuals and their property from violence, theft, fraud, and fire. They enforce most of the public police-power regulations that affect public health, safety, morals, and general welfare. Through their educational, planning, and other activities, they have considerable opportunity and power to influence individual attitudes and actions regarding land resource use.

Public land policies sometimes call for international agreements such as those

---

[14] *United States* v. *Butler et al.*, 297 U.S. 1 (1936).

that govern the development of power and water storage facilities on the Rio Grande or control pollution in the waters of the Great Lakes. Interstate compacts such as the Colorado River compact are occasionally needed to deal with interstate problems. Regional watershed approaches such as that used by TVA are also used at times. Within the states, counties and minor civil divisions sometimes work together on drainage and water management problems or in planning the coordinated development of metropolitan areas. *Ad hoc* units of government are also set up at times to deal with the particular problems that arise with the establishment of grazing, conservation, drainage, irrigation, flood control, and levee districts. These special districts ordinarily disregard political boundaries and include only those areas affected by the land problem in question.

Law and the Legal System

Law has been defined as that body of rules and regulations recognized as binding by citizens and nations. These rules and regulations represent an easily visualized type of collective action in control of individual behavior. They are important because they set the legal boundaries within which accepted individual and group behavior takes place. Because of its scope, "law is an all-pervading part of our social structure" and "there is no moment in our lives when our actions or inactions are not in some way subject to legal valuation."[15] As Barnes observes,

> Laws and lawyers are today the most important directive element in our civilization. Our technique of production, transportation, and communication may be determined and controlled by science and machinery, but our institutional life is dominated by law and lawyers. . . . Ours is as much a lawyer-made civilization, on its institutional side, as the civilization of Assyria and Rome was a military one, and that of the Middle Ages a religious one.[16]

What we ordinarily recognize as law really comes from three sources: (1) from statutes, ordinances, and administrative regulations; (2) from established customs that have gained the sanction of legal authority; and (3) from judicial interpretations and court decisions. Altogether, we probably have several million laws. Some have been passed by Congress and state legislatures, many more by city councils, county boards, and voters themselves. Others involve executive orders and administrative rules laid down by various public agencies. Yet as important as these examples are, one must not lose sight of the fact that a considerable proportion of our law is court-made.

From a historical standpoint, the beginnings of law are found in the social customs accepted and enforced by primitive peoples. As civilization developed, these social usages often became the basis for written law. Sometimes their substance was enacted into statutes. More often they were accepted as a type of prece-

---

[15] J. H. Beuscher, *Farm Law in Wisconsin* (Appleton, Wis.: C. C. Nelson, 1951), p. 1.
[16] Barnes, *op. cit.*, p. 354.

dent and became part of the law because of their recognition and acceptance in judicial decisions. Eventually, many judicially sanctioned customs found their way into various codifications of law such as the Code of Hammurabi (2000 B.C.), the Justinian Code (A.D. 529), and the Code Napoleon (A.D. 1804). But even when codified, the real meaning of these laws has usually depended on judicial interpretations.

This evolutionary pattern describes the development of the legal system accepted in most English-speaking countries. Units of government within these countries have their constitutions, statutes, ordinances, and regulations—many provisions of which are based on custom. Some of them have codified portions of the prevailing legal theory advanced in judicial decisions. But most still look to the English common law for significant portions of their ruling law. This is particularly true of laws relating to land resources. As Beuscher indicates,

> Almost all of the law of contracts; practically all of the law of torts (intentional or unintentional injury to person or property); much of the law of real estate and most of the law of agency is based, not on statutes, or administrative legislation or municipal ordinances but on court opinions.[17]

The English common law on which much of our legal system is based involves court-made law. It is found not in the statute books but rather in leading court opinions and in legal commentaries and compendiums. Many common-law principles go back to the customary practices of medieval England. But while the common law places heavy emphasis on precedents established in times past, it nevertheless represents a dynamic, flexible body of legal doctrine. Its framework is such that it can easily adjust to new situations and changing conditions.

The common-law approach, with its acceptance of individual decisions as precedents for future action that thereby become rules of the game, is widely used in our society. It is frequently applied in individual households and firms as well as in public actions. It provides a workable means for resolving conflicts of interest. In their use of this approach, courts often make far-reaching decisions that have a telling impact on various aspects of economic and social life as well as on legal institutions. When the Supreme Court reaches decisions on subjects such as racial segregation, antitrust regulations, and land-use practices, it really acts as "an authoritative faculty of political economy for the United States."[18]

It should be noted in passing that the legal system accepted in English-speaking countries is not the only system of law that prevails in the world today.[19] Eight important systems of law are now in use.[20] These systems have developed from dif-

---

[17] Beuscher, *op. cit.*, p. 14.

[18] Commons, *op. cit.*, p. 712.

[19] The English common law is accepted in all of the United States except Louisiana, which follows the Code Napoleon. Various aspects of Spanish law are accepted in some of the Southwestern states.

[20] Sixteen important systems of law have been developed during recorded history. Of these the Egyptian, Mesopotamian, Hebrew, Greek, Roman, Celtic, maritime, and canon systems

ferent customs, circumstances, and attitudes. They differ in content and coverage, in their philosophy and points of view. Some, such as the English system, emphasize individual rights while others exalt the state and stress an absolutist approach. With these differences, it is only natural that the legal concepts of rights and responsibilities accepted in some countries should appear quite foreign to the thinking of others.

Effects of law upon land and resources. Hundreds of examples can be cited to illustrate the imprint of law and legal institutions on land resources. Constitutions involve laws that control government. They allocate legal powers concerning the use of land resources between units of government. They also contain protective clauses, such as the "due process" and "equal protection of the law" clauses of the Constitution of the United States, which safeguard personal and property rights against arbitrary actions.

Governments have laws and ordinances that affect the acquisition of real property, registration of land titles, and leasing and mortgaging of properties. Transfers of property from generation to generation are carefully regulated by legal procedures. Landowners are protected by laws concerning property boundaries, trespassing, and adverse possession. Operators are directly affected by taxation and police power measures, laws that create drainage or conservation districts, and regulations such as those that require produce grading or fair trade practices.

Most laws affecting real estate resources involve the distribution or control of the rights and responsibilities individuals and groups have in property. This is true of the laws of real property. It applies to an individual's right to personal liberty, freedom to own property and freedom from the slavery or peonage rights others could hold in one's person. It also applies in varying degrees to the laws of torts, agency, and contract. The right individuals have to claim damages for injuries, for example, often has a tempering effect on the way in which land resources are used. Real estate use can also be affected by laws that define the powers of agency that can be delegated to others to act in one's behalf and by the emphasis given to fulfillment of contracts.

Custom and Habit

Law and government represent organized forms of collective action. As such, they have important impacts on individual behavior. Yet their influence is often matched, if not exceeded, by the less formal controls exerted by custom and habit. Like law, custom represents an accepted way of doing things. It can provide the basis for new laws; but it differs from law in the extent to which it has been formalized and to

---

no longer operate as separate systems. Wigmore classifies the present systems as Anglican, Romanesque, Germanic, Slavic, Chinese, Hindu, Japanese, and Moslem. In addition to these classifications, tribal customs still rule in many areas where no organized system of law has developed. Cf. John H. Wigmore, "A Map of the World's Law," *Geographic Review*, 19 (1929), 114-20; also Wigmore, *A Panorama of the World's Legal Systems* (St. Paul, Minn.: West Publishing, 1928).

which it can be enforced. The areas of collective action covered by custom and law often overlap but they are by no means contiguous. Laws cover many matters not affected by custom; and customs in turn involve many practices not as yet covered by law.

Most customs start as rational decisions. Farmers may find that a certain cultural practice fits their conditions. They accept it, and others follow their example. Soon this precedent for action becomes a working rule of the local society. Others coming later may accept this and other customary practices without question because of the prestige and status they have acquired with age. Even when people are inclined to question the authority ascribed to the "dead hand of the past," they often accept customs because they represent part of the system of collective controls into which they were born. As Commons has observed,

> Individuals begin as babies. They learn the custom of language, of cooperation with other individuals, of working towards common ends, of negotiations to eliminate conflicts of interest , of subordination to the working rules of the many concerns of which they are members. They meet each other, not as physiological bodies moved by glands, nor as "globules of desire" moved by pain and pleasure, . . . but as prepared more or less by habit, induced by the pressure of custom, to engage in those highly artificial transactions created by the collective human will. They are . . . found where conflict, interdependence, and order among human beings are preliminary to getting a living. Instead of individuals the participants are citizens of a going concern. Instead of forces of nature they are forces of human nature. Instead of mechanical uniformities of desire . . . , they are highly variable personalities. Instead of isolated individuals in a state of nature they are always participants in transactions, members of a concern in which they come and go, citizens of an institution that lived before them and will live after them.[21]

People are rational beings. An ability to reason and plan separates them from lower forms of life. Yet even though they are capable of deep thought, much of their activity is rooted in habits of thought and action. When they must, they can compute marginal rates of substitution for different goods and determine the approximate value of the marginal utilities they receive from various expenditures. In countless cases, however, they avoid the need for these calculations by following routine patterns of action already established by custom or habit. Most of their routine activities are rooted in earlier deliberate decisions. On a day-to-day basis, they read the same newspapers, eat given types of food, brush their teeth, drive to work along the same routes, and do a host of other things mostly as a matter of habit.

Custom and habit affect land resources in many different ways. Most Americans have a customary preference for varied diets, which are rich in animal products. As a result, large areas of agricultural land are used to produce forage for livestock, whereas smaller areas are needed for food crops such as cereals and potatoes. Cus-

---

[21] Commons, *op. cit.*, pp. 73–74.

tomary consumption of products such as tobacco, coffee, tea, coke, and alcoholic beverages has favored the diversion of considerable areas to particular uses. Without our acceptance and habitual use of these products, the two million acres planted to tobacco in the United States could be used for other purposes. Brazil and Ceylon would lose their leading exports, and Milwaukee would need something else to make it famous. Producers of the raw materials used for these products would lose their markets, while the urban areas employed for their processing and sale would be available for other uses.

Tastes in clothing also have important effects on land resource use. In times past, the average citizen usually accepted a simple wardrobe, perhaps a bearskin, loincloth, the simple tunic of ancient Rome, or the warmer homespuns of the early American settler. With the opportunities provided by modern society, we have turned to the customary acceptance and use of colorful and elaborate wardrobes. As a result, large areas must be used for the production and processing of cotton, wool, and flax. Natural resources such as pulpwood and coal are utilized in the manufacture of rayon, nylon, and other synthetic fibers. Fur farms are required to fill the demand for the furs used on articles of adornment. New markets have also emerged for leather products, bird feathers, coral, and semiprecious stones.

Many other examples can be used to illustrate the effects habitual modes of life have on the demand for land products. The customary use of the daily newspaper, for example, requires the production of tremendous quantities of wood pulp. Long-accepted attitudes concerning housing facilities call for the use of billions of tons of building materials. Great quantities of coal, oil, and natural gas are used each year in the customary heating of homes. The customary use of areas around homes for lawns and landscaped grounds requires the use of substantial acreages for this purpose plus the use of additional areas for the production of nursery stock.

Customs can have direct as well as indirect effects on the ownership and use of land resources. Farmers may cling to customary cultural and crop-rotation practices, even when improved practices are brought to their attention. Sometimes they refuse to change because of the proven value of their practices or because "what was good enough for my father is good enough for me." On other occasions, particularly in those areas where human life is supported by a slender thread, decisions are motivated by hopes for survival, operators may hesitate to accept the risks associated with suggested shifts to new and (for them) untried methods.[22] Rigid adherence to custom can lead to inefficiencies in production and losses of potential income not only for farmers but also for builders, artisans, merchants, miners, foresters, and other users of land resources.

Property inheritance, rental, and ownership arrangements are also affected by custom. Our emphasis on equal division of estates among heirs of equal relationship, for example, is based as much on custom, as was the system of primogeniture endorsed by earlier societies. Similarly, our acceptance of particular rental arrange-

---

[22] Cf. Herrell DeGraff, "Some Problems Involved in Transferring Technology to Underdeveloped Areas." *Journal of Farm Economics*, 33 (November 1957), 700.

ments and the high regard with which we view home ownership owe much of their basis to traditional habits of thought.

Religious Institutions

Although religious institutions are primarily concerned with spiritual matters, they nevertheless have important effects on economic life and the uses made of land resources. With real estate, this influence usually takes one of three forms: (1) ownership or operation of land by religious bodies, (2) claims of a church or religious body to the income of land, or (3) religious beliefs affecting land-use practices.

Church and religious bodies have owned or controlled substantial property holdings throughout most of the world's history. One of the most extreme examples existed in ancient Egypt, where the land belonged to the pharaoh, a quasi-divine ruler, who administered it with the aid of his priests and nobles. Church ownership of property and domination of temporal affairs were also important in Western Europe throughout the Middle Ages. During this period, the Roman Catholic church acquired numerous properties by gift and will and became the principal landowner in many countries. Almost a third of the landed property in England, for example, was held by the church during the thirteenth century.

Churches and religious bodies own considerable real property in the United States. But their share of the total is small, and the problem of church ownership is not considered serious. Important land-use problems do stem, however, from the location of churches and synagogues at sites of high commercial value near the 100 percent spots of downtown business districts. The preemption of these sites for religious uses, together with the tax-exempt status of these properties, naturally discourages their redevelopment for what many people regard as their highest and best economic uses. It may be noted, however, that cathedrals and other religious centers have often provided the focal points around which many cities have developed.

The extent of the claim religious bodies have to the income derived from land resources varies over a wide range. When a church owns a house or an office building, it naturally has first claim on the income or rent derived from the use of its property. With other properties, a church may claim tithes, or its claim for support may be limited to the voluntary contributions of its members. During biblical times and beginning again during the Middle Ages, tithes equal to one-tenth of the returns from land were often collected by governments for the support of the state church. These collections have for the most part been discontinued. Some European governments, however, still support their official churches with funds collected from taxes that affect church members and nonmembers alike. Tithes also are contributed voluntarily by the members of some religious groups.

Religious beliefs also can affect the use of land resources. Many possible practices are discouraged in tribal societies because they are regarded as taboo. Some of these taboos, such as those against eating certain kinds of food or working on certain days of the week, have carried over to the present day and have indirect effects on land resource use. Another type of taboo is accepted in India, where the Hindu belief in transmigration of souls has resulted in the toleration of large local populations of cattle and monkeys. The refusal of local people to keep the numbers of these animals in check has at times contributed to losses in agricultural productivity.

In ancient times particular sites were often regarded as holy places. Even today, some of these sites, such as Rome, Jerusalem, Mecca, and the Ganges River are focal points for pilgrimages. Burial grounds also have been maintained for religious reasons; and significant areas of potentially productive land in many countries are retained in this use. Another example involving the sanctity of earth is suggested by the opposition of some Tibetans to mining operations for the religious reason that removal of minerals will disturb the spirits of the earth and result in losses of soil fertility.

Among the ancient Hebrews, land was considered a sacred possession of Yahweh (God). Property sales to foreigners were frowned on because land was not to be alienated from Yahweh's chosen people. Another Hebrew belief called for periodic years of complete rest. During the jubilee years, all cropland was supposedly left fallow.

Most early civilizations had agrarian deities. Sacrifices were offered and special rituals held to ensure favorable crop planting and growing conditions. Festivals were held to celebrate harvests, and feast days were held throughout the year. Agricultural operations in ancient Rome were complicated by forty-five different festival days each year.[23] This picture has changed in most countries. But priests are still asked to bless the fields and fishing grounds, and religious festivals still punctuate the lives of many people. The adornments of some of these festivals, such as the popular use of Christmas trees, have even given rise to significant land uses.

Many of the religious beliefs that have influenced our use of land resources in times past have been discarded. Yet beliefs such as the golden rule, respect for authority, and faith in the future still have strong impacts on human behavior. Two other important land resource concepts—the stewardship concept of land use and the goal of family farm and urban home ownership—also have religious roots. The *stewardship* concept—the belief that each generation of operators should turn the land resource base over to the next in as good condition as it is received—has positive implications for the development of a sound conservation philosophy. Home ownership—the idea that everyone should have an opportunity to enjoy life "under his vine and under his fig tree"—on the other hand, is already accepted as a popular goal of land policy in many countries.

PERSONAL AND HOUSEHOLD CONSIDERATIONS

In addition to the institutional factors listed above, economic behavior is affected to a considerable extent by personal and household considerations. These considerations include factors such as one's attitudes and goals, ability to work, and family obligations that cause one to differ from the prototype of the economic man. Some

---

[23] Cf. Barnes, *op. cit.*, p. 74.

of these factors bear the imprint of institutions such as the family, education, religious systems, and our cultural heritage, whereas others are tied more to the life cycle of man. Regardless of how they are classified, these factors are closely associated with institutional factors in the sense that they help to provide the society-made conditions under which economic activity takes place.

The importance of personal and household considerations can best be illustrated with the example of the economic man. As an operator, this remarkable character is blessed with perfect foresight and knowledge. As a worker, he is ageless, immune to sickness and worry, and always in possession of his fullest physical and mental powers. As an individual, he is not troubled by family or group obligations, has a single-minded devotion to the principle of profit maximization, and always strives to attain the highest possible level of economic efficiency in his operations.

This idealized assumption has its rightful place in economic analysis. Under real-life conditions, however, it is foolish to assume that average operators will always act like the economic man. Personal attitudes, choices, and goals always have important impacts on individual economic behavior. And personal factors can make programs that pass the test of economic practicability unacceptable on an individual level in much the same way that institutional barriers frequently prevent their successful operation on a larger scale.

Most people are motivated by nonmonetary as well as by monetary goals. Wide differences exist in the extent to which different people emphasize these two sets of goals. Some business operators approach the concept of the economic man, particularly during business hours. But Ebenezer Scrooge is one of the few people who have ever given full allegiance to the profit-maximization goal. Somewhere along the scale of their income production possibilities, most operators substitute welfare and nonmonetary goals for at least part of their interest in additional economic returns. This substitution process often causes them to forgo opportunities to make more money so that they will have more time to relax, go fishing, or play golf. In similar fashion, it often prompts them to take time and money they could use to advantage in their businesses to indulge in charitable, civic, cultural, political, religious, or other activities.

An operator's success or failure in a business can often be explained in terms of attitudes toward work, willingness to adapt to change, and ability to get along with other people. To a considerable extent, success is also determined by one's capacity and ability to perform tasks. Many operators fail because they are not "cut out" for their particular jobs or enterprises. Others fail to realize their potentialities because of sickness, ill health, or sometimes sheer laziness.

Individual economic behavior is often conditioned by group obligations. Family responsibilities and community ties frequently discourage operators from migrating to areas of greater economic opportunity. At times, they require large portions of an operator's time—as when one personally cares for disabled members of a family or gives considerable time to service organizations—and leave a minimum of time for normal economic pursuits.

These obligations can have a severe effect on the operator's ability to accumulate and reinvest capital. Young business operators and farmers usually have far less working capital than they could use to best advantage. Yet their heaviest family obligations ordinarily come at the very time when they are most in need of additional money for land, equipment, and other input factors. As operators, they are conscious of their capital needs and of the fact that it "takes money to make money." Many skimp on personal and family expenditures so that they may use maximum portions of their earnings as production and working capital. Yet when their disposable incomes are low, they usually give first priority to shoes for their children or medicine for their mates, even though this choice may keep them from operating in the so-called rational zone of economic action.

—SELECTED READINGS

Barnes, Harry E., *Social Institutions* (Englewood Cliffs, N.J.: Prentice-Hall, 1942).

Clark, John Maurice, *Economic Institutions and Human Welfare* (New York: Knopf, 1957).

Ely, Richard T., *Property and Contract in Their Relation to the Distribution of Property* (New York: Macmillan, 1914), Book I, chaps. 1 and 2.

Gruchy, Allen G., *Modern Economic Thought* (Englewood Cliffs, N.J.: Prentice-Hall, 1947).

Hite, James C., *Room & Situation: The Political Economy of Land-Use Policy* (Chicago: Nelson-Hall, 1979), chaps. 4 and 5.

Steiner, George A. *Government's Role in Economic Life* (New York: McGraw-Hill, 1953), Part I.

Wehrwein, George S., "Institutional Economics in Land Economic Theory," *Journal of Farm Economics*, 23 (February 1941), 161–70.

12

PROPERTY
IN LAND RESOURCES

Real estate resources are more than a strictly physical factor of production. From a cultural, economic, legal, and social point of view, they are "an element of nature inextricably interwoven with man's institutions."[1] People covet and want land resources for their personal use and satisfaction. To secure this end, institutional arrangements have been devised that permit individuals and groups to exercise the privilege of owning, possessing, and utilizing specified real estate resources while excluding others from exercise of these same rights. The systems of rights represented by these arrangements provide the basis of the concept of property.

Property rights play an ever-present role in determining what people may or may not do with land resources. Their importance is such that some understanding of the rights individuals and groups hold in real estate is essential if one is to fully comprehend human behavior and conduct with respect to land. The discussion that follows focuses on the rights people hold in real estate. It starts with a general commentary on the nature, characteristics, and scope of property rights and then proceeds to a more-detailed examination of the various types of interests individuals and society have in real property, in water, in air, and in subsurface resources.

[1] Karl Polanyi, *The Great Transformation* (New York: Farrar and Rinehart, 1944), p. 178.

NATURE AND SCOPE OF PROPERTY RIGHTS

Property is a complicated legal concept. Many people think of it in terms of objects that can be owned or possessed. In a legal sense, however, property consists not of objects but rather of "man's rights with respect to material objects."[2] As one court has held:

> The term "property" may be defined to be the interest which can be acquired in external objects or things. The things themselves are not, in a true sense property, but they constitute its foundation and material, and the idea of property springs out of the connection, or control, or interest which according to law, may be acquired in them, or over them. This interest may be absolute when a thing is objectively and lawfully appropriated by one to his own use in exclusion of all others. It is limited or qualified when the control acquired falls short of the absolute.[3]

Generalizing from these definitions, one might describe *property* as "the exclusive right of possessing, enjoying, and disposing of a thing"[4] or as "the exclusive right to control an economic good."[5] Our concept of property rights is both broad and diverse. It ranges from complete ownership to the more limited rights one may hold with an easement or under a lease. In more eloquent terms, it may be noted that

> property is a euphonious collocation of letters which serves as a general term for the miscellany of equities that persons hold in the commonwealth. A coin, a lance, a tapestry, a monastic vow, a yoke of oxen, a female slave, an award of alimony, a homestead, a first mortgage, a railroad system, a preferred list and a right of contract are all to be discovered within the catholic category. Each of these terms, meaningless in itself, is a token or focus of a scheme of relationships; each has its support in sanction and repute; each is an aspect of an enveloping culture. A Maori claiming his share of the potato crop, a Semitic patriarch tending his flock, a devout abbot lording it vicariously over fertile acres, a Yankee captain homeward bound with black cargo, an amateur general swaggering a commission he has bought, an adventurous speculator selling futures in a grain he has never seen and a commissar clothed with high office in a communistic state are all men of property. In fact, property is as heterogeneous as the societies within which it is found; in idea, it is as cosmopolitan as the systems of thought by which it is explained.[6]

[2] C. Reinold Noyes, *The Institution of Property* (New York: Longman, 1936), p. 353.

[3] *Griffith* v. *Charlotte et al.*, 23 S.C. 25, 38 (1884).

[4] *McKeon* v. *Bisbee*, 9 Calif. 137 (1858).

[5] Richard T. Ely, *Property and Contract in Their Relation to the Distribution of Wealth*, Book I, (New York: Macmillan, 1914), p. 101.

[6] Walton H. Hamilton and Irene Till, "Property," *Encyclopedia of the Social Sciences* (New York: Macmillan, 1933), 12, 528-29.

Attributes and Characteristics of Property

Property has many important characteristics.[7] It is an attribute of human beings, not of chattels.[8] It involves rights to the use of material things, not personal rights or liberties. It differs from free goods in the sense that it involves only appropriable objects of value that people can possess. Furthermore, it is an *exclusive* not an *absolute* right. Individuals can hold property rights alone or share them with others to the exclusion of all other persons. But these rights are always subject to the supervision as well as the protection of a sovereign power. In our society, the existence of property rights presupposes the presence of (1) an owner together with other persons who can be excluded from the exercise of ownership rights; (2) property objects that can be held as private or public possessions; and (3) a sovereign power that will sanction, and if necessary protect, the property rights vested in individuals or groups.

Property involves exclusive rights. These rights obviously cannot exist until there is both an owner to possess and use the object in question and other interested persons who can be excluded from possession and use. Property, as we know it, simply does not exist in areas where there is no population and no outside claimants. Nor does it exist in those isolated instances in which a single user, such as Robinson Crusoe, may be present. Property rights come into existence only when two or more people compete for the possession and use of some object and need develops for the allocation of recognized rights between them. In this sense, the concept of property involves more than a simple relationship between persons and things. It involves relationships among people regarding their rights to use and to exclude others from the use of particular objects.

Before property can exist there must be a property object. These objects usually involve material things, though they may involve quasimaterial items such as franchises, patents, copyrights, or industrial goodwill.

Two additional attributes of property objects are their appropriability and value. Before anything can be classed as a property object, it must be capable of appropriation. The lands beneath the deeper portions of the ocean and beneath the polar

[7] For other descriptions of these characteristics, cf. Noyes, *op. cit.*, chaps. 4–6; and Ely, *op. cit.*, Book I, chap. 5.

[8] People frequently speak of a bird's cage or a dog's bone. These are merely descriptive terms because the property in each case belongs to the master, not to the chattel. Sentiment sometimes causes people to treat chattels and physical objects as though they had human rights. Thus one occasionally reads of an estate being left to a pet parrot or a favorite cat. Actually these estates are held in trust for the support of these creatures, and they do not exercise true ownership rights. Ripley's "Believe It or Not" has featured stories of a facelike rock formation at Ploumanach, France, which was legally conveyed to itself and of the bridge of Pre-St.-Didier in Italy which was made the legal property of two trees that stand like sentinels at one end of the bridge. These examples represent cases in which ownerships without legal foundation are tolerated as long as society respects the grantor's sentiment and withholds its power of taxation. The position of society in these cases is illustrated by the Pennsylvania example of a 600-acre forest, which was deeded to God but which reverted to the state for nonpayment of taxes. Cf. *American Forests*, February 1931, p. 112.

ice caps are not property because they have not as yet been appropriated for human use. Free goods such as air and ocean water can be appropriated. But in their natural state, they are not regarded as property objects because no one can enforce exclusive rights over their use.

Our supply of property objects has come from two primary sources—the capture of free goods and the creation of new goods through the processing of existing resources. People ordinarily try to capture, develop, or produce things that have value. By accident, or as a byproduct of their activities, they sometimes produce things of negative value such as slag from mines. They may also continue to hold objects that have lost their value. But as a general rule, property objects have value. Otherwise, owners would have little incentive for the continued retention of their property rights.

As a final requirement, the existence of property implies the assent or sanction of a sovereign power vested with both authority and ability to protect the rights of its subjects. People have been able on various occasions to acquire and hold objects they desire through cunningness or sheer force. By these means they have acquired possession; but they have not acquired property rights. Possession may be defended by strategy or force over long periods of time. But property rights, and with them the right of protection against those one would exclude from the holding or use of a property object, arise only when a sovereign authority—the family, clan, tribe, or the state—recognizes and enforces one's exclusive right of possession.

The dependence of the property concept on the protection afforded by sovereign powers is best illustrated by examples of what happens when this power is weakened or destroyed. Children often appropriate the toys or possessions of others. Gangsterism and lawlessness sometimes break out when the enforcement powers of government appear weak or ineffective. Similarly, the breakdown of authority that comes with armed conquest often results in some appropriation, "liberation," or pillaging of private possessions by the personnel of the conquering armies.

Bundle of rights. Property involves several distinct interests or rights, which can be held separately and which when taken together represent a "bundle of rights." The largest bundle of rights a private owner can hold in landed property is known as complete ownership or as ownership in *fee simple.*[9] Fee simple owners hold a bundle of separable property sticks or rights, (see Figure 12-1). They have the right to possess, use, and within reason exploit, abuse, and even destroy their land resources. They can sell land with or without deed restrictions that affects its future use. They can give it away, trade it for other things, or devise it in any of a number of ways to heirs.[10] They can lease use rights to others; mortgage their property or

---

[9] The term *fee* is used in English law to signify an estate of inheritance (interests in property that can be passed on to one's heirs), as contrasted with an estate that can be held only for life. There are three major types of fee estates: estates in fee simple or fee simple absolute, which connote the most complete possession of private ownership rights permitted in our society; and the conditional fee estate; and fee tail estate. The latter two are described later.

[10] In legal parlance, property owners can *bequeath* personal property through use of a will but they *devise* real property.

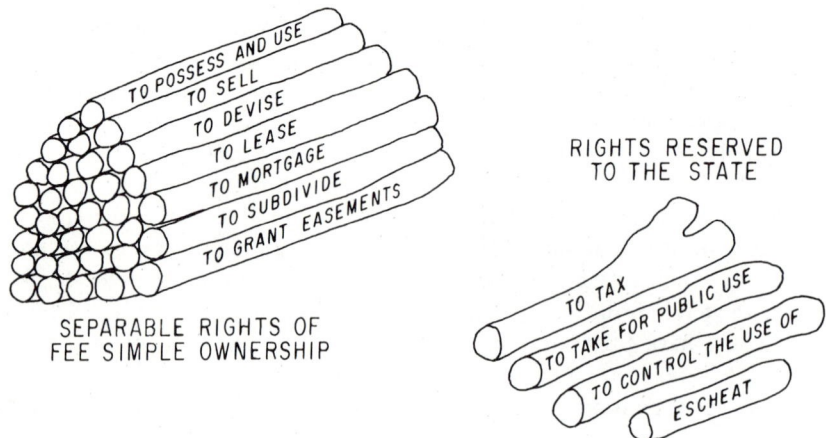

FIGURE 12-1. The bundle of rights in property.

permit liens to be established against it; subdivide their holding or grant easements for particular uses; enter into contractual arrangements involving its development, use, or disposition; and exercise these rights, as long as they have not disposed of them, to the exclusion of all other persons.

Fee simple ownership is one of the broadest and most complete concepts of property ownership yet developed. Yet it must be recognized that the fee simple owner holds exclusive, not absolute, rights. Ownership rights are always limited and conditioned by the overall interests of society administered by the state. Because of their public nature, four important sticks are never included in the fee simple bundle of rights. These include the public rights of taxation, taking for public use, regulation, and escheat.

Some imperial rulers in times past have claimed absolute rights of property ownership. The nearest approach to absolute ownership in our system is found with the landholdings of the state and federal governments. Since these owners exercise the powers of government, it may be argued that they hold all the rights of property and thus possess absolute ownership. The rights they hold, however, are definitely limited by public opinion and by various reservations of public economic and social policy.

Qualified and common property. Property rights can be asserted only over those objects that individuals or groups can appropriate to exclusive possession. This means that most objects can be classified as either free goods or property. An intermediate classification applies to those free goods that can under certain circumstances be reduced to private ownership. These objects, often described as "qualified property," include such items as wild game and fish, wild fur-bearing animals, dogs, and water.

Wild creatures such as deer, rabbits, pheasants, and fish ordinarily have the status of free goods. Most states now have laws declaring these creatures the property of the state.[11] This public proprietary interest, however, is asserted for the purpose of regulating and controlling the administration and private taking of these resources, not for their exploitation or use as public property. Individuals can acquire property rights for these creatures if they kill, catch, or capture them in compliance with the licensing provisions and other regulations set up by individual states.

Wild animals valued for their fur are usually treated in much the same way as game and fish. They exist as free goods in nature but can be trapped or captured subject to public regulations. The water found in its natural state in streams, lakes, and the ocean might also be treated as a free good. As a New York court has observed: "Water, when reduced to possession, is property, and may be bought and sold and have market value, but it must be in actual possession, subject to control and management. Running water in natural streams is not property, and never was."[12]

Most qualified properties along with some resources held in public ownership can also be described as common properties. This grouping includes resources such as air; water; ocean and freshwater fisheries, game animals and birds; and public grazing, recreation, and wilderness areas. Their key characteristics are that (1) they are used in common by members of communities and society, (2) individuals and groups have historically exercised free and unlimited rights of entry for their use, and (3) no user enjoys the privilege of excluding others from their use. High values are associated with many of these resources. But unrestrained rights of entry and increasing competition between users can easily result in overutilization and exploitation. Public regulations have accordingly been adopted in many instances to limit their utilization to levels at which their resource values can be regenerated and maintained.

---

[11] In two key decisions, the U.S. Supreme Court has indicated that these creatures are free goods even when claimed in state ownership. The question of state title to migratory water fowl was raised in *Missouri* v. *Holland*, 252 U.S. 416 (1920). Justice Holmes observed on this occasion; "No doubt it is true that, as between a state and its inhabitants, the state may regulate the killing and sale of such birds, but it does not follow that its authority is exclusive of paramount powers. To put the claim of the state upon title is to lean upon a slender reed. Wild birds are not in the possession of anyone; and possession is the beginning of ownership. The whole foundation of the state's rights is the presence within their jurisdiction of birds that yesterday had not arrived, tomorrow may be in another state, and in a week a thousand miles away." In *Toomer* v. *Witsell*, 334 U.S. 384 (1948), Chief Justice Vinson observed that "the whole ownership theory, in fact, is now generally regarded as but a fiction expressive in legal short-hand of the importance to its people that a State have power to preserve and regulate the exploitation of an important resource." According to some observers, this decision delivered "a death-dealing blow . . . to the doctrine of state ownership of fish. The states were stripped of their proprietary rights in game and fish and were told that all they had was merely the police power to regulate their taking." Cf. Nicholas V. Olds and Harold W. Glassen, "Do States Still Own Their Fish and Game?" Michigan State Bar Journal, 30 (April 1951), 16–23.

[12] *City of Syracuse* v. *Stacey*, 169 N.Y. 231, 245 (1901).

Basis of the Concept of Property

Most authorities agree that our concept of property rights is really the outgrowth of a long period of evolutionary development.[13] Among the primitive and nomadic societies from which our culture sprang, land resources were at first regarded as free goods to be used at will. Group ownership rights were gradually recognized for particular types of land such as burial grounds, sites of religious significance, salt and mineral deposits, and springs and watering holes. Private and group ownership came later for tillage, grazing, and hunting areas. Even then, these lands were usually held in common at first under communal or tribal ownership.

Development of fee simple ownership. The concept of fee simple ownership now accepted in most English-speaking areas evolved from the village and feudal tenure systems found in Western Europe during the Middle Ages.[14] Most of the nomadic native tribes in these areas settled down to a village economy during Roman times.[15] The residents of these villages were primarily dependent on agricultural pursuits. Ordinarily they used an open field system with the lands around the villages. Under this system, all the land was usually held in common ownership—at least at first—and each family shared a portion of each field.

With the rise of feudalism, most villages came under the political control of overlords who frequently converted them into manorial or feudal estates. Actual ownership of the land in these estates was vested in an overlord or king, usufructuary rights were held by the villagers, serfs, and villeins. These rights varied considerably. Some land occupants enjoyed relatively free tenure. But most fields were operated by villeins, who were born to their status and who exercised tillage rights subject both to customary obligations to the lord and to his arbitrary will.

[13] Several theories have been expounded at various times regarding the origins of and justification for property rights. Important among these are the legal, occupancy and possession, gift of God, natural right, social contract, human nature, labor, and general welfare theories of property. Cf. Ely, *op. cit.*, 11, pp. 531–51; and Edmund Whittaker, *A History of Economic Ideas* (New York: Longmans, Green, 1940), pp. 175–241.

The more relevant aspects of the first seven of these theories may be combined in a general welfare or social theory of property. With this approach, one may argue that society is the true fount of property rights. Society allows individuals (or those who count in the eyes of the ruling group) to acquire, exercise, and maintain property rights and thereby maximize their personal satisfactions, because this procedure usually enhances the prevailing concept of "social welfare." In an ultimate sense, however, society always retains its right to regulate the allocation and distribution of property rights; and as the interests and goals of society change, the institution of property also changes.

[14] For a more detailed discussion of this subject, cf. Marshall Harris. *Origin of the Land Tenure System in the United States* (Ames: Iowa State College Press, 1953): Donald R. Denman, *Origins of Ownership* (London: Allen & Unwin, 1958); V. Webster Johnson and Raleigh Barlowe, *Land Problems and Policies* (New York: McGraw-Hill, 1954), chap. 2; and Roland R. Renne, *Land Economics*, 2nd ed. (New York: Harper & Row Publishers, Inc., 1958), pp. 315–39.

[15] Private property rights were recognized in the ancient world, and a relatively full concept of property rights was developed under Roman law. Much of the fullness of this concept was lost, however, with the breakdown of the Roman Empire and the rise of the feudal system.

A trend toward freer tenure conditions in England began with the gradual weakening of the feudal system during the thirteenth and fourteenth centuries.[16] Military service and other incidents of feudalism were relaxed as conditions of proprietorship. Operators gradually acquired the privilege of selling or passing their holdings to heirs without the approval and payment of dues to an overlord. Owners also acquired the right to exclude landless villagers from the use of land, a development that precipitated the "enclosure movement." These events brought an end to villeinage tenure in many parts of England during the fifteenth century and its complete extinction by the seventeenth century. With this change, land operators became owners or tenants, and landless villagers sought other types of employment.

Development of property rights in other areas. England gave up the feudal system gradually and at an early date. Different patterns were experienced in other areas. France retained its feudal system until the French Revolution; and vestiges of feudalism remained in many parts of Central and Eastern Europe until the uprisings of 1848, the emancipation of the Russian serfs in 1862, and the Armistice of 1918. Feudalistic and semifeudalistic practices persisted in many other parts of the world until they were ended by land reform programs after World War II.

As one might expect, wide variations exist between the property right concepts developed in different areas. Many countries have developed relatively full concepts of private property rights comparable with those enjoyed with fee simple ownership. But many of the world's people have never really experienced full ownership rights in land. Some have held to semifeudalistic practices under which most of the land is held by a limited number of often absentee landlords. Others have emphasized village communal ownership, while those within the communistic orbit have shifted to state and collective ownership.

TYPES OF INTERESTS IN LANDED PROPERTY

Property can be classified in many different ways. From a physical standpoint, property objects may be regarded as mobile or immobile, tangible or intangible, or as types of objects such as a student's clothes, furniture, books, and car. In an economic sense, they can be divided into production goods and consumer goods. They may also be classified as real property and personal property; as public and private property; or as properties held by individuals, partnerships, and corporations.

Some of the most important classifications of property interests involve the distribution and sharing of the bundle of rights people hold in land. With this approach, emphasis may be focused on the (1) types of estates or interests held in land, (2) layers of rights, (3) number of owners, (4) conditions of holding, (5) duration of interests, and (6) time of enjoyment. Emphasis is focused here on the leading types of estates people hold in land, after which brief attention is given to the

---

[16] Cf. J. M. W. Bean, *The Decline of English Feudalism. 1215-1540* (Manchester, U.K.: Manchester University Press, 1968).

other five classifications and to the general nature of the interests society has in landed property.

Leading Types of Estates in Landed Property

The rights and interests one holds in the ownership, possession, or control of property are often described as one's *estate*. Estates vary in scope from fee simple ownership to estates of remote or negligible importance. The most important estates involve the interests held by owner-operators, holders of life and remainder estates, landlords and tenants, mortgagors and mortgagees, and the givers and holders of land contracts. Lawyers ordinarily give detailed consideration to the nature, characteristics, rights, and responsibilities associated with each of these estates. It is sufficient here, however, to briefly identify each of the leading types of property interests.[17]

Complete or *fee simple ownership* represents the highest combination of rights a person can hold in landed property. Most owner-operated properties are considered as held in fee simple. This is true even though many ownerships are affected by minor subtractions of rights such as a powerline easement or a deed restriction or covenant.

Easements involve rights held by others to use one's land for special purposes. A utility company, for example, may hold an easement that permits it to run its utility lines above or beneath the surface of one's property and gives its workers a right of access for servicing these lines. Easements can involve a wide variety of privileges such as the right to encroach on one's airspace, cross and transport goods across one's property, drain water across one's land, or compel a property owner to maintain a share of a common driveway. Easements can be created by oral or written agreement or sometimes by implication. They can be acquired by purchase, deed reservation, gift, condemnation, or adverse use and possession throughout the prescriptive period recognized by law. They "run with the land" when property is transferred to new owners. And they cannot be revoked except by sale, release, abandonment, or condemnation.

Property owners are frequently affected by specific provisions in their deeds which limit the scope of their ownership rights. Sometimes these provisions involve

---

[17] The following discussion is limited to the presentation of a thumbnail sketch of the principal rights people hold in landed property under the American and English system of law. For more-detailed popular treatments of this subject, cf. Robert Kratovil and Raymond J. Werner, *Real Estate Law*, 8th ed. (Englewood Cliffs, N.J.: Prentice-Hall, 1983), or the appropriate chapters in most standard textbooks dealing with real estate principles. Readers who want a working knowledge of real property law will also find it desirable to study some of the standard legal sources in this field. For examples of these, cf. Curtis J. Berger, *Land Ownership and Use: Cases, Statutes, and Other Materials* (Boston: Little, Brown, 1968); Richard R. Powell, *The Law of Real Property*, 7 vols. (New York: Matthew Bender, 1954); George W. Thompson, *Commentaries on the Modern Law of Real Property*, 12 vols. (Indianapolis, Ind.: Bobbs-Merrill, 1940); Herbert T. Tiffany, *The Law of Real Property*, 6 vols. (Chicago: Callaghan, 1939); or A. James Casner *et al., American Law of Property*, 7 vols. (Boston: Little, Brown, 1952).

deed reservations that reserve mineral rights, timber-cutting rights, rights-of-way, or other comparable privileges to the grantor. Reservations of this type often create easements against the property. *Deed restrictions and covenants* are used to impose private controls over the future use of land. Building lots in most residential subdivisions, for example, are now sold with the stipulation that the sites must be used for single-family residential purposes and that the buildings be located in conformance with prescribed specifications. Deed restrictions may also be used to forbid the future use of premises for particular purposes such as the sale of alcoholic beverages.

Deed restrictions and covenants are used to secure specific ends desired by grantors or grantees. They may run indefinitely, for definite periods, or they may be limited by statute. Ordinarily, they are legally enforceable as long as they do not run counter to public policy.[18] Restrictive covenants can be enforced by court orders or injunctions issued against persons who would violate the covenants or by personal actions for damages against violators. Unlike covenants, deed restrictions usually contain forfeiture or reverter clauses that provide for the forfeiture of properties and their reversion to the original grantors whenever restrictive conditions are broken. Thus owners may grant a site for a church or sell commercial property for specified noncompetitive uses with the stipulation that the ownership rights will revert to them, their heirs, or assignees should the land ever be used for other purposes. The ownership rights held by the grantees in these cases are called *determinable, base,* or *qualified fee* estates.

Another type of limitation on the rights of ownership occurs with the entailment of estates. This practice has been discontinued. Yet in some American colonies and for a long period in England, property owners were free to entail their estates by specifying that they could be handed down only to "heirs of the body." Under this system of *fee tail* estates, the owner in each succeeding generation (usually the oldest son of the previous owner) had the right to possess and enjoy the property but could not sell nor dispose of it to persons other than the heir next in line of succession.

In contrast to estates held in fee, many properties are held as *life estates.* With this arrangement, life tenants can enjoy, possess, and use properties throughout their lifetimes. They can lease their estates to others; and assuming they can find someone willing to risk such a venture, they can mortgage or sell their interests for the duration of their lifetimes. But these rights exist only for the duration of their lives. At their death, the estates revert to the grantors, their heirs, or assignees, or pass to designated remaindermen. A comparable but less common arrangement known as an estate *pur autre vie* exists when one's rights are limited to the lifetime of some other person or persons. A son-in-law, for example, may hold an estate of this type during the lifetime of his wife.

---

[18] Some older deeds to residential properties contain restrictive convenants against occupancy or ownership by persons not of the Caucasian race. Racial covenants of this type are not enforceable in the United States because they run counter to public policy. Cf. *Shelley* v. *Kraemer*, 334 U.S. 1 (1948); and *Barrows* v. *Jackson*, 346 U.S. 249 (1953).

Life estates ordinarily fall into two classes: (1) conventional life estates created by deeds, wills, and other contracts; and (2) legal life estates authorized by law. Some life estates are set up during the lifetime of the donor with the donor holding a *reversion* interest. Most conventional life estates, however, are established by will, as when a man leaves his widow a life estate in his property with the provision that their children hold a *remainder* interest, which will vest them with the estate at her death. The principal legal life estates involve the rights of dower, curtesy, and homestead.

Closely analogous to life estates are the rights involved in *trusts*. Trusts involve contractual arrangements under which properties are held and administered by trustees for the benefit of specific beneficiaries. As is the case with life estates, a *living trust* can be established during a property owner's lifetime, while a *testamentary trust* can be provided by will. *Business trusts* are sometimes established and operated for investment and other business purposes, and *land trusts* are occasionally used by some operators as a means of concealing property ownership while they benefit from the acquisition or control of properties.

Twenty-two states recognize the *dower* right of a wife to a life estate in one-third of a husband's property at the time of his death. This share may be more or less than the widow receives under the husband's will or as her statutory share under the laws of descent. Since she cannot take both shares, she is ordinarily required to choose between her dower right and the shares she would otherwise receive.

Some states recognize a dower right of a husband to the property of his deceased wife. Twelve others give a husband a *curtesy* right to a life estate in the real property owned by the deceased wife during the marriage. This right is analogous to the right of dower except that it usually applies to all rather than merely a third of the wife's property. Another arrangement similar to dower and curtesy exists in the eight community-property states of the West and Southwest. Husbands and wives in these states can hold property separately or together as community property. Whenever a husband or wife dies, the community property is divided into halves and the survivor receives his or her share plus one-half of the deceased spouse's share. This share passes in fee simple rather than as a life estate.

Forty-one states recognize a real property concept known as the *homestead*, to which particular homestead rights apply. Homesteads are usually defined as a portion of the holding, limited both as to total area and value, owned and occupied by families as their home. In rural Michigan, for example, property owners hold homestead rights to their dwelling and not more than forty adjoining acres of farmland with an exempt property value of not more than $2,500. Homesteads of this type are exempt from forced sale for debt and cannot be mortgaged or sold during the lifetime of the husband and wife without the consent of both parties. After a husband's death, the widow and children can continue to occupy the homestead for the duration of the widow's life or so long as the children are minors without regard to the provisions of the husband's will or the claims of his creditors. Homestead rights cannot be sold, but they can be lost through the remarriage of the widow or by abandonment.

Most people are familiar with the general division of rights that takes place with the leasing of properties. A *lease* is the relationship created by a contract that gives a tenant or lessee the right to possess and use property held or owned by a landlord or lessor. Leases can run for given periods of time or continue indefinitely by mutual consent. Rent or some other consideration is normally paid by the tenant to the landlord or his or her agent as a condition of the lease. At the expiration of the lease, it is ordinarily expected that the tenant will return the premises to the landlord in approximately the same condition as they were received less normal wear and tear or any damage caused by the elements. Throughout the period of the lease, the landlord normally has no right to enter upon the property without the tenant's permission and has no right to interfere with the tenant's use of the property unless these rights are reserved to the landlord in the lease. Under a leasing arrangement, a tenant is said to have a *leasehold* estate, while the landlord retains a reversion interest.

A *mortgage* represents a conveyance of landed property by a borrower (mortgagor) to a lender (mortgagee) as security for payment of a debt, with the provision that the conveyance is to be void if the debt is paid in the manner and period prescribed. In early England, the term *mort gage* meant death pledge. Mortgagors turned properties over to their mortgagees; and the latter not only enjoyed the use and the income from these properties throughout the mortgage period but also acquired full ownership rights if the mortgagor failed to repay the debt on the due date. Mortgagors now retain their properties throughout the mortgage period; and if they default in their payments, they can remain in possession during redemption periods of up to eighteen months while the mortgagee initiates foreclosure proceedings. Also, if a property is foreclosed, the mortgagee has a valid claim only to the outstanding value of the loan plus interest, not to the entire property.

The exact distribution of property rights between the mortgagor and the mortgagee depends on the details of the mortgage agreement and the laws under which the mortgage is prepared and filed. In "title-theory" states such as Arkansas, "legal title passes, at law, directly to the mortgagee, subject to be defeated by the performance of the conditions of the mortgagee." In contrast, "lien-theory" states such as Oklahoma hold that "a mortgage does not vest any estate in the mortgagee, but is merely a security operating as a lien or incumbrance on the property." Still other states accept variations of these two theories. Mississippi, for example, treats the mortgagor as "owner of the legal title of the property conveyed" except that "upon a breach of the conditions of a mortgage, the legal title becomes absolute in the mortgagee."[19]

Land or *purchase contract* arrangements are accepted in some areas as a popular means by which buyers with limited capital may acquire rights to property. These

[19] Leonard A. Jones, *A Treatise of the Law of Mortgages of Real Property*, 8th ed. (Indianapolis, Ind.: Bobbs-Merrill, 1928), Vol. 1, secs. 21, 40, and 52.

contracts resemble mortgage transactions but they differ in the types of estates created. Givers of land contracts can gradually build up their equities in the properties they are buying to a point at which they can convert them into mortgages or even become full owners. However, as long as they operate under a land contract, the titles to the properties remain with the holders of the contracts. Buyers under these conditions have a right to possess and use the properties. But these rights together with their equities in the properties can easily be forfeited without need for foreclosure proceedings if they default on their payments.

Another rights-sharing arrangement involves the use of liens. A *lien* is a right enjoyed by certain classes of creditors (including mortgagees) to require, if necessary, the sale of a debtor's property to satisfy a debt or charge. Three principal types of liens can affect property ownerships. These include (1) mechanic's liens for charges associated with the use of labor and building materials; (2) tax liens for the payment of delinquent property, inheritance, gift, or income taxes; and (3) judgment liens, which result from court actions. A properly filed lien always poses a threat to an owner's continued use of property. It places a cloud on the title that affects one's ability to sell or secure mortgage financing; and as long as it stands, legal action can be taken to force the foreclosure of a property to pay the claims against it.

Other Classifications of Property Interests

It is often convenient to classify property rights as layers of rights—surface, suprasurface or air rights, and subsurface rights. Most discussions of property deal primarily with rights to surface land. But the rights one holds in surface waters, air and the space above surface holdings, and minerals and other subsurface resources represent significant aspects of the property concept. These rights, which are discussed later in this chapter, can be and often are separated from the bundle of surface rights held in land.

Other criteria are also used in the classification of the legal interests individuals hold in landed property. Important among these are the classification of estates by number of owners, conditions of holding, duration, and time of enjoyment.

Number of owners. Estates held in landed property can be classified by number of owners into three general groups: (1) resources held as the common property of all the members of a community or society, (2) properties held as undivided interests by two or more co-owners, and (3) properties held in severalty by single owners.

Primitive societies have usually treated land resources as the *common property* of the village or group. Each individual or family in these societies has enjoyed use rights in these resources, but no single individual has had a recognized ownership right that can be leased, mortgaged, sold, or devised to others. This system of ownership has for the most part broken down; and most properties are now held *in severalty*—that is, in separate individually controlled ownerships. With this change, the term *common property* is now used mostly to describe the remaining

qualified properties in which members of society hold common use rights. Vestiges of the common-property concept still remain, however, with the Mexican *ejidos*, the land holdings of several American Indian tribes, and some mountain pastures in Central Europe. Common grazing areas were a familiar feature of many early American village settlements.

Properties held by two or more persons in *undivided ownerships* are normally held under one of four arrangements: tenancy-in-common, joint tenancy, tenancy-by-the-entireties, or community property. When ownership interests are held under *tenancy-in-common*, each party owns an undivided share of the property. An owner may sell this undivided interest or dispose of it by will; otherwise it passes to his or her legal heirs.

With *joint tenancy*, two or more persons hold joint ownership of a property, each with rights of survivorship to the interests of the others. This means that if one owner dies, his or her rights pass automatically to the surviving owner(s) rather than to his or her heirs or devisees as would be the case with tenancy-in-common.

Joint tenancies cannot be created without the inclusion of an express statement to this effect in the deed that confers title. Whenever deeds are made out to two or more persons, not husband and wife, without stipulations concerning the type of tenancy created, the grantees acquire title as tenants-in-common. Some states do not recognize a right of survivorship unless a deed specifies that the title has passed to "joint tenants with the right of survivorship." Joint tenancies can be dissolved by the mutual consent of the owners or by the action of a single tenant should he or she request court partition of the property. They also are broken and become tenancies-in-common when a tenant sells his or her undivided interest. This situation stems from a legal rule that requires joint tenants to share the same interest, acquired at the same time and in the same deed, and held throughout in the same undivided possession.

Twenty-nine states treat husbands and wives as legally the same person in matters involving co-ownership of property. A husband and wife in these cases cannot hold property as tenants-in-common or as joint tenants. Instead they hold their ownership rights as *tenants-by-the-entireties*. This co-ownership arrangement is similar to joint tenancy, except that neither party can break it without the other's consent. Also when the husband or wife dies, the surviving spouse takes the entire ownership, not by right of survivorship, but under the terms of the original title. In some states, any deed or will of real property to a husband and wife creates a tenancy-by-the-entireties. Some others do not recognize this type of ownership unless it is expressly provided for in the deed. Still others have acted to abolish tenancies-by-the-entireties, usually by converting them into tenancies-in-common.

In community-property states, husbands and wives can hold any property owned at the time of their marriage or acquired by gift, will, or inheritance during marriage as separate property. All other property acquired during marriage is shared equally as community property. With this doctrine, both parties share in the ownership of real property acquired during marriage regardless of whether the deed is made out to the husband, wife, or both.

Conditions of holding an estate. Most estates are held in fee simple absolute without conditions as to holding. When conditions are specified, they fall into two categories: (1) conditions *precedent*, which involve requirements that must be met or events that must occur before an estate will vest, and (2) conditions *subsequent*, which involve events or types of action the nonperformance of which will defeat estates already vested. Conditions precedent are involved whenever a will provides that property shall vest with a given heir on a specified birthday, when he or she marries, or on the death of a life tenant. Other examples may involve the promise of a father to set a son up in business when he graduates from college, or the agreement of a holder of a land contract to exchange the contract for a mortgage once buyers build their equity up to the 25 percent level.

Estates granted subject to conditions subsequent continue only until the happening of certain events or "during," "while," or "so long as" the grantee complies with certain conditions. A will may provide that an heir shall enjoy an estate for life, or perhaps only until "John becomes of age," or so long as one cares for an invalid relative. Similarly, a father may grant property to a son--in-law to be held so long as he remains married to the grantor's daughter. In other examples, a tenant may retain a leasehold as long as rent is paid and the property is maintained and not subleased to others; while a mortgagee will surrender rights held under a mortgage once the debt is paid.

Various conditions can be attached to the holding of estates. But courts usually insist that provisions be made for the grantor's reentry into the rights of the estate if actual reversion is to take place. Also, the conditions attached to the vesting of ownership must be possible of fulfillment. Most courts, for example, would not accept a condition that an heir must marry a given person before an estate vests if this person died before the marriage could take place. Generally speaking, courts do not favor restrictions or conditions that may defeat estates that have already vested in fee.

Duration of estates. From the standpoint of duration, estates can be classified into six groups: estates in fee, for life, for years, from year to year, at will, and at sufferance. Estates held in fee simple are not limited as to duration. Owners have the right to select their successors in ownership and to use deed or other restrictions to qualify the interests they sell, devise, or otherwise convey to the next owner. Life estates involve a period of shorter duration because they are limited to the holder's lifetime. Both of these types of interests involve freehold estates.

The classification of estates for years, from year to year, at will, and at sufferance involves leasehold estates. A *tenancy for years* is created whenever a lease specifies the time period during which a tenant is to possess and use a landlord's property. When tenants originally rent a property for a year and at the end of that period hold over and remain in possession of the premises, landlords can either evict them or hold them for another year's rent. Some states have abolished or modified the rule that makes tenants liable for another year's rent if they voluntarily hold over beyond the expiration of their lease. In some of these states, the holdover tenant is treated as a tenant from month to month or as a tenant at will. If a landlord chooses the second course or accepts rent from the tenant, a *tenancy from year to year* (or *from month to month* with most residential properties) is created. Once this type of tenancy has been established, advance notice is usually required for its termination.

Whenever a tenancy continues for an indefinite period subject to termination by either the landlord or the tenant on short notice. a *tenancy at will* exists. Some states require the minimum periods of notice for termination. This type of leasehold involves leases for indefinite periods, which can be terminated at the will of either party, at the sale or conveyance of the landlord's interests to others, or at the death of the landlord or tenant.

Holdover tenants who continue to occupy a landlord's premises after the expiration of their leases become *tenants at sufferance.* As long as the landlord does not consent to their continued occupancy and does not accept rent, he or she can terminate the tenancy at any time and treat the occupants as trespassers. Tenants who hold over in this fashion can be held liable for penalty rent.

Time of enjoyment of estate. Estates can be enjoyed either now or in the future. Those that are enjoyed at the present time must be held in current possession. Examples include estates in fee, life estates, and leaseholds held by tenants. In contrast to these types of estates, reversion, remainder, and executory interests cannot be enjoyed until some future date.

Owners retain an *estate in reversion* whenever they grant interests in property to others with the provision that the interest will revert to them at the expiration of the grant. Landlords retain this interest when they lease their premises to tenants. People who give life estates to others during their lifetimes may also retain a reversion interest. Much the same situation exists with grantors of determinable fee estates, though the uncertainty of reversion in these cases gives grantors only a possibility of reversion.

Estates in remainder are created when interests are conveyed to a person to take effect on the termination of a prior estate. This type of estate with its promise of future rights of enjoyment usually accompanies grants of life estates. A husband, for example, may leave his widow a life estate in a property while he vests a nephew with a remainder interest, which entitles him to the property upon her death. Owners have at times used their right to name remaindermen as a means of entailing their estates in perpetuity. Most jurisdictions now limit this possibility with rules against perpetuities.

Future interests that vest after a given period of time or after the happening of some particular event are known as *executory interests.* These interests are similar to remainder rights except that they need not follow the termination of some prior estate.

Social Controls over Landed Property

Society has an inherent interest in all arrangements involving the ownership and use of landed property. This interest exists because of (1) the original role society plays in granting, recognizing, and protecting property rights; (2) the economic and social significance of property in our daily lives; and (3) the overall responsibility society has for maximizing social returns both now and in the future.

Except for the countries behind the Iron Curtain, most nations now favor the private ownership of most types of land resources. This private ownership involves exclusive rights of possession. But these rights are exclusive, not absolute; and they are held subject to certain social controls. Some of this social control is informal and is exerted through customs, tradition, religious and moral restraints, education, and public opinion. On the more formal side, private property rights are also conditioned and limited by governmental actions that interpret and implement the will of society.

At the present time, the governments of Canada and the United States have five important formal powers they can use to direct and control the uses made of land resources. Three of these powers—the police, eminent domain, and taxation powers—involve social controls over landed property, while the spending and proprietary powers represent auxiliary powers that governments can use to achieve particular objectives in land use.[20]

RIGHTS IN WATER

Water rights were identified as an important legal concern in the arid farming areas of the Middle East at an early date. Availability of plentiful supplies of water relative to demand in many other parts of the world, however, favored the widespread treatment of water as a free good until long after surface ownership rights were defined in land. This situation has changed during the last two centuries as increasing competition and conflicts of interest have called for specific working rules concerning the who, how, when, and where of water rights.

Different legal doctrines now apply regarding the rights individuals have in the use of water from different sources. For discussion purposes, these sources can be classified into four principal groups: (1) ocean waters; (2) *diffused surface waters* (from rain, melting snow; or possible waste water from irrigation works) found either standing in natural depressions, bogs or marshes, or flowing vagrantly over land while enroute to some watercourse, lake, or pond; (3) *surface waters* found in lakes, ponds, rivers, streams, and springs; and (4) subsurface or *ground waters*, which occur either as flowing water in defined subterranean channels or as diffused percolating waters.

Ocean waters are generally regarded as a more or less free good that individuals can use for navigation, fishing, recreation, and other purposes subject to certain national and international regulations. Among the countries and states that accept the English system of law, diffused surface waters, together with water in the soil, are normally regarded as the property of the landowner. The principal water-rights

---

[20]The use of these powers is discussed in more detail in Chapter 18. Donald R. Denman, *Land in the Market* (London: Institute of Economic Affairs, 1964), pp. 35–43, lists the public means for controlling property in the United Kingdom as sanctions, levies, prescribed maxima, forced sale, land nationalization, and compulsory reallocation.

problem with this resource concerns the drainage of unwanted waters. More complicated water-rights issues are associated with the use of surface and ground waters. Surface water rights are governed by the riparian rights doctrine in most of the more humid areas of the United States. In the more arid regions of the West, constitutional and statutory provisions require application of either an appropriation or a modified riparian doctrine. Comparable differences occur in the doctrines that govern the allocation and use of ground waters.

Water rights concepts similar to the American riparian and appropriation doctrines are accepted in several other nations. Most nations, however, emphasize administrative arrangements which permit governments to exercise varying powers to grant, revoke, and withhold water-use permits and concessions. Acceptance of reciprocal obligations are often associated with the receipt of use permits.[21]

Riparian Doctrine[22]

With the common-law riparian doctrine, all landowners whose properties are bounded or traversed by a river, stream, spring, or natural body of water have riparian rights. These owners have a right to use those waters to which they are riparian for domestic and household purposes, for watering their livestock, for navigation, for the generation of power, for fishing and recreation purposes, and for certain other uses.

Strictly interpreted, the riparian doctrine grants riparian owners a right to have water flow by or through their lands undiminished in quantity, unchanged in quality, and undisturbed in time of flow except for its use by upper riparian owners for domestic purposes and for the watering of livestock. As this statement suggests, riparian owners have usufructuary but not proprietary rights in the water that flows by their land. They can use the water for a variety of purposes. But except for domestic and stock watering purposes, they have no right to divert or take more water from a stream or lake than they expect to return to it.

Few state courts now hold to this strict "natural flow" doctrine with its limitation of riparian rights to "ordinary and natural" uses. Most of them have substituted a reasonable-use doctrine, which permits the diversion or taking of some waters for "extraordinary and artificial" uses. These uses may involve the consumption of all or part of the water taken and thus preclude its complete return to the stream. With this modification, riparian waters can be used for municipal, industrial, irrigation, and other "extraordinary" uses as long as an adequate supply of water remains available to meet the "natural" needs of other riparian owners.

It should be noted that the right to take water for "extraordinary" purposes is a limited right, which can be exercised only with reasonable regard for the equal

---

[21] Cf. Ludwik A. Teclaff, *Abstraction and Use of Water: A Comparison of Legal Regimes* (New York: United Nations Department of Economic and Social Affairs, 1972).

[22] Cf. Wells A. Hutchins, *Water Rights Laws in the Nineteen Western States*, U.S. Department of Agriculture Misc. Publ. No. 1206, 1971, pp. 38–64; and Roy E. Huffman, *Irrigation Development and Public Water Policy* (New York: Ronald Press, 1953), pp. 37–39.

rights of other riparian owners. An upper riparian owner can take the entire flow of a stream for "natural" riparian uses such as the watering of cattle. Water can be taken for industrial or irrigation uses, however, only as long as sufficient water remains to care for the "natural" uses of lower riparian owners. Whenever lower riparian owners suffer material injury because of upper diversions for "extraordinary" purposes, they can request court action to enforce their right to a continued flow of a stream.

Riparian rights are limited to riparian land. This means that nonriparian owners have no rights to the use of riparian waters—except for their rights as citizens to use public waters, beaches, and fishing sites. Riparian owners enjoy comparable use rights regardless of the extent of their property frontages along streams or lakes. When riparian waters are used in connection with land (as with irrigation), the right of use applies only to holdings that lie within the watershed of a stream, lake, or pond and are contiguous to or abut on these waters. In many jurisdictions, the riparian right "extends only to the smallest tract held under one title in the chain of title leading to the present owner."[23] This means that riparian rights do not always apply to lands added to a riparian owner's holdings if the lands added have at some time been parts of nonriparian holdings.

Riparian owners ordinarily have the right to change their points of use and diversion and also the right to use dams to retain water if their exercise of these rights does not cause injury to others. They also can reserve or separate riparian rights from land when it is granted or conveyed to others. As a rule, however, riparian rights pass with conveyances of riparian land unless the grantor specifically reserves or excepts the right from the conveyance. Owners of water frontage who sell rights of way for a highway between their homes and a lake, for example, may reserve their riparian right and thus retain rights of access to the lake.

Appropriation Doctrine

With the settlement of the arid lands of the American West, it soon became apparent that water is a strategic resource and that its most beneficial use sometimes calls for outright appropriation. This situation was clearly recognized by the Mormon pioneers who appropriated surface waters without regard for riparian rights when they started their irrigation of the Great Salt Lake valley in 1847. Similar practices were soon applied on a larger scale by the early gold miners in California. An appropriation doctrine was accepted in both of these areas as a matter of expediency, primarily because the riparian doctrine did not serve the best interests of the settlers.[24]

[23] Hutchins, *op. cit.*, p. 46.

[24] From a legalistic standpoint, the beginnings of the appropriation doctrine are often traced to the Spanish and Mexican law applied in the Southwest prior to the Mexican cession of this area to the United States in 1848. However, as Hutchins points out: "The appropriation principle in the form in which it is now recognized throughout the West—embodying the essential element of priority—is not traceable to Mexican laws and customs, but sprang from the require-

Under the appropriation doctrine, both riparian and nonriparian owners can file claims to divert water from streams or other bodies of water as long as their claims do not conflict with prior claims on water from the same source. As successive claims are filed, a system of priorities develops. This system vests each claimant with a recognized exclusive right to take water up to the amount of his or her claim for beneficial use, provided there is sufficient water to satisfy all claims of higher priority.

The appropriation doctrine has the following basic features:

1. It gives an exclusive right to the first appropriator; and, in accordance with the doctrine of priority, the rights of later appropriators are conditioned by the prior rights of those who have preceded.

2. It makes all rights conditional upon beneficial use—as the doctrine of priority was adopted for the protection of the first settlers in time of scarcity, so the doctrine of beneficial use became a protection to later appropriators against wasteful use by those with earlier rights.

3. It permits water to be used on nonriparian lands as well as on riparian lands.

4. It permits diversion of water regardless of the diminution of the stream.

5. Continuation of the right depends upon beneficial use. The right may be lost by nonuse.[25]

The essence of the prior-appropriation doctrine is aptly summarized by the catchphrase "first in time, first in right." As this maxim suggests, the key feature of the appropriation doctrine involves its recognition of priorities in appropriative rights. The operation of these priorities can be illustrated by an example of a stream, such as that depicted in Figure 12-2, along which several claims for water have been filed.

Let us assume that the first and sixth claims involve riparian users who take water to irrigate relatively small acreages, that the second and fifth priorities are held by irrigation projects along the stream, that the fourth priority involves a large irrigation project located several miles away from the stream, and that the third priority is held by a downstream hydroelectric plant which claims a definite minimum flow of the stream at all times.

Regardless of their location along the stream, these claimants enjoy water rights only in accordance with the priority of their claims. The first and second claimants may expect to take water up to the full amount of their claims every year. After

---

ments of a mining region for protection in the use of water supplies needed to work the mining claims. . . . The miner's customs became law, adaptable to diversions of water for irrigation as well as for mining purposes; and it is the specific principles there developed under the exigencies of that environment, rather than the less widely known principles of Mexican appropriation law and custom, that have been adopted by legislation and court decisions and are now a part of the water codes throughout the West." Hutchins, *op. cit.,* pp. 67–68.

[25] Roy E. Huffman, *Irrigation Development and Public Water Policy*, Copyright 1953, The Ronald Press Co. (New York: Ronald Press, 1953), p. 43.

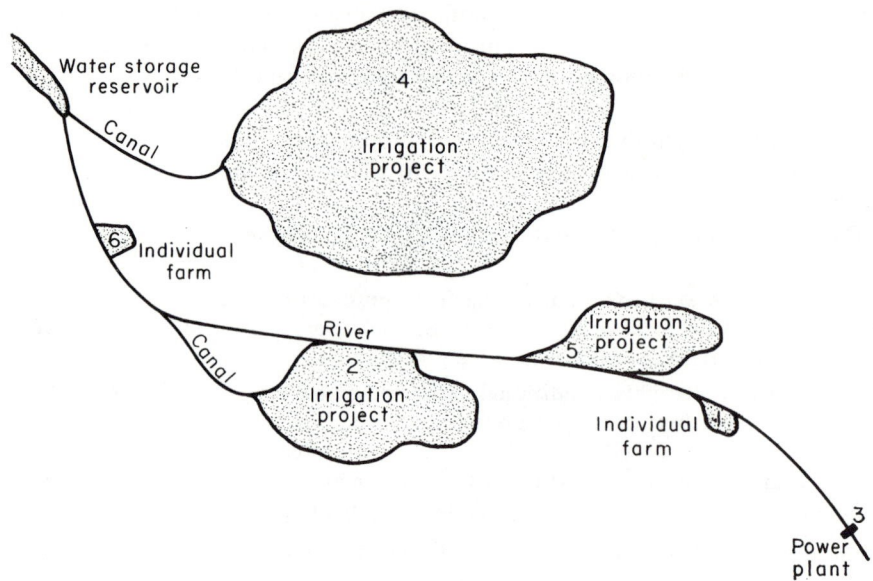

FIGURE 12-2. Example showing priorities in appropriative rights under the appropriation doctrine.

them, the power company can insist on a minimum natural flow of the stream, which may take all the flow during the summer months. Since the priorities of the first three claimants may entitle them to the full flow of the stream during the drier months, the remaining claimants can find their rights cut off when they experience the greatest need for irrigation water. They can remedy this situation by building an upstream storage reservoir that will hold seasonal flood water for later use.

Additional demands involving possible claimants along the river, on a tributary stream, or upstream from the reservoir could arise after the allocation of all of the surplus water in the reservoir. Since the normal summer flow would be covered by prior claims, they would be able to get water only by helping to finance the storage of additional seasonal surplus waters, perhaps on a tributary stream, which could be used as trade-offs for the water they hope to secure.

The appropriation doctrine is now accepted to a greater or lesser degree in eighteen Western states. The eight intermountain states—Montana, Idaho, Wyoming, Colorado, Utah, Nevada, Arizona, and New Mexico—and Alaska have completely abrogated the riparian doctrine in favor of the appropriation doctrine. Oregon also can be listed with these states because it too has abandoned the riparian doctrine except for its application in a few early cases in which riparian rights were established for beneficial uses. These states are sometimes said to adhere to the "Colorado doctrine," so named because the exclusive appropriation doctrine was prescribed in the constitution of that state when it was admitted to the union in 1876.

Throughout the remaining eight Western states—Washington and California on the West Coast and North Dakota, South Dakota, Nebraska, Kansas, Oklahoma, and

Texas on the eastern fringe of the arid and semiarid West—the appropriation principle is used in combination with the riparian-rights doctrine. Each of these states has some humid or subhumid areas, and most of them have superimposed the appropriation doctrine on an underlying riparian doctrine. Variations exist between these states; but in general they have all modified the riparian doctrine by restricting the claims of riparian owners while at the same time recognizing appropriation rights to reasonable amounts of water used for beneficial purposes.[26] California was the first of these states to spell out this *modified riparian doctrine*. For this reason, this approach is often referred to as the "California doctrine." Several humid-area states have modified the riparian doctrine in various respects. Mississippi enacted legislation that accepts aspects of the appropriation doctrine in 1956. Iowa adopted a ten-year permit law in 1957. Hawaii, Maryland, Minnesota, North Carolina, and Wisconsin also have laws that limit the riparian doctrine.

Unlike the riparian doctrine, operation of the appropriation doctrine calls for a certain amount of administration. Administration is involved in filing and recording claims, checking water gauges to make sure that appropriators do not exceed their rights, and in the action of courts and administrative boards in defining and adjudicating water rights. Some of the most notable administrative differences involve methods of filing claims. Wyoming, for example, claims ownership of the water found in streams, and all water rights must be filed with and granted by the state. Some other states in times past have permitted water users to file claims simply by posting a notice at the point of diversion and then filing copies of this notice with the appropriate county officers. This approach led to considerable confusion and the filing of more claims on some streams than could possibly be filled. Most states now require the filing of appropriation claims with some state agency.

Use of Ground Waters

Wells have been used as a source of domestic water supplies since the beginnings of history. Throughout this period rights for the use of ground waters have usually been taken for granted. The increased taking of this resource in recent decades for domestic, municipal, industrial, irrigation, and other uses has given rise to important questions concerning both ground-water rights and ground-water conservation.

From a legal standpoint, ground waters may be divided into two groups: underground streams and percolating waters. Most courts apply the same water-rights doctrine to underground streams that they apply to surface waters. Thus if an underground stream is found in a riparian-doctrine state, its waters are normally subject to the riparian rights recognized in that state. Likewise, if the underground

---

[26] For a more-detailed discussion of the water-rights doctrines accepted in these states, cf. Hutchins, *op. cit.*: and *Water Resources Law*, Vol. 3 of the Report of the President's Water Resources Policy Commission, 1950; Frank J. Trelease, Harold S. Bloomenthal, and Joseph H. Geraud, *Cases and Materials on Natural Resources Law* (St. Paul, Minn.: West Publishing, 1965), Part I; and Jacob H. Beuscher, *Water Rights* (Madison, Wis. College Printing & Typing Co., 1967).

stream is located in an appropriation or modified riparian doctrine state, its waters in most cases are available for appropriation.

Percolating waters—those waters below the surface that are not confined to any channel—are affected by four legal doctrines: the doctrines of (1) absolute owner-ship, (2) reasonable use, (3) correlative rights, and (4) prior appropriation rights. The first of these doctrines, the common-law rule of absolute ownership, was originally accepted in most areas that accept English law. This doctrine recognizes a landowner's right to all the water on or under the surface of his or her land as long as it is not part of a definite stream. As one court has complained, this doctrine "affirms the right of the owner of land to sink wells thereon, and use the water therefore, supplied by percolation, in any way he chooses to use it, to allow it to flow away, even though he thereby diminishes the water in his neighbor's wells or dries them entirely, and even though in so doing he is actuated by malice."[27]

The obvious lack of justice associated with unreasonable withdrawals of ground water under the absolute ownership rule has caused the courts in most states to qualify this rule. Out of these actions has come the American doctrine of *reason-able use*. This doctrine recognizes that landowners hold joint rights in the use of common ground waters. Individual owners are permitted considerable freedom in their use of percolating ground waters on overlying lands. But the rule of reasonable use may be interpreted to prevent wasteful, malicious, or other unreasonable uses of water, particularly if these uses have a harmful or injurious effect on others. Owners who suffer injuries from ground-water depletion can in some cases secure legal action to prevent the pumping of ground waters for sale or for use on non-overlying lands. Under some circumstances, they may also seek and secure damages from persons whose pumping activities cause them injury.

The doctrine of *correlative rights* is applied in California and some other Western states and represents an outgrowth of the rule of reasonable use. With this doctrine, landowners who are drawing water from a defined ground-water basin have coequal rights in the use of the water for reasonable purposes on or in connection with their overlying lands. Their rights can be limited to an equitable proportion of the total supply any time the supply appears inadequate to meet the needs of all the owners. Every owner could thus be required to reduce water consumption by 20 percent, for example, if it were determined that ground-water was being pumped out of the basin 20 percent faster than it was being recharged.

A few states in the West apply the prior-appropriation doctrine to percolating ground waters. This doctrine permits use of ground waters on a first-come, first-served basis. It also permits the appropriation of ground waters from overlying lands not owned by an appropriator for use on lands that do not overlie the source of the water supply. Problems have arisen with this approach, however, because of "the practical difficulty in identifying such waters and proving their characteris-tics."[28]

---

[27] *Schenk* v. *City of Ann Arbor*, 196 Michigan 75, 82 (1917).

[28] Hutchins, *op. cit.*, p. 161.

With the growing use of ground waters for municipal, industrial, and irrigation purposes, the problem of ground-water conservation and use is becoming increasingly critical in many localities. As these problems are recognized and publicized, more and more attention will be given to the development of regulations affecting ground waters. In some Eastern states, these problems may call merely for a greater exercise of the police powers of these states in regulating the taking or use of ground waters. In other areas, they may call for the enactment of comprehensive ground-water codes. Several Western states have already faced up to this problem. New Mexico, for example, has a highly regarded ground-water code, which asserts public ownership of the ground waters of the state and provides for their apportionment in the public interest.

Rights Affecting Drainage

Emphasis has been given thus far in this discussion to the rights individuals and groups have to take and use water for various purposes. A different type of problem arises with the rights landowners have to drain surplus waters from their lands or to construct levees that will prevent unwanted waters from encroaching on their properties.

Most states have constitutional or statutory provisions that authorize the establishment of drainage or levee districts together with the construction and maintenance of drainage and levee works. In addition to these provisions, most of the more humid states have enacted laws covering drainage problems. These laws tend to follow one or the other of two alternative doctrines. They either accept the common-law or so-called common-enemy rule or they follow the civil-law rule.[29]

The common-law rule treats both flood waters and unwanted runoff waters as a common enemy against which landowners have a recognized right to protect their lands. Some courts have modified this rule to permit the drainage of upper lands onto lower lands if no appreciable damage occurs. Owners of upper lands, however, have no lawful right to drain unwanted waters onto lower lands or to use tile, drainage ditches, or other man-made means to discharge drainage waters into a natural stream if their actions cause injury to lower owners.

Unlike the common-enemy rule, the civil-law rule recognizes the right of upper landowners to have flood and runoff waters flow naturally to lower lands. Lower owners cannot obstruct or refuse to receive the natural runoff from upper lands. But they can object whenever upper landowners hold back water for later release, use terraces or dikes to concentrate the runoff in given locations, or use ditches to facilitate faster runoff.

[29] Cf. Bernard A. Etcheverry, *Land Drainage and Flood Protection* (Stanford, Calif.: Stanford University Press, 1940), pp. 282–84. This author indicates that the common-law rule has been adopted in Arkansas, Connecticut, Indiana, Kansas, Maine, Massachusetts, Minnesota, Missouri, Nebraska, New Hampshire, New Jersey, New Mexico, New York, Oklahoma, South Carolina, Virginia, Washington, and Wisconsin. The civil-law rule applies in Alabama, California, Georgia, Illinois, Iowa, Kentucky, Louisiana, Maryland, Michigan, North Carolina, Ohio, Pennsylvania, and Texas.

Courts in some civil-law-rule states have held that landowners have a right to natural drainage, but that they have no right to increase the flow of water from their lands to those of lower owners. Upper owners may also be allowed to speed up the drainage of their lands by artificial means, as long as they have an outlet in a watercourse or natural depression and as long as they do not change the natural direction of water flow.

Both rules allow upper owners to acquire drainage rights over the lands of lower owners through the use of easements or by maintaining adverse uses over the authorized prescriptive period. They also recognize that different situations exist whenever drainage or levee districts are created. Landowners living within drainage districts, for example, have a lawful right to use tile, ditches, and other man-made devices when they discharge drainage waters into artificial watercourses, that is, into public drains or natural watercourses that have been improved for drainage purposes.

Property owners living within drainage and levee districts are usually subject to special assessments for drainage and levee districts works, even when they receive few, if any, benefits from these works. In the absence of districts, upper landowners covered by the common-law rule can often be assessed for downstream improvements necessitated by their use of artificial drainage methods. Upper owners covered by the civil-law rule are usually exempt from special assessments of this type. Owners of lower lands can usually claim damages from upper owners for just cause. In many jurisdictions, they can also secure injunctions against future uses of artificial drainage measures by upper owners if the expected extra water flow caused by these measures threatens injury or damage to the lower owners.

Public Interests in Water

Important powers concerning water are also held by the federal and state governments. The federal Constitution clothes Congress with the power "to regulate commerce with foreign nations and among the several states" and with the power "to dispose of and make needful regulations respecting government property." These two grants of power—commonly referred to as the "commerce power" and the "property clause"—underlie most of the interest the federal government has in water resources. But important powers affecting the administration, use, and development of water resources may also be implied from the authority vested in Congress "to provide for the common defense," "to provide for the general welfare," and to pass on interstate compacts. Other federal powers over water are associated with the federal treaty-making power and the original and exclusive jurisdiction the Supreme Court has over controversies between states.

Under the commerce power, Congress is responsible for the control of navigation and has full jurisdiction over navigable rivers. Because of the scope of this power, considerable importance attaches to the meaning of "navigable waters." The English common law limits the legal concept of navigability to ocean waters and to inland waters affected by the ebb and flow of tides. This doctrine was accepted for a while in the United States and then expanded to cover those inland waters that are navigable in fact. The concept of legal navigability has been further expanded to include all those rivers and lakes that are used, are susceptible of being used, or with reasonable improvements can be used in interstate commerce.[30] Moreover, it appears that "the commerce jurisdiction of Congress may be appropriately invoked both as to the upper nonnavigable reaches of a navigable waterway and as to its nonnavigable tributaries, if the navigable capacity of the navigable waterway is affected or if interstate commerce is otherwise affected."[31]

In the exercise of its power over commerce and navigation, "Congress has enacted legislation governing erection of dams, bridges, dikes, causeways, piers, wharves, and other structures; the removal of sunken vessels; the deposit of refuse materials; the operation of drawbridges; the use, administration, and navigation of waterways; the deposit of oil in coastal waters, and other protective measures."[32] Beyond this, the federal government has licensed private power plants on navigable waters, required the removal of unauthorized structures regarded as obstacles to navigation, built storage dams and locks, deepened channels, improved harbors, acquired private canals for public use, and developed its own water transportation facilities. Going farther, Congress has declared it national policy to work to the end that the nation may "fulfill the responsibilities of each generation as trustee of the environment for succeeding generations" and "assure for all Americans safe, healthful, productive, and esthetically and culturally pleasing surroundings." In so doing, it has broadened the interpretation of the commerce and navigation powers to give the federal government constitutional authority to work with the states on programs for protecting, maintaining, and enhancing the quality of air and water resources.[33]

In addition to its commerce power, the federal government has used its proprietary power under the "property clause" to undertake resource development programs such as those carried on under the Reclamation Act of 1902. With this power, the federal government can acquire and condemn property rights; it can build

---

[30] In the *Daniel Ball* case, the Supreme Court held: "Those rivers must be regarded as public navigable rivers in law which are navigable in fact. And they are navigable in fact when they are used, or are susceptible of being used, in their ordinary conditions, as highways for commerce, over which trade and travel are or may be conducted in the customary modes of trade and travel on water." *The Daniel Ball*, 10 Wallace 557, 563 (1870). Later in the *New River* case, it held further that "the power of Congress over commerce is not to be hampered because of the necessity for reasonable improvements to make an interstate waterway available for traffic." *United States* v. *Appalachian Electric Power Co.*, 311 U.S. 377 (1940).

[31] President's Water Resources Policy Commission, *A Water Policy for the American People* (Washington, D.C.: Government Printing Office, 1950), p. 278.

[32] *Ibid.*, p. 283.

[33] Cf. declarations of policy goals in National Environmental Protection Act of 1970 (42 U.S.C.A. § 4331) and Federal Water Pollution Control Act Amendments of 1972 (33 U.S.C.A. § 1251). This broad interpretation has been accepted by the courts. Cf., for example, *Avoyelles Sportsmen's League, Inc.* v. *Brown*, 715 F. 2d 897 (1983).

irrigation, flood-control, and multipurpose dams and generate and sell electrical power as an incident of ownership.

Broad powers concerning water are also exercised by the states. Each state has the power, within constitutional limits, to prescribe the rules that govern the use of surface and ground waters within its jurisdiction; and some have asserted proprietary interests in water by declaring their ownership of the surface or ground waters found within their boundaries. States can engage in water resource developments; they can exercise their police powers to control water pollution, prescribe measures affecting the supply and treatment of municipal water supplies, regulate the pumping of ground waters, and authorize flood-control measures and drainage works.

Every state has a perpetual and inalienable responsibility for safeguarding the rights of its citizens in the use of public waters. The full nature of these rights is subject to court interpretation, but most courts agree that these public rights include the privilege of individuals to use public waters for such purposes as navigation, fishing, fowling, bathing, skating, and the enjoyment of scenic beauty. How far states go in protecting these rights depends on court interpretations as to what constitutes "public waters" or waters subject to state regulation.

State courts frequently use state definitions of "navigability" as the criterion for determining whether a lake or stream is a "public water" that is subject to state regulation. Some of the Atlantic Coast states hold to a saltwater test of navigability, which treats only those waters affected by the ebb and flow of tides as navigable. Others have used a sawlog test, which holds that streams that can or have floated a log or boat are navigable. Between these two extremes, states such as Texas hold that a stream is navigable only as far as it retains an average width of 30 feet, Mississippi specifies that a navigable stream must be deep enough for any thirty consecutive days to float a steamboat large enough to carry 200 bales of cotton.

Some of the most important differences in definitions concerning "public waters" relate to the distinction between public and private lakes or ponds. Lakes or ponds connected with navigable streams are ordinarily regarded as public waters. They may also be so regarded if they cover more than a prescribed area, if they were meandered in the original public survey, or if owners have consented to the public planting of fish in what was their private water.[34]

Two problems occasionally arise with the exercise of public rights to public waters. The first of these involves the ownership of the ground underlying public waters; the second concerns rights of access. Most states have retained public ownership of the land beneath their public waters. But some have surrendered the ownership of these lands to riparian owners. Questions sometimes arise in these states concerning rights anglers and others have in the use of waters that flow over private bottom lands. In a leading case on this point, one state court has held that

---

[34] Cf. Leighton L. Leighty, "The Source and Scope of Public and Private Rights in Navigable Waters," *University of Wyoming Land and Water Law Review*, 5 (1970), 391–440; and 6 (1971), 459–90.

anglers can wade upstream, fishing as they go and contrary to the posted warnings of the riparian owner who owns the bottom of the stream, without being guilty of trespass as long as they do not go on the owner's upland.[35]

Lack of access rights keeps many people from exercising their rights to the use of public waters. Several states have recognized this situation and have taken specific steps to acquire riparian sites that give citizens access to public waters for fishing, bathing, boating, and other recreation uses.

Another major concern of states and the federal government is the need for controlling water pollution. Discharge of untreated sewerage and wastes into lakes, streams, and the ocean was accepted for a long period as a convenient and inexpensive way of disposing of unwanted materials. This practice, which appeared both logical and relatively harmless when few people lived along public waters, became unacceptable from a health standpoint once modern medicine discovered its impact on the incidence of disease. It also became aesthetically objectionable as more and more industries and cities turned to this means for disposing of wastes, as the volume of wastes began to tax the dilution and purifying capacities of the available water supplies, and as problems of deteriorating quality made many public waters unattractive or unacceptable for recreation uses.

Treatment of public water supplies and municipal wastes was pushed for many years as the most practical way of dealing with the water pollution problem. Since the enactment of the Water Quality Act of 1965, considerable progress has been made in cleaning up the nation's waters. Water quality standards have been established for different water uses in all the states, and strong steps have been taken to eliminate sources of pollution and enhance the quality of the water in most lakes and streams.

AIR AND SUBSURFACE RIGHTS

The property rights individual owners hold in land are sometimes visualized as an inverted pyramid that starts at the center of the earth and extends upward through the surface boundaries of the owner's holding to the highest heavens. As this concept suggests, landed property rights can be divided into layers of rights—air rights, surface rights, and subsurface rights. From a legal standpoint, each of these layers can be held separately from the others and each has its particular characteristics.

Air Rights

Until recent years, our courts have almost universally accepted the principle that surface owners hold all the rights to the column of airspace above their surface holdings. This doctrine has been modified to permit public use of the air for air

[35] Cf. *Collins* v. *Gerhardt*, 237 Michigan 38, 211 N.W. 115 (1926).

travel. But individual owners still have exclusive rights to all the airspace above their land they can occupy and use. As one court has held:

> The air, like the sea, is by its nature incapable of private ownership, except insofar as one may actually use it. . . . We own so much of the space above the ground as we can occupy or make use of, in connection with the enjoyment of our land. This right is not fixed. It varies with our varying needs and is co-extensive with them. The owner of land owns as much of the space above him as he uses, but only so long as he uses it. All that lies beyond belongs to the world.
>
> When it is said that man owns, or may own, to the heavens, that merely means that no one can acquire a right to the space above him that will limit him in whatever use he can make of it as a part of his enjoyment of the land. To this extent his title to the air is paramount. No other person can acquire any title or exclusive right to any space above him.
>
> Any use of such air space by others which is injurious to his land, or which constitutes an actual interference with his possession or his beneficial use thereof, would be a trespass for which he would have remedy. But any claim of the land owner beyond this cannot find a precedent in law, nor support in reason.[36]

Under our accepted concept of air rights, landowners can claim trespass whenever telephone wires, limbs of trees, or overhanging parts of buildings based on adjacent properties project into their columns of airspace. Easements affecting air rights are often purchased by utility companies and others. These rights may also be secured by deed reservations or by adverse possession. In exceptional cases, they may be sold or leased to others. Airspace above railroad tracks in downtown Chicago and New York, for example, has been sold together with rights of support for commercial developments such as the Chicago Merchandise Mart.

Surface land ownership carries with it a right of access to sunlight, air, and rain. Landowners hold these rights within their own columns of airspace. Their rights may be limited, however, by structures placed on adjacent lands. With the doctrine of "ancient lights" accepted in Great Britain, building owners who have enjoyed access to light and air from their windows for a twenty-year period have a prescriptive right to retain this access even though it limits the rights of adjacent owners. This doctrine is not accepted by the American courts. Property owners in the United States can put up buildings that completely shut off the air and light received by abutting windows on adjacent properties. Problems of this type are often averted by easements, deed restrictions, zoning ordinances, and building setback regulations.

Some of the most important problems involving air rights arise because of the movement of undesirable odors, smoke and chemical substances, rain clouds, and aircraft across property lines. Individuals and communities that suffer because of offensive odors caused by slaughterhouses, glue factories, or other comparable

---

[36] *Hinman* v. *Pacific Air Transport*, Ninth Circuit Court of Appeals, 84 F. 2d 755 (1936).

property uses can ordinarily take legal action to have these uses modified or stopped. Courts have often held, however, for the offending users—particularly when their prospective losses seem to outweigh promised social gains—if the offensive uses have continued for a long period and if the protesting individuals have acquired their properties since the beginning of the offensive use.

Urban and rural areas have been plagued with smog and air pollution. These are old problems that have been aggravated in recent times by the addition of exhaust fumes from automobiles, smoke and chemical wastes from manufacturing and smelting plants, and pollutants generated by incinerators and home heating systems and by the role these pollutants play in causing acid rain. Public concern with the air pollution problem prompted the passage of the Clean Air Act Amendments of 1970. This law instructed the Environmental Protection Agency to work with the several states in establishing and enforcing ambient air quality and emission standards. These standards limit the rights of individuals and others to discharge undesired wastes into the atmosphere. Prior to the enactment of these regulations, legal action could be taken against polluters when it was found that their actions interfered with the legitimate rights of others. Before granting damages or issuing injunctions against pollution practices, however, the courts usually weigh the equities on each side and give the offending parties an opportunity to install pollution-abatement facilities.

No definite customs or legal rules have as yet been developed concerning ownership rights to atmospheric moisture. But the possibility of artificial rainmaking has prompted some concerns with atmospheric water rights. At present, it appears that landowners have no more of a property right in the clouds that cross their land than they have in the birds that fly overhead. Until specific rules are developed, rainmakers can treat atmospheric moisture as a free good. As the techniques used to cause rain to fall in specific areas become more exact, however, possible legal grievances will rise both from persons deprived of the benefits of rainfall and from those who suffer damage from unwanted rainfall.

Congress declared in the Air Commerce Act of 1926 that the United States has "complete and exclusive national sovereignty" over the nation's airspace. This modification of the common-law doctrine of air rights was necessary to permit commercial air navigation without countless trespass suits. Zoning regulations and easements can be used to keep property owners from erecting high structures that may impede safe air travel. But with the advent of low-flying planes, an increasing volume of air traffic, ear-splitting takeoffs, and nerve-racking sonic booms have brought complaints about the adverse effects of air traffic on property values. The Supreme Court has recognized this problem and held that property owners can claim damages for actual losses caused by low-flying planes. In a leading case involving a claim for damages on a chicken farm located near an air base, the Court observed that

the path of glide for airplanes might reduce a valuable factory site to grazing land, an orchard to a vegetable patch, a residential section to a wheat field.

Some value would remain. But the use of the airspace immediately above the land would limit the utility of the land and cause a diminution of its value.

We have said that the airspace is a public highway. Yet it is obvious that if the land owner is to have full enjoyment of the land, he must have exclusive control of the immediate reaches of the enveloping atmosphere. Otherwise buildings could not be erected, trees could not be planted, and even fences could not be run. The principle is recognized when the law gives a remedy in case overhanging structures are erected on adjoining land. The land owner owns at least as much of the space above ground as he can occupy or use in connection with the land.... The fact that he does not occupy it in a physical sense— by the erection of buildings and the like—is not material. As we have said, the flight of airplanes, which skim the surface but do not touch it, is as much an appropriation of the use of land as a more conventional entry upon it.[37]

Subsurface Rights

Landowners hold rights to the minerals and other materials found beneath the surface of their land as well as to the land surface itself. Surface and subsurface rights are usually held and conveyed together. But they can be divided and held separately, either by the sale, devising, or leasing of mineral or oil and gas rights to others or by deed reservations that provide for the grantor's retention of subsurface rights. Although leases, purchases, and reservations of these rights are fairly common, their separation from surface rights often complicate the ownership and mortgage credit status of surface owners. Some states now provide for the separate taxation of subsurface rights when they are known to have economic value and for their reunification with the rights held by surface owners if they are not exercised or rerecorded within given time periods.

Rights to develop and use subsurface resources are frequently conditioned by the rights of other property owners. An owner who plans to excavate a basement, for example, must respect the subjacent rights of adjacent property owners to have such side support as is necessary to keep their properties from caving into excavated areas. In like manner, holders of mineral and oil and gas rights must respect the rights of surface owners when they exercise their right of surface access in getting to and from their properties.

Mineral rights. Landowners in the United States have a recognized right to take the minerals found on or under their surface holdings. When mineral rights are held separate from surface rights, owners of the mineral rights have recognized rights or easements over the surface. They can make mineral explorations, sink shafts, and build such roads and railroad tracks over the surface as may be needed to transport supplies to the mine and carry minerals to market. At the same time, holders of the surface rights ordinarily have a right to continued support. This means that if there is no contractual arrangement concerning possible land disruption or subsidence, mining operations must be carried on in such a way as to prevent the land surface from sinking or collapsing.

Unlike the United States, the governments of many nations reserve mining rights

[37] *United States* v. *Causby*, 328 U.S. 256 (1946).

to themselves. This situation prevails in continental Europe, in Mexico, and in South America.[38] Among the English-speaking countries, royalties representing specified percentage shares of certain minerals were long reserved to the crown. This practice was carried over in the early land legislation of the United States. The Ordinance of 1785 provided that the federal government should receive one-third of the gold, silver, lead, and copper found in the lands granted from the public domain.

This federal royalty system was applied in the Missouri and Upper Mississippi lead regions where it broke down because of lackadaisical enforcement during the 1840s.[39] No effort was made to collect royalties from the gold miners in California. Miners were free to prospect all over the public domain and frequently rushed to new "strikes" in unsettled areas where public lands had not yet been surveyed for sale. With no enforcement of government regulations, they staked out and operated individual claims in accordance with miners' rules. These rules spread from community to community and held "that discovery and development of a mine were the foundation of a property right in it."[40] Congress finally legitimized these local mining rules, abolished the royalty system, and opened the public domain for free mining in its mining act of 1866.

Private rights in mineral deposits have been modified in two important respects by federal legislation. The federal mining law of 1866 gives the owner of a vein or lode of ore extralateral rights—the right to follow a claim beyond the boundaries of one's surface ownership—if the vein or lode has its apex within the owner's holding. A second modification of far-reaching consequence is contained in the Atomic Energy Act of 1946, which declares all fissionable material, now or hereafter produced, to be the property of the Atomic Energy Commission. Individuals are encouraged to discover and develop uranium deposits; but those who possess and use the ores must be licensed to do so by the Atomic Energy Commission (now the Nuclear Regulatory Commission).

Oil and natural gas rights. Rights to oil and natural gas are treated in much the same way as mineral rights.[41] Like ground waters, oil and natural gas are migratory resources. They occur in underground basins that underlie substantial areas and

---

[38] Cf. Rudolf Isay, "Mining Rights," *Encyclopedia of the Social Sciences* (New York: Macmillan, 1932), X, 513–17.

[39] Cf. James E. Wright, *The Galena Lead District: Federal Policy and Practice, 1824–1847* (Madison: State Historical Society of Wisconsin, 1966); and Robert W. Swenson, "Legal Aspects of Mineral Resources Exploitation," in Paul W. Gates, *History of Public Land Law Development* (Washington, D.C.: Government Printing Office, 1968), pp. 701–16.

[40] Isay, *loc. cit.* p. 517.

[41] For more detailed discussions of the nature of oil and gas rights, cf. L. A. Parcher, John H. Southern, and S. W. Voelker, *Mineral Rights Management by Private Land Owners*, Great Plains Agricultural Council Publication No. 13, Oklahoma Agricultural Experiment Station, 1956; Stanley W. Voelker, *Mineral Rights and Oil Developments in Williams County, North Dakota*, North Dakota Agricultural Experiment Station Bulletin 395, 1954; Howard R. Williams, Richard C. Maxwell, and Charles J. Myers, *Cases and Materials on the Law of Oil and Gas* (Brooklyn, N.Y.: Foundation Press, 1956); and Frank J. Trelease *et al., Cases and Mateials on Natural Resources Law* (St. Paul, Minn.: West Publishing, 1965), Parts II and III.

often numerous ownership holdings. Landowners seldom know how much oil or natural gas underlies their land; and they cannot keep these resources from flowing to the ownership of others, particularly when others drill deep wells for their capture.

Because of the migratory nature of these deposits, owners of oil and natural gas rights do not acquire title to these resources until they actually capture them. This situation has put a definite premium on the early tapping and capture of these resources. Landowners who have struck oil have often drilled offset wells around the borders of their property to ensure a maximum take on their part. This practice forces the owners of adjacent properties to either join in the mad scramble for oil or risk the loss of their share of the possible profits. As one might expect, the acceptance of the "rule of capture" has resulted in wasteful competition. Unneeded wells have been drilled; and these wells have reduced oil reservoir pressures and have contributed to the too-rapid depletion of many petroleum deposits.

All of the leading oil-producing states now have conservation laws that prohibit avoidable wastes. These states fix and regulate production quotas for individual wells and thus reduce the underground waste that can result from overly rapid depletion. They may also require ten-, twenty-, or forty-acre units as the minimum spacing for new wells. Unitization—the development of complete oil fields by one management under a unified drilling and production program—also has been suggested as a conservation measure. This approach is easily applied only in those cases in which an entire oil field is controlled by one company.

Widescale leasing of oil rights in areas believed to overlay petroleum deposits provides another means by which wasteful competition in the drilling of oil wells can be prevented. With the leasing approach, surface owners who lease their oil rights receive nominal rental payments together with the promise of royalties if petroleum is discovered and pumped on their land. Since the lease rights are usually held by a limited number of oil companies, this system often results in the drilling of fewer wells.

—SELECTED READINGS

Berger, Curtis J., *Land Ownership and Use: Cases, Statutes, and Other Materials* (Boston: Little, Brown, 1968), pp. 113–244.

Ely, Richard T., and **George S. Wehrwein**, *Land Economics* (Madison: University of Wisconsin Press, 1964), chap. 5. Originally published by Macmillan, 1940.

Fisher, Ernest M., and **Robert M. Fisher**, *Urban Real Estate* (New York: Holt, Rinehart & Winston, 1954), chap. 4.

Hutchins, Wells A., *Water Rights Laws in the Nineteen Western States, Selected Problems in the Law of Water Rights in the West*, U.S. Department of Agriculture Misc. Publ. No. 1206, 1971.

Kratovil, Robert, and Raymond J. Werner, *Real Estate Law*, 8th ed. (Englewood Cliffs, N.J.: Prentice-Hall, 1983).

Ratcliff, Richard U., *Urban Land Economics* (New York: McGraw-Hill, 1949), chap. 1.

Renne, Roland R., *Land Economics*, 2nd ed. (New York: Harper & Row Publishers, Inc., 1958), chaps. 15 and 16.

Schmid, A. Allan, *Property, Power, and Public Choice* (New York: Praeger Publishers, 1978), chaps. 1–3.

13

ACQUISITION AND TRANSFER OF OWNERSHIP RIGHTS

Some of our most important real estate resource problems stem directly from the arrangements under which people hold and share property rights. These problems provide the subject matter of *land tenure*—a concept that involves the many relationships established among people that determine their varying rights to control, occupy, and use landed property. Land tenure concerns the many ways in which persons, corporate bodies, and governments share in the bundle of property rights. It is also concerned with the time periods during which these rights are held.

Emphasis is given in this chapter and in the three subsequent chapters to four major aspects of land tenure. This chapter deals with the acquisition and transfer of ownership rights, Chapter 14 with leasing arrangements, Chapter 15 with real estate credit problems, and Chapter 16 with society's right to tax private property. Emphasis in all of these chapters is on tenure conditions in the United States.

ACQUISITION AND MAINTENANCE OF OWNERSHIP

Most people have a strong desire for property ownership, particularly for home ownership. This desire has some basis in our biological nature.[1] It is also fostered by (1) traditional attitudes and sentiments favoring ownership, (2) the cultural approval of society, and (3) the promotional efforts of groups that develop and sell properties.

There is nothing new about this desire for ownership. It was a recognized factor in ancient times. And it has been nurtured through the centuries both by the average

[1] Cf. Robert Ardrey, *The Territorial Imperative* (New York: Dell Pub. Co., Inc., 1966).

person's limited opportunities to own land and by the social status, economic and political power, and other privileges so often associated with the ownership of property.

Much of our emotional sentiment favoring home ownership in the United States and Canada can be traced to the attitudes of the land-hungry settlers who helped develop these countries. These settlers agreed with Arthur Young that "the magic of property turns sand to gold" and with Thomas Jefferson that "small landholders are the most precious part of the state."[2] This passion for ownership and the tendency of our forebearers to identify landownership with economic, political, and social prestige have been somewhat diluted with the maturing of our national economy. But we still endorse the ideas of home ownership, family owner-operatorship of businesses, and wide dispersion of ownership rights as high-priority goals in public policy. And we still look somewhat askance at anyone who dares to question the desirability of the ownership ideal.

With the ownership goal playing the role it does, the strategic question in many minds is not so much Should I become an owner? but rather, How can I become an owner? Emphasis is given here to four landownership issues: (1) acquisition of land titles, (2) costs of landownership, (3) getting the most for one's money, and (4) risks to successful ownership.

Acquisition of Land Titles

Acquisition of rights to landed property involves a legal as well as an economic process. Prospective owners must use legally sanctioned methods in acquiring title to the properties they want and thereafter take appropriate actions to support and maintain their property rights.

Methods of acquisition. Ownership in land can be acquired in any of ten different ways. It can be acquired through (1) patents or grants from the government, (2) private grants by deed, (3) grants by devise, (4) acquisition under the laws of descent, (5) dedication, (6) eminent domain, (7) forfeiture, (8) adverse possession, (9) accretion, and (10) escheat.

While ownership is seldom acquired through *patents* from government at the present time, every land title in the United States can supposedly be traced back through an unbroken chain of ownerships to an official grant of title from the government or crown. Most of these patents were originally acquired through grants to royal favorites; purchases at public sales; or through homestead, military bounty, or public-improvement grants. The titles held by governments can in turn be traced back through purchase and treaty arrangements to the claims European sovereigns asserted to portions of the New World by right of discovery and to the actions of the federal government in extinguishing the claims of the original Indian population.

Most landed properties now pass from owner to owner by the private grant of a

---

[2] Andrew A. Lipscomb, ed., *The Writings of Thomas Jefferson*, Monticello edition (Washington: Thomas Jefferson Memorial Association, 1904), IX, 18.

deed. The *deed* in this case represents a properly drafted, written statement by which an owner conveys the rights he or she has in a given tract of land to someone else. Deeds are used with sales of properties and also with transfers by gift or trade.

Real properties are often acquired by inheritance. This process usually takes one of two forms. Deceased persons who leave a valid and effective will are said to die *testate*, and their real property passes by *devise* to their designated heirs. When no valid will is known to exist, the deceased is said to die *intestate*, and the estate is divided according to the *laws of descent*. The overall importance of inheritance as a method of ownership acquisition is suggested by the findings of a nationwide survey of landownership in 1978. This study found that 10.8 percent of the ownerships and 19.1 percent of the acreage reported as owned had been acquired from gifts or inheritances. An additional 13.6 percent of the ownerships and 14.7 percent of the land were acquired through purchase from relatives. Elements of gifts or inheritances were involved in a fourth of the farm transfers between 1966 and 1982.[3]

Dedication represents a less frequently used method of ownership acquisition. This process takes place when private landowners make specified areas available for public use. Dedication may result from an owner's implied offer and the public taking of land for use as a park or road; or it may result from an official statement dedicating an area to public use. Official dedications of specified areas for streets, alleys, and other public uses are found in most subdivision plat recordings. These dedications do not become final until they are officially accepted by the units of government that must bear the responsibility for their administration and maintenance.

Lands needed for public purposes can also be acquired under the power of *eminent domain*. This power to acquire needed sites by condemnation is limited to public, quasi-public, and specified private agencies and groups. Its use is limited to the acquisition of lands for public purposes; and reasonable compensation must ordinarily be paid for the lands taken.

Landed property is occasionally acquired by *forfeiture*, either through mortgage foreclosure proceedings or through reversion to some unit of government for non-payment of taxes. Most of the properties lost through forfeiture are offered for purchase at public mortgage foreclosure or tax sales. The highest bidders at these sales ordinarily receive either a deed for the properties they buy or a certificate of deed, which conveys ownership rights if the forfeiting owner does not redeem his or her ownership within a specified redemption period. Some states have mortgage foreclosure arrangements that permit mortgagees to take title to delinquent properties without public sale. Several also have legislation that permits the automatic bidding off of tax titles to the state or county at the time of tax sale.

Despite our faith in recorded titles, property rights can be lost through a significant though seldom-used property acquisition method known as *prescription*

[3] Cf. James A. Lewis, *Land Ownership in the United States*, U.S. Department of Agriculture, Economics, Statistics, and Cooperatives Service Agriculture Information Bulletin No. 435, 1980.

or *adverse possession*. This method permits one to acquire specific rights in a property or even the title itself through continuous use of a property in an open and visible manner, without the owner's expressed permission, while always claiming the right of use or ownership throughout the prescriptive period set by law. Acquisition by adverse possession can involve specific interests such as the right to drain water or use a road across another person's land. This method can also be used to acquire title to unused or abandoned lands and to clear possible clouds against one's title. Most states require that a use continue for periods ranging from ten to twenty years before one can acquire title by adverse possession. Periods of as little as five years are permitted in some states when this process is used to remove color of title.

Ownership rights can be acquired by *accretion* in two different ways. A common example occurs when a property bounded by a watercourse has surface land added to it by action of the stream. A second type of accretion occurs when a house or some other improvement is added to one's land without one's consent. If a house is built on one's lot by mistake, the landowner acquires title by accretion and is under no obligation to pay its cost of construction.

A final method of property acquisition known as *escheat* involves the reversion of estates to the state whenever a person dies intestate without known heirs who are eligible to receive the property under the laws of descent. Escheat was a jealously guarded power under feudalism. It was often a highly remunerative power because properties reverted to overlords or to the crown when owners were adjudged as guilty of felonies or of treason as well as when they died without heirs. It provides a smaller source of public revenues today, but it is used by most states to claim abandoned properties such as forgotten bank accounts or unexercised options to claim properties.

Need for a clear title. Regardless of how they acquire property, owners naturally want and expect security in the continued possession and exercise of their rights of ownership. To secure this end, they should always make certain that they secure clear and merchantable titles and follow the proper procedures in registering their evidences of ownership. With the usual purchase transaction, they should insist that they receive a valid deed, that the deed is duly recorded, and that the grantor conveys all of the expected rights of ownership.

Deeds can be classified into four different types according to the rights they convey. A *warranty deed* contains covenants of title guaranteeing the conveyance of a clear title that is unencumbered except for those claims specifically stated in the deed. Should any of these covenants be violated by the later discovery of claims against the title, the grantor can be held liable for damages. Under *special warranty deeds*, grantors assume less risk because they covenant the title only against the lawful claims of persons who have claims arising during their period of ownership. A third type of deed known as a *bargain-and-sale deed* conveys land without any warranty of title. Still another type, known as a *quitclaim deed* is used to convey whatever interest a grantor may have in real estate rather than the land itself. Quitclaim deeds are frequently used to clear up possible claims against titles.

The requirements of a valid deed vary somewhat from state to state. It must list the name of the grantor (who must be a competent person), the name of the grantee, a recital of the consideration given for the property, a legal description of the property, a statement of conveyance, and the signature of the grantor.[4] The recital of consideration ordinarily lists the actual sales price when the property is conveyed by a corporation, trustee, or executor. In cases of transfers between individuals, the amount of the sales price is usually cloaked by some phrase such as "one dollar and other valuable considerations." The requirement that there be a recital of consideration does not prevent owners from giving property to others. A gift or sale for a nominal consideration can be invalidated, however, if it can be shown that this means of disposal was used to prevent satisfaction of the claims of a grantor's creditors. A deed may also contain the date of transfer, the addresses or some other identification of the grantor and grantee, a warranty of title, statements concerning outstanding mortgages and other encumbrances, a statement of possible restrictions and conditions, a waiver of dower and homestead rights, signatures of witnesses, an official acknowledgment by the grantor before a notary public or other public official, the grantor's official seal, and revenue stamps.

Once prepared, a deed must be delivered to the grantee (usually during the lifetime of the grantor) and be accepted before it becomes valid. This action makes the transfer of ownership effective as far as the two parties to the deed are concerned. For official purposes, however, the deed should then be recorded with the public recorder or register of deeds and thus become a matter of public record.

Possession of a valid deed or other legal instruments showing conveyance of ownership gives an owner evidence of title.[5] Unfortunately, this does not necessarily mean that the owner has a clear or merchantable title. To have a clear or merchantable title, one must acquire and still possess all the property rights one claims and be able to support these claims with legal evidence of the extinction or severance of the interests previous owners and others have held in the property.

Owners have a perfect title when they can show that every item in the chain of legal actions involving their properties going back to the time of their original patent from the government has been properly handled and recorded without error. Titles of this type are a rarity in most older settled areas. Somewhere in its

---

[4] Cf. Robert Kratovil, and Raymond J. Werner, *Real Estate Law*, 8th ed. (Englewood Cliffs, N.J.: Prentice-Hall 1983), chap 7; and Bruce Harwood, *Real Estate Principles*, 2nd ed. (Reston, Va.: Reston, 1980), chap. 5.

[5] Official evidences of ownership are also needed with most properties acquired by methods other than the granting of a deed. Homesteaders on the public domain needed patents of ownership from the government before they could fully exercise their rights of ownership. The rights of heirs operating an estate are often in doubt until the estate is probated and an official determination of heirs is made. Specified procedures are required when a unit of government acquires land ownership by eminent domain, tax reversion, or escheat or when a mortgagee acquires property by foreclosure. Rights acquired by adverse possession must be legally recognized and recorded in the owner's name before they can be transferred to others. Private dedications of land areas for public use should be made a matter of public record, even though this is not strictly necessary.

history, almost every title involves some minor errors or defects such as a faulty or incomplete property description, a wife's failure to waive her dower interest, or an owner's failure to register the release of a mortgage. These defects can usually be cleared by reasonable interpretation or by use of affidavits and quitclaim deeds to provide a merchantable title. Titles involving minor errors or omissions are usually regarded as imperfect and should be cleared before they are transferred, even though they can often be accepted with little risk to the grantee.

Titles with major errors or omissions can be described as incomplete or defective. Time, expense, and litigation are often needed to clear these defects. These costs and the possible claims others may have against the title make it prudent policy for buyers to refuse to accept defective titles until the necessary steps have been taken to make them merchantable.

Title examinations. Since few buyers or sellers have the training needed to examine and vouch for the merchantable nature of their land titles, it is usual practice to enlist the services of an attorney for this purpose. The attorney conducts a title examination in which all legal actions affecting the property are reviewed before submitting an opinion concerning validity of a title.

The title examination process ordinarily involves the preparation of an *abstract of title* by the lawyer, some public official, or a private abstracting company. This abstract represents a chronological summary of all the recorded legal actions—data on deeds, wills, mortgages, liens, judgments, tax sales, and so on—that involves one's title to property. An abstract provides no guarantee of title. But it does provide a condensed history of the legal actions affecting a property, which will usually disclose any errors or omissions in the official records that could cloud the title.

In their title examinations, lawyers sometimes check every item back to the original patent from the government.[6] On other occasions, they may limit their study to those entries of record made since some prior title examination. Following these examinations, they prepare a statement known as a *title opinion* (or in some states as a *certificate of title*), in which they indicate that they have examined the title and believe it to be merchantable or not merchantable. In the latter case, the opinion usually lists specific errors, omissions, or other factors that cloud the title.

Even under ideal conditions, a title search will sometimes fail to disclose hidden defects such as a forged signature on a deed or a grantor's minority, possible insanity, or failure to indicate that he or she was married. This possibility, together with the normal costs of clearing titles and the tendency of some attorneys to raise technical objections to details that would ordinarily pass as merchantable, has caused many buyers to seek other means for guaranteeing their titles. The two most

[6] Some states have simplified this process by enacting marketable title statutes, which invalidate old claims after a specified time period. These laws make it unnecessary for the title examiner to go back more than the specified number of years (forty years under the Michigan law).

important of these involve the use of title insurance and registration under the Torrens system.

Title insurance.   Title insurance is now available in most areas. The companies that offer this insurance naturally refuse to insure defective titles. But if they find no serious defects in their title examinations, they usually issue insurance that guarantees titles against possible defects or claims up to a specified policy value. With these policies, the insurance company agrees to defend at its own expense any lawsuits attacking the validity of a title and to reimburse owners up to the amount of their policies should they be ejected from ownership.

Unlike most other insurance, a title insurance policy involves a flat fee that protects owners for their full period of ownership. Title insurance policies vary in their details and are not ordinarily transferable. They apply only to defects of title that exist prior to the date of the policy; and they usually involve exceptions and conditions that limit the company's liability. Title insurance is widely used both because of the protection it gives and because it is required by most real estate credit lending agencies.

The Torrens system of title registration.   An alternative to our usual system of land titles is provided by the Torrens system of title registration. This system calls for updating land titles, creating indefeasible titles warranted and guaranteed by the state, official registration of titles rather than the registration of mere evidences of titles, and eventual elimination of the time-consuming and frequently expensive practice of clearing land titles.[7] To get a title registered, a landowner must first file an application for this purpose together with a complete abstract of title for the property with the proper public official. The title is examined, and official notices are sent to all persons who appear to have any interest in the property, advising them of the owner's action and of their right to contest the owner's claim to title. Special court proceedings are then conducted to establish the validity of title.

If the court finds the title defective, it naturally refuses to take further action until the title is cleared. If it finds that the owner has an acceptable title, it issues a certificate of title and orders the title registered in an official record. This registered title is warranted by the state and is considered indefeasible in the sense that it cannot be nullified by possible past claims. For this service, the owner must pay a fee that covers the cost of establishing the title plus a contribution to an indemnity fund, which is set up to pay off possible future claims for damages.

Once a title has been registered, all future legal actions involving the property must be noted on an official registration sheet maintained for the property; and changes in title can come only with the issuance of new certificates of ownership. With this procedure, future searches of title become unnecessary, and transfers can be made speedily at a nominal cost.

The Torrens system was first used in South Australia in 1858 and has since been accepted in many countries. It has been used with some success—often as a means

---

[7]Cf. Harwood, *op. cit.*, chap. 6.

of last resort—to clear land titles in the United States. Its use has been authorized on a permissive basis by nineteen states. The manner in which it has been used together with the cumbersome and expensive nature of the initial registration process and opposition by groups with vested interests in the present system have discouraged its popular acceptance. Proponents of the Torrens system argue that its full benefits can be realized only with a mandatory system of title registration.

Costs of Real Estate Ownership

When one mentions the costs of real estate ownership, most people think of the capital outlay needed to purchase property. This investment cost is only one element in the total cost picture. Yet it often represents the most important single hurdle in the acquisition process.

The purchase price or market value of any given property always depends on the characteristics of the property, its location, and the supply and demand situation at the time of sale. Other things being equal, an eight-room house will cost more than a five-room house; and a 200-acre farm will cost more than one having 120 acres. In like manner, a house located at a prestigious address in a large city will have a higher sales value than a house of similar design in a small village; and farmland located near a good market will sell for more than comparable land at a more remote location.

Time of purchase is also important because properties normally command higher prices during periods of business prosperity than during recessions. The nature of these shifts in market value can best be illustrated by examples based on national indices of farm and residential real estate prices. (See Figures 10-1 and Table 10-5.) An average farm valued at $10,000 in 1912 was worth $17,520 in 1920 but only $7,520 in 1933 and then increased to a value of $17,780 in 1950, $30,381 in 1960, $50,908 in 1970, and $177,749 in 1980. Meanwhile, a residence that could have been built for $5,400 in 1914 would have cost $10,000 in 1920, $6,700 in 1932, $17,600 in 1950, $22,200 in 1960, $30,600 in 1970, and $69,800 in 1980.

Fortunately for the average buyer, it is not necessary for one to save the entire purchase price of a property before one can become an owner. Credit arrangements make it possible for buyers to acquire ownership with minimum down payments ranging from practically nothing with some loans up to 50 percent or more of the purchase price with others. This use of credit definitely opens the door of ownership to people who could not otherwise afford to buy properties of their own. But while use of credit makes it possible for buyers to shift much of their problem of providing an initial capital outlay of investment funds to the future, it has the disadvantage of adding debt service charges to the periodic, ongoing costs of ownership.

A buyer's capital investment outlay—the down payment plus any periodic payments of principal required by a credit arrangement—really represents an investment that can be recovered through later sale. The actual costs of property ownership, which cannot be recouped, include (1) fees for securing mortgage credit; (2) acquisition closing costs, (3) debt-service charges; (4) allowance of interest on one's equity in the total value of the property; (5) taxes and special assessments; (6) fire and other property insurance; (7) repair and maintenance costs; (8) allowances for depreciation and obsolescence; and (9) a charge for other items such as utilites, fuel, and yard work, which the buyer may or may not have paid as a tenant.

These costs vary considerably depending on individual circumstances. If the buyer can secure mortgage credit from a private lender or local lending agency, the cost of obtaining credit may be limited to a nominal charge for the preparation and filing of the mortgage. But if one deals with a mortgage broker, it is common practice to pay a loan origination or placement fee—often priced in terms of a discount of "points" from the face value of the mortgage—for the broker's service in placing the mortgage. In addition lending agencies usually require a series of charges for a property appraisal and survey, a title examination, title insurance, and the preparation and filing of the mortgage. These charges together with other closing costs for property inspection, escrow, recording, and attorneys fees can add between 2 and 15 percent to the actual purchase cost of ownership.[8]

Debt-service charges vary with the size of one's mortgage, the interest rate, the length of the repayment period, and the amortization provisions. Most real estate loans in 1982 were financed at interest rates ranging from 12 to 16 percent and were amortized over periods ranging from ten to thirty years. Even with lower interest rates, debt-service charges can represent a substantial cost of ownership. With a 6 percent interest rate and a standard amortized loan, a borrower pays $1,719.43 for every $1,000 borrowed and repaid over a twenty-year period and $2,158.38 for every $1,000 borrowed and repaid over a thirty-year period. At 9 percent, these payments rise to $2,159.34 and $2,896.64 and at 12 percent to $2,642.40 and $3,703.32 for the two respective time periods.

Owners frequently fail to consider the opportunity costs associated with the equity interests they hold in properties they own. From an economic standpoint, an ownership cost should always be assigned to the income one could receive by investing one's money in some alternative enterprise. The interest rate applied in calculating this cost varies with one's alternative investment opportunities. As a bare minimum, one should allow a rate of return at least equal to the interest rate paid on savings accounts or government bonds.

Property taxes and special assessments vary from year to year and from community to community. On an annual basis, these costs ordinarily range between 1.5 and 4 percent of the property's value. Fire and other property insurance costs

---

[8] Mortgage placement fees vary over a wide range depending on one's choice of a lender, the supply and demand situation with mortgage money, and the difficulties one may encounter in placing a mortgage. Typical placement fees in 1980 ranged from 2 to 12 percent of the amount of the mortgage, with most home buyers paying in the 4 to 8 percent range. The Federal Housing Administration reported average closing costs of $1,049 for new houses and $1,054 for existing houses for its insured loans in 1981. These costs represent 1.8 and 2.3 percent of the average appraised values of the respective houses and do not include points or fees paid for placement of the mortgages.

also vary somewhat but ordinarily range from one-eighth to one-half of 1 percent of the insured value of the property each year.

Wide variations also apply with repair and maintenance costs. These costs are usually lower with new buildings than with older structures; and they are always higher in years when one installs a new furnace or puts on a new roof than in years when no major repairs are made. Owners of relatively new properties should allow around 1 to 1.5 percent of the value of their buildings and improvements for repairs and maintenance costs. Owners of older properties should raise this annual cost outlay to sums representing 2 to 2.5 percent of the value of their buildings.[9]

Since building improvements tend to deteriorate with passing time, allowances must also be made for their depreciation and eventual replacement. This charge varies with the quality of the original structure, the wear and tear it receives, and owner practices in caring for property. A depreciation charge of from 1.5 to 3 percent of the value of buildings and other improvements should be allowed each year unless offset by expenditures for improvements and property renewal. In many instances, a charge for obsolescence should also be listed among the costs of ownership. This cost is particularly important when property values decline because of neighborhood blight or because of the effects of changing tastes and new technology in outdating older structures.

A final group of ownership costs involves a miscellaneous assortment of expenditures that prospective owners may or may not have paid as tenants. Apartment dwellers often have charges for electricity, gas, heat, water, and garbage and trash disposal included in their rents. They also benefit from the yard work and janitorial services normally provided by management. The costs associated with these services represent significant items in most household budgets. Buyers who fail to reckon with these costs at the time they make their ownership commitment often end up with unbalanced household budgets and with a more healthy regard for the costs of ownership.

Investment alternatives. Once operators visualize the costs of property ownership, it is appropriate to ask: Is it really wise to become an owner? Their answers to this question are often influenced by sentimental considerations, by desire to own, to please their families, or to free themselves from dependence on a landlord and by the extent to which they view property ownership as a consumption expenditure rather than as a production investment. If they want to examine their ownership acquisition decisions in economic terms, they can easily evaluate their investment alternatives and weigh their expected costs of ownership against the costs they would bear as nonowners.

When alternative investment opportunities are evaluated in this manner, operators often find that they can realize higher economic returns by using their available capital for purposes other than acquisition of ownership. Farm tenants in areas of

---

[9] For other discussions of this topic, cf. Leland J. Gordon and Stewart M. Lee, *Economics for Consumers*, 7th ed. (New York: Van Nostrand Reinhold, 1977), chap. 17; and John P. Dean *Home Ownership: Is It Sound?* (New York: Harper & Row Publishers, Inc., 1945).

high land values, for example, frequently find that it is more profitable to invest their savings in machinery and livestock than in land because they can make more money as well-equipped renters on larger or more productive farms than as owner-operators of the smaller or less-productive units they could afford to buy.[10] Operators of grocery stores, ready-to-wear shops, and other businesses often covet their mobility as tenants and argue that it is far better to use capital to expand their scales of operation than to purchase their sites of business. In a similar vein, people who work for others may also find it economically better to live as tenants and invest their savings in other ventures than use them to acquire ownership.

With other circumstances, property ownership can represent a wise investment for the prospective owner. If a buyer expects to become an owner anyway, he or she can often benefit by acquiring the rights of ownership now rather than later. With liberal use of credit, ownership can be acquired with a relatively small down payment, and one's equity can be built up with payments not greatly different from rent. If one finds it hard to save money, capital accumulations will come with the system of forced savings dictated by one's mortgage or land contract payment schedule. By investing in a home or business, one can pamper an investment along and treat it as a personal savings account that can be converted into cash at some future date through the mortgaging, sale, or leasing of one's interests.

To buy or not to buy. Prospective buyers should carefully evaluate and compare the costs and advantages associated with their ownership and renting alternatives. Like the prospective investor who weighs the estimated benefits and costs associated with alternative investment opportunities, business operators should compare the economic opportunities available to them as owners and tenants. They should not plan to buy a business site unless the site offers them economic opportunities at least as good as those they could enjoy as tenants. In similar fashion, prospective buyers of residential properties should carefully consider their expected economic benefits and costs as well as the consumer satisfactions associated with their opportunities for buying and renting.

To illustrate the economic cost portion of this evaluation process, one might assume the cost comparisons of a married couple who rent an unfurnished house for $550 a month, no utilities included. They have a salary income of $30,000 a year, enjoy job security, anticipate no marked increase or decrease in income in the near future, and have $20,800 in a savings and loan association savings account drawing 7.5 percent interest. Their living quarters are adequate but they would like to own their own home. They have found a house they like, which they can buy for $60,000 plus an estimated $800 for closing costs. They can borrow the additional $40,000 they need at 11.5 percent interest on a twenty-five-year mortgage with monthly payments of $406.59. The house is twelve years old and in good condition but will need repairs in the future.

---

[10] Cf. T. W. Schultz, "Capital Rationing, Uncertainty, and Farm Tenancy Reform," *Journal of Political Economy*, 48 (June 1940), 309–24; also Robert E. Pritchard and Thomas J. Hindelang, *The Lease/Buy Decision* (New York: AMACON, 1980), chaps. 1 and 8.

With this combination of circumstances, they find that their initial costs would include an outlay of $800 for closing costs plus any payments they make for repairs or changes in the house before they move into it. Their probable future costs of ownership can be visualized as in Table 13-1. These costs include not only the actual cash outlays for mortgage debt service, property taxes and insurance, and

TABLE 13-1. Hypothetical Example Illustrating Costs of Owning a Residential Property in the First and Twenty-Sixth Year after Purchase*

COST ITEM AND RATE OF CHARGE	1ST YEAR AFTER PURCHASE	26TH YEAR AFTER PURCHASE
Debt service (decreasing from one-twelfth of 11.5 percent of $40,000 in first month to nothing in 26th year)	$4,566	None
Property taxes at 2.5 percent of market value of $60,000	1,500	$1,500
Property insurance (0.25 percent of building value of $48,000)	120	120
Repairs and maintenance (rising from 2 percent of $48,000 in first year to 2.5 percent in 26th year)	960	1,200
Annual cash outlay for ownership	**$7,146**	**$2,820**
Depreciation at 2 percent of building value of $48,000	960	960
Interest on owner's equity in total investment (increasing from 7.5 percent of $20,800 in first year to 7.5 percent of $60,800 in 26th year)	1,560	4,560
Total annual gross costs	**$9,666**	**$8,340**
Federal income tax benefits		
Savings from owners' option for itemizing portion of their property taxes and interest payments that exceed their maximum standard deduction of $3,400	613–1,395	0–345
Savings from exemption of imputed interest on equity funds from taxation (23 percent of $1,560 in 1st year to 23 percent of $4,560 in 26th year)	359	1,049
Total federal income tax benefits	$972–$1,754	$1,049–$1,394
Net cost of home ownership	$7,912–$8,694	$6,946–$7,291

*Example assumes a husband and wife with two children with a family income of $30,000, $20,800 in savings invested at 7.5 percent interest, and living as tenants in a house that rents for $550 a month. The house they are considering is twelve years old, is generally comparable with their present house, and can be purchased for $60,000 plus $800 in closing and incidental costs. They can borrow $40,000 at 11.5 percent interest on a twenty-five-year mortgage. Their payments will come to $406.59 a month and will total $117,977 in twenty-five years. Family is in the 23 percent federal income tax bracket.

repairs and maintenance but also the additional allowances they should make for depreciation and imputed interest on their equity funds. Their total ownership costs come to $9,666 in the first year and $8,340 in the twenty-sixth year and can be compared with an alternative annual cost of $6,600 if they continue as tenants without a change in rental rates.

Their case for purchasing the house is strengthened when allowance is made for the federal income tax benefits they can claim as home owners. Two types of tax benefits are available: (1) payments of mortgage interest and property taxes can be itemized as deductions from taxable income, and (2) the imputed interest on their equity funds would be exempt from taxation. As taxpayers in the 23 percent federal income tax bracket in 1983, they would have had tax benefits of between $613 and $1,395 (depending on the amount of their other deductions) in the first year and between $1,049 and $1,395 in the twenty-sixth year of ownership. These benefits would improve their case for buying but would still leave them with strong economic reasons for feeling that they could save money by remaining as tenants.

From an economic view, buyers should also consider the effects of possible monetary inflation or deflation on their purchase decisions. They may anticipate a steady increase in income that will permit them to allocate larger sums to housing expenditures. They may also anticipate an inflationary trend in the nation's economy that will boost both the market value of the house and the amounts they would be required to pay as contract rent should they remain as tenants. Should housing prices increase as much in the next twenty-five years as in the 1955–1980 period (see Table 10-2), the market value of the house they could purchase for $60,000 now could rise to $175,000 by the end of the mortgage period. In contrast, a serious depression such as that of the early 1930s could cause property values to drop below their present levels.

Blanket conclusions cannot be drawn from the example presented in Table 13-1, because the circumstances associated with various purchase decisions differ. Cost comparisons of this order can serve a useful function in assisting would-be buyers to think through their cost commitments. Lest one take this approach too seriously, however, it should be noted that prospective buyers can easily make erroneous assumptions about future price, cost, and income developments. One must agree with Ratcliff that

> whether or not the purchase of a home in a given case is or is not a "good investment" cannot finally be determined until the investment is liquidated or shifted by sale. Not until the end of the ownership is it possible to make a full accounting. If the home were purchased when prices were depressed and sold during inflation or even in a normal market, the result would be favorable. If conditions were reversed, so would be the result.[11]

Finally, it should be stressed that the decision to buy a home is as much a consumer decision as it is an economic investment. Buyers hope they can resell their

[11] Richard U. Ratcliff *Urban Land Economics* (New York: McGraw-Hill, 1949), p. 109.

properties at a later date for as much as or more than they pay for them, but meanwhile the properties provide them with shelter and opportunities for the enjoyment of life. Buyers usually upgrade their housing when they acquire ownership. Many buy homes more for the community status, feeling of independence, or other satisfactions they expect than for strictly economic investment purposes. Should they dislike their landlord, want to move to the suburbs or to a more prestigious address, or feel that their life style requires homeownership, they have good reasons for buying. Viewed in this way, expenditures for homeownership can be justified in much the same way as expenditures for a deluxe dinner, theater tickets, a television set, or a new car. On the other hand, if one's job calls for frequent transfers, if the decision to buy requires a longer ride to work or the purchase of an additional automobile, if the buyer dislikes yard and home repair work, if one demands more than a 7.5 percent return on one's equity or has limited savings, a good case can be made for remaining as a tenant.

GETTING THE MOST FOR ONE'S MONEY

Assuming that one has decided to become an owner and has determined the price range of the property he or she can buy, the next major problem is that of getting the most for one's money. Operators who are seeking a productive property should look for one that fits their specifications and should make certain that it has sufficient production potential to warrant its cost. If they are seeking a house, they will usually limit their choice to certain communities and then decide whether to buy a new house, an older house, or a residential lot on which they can have a house built with a floor plan of their own choosing.

Building a House

Both building knowhow and bargaining ability are needed by those who would secure the maximum in consumer satisfactions and property values from the houses they build. Some couples have the necessary experience, courage, and imagination to design and build their own houses. Most, however, need competent advice and should discuss their plans with and seek the assistance of architects, builders, or other qualified individuals. Even with these advisors, it is important that they have clearly developed ideas as to what they want and what they can afford. Ten general rules can be advanced as guides to help those who plan to build houses:

1. *Choose a desirable neighborhood.* Prospective buyers should visit the areas where they might build, talk to their possible neighbors, consider the income class of the area, and make inquiries about schools and public services and about zoning and building restrictions. They should ask: Is this where we want to live? Is this where we can afford to live? and What will happen to property values in this neighborhood in the next twenty years?

2. *Select a serviceable lot.* Careful consideration should be given to the frontage and size of the lot, its topography and drainage, surface and subsurface soil conditions, the presence or lack of trees and other natural features, its market price, any outstanding or expected charges for public improvements, and any easements or building restrictions that may adversely affect its use. Observations should be made concerning orientation with respect to neighboring houses and yards, to the sun, to prevailing winds, and to possible scenic views. Buyers should ask how well various lots are adapted to the house design or plan they have in mind.

3. *Orient the house to make best use of the lot.* Houses should be sited to permit optimum use of the size and shape, terrain, and orientation of the lot with respect to the sun and wind, other houses and yards, and possible scenic views. Care should be taken to provide adequate areas for possible future additions to the house, off-street parking, privacy on patios or terraces, children's play space, and possible gardens and planting areas.

4. *Insist upon good architectural style.* People differ in their architectural tastes, and no single style is favored by all buyers. Couples should select a style that pleases them and one that does not clash with the styles of other houses in the neighborhood. Regardless of whether they design their own house, copy someone else's plan, or engage the services of an architect, they should insist on good clean architectural lines and form. Inside and out, house plans should provide for good scale, proportion, and texture and quality workmanship. Avoidance of fads and insistence on good taste provide the best guides to the choice of a housing style that will retain its future resale value.

5. *Select a desirable floor plan.* Floor plans vary with individual tastes, family needs, architectural styles, and total cost outlays. Regardless of cost, emphasis should be placed on good traffic circulation. All rooms and parts of the house should be readily accessible to one another. Adequate space should be allowed for placement of furniture and for storage; and odd-shaped rooms and cut-up designs should usually be avoided.

    Families building small homes should seek all the comfort, convenience, and privacy they can get from a limited amount of floor space. Hall space is usually minimized; the front door ordinarily opens on the living room; there may be no dining room; a modern kitchen with family eating space is a must; some rooms can serve a double purpose; and storage space is often limited. Sound planning with the small house calls for good traffic flow and provisions that permit expansion of family living into outdoor areas during the warmer months.

    Families building larger houses will logically want additional features such as a separate entrance hall, larger living areas, a dining room, a family room, perhaps a study or den, extra bedrooms and bathrooms, and an attached garage or carport. Still other features such as recreation rooms, work and hobby rooms, greenhouses, swimming pools, and servants' quarters can be added if the family's budget permits.

6. *Keep the floor plan functional but livable.* Regardless of architectural style, house plans should be designed for ease of maintenance and enhancement of

family living. Provisions should be made for the installation of modern equipment and utilities, for providing storage areas, and for encouraging family interests such as music, hobbies, television viewing, or the displaying of collected objects. The house should be easy to care for and maintain but still reflect the interests and personalities of the family members.

7. *Keep the plan flexible.* Good planning should allow flexibility in future use. Rooms can be designed to fulfill any of several uses and in this way care for the changing needs of the family that first lives in the house and possible future residents. Plans also can facilitate the later addition of bedrooms, more living space, porches, or garages.

8. *Ensure privacy in the house and yard.* Most people demand privacy in their sleeping and dressing quarters, and more people are taking steps to secure privacy in their use of indoor and outdoor living areas. Walls, plantings, and fencing can be used effectively to screen out undesirable views and outside viewers and to ensure privacy in both indoor and outdoor living spaces.

9. *Plan the house for comfort, health, and safety.* First-class housing calls for more than just a good floor plan. Houses should be well lighted, both during the day and at night. Provisions are needed for adequate electric wiring, plumbing, insulation, and energy-efficient heating and cooling facilities. Care should also be taken to minimize possible accidents and fire hazards.

10. *Watch your budget.* Buyers must keep a constant watch on their budgets. It is easy to plan for more than one can afford. A bigger and better-appointed house is always a tempting prospect, but it often represents a larger investment than families can carry without scrimping on other necessary expenditures. Buyers must also remember that while personalized features may appeal to them, they may be of little value to others. Families that plan to resell their houses in the future must expect to subsidize those extras and special features that vary much from conventional taste.

Buying an Already Constructed House or Condominium

Buyers who are interested in the purchase of an existing house or condominium have many of the same problems as those who build their own houses. They may prefer to buy an existing structure because they need housing in the immediate future, because no vacant lots are available in the neighborhood where they want to live, or because they can see a house but have trouble visualizing the product of a blueprint. In their search for a good buy, they must carefully evaluate the merits and weaknesses of the neighborhoods, building sites, and structures they are considering.

When possible, buyers should inspect the properties they are seriously considering more than once, see them in the daylight as well as during the evening, walk around the neighborhood, and allow themselves sufficient time to make judicious choices. The decision to buy should not be hurried; and prospective buyers should repeatedly ask the questions: Is this property worth the asking price? Can we afford it? Does it fit our needs? Is it for us?

Attention should be given to the overall appearance, floor plan, and adequacy of the house as well as to its state of repair. Much can be determined about the condition of a house by examining its external and internal features. Outside, one should look for structural defects such as a sagging roof line, missing shingles, leaning chimneys, and other evidences of deterioration or wear and age. Inside, one should check the condition of the walls, ceilings, floors, doors, windows, basement, and attics and look for evidence of water or termite damage. Careful inspection is needed of the kitchen, any included equipment, the plumbing and heating systems, and the adequacy of the electric wiring. Inspection of these features poses no problems for those with practical experience in work with housing quality. Inexperienced buyers are advised to seek the assistance of an unbiased appraiser, builder, broker, or engineer who can help them judge the quality and value of used properties.

A second important issue in the purchase of existing properties relates to the problem many buyers have in choosing among their alternatives. Frequently, they will narrow their choices to two, three, or four properties that meet their needs but will find it difficult to make a final choice among them. Unlike most shoppers' goods, their alternatives are not near-perfect substitutes, and they cannot be lined up alongside one another for easy comparison. Ordinarily, the properties are located at some distance from one another and are visited at different times, sometimes on different days.

When buyers find themselves in this position, it is often a good policy to write down certain basic sales information about each of the properties being considered and prepare rating sheets that can be used to evaluate and compare the separate housing features one regards as important. With this technique, one can avoid the common pitfall of being overly impressed with some features while ignoring others and of usually favoring the property that was viewed last.

An application of this technique can be illustrated with the example of a couple who plan to buy a two-bedroom condominium apartment and who have narrowed their choice to three properties. For comparison purposes, they assemble data on each unit about living room size and layout, closet space, interior square footage, patio or balcony size, market price, price per square foot, annual property taxes, monthly maintenance charges, and credit arrangements offered by the seller.

In addition to this comparison, they may also rate selected features on a scale from one to five. One group of comparisons may emphasize such outside and common features as neighborhood quality, grounds, parking areas, outside appearance of buildings, access to shopping areas, entrance halls, elevators and halls, common rooms, recreation facilities, security system, availability of extra storage facilities, and quality of maintenance. A second group may call for rating each unit on its floor plan, traffic flow, cross ventilation, spaciousness, quality of foyer, living room, kitchen layout and equipment, bathrooms, closets, balcony size, balcony view, view from apartment windows, and outside noise.

With this property-rating technique, prospective buyers can give parallel consideration to those features they identify as being wanted in a house or condominium unit. Comparisons of these personal ratings can facilitate rational choices as to which unit offers them the most for their money.

Risks to Successful Ownership

Once ownership has been acquired, owners have the problem of maintaining their ownership rights. Success in this regard is conditioned by the supply of capital and resources they have at their command, the ability they show in managing their resources, and by their exposure to various risk factors. Operators who have the resources and managerial skill needed for making productive use of their business sites ordinarily have little trouble in maintaining their ownership status. Similarly, those who have adequate sources of outside income to pay for their homes also experience little difficulty in this regard. Serious problems often arise, however, when owners try to operate on a "shoestring" basis with too little capital, when their income is inadequate to meet their costs of ownership, when their ownership equity is reduced or wiped out by an unexpected decline in property values, or when plans are adversely affected by family considerations.

Risks involving overcommitments. Major risks to successful ownership are encountered when unwary buyers agree to higher schedules of costs then they can rightly afford. Buyers do not rationally decide to overcommit themselves. They stumble into this situation, usually through bad judgment or overoptimism. Those who allow themselves to be pressured into unfavorable ownership situations or who blindly agree to higher schedules of ownership costs than they can afford must charge their decisions to bad judgment. Misfortune sometimes strikes even those who carefully consider their prospective costs and returns as owners. This is particularly true when an optimistic buyer acquires ownership, only to be victimized by an unfortunate time of purchase, unfavorable conditions of purchase, unpredicatble property deterioration, or unstable neighborhood change.[12]

As this statement suggests, overoptimism can lead to overcommitments when one buys a property and assumes a mortgage under boomtime conditions and later experiences a decline in both income and property value. It occurs again when one amortizes mortgage debt over too short a time period and thus takes on a heavier debt load than can be conveniently carried. It occurs when one buys a building in good faith only to find that it is a jerry-built structure or that it has hidden defects that result in rapid deterioration. Overoptimism can also lead to problems when the value of one's property suddenly declines because of the devastating effects of a drought or an earthquake, because of the loss of its principal market, or because of the sudden blighting of a neighborhood.

Most people are guilty of occasional decisions involving bad judgment or overoptimism. Common sense, careful consideration of one's alternatives, refusal to be stampeded into rushed decisions, and a willingness to seek and take impartial advice provide the best deterrents to bad decisions. These factors can temper overoptimism and at the same time provide a rational basis for the average investor's faith and hope in the future.

Overoptimistic buyers often find it difficult to recognize the limits within which

[12] Cf. Dean, *op cit.*, p. 75.

they can operate. With productive properties, they should never offer a higher price than that justified by the property's productive potential. They should protect their future equities in the properties they buy by making a substantial down payment and by making certain that they can borrow sufficient capital to cover their normal operating expenses.

Three rules of thumb have been advanced to guide home buyers in their ownership-acquisition decisions. Several studies of family expenditures in the United States indicate that the average medium-income family spends approximately one-fourth of its income for housing and household operations.[13] This proportion is often cited as a good index of the portion of one's disposable income that should go for housing purposes. Families with no children and families with high incomes can naturally afford to spend higher portions of their income on housing.

A second rule of thumb proclaims that families should spend no more than 2.5 times their annual disposable income on housing. This rule is also flexible. Families with considerable savings can safely spend more for a home than families with limited savings. Similarly, childless couples can usually justify higher expenditures than large families; and families that get a great deal of enjoyment out of their homes can often pay more than families that want to be on the go. Buyers have probably ignored this rule more than they have honored it since World War II. Yet it still provides a conservative index of how far buyers can safely commit themselves in acquiring home ownership.

A third rule of thumb asserts that residential properties should have market values equal to approximately 100 times their monthly rental rates. This rule fell into disuse during World War II, when property values were allowed to increase while contract rental rates were subject to public rent controls. Rental rates have since increased but in most cases rose at slower rates during the 1970s and early 1980s than property values. As a result, many properties can be rented for considerably less than 1 percent of their market value each month. This situation can create problems because, under stable market conditions, owners of residential rental properties must receive gross rental returns of 12 percent or more of the value of their properties annually if they are to realize fair returns on their investments.[14]

Influence of personal and family factors. Experience shows that property owners themselves often constitute a major risk to successful ownership. A person's prospects for success as an owner frequently reflect (1) personal and

[13] Cf. U.S. Bureau of the Census *Statistical Abstract of the United States 1982–83* (Washington, D.C.: Government Printing Office, 1982), p. 466.

[14] When 2.5 percent of the residential property's value is allowed for property taxes and insurance payments, 2.0 percent for repairs and maintenance, and 2.0 percent for depreciation and obsolence, owners must receive 12 percent of the value of the property as contract rent if they are to realize a 5.5 percent return on the investment value of their property. Owners have been willing to accept lower rates of return only because (1) most of their properties were acquired at less than current market prices, and (2) their internal rates of return have involved contract rents plus annual increments of appreciation in property values.

family attitudes regarding the obligations of ownership, (2) one's relative youth or maturity, (3) changing family needs, (4) size and fluctuations in the family income, and (5) family mobility.

An unhappy or discontented owner often ends up as an unsuccessful owner. Insufficient housing space and business holdings of inadequate size are common causes of dissatisfaction. Beyond this, discontent finds its seeds in many places. Sometimes it springs from a buyer's underestimate of the costs and obligations of ownership or overrating of the satisfactions to be derived from ownership. Sometimes it has roots in a buyer's feeling that he or she was pressured into buying by family insistence, a salesman's slick talk, or possibly by the seller's misrepresentation of certain facts. At times, it also results from factors such as envy of one's neighbors, ill health, family discord, a death in the family, unemployment, alcoholism, or the buyer's disdain for work.

Considerable significance can be associated with the buyers' ages and the stage of their family cycle at the time they become owners. Buyers who are in their forties usually have advantages over younger buyers. They should have more money saved for a down payment. Their ideas on what they want are ordinarily more set. And since the size of their families has already been determined in most cases, they know the size of the house they need or, in the case of a farm, how much family labor they can count on in the future.

Unlike older buyers, young owners have a longer period of productive work ahead. Yet their ideas about what they want in a home or a business property are often more subject to change. They also find it difficult to plan for the future because they always have the problem of trying to match the size of their house to the estimated future size of their families. In this process, they sometimes end up with larger and presumably more expensive houses than they need, while at other times they may find that they must choose between adding additional bedrooms to their houses or selling them so they might move to larger houses.

Income and job security also can become risk factors, particularly during recession periods. Most buyers look forward to continuous employment and steady or rising incomes when they acquire ownership. Past experience shows that this hope is not always justified. The thousands of ownerships sacrificed through mortgage foreclosures and tax forfeitures during past recessions underline the need for steady incomes to support continued ownership.

Family mobility represents still another risk to successful ownership. This factor often takes the form of frequent moving from one apartment, house, or farm to another—usually because of dissatisfaction with one's present quarters and a continuing search for "greener pastures." Many owners as well as tenants indulge in this moving practice, often with considerable wastage of resources. A reverse phase of the mobility problem can operate during recession periods when owners feel that they are tied to present jobs and present locations because their life savings are sunk in a home that they can sell only at considerable discount. During the Depression of the 1930s, Charles Abrams observed that "universal home ownership fits ideally into the industrial scheme: for once fixed to his investment, the home owner makes a most satisfactory and reasonable worker, quick to yield in industrial disputes, anxious to maintain the status quo by doing almost anything to preserve his nominal ownership of the 'equity' into which he has placed his savings, even where prodigious sacrifices of food, clothing, or comfort are required."[15]

As these risk factors suggest, the ownership ideal should not be held out as a boon to all people. Business and home ownership has its satisfaction for those people who can afford it and for those who can maintain their ownership rights once they have acquired them. But for some people, the risks of successful ownership outweigh the advantages. For these people, it may be wise policy to discourage ownership and thereby avoid the heartaches and wasting of resources that can accompany cases of unsuccessful ownership.

TRANSFERRING OWNERSHIP RIGHTS

Sooner or later every privately owned property changes hands. Sometimes these transfers occur only once in a generation; sometimes on a more frequent basis. In most years, between 4 and 6 percent of the nation's single-family houses and farms are transferred to new owners. Approximately 90 percent of these transfers in most years involve deeds with the purchase of either existing properties or of lots on which houses or business structures can be built. A second most important method of transfer involves inheritance arrangements with transfers by devise or under the laws of descent.[16] Patents from government were important prior to 1920 while the government was still disposing of its public domain. Mortgage foreclosures and tax forefeitures also have been important in Depression years such as 1933 when they affected 5.4 percent of the nation's farms and 252,400 nonfarm properties.

Emphasis is given in this section to the practices associated with the two major methods for transferring ownership rights. Consideration is given first to transfers by deed and then to inheritance arrangements.

Transfers by Deed

Some transfers by deed involve outright gifts, but most involve purchase-and-sale arrangements. Before sellers can enter into this type of arrangement, they must first find a prospective buyer. In some instances, buyers seek out the owner with offers to purchase a property. More often, owners must take the initiative. They may approach an acquaintance with a proposal that they buy the property. They may advertise the fact that their properties are for sale. They may list them with a real estate broker, or offer them for sale at a public auction.

[15] Charles Abrams, *Revolution in the Land* (New York: Harper & Row, Pub., 1939), pp. 48-49.

[16] Lewis, *op cit.*, p. 30, found in his survey published in 1980 that 8.3 percent of the land-ownerships in the United States and 16.7 percent of the privately owned land had been acquired through inheritances.

Once a prospective buyer and the seller are brought together, the buyer usually examines the premises and inquires concerning the owner's terms of sale. The prospective buyer can then decide whether to buy the property. If prospective buyers are interested but do not like the owner's terms, they may bargain for more favorable conditions. In this bargaining process, buyers usually try to get the property at a lower price. They can also bargain for the inclusion of "extras" in the sales price; for the provision of specific improvements; for an earlier possession date than the seller has specified; or, if the seller is financing the sale, for a lower down payment or more favorable credit terms. With each of these items, sellers can accept or reject the buyer's offer; or if they wish to bargain further, they may make counteroffers, which may lead to a compromise price.

As soon as the buyer and seller definitely agree on the terms of sale, both parties are committed to the fulfillment of their portion of the sales agreement.[17] There is no formal requirement that the details of these agreements be committed to writing; but it is normal procedure, particularly when the purchase agreement involves a substantial consideration or is made through a broker, for the buyer and seller to sign a *purchase-and-sale contract*.[18] This contract provides legal evidence of the fact that a purchase agreement has been made; it itemizes the terms of sale so that there will be no future misunderstandings owing to faulty memories; and it protects both the buyer and the seller from a possible change of intentions on the other's part. Problems can arise when buyers or sellers fail or refuse to comply with their obligations under a purchase-and-sale contract. Under these conditions, the offending party can usually be sued either for specific performance of the contract or for breach of contract and damages.

No definite form is required in purchase-and-sale contracts. But a well-drafted contract will (1) identify the buyer and seller, (2) provide an accurate description

---

[17] For more-detailed discussions of the land sale and title transfer process, cf. Kratovil and Werner, *op. cit.* chaps. 11–14; and William M. Shenkel, *Modern Real Estate Principles* (Dallas, Tex.: Business Publications, Inc., 1977), chaps. 7 and 17.

[18] Three other types of agreements—usually made in writing—may also be used in the selling process. When sellers list property with a broker, they ordinarily sign a *listing contract*. This contract lists the names of the seller and the broker, a description of the property to be sold, the owner's sale price and other terms of sale, the duration of the broker's agency to sell the property, the broker's commission, and usually an agreement concerning whether the broker has exclusive sales rights.

When a buyer deals directly through a broker rather than with the seller, the "offer to buy" is usually written up in a document known as a *binder*. This document specifies the terms under which the buyer will purchase the property. If the seller accepts these terms by signing the binder the sales agreement is then formalized in a purchase-and-sale contract. Buyers often limit their offers by making them subject to their ability to secure adequate financing.

A third type of agreement known as an *option* may be used when buyers want to hold a purchase opportunity open while they analyze their alternative opportunities, act to secure needed financing, or sell another property. With this arrangement, sellers receive a consideration for which they agree to sell their property to the optionee at a fixed price any time during a specified period of time. If the optionee does not buy the property during this period, the option expires and the owner is again free to sell to anyone.

of the property being sold, (3) list the sales price and any conditions of sale, and (4) bear the signatures of the parties to the sale. In addition, it may specify the approximate "closing date" when the ownership title is to be transferred to the new owner and the date when the seller is to vacate the property so that the buyer can take possession.

At the time a purchase-and-sale contract is signed, buyers are normally expected to deposit a sum of money with the broker or a third party to indicate their good faith in going through with their purchase offers. This deposit—usually between 5 and 10 percent of the purchase price—is known as *earnest money*. If buyers follow through with their acquisition plans, this sum is credited toward the purchase price. Should they fail to carry out their purchase agreement, this money can be used to compensate the broker and the seller for losses they may incur because of a buyer's failure to take title.

Since buyers are always interested in securing good titles to the properties they buy, it is assumed that sellers will provide buyers with merchantable and unencumbered titles. Purchase-and-sale contracts often specify that buyers will take the property subject to existing easements and restrictions, that they will recognize the rights of tenants under existing leases, or that they will take over existing mortgages. In the absence of provisions of this type, sellers must be ready to grant a warranty deed and to clear titles of any easements, restrictions, mortgages, liens, tenancies, or other claims that might limit their ability to convey merchantable and unencumbered titles. If sellers fail in this regard, buyers are under no obligation to accept the titles.

Buyers ordinarily insist on an examination of title before the "closing date." As a part of this process, sellers make their documents regarding the property—the deed, abstract of title, opinion as to title, title insurance policy, and property surveys—available to buyers or their agents for examination. Any omissions in the record or clouds against the title that might be detected at this point are ordinarily cleared at the seller's expense.

Some contracts specify a definite closing date when the deed will be granted and the title will actually pass to the new owner. When not mentioned in the contract, the closing transaction is usually scheduled within a reasonable period after the buyers have had an opportunity to check the validity of the titles they are to receive. This transaction can involve the simple exchange of a buyer's personal or certified check for the seller's deed. Sometimes the settlement process includes arrangements for prorating certain costs—property insurance premiums paid by the seller, property taxes for the year, value of fuel on hand, and so on—between the buyer and the seller. And it is frequently complicated further by a need for several near-simultaneous transactions involving the buyer and seller and their respective mortgages.

A closing transaction involving a mortgaged property often calls for a meeting of nine different parties: the husband and wife who are selling the property, the husband and wife who are buying it, the listing broker, the broker who showed the property to the buyers, the seller's mortgagee, the buyer's mortgagee, and a lawyer.

The sellers must be present to accept payment for the property and to sign the deed in which they convey all their rights—including dower and homestead rights—to the grantee. The buyers are present to pay for the property, to accept the deed, and to mortgage the property as soon as they acquire title. The two brokers are present to represent their clients and to collect their commission. The sellers' mortgagee (or agent) is present to accept payment of the mortgage and to sign a mortgage release. The buyers' mortgagee (or agent) is present to advance the funds the buyers need for the transaction and to receive the new mortgage. A lawyer is present to handle the sale documents.

Properties are frequently transferred in *escrow*. Under this arrangement, buyers and sellers designate a third party who acts as an escrow agent. These parties then sign an escrow agreement, which specifies the conditions each must fulfill to close the sale. Sellers then deliver their deeds to the escrow agent to be delivered to the buyers once they have made their agreed-upon payment for the property; and buyers turn their payments over to the escrow agent to be delivered to the sellers as soon as they provide a merchantable and unencumbered title to the property. Once the terms of the escrow agreement are met, the seller receives payment while the buyer receives a deed, which can be registered in the buyer's name.

Goals in Property Inheritance

Talk about property inheritance and almost everyone perks up their ears. Property owners wonder how they might best handle their estates; prospective heirs wonder how they will make out; neutral bystanders are interested in who will get what and when. From almost every angle, inheritance makes an interesting and intriguing subject. This situation arises both because of the variety of differences between individual cases and the atmosphere of uncertainty regarding future developments. A flavor of intrigue is also added by occasional rumors of underhanded dealings, and by the general hush-hush treatment that surrounds many inheritance plans.

Most people agree that the question of how property is transferred and who receives it deserves careful consideration. Unfortunately, this subject is not one that we discuss freely at the breakfast table. As long as parents are not under pressure, they ordinarily prefer to leave their plans fluid and unjelled so that they can be adjusted to changing situations. As a result, most property owners tend to postpone their decisions relative to inheritance arrangements. Even when they have made definite plans, they frequently hesitate to discuss the details with their prospective heirs.

Children and other heirs, on the other hand, are usually just as hesitant in asking questions about the details of future inheritance arrangements. Rather than give the appearance of being vulturous or "grabby," many prospective heirs seem content to operate on faith and hope, with no definite assurances concerning their future position as heirs. This atmosphere of uncertainty is sometimes sobering and wholesome. But just as often—and particularly when one of the prospective heirs has assumed responsibility for operating the family business or for looking after the parents—this situation may involve injustices and present a barrier to the efficient management and operation of a family business or farm.

Four important goals should be stressed in planning inheritance arrangements. These goals call for (1) minimizing the cost, time, and trouble involved in estate settlements; (2) maintaining productive properties as going operating units; (3) safeguarding the security of the parents; and (4) securing fair treatment of the heirs.

Most families want to hold their property inheritance costs to a minimum. At the same time, they want to avoid time-consuming delays in estate settlements; and they would like to forestall misunderstandings and quarrels between heirs. Attainment of these objectives cannot always be assured in advance. Family quarrels could often be avoided, however, if parents would discuss their inheritance plans with their heirs and join with them in agreements concerning the shares each heir should expect.

Divisions of the physical assets of going businesses between two or more heirs can lead to production inefficiencies and sometimes their demise as economic going concerns. Good reasons sometimes exist for breaking up large business holdings, particularly when they involve sizable acreages or more than one operating unit. Except when holdings involve readily separable units, however, it is usually in the best interests of heirs, local communities, and society at large to work out inheritance arrangements that will permit complete economic units to pass as going concerns. The entire inheritance need not go to one person. Joint ownership arrangements may be considered; some heirs may receive cash or other properties; or arrangements may be worked out to have the recipient of a farm or other business pay off the interests of other heirs over a suitable period of time.

Property owners ordinarily give considerable attention in their inheritance plans to the welfare and well-being of their prospective heirs. They recognize that heirs who have reached the age at which they can operate on their own can often benefit from the productive use of their expected inheritance, and that the value of this opportunity can decline if years pass before its receipt. As long as parents can afford to do so, they often provide their children with various types of help in getting started. Some children are sent to college, some receive money to start businesses or to provide an initial payment on a home. Children may be taken into the family business, and parents can work out intrafamily business transfer arrangements that permit heirs to take over a family business while they are still in their physical prime. Yet despite their good intentions, parents must also consider their own future welfare. Before they transfer their properties, they should consider the satisfactions they associate with retention of managerial control over their properties; the effects of their actions should they outlive their children; and the need for maintaining financial reserves to offset the eventualities that may arise with inflation, high medical expenses, or depressed business conditions.

As a final goal, serious consideration should be given to the equitable treatment of heirs. Our inheritance laws favor equal treatment of children. "Equal treatment," however, can involve inequitable treatment. Almost every community has examples of sons or daughters who have operated the family business for their parents, or who have cared for them during their declining years, while brothers and sisters have been free from this responsibility. In these cases, it is only fair that the heirs who have stayed at home or those who have cared for their parents should receive special consideration. When some heirs have received special financial help for their education, the launching of businesses, or other purposes, these grants should also be considered in the determination of what is and what is not equitable treatment.

Alternative Inheritance Arrangements

Property owners can use any of several alternative arrangements in transferring their ownership rights to their normal heirs. They can give their property away. They can arrange to sell it to one of their heirs at the market price or at a reduced price. Sometimes they use contractual arrangements to turn their ownership rights over to other parties during their lifetime or at the time of their death in return for special services or support. Owners can hold their ownership rights in joint tenancy with rights of survivorship with another person. They may draw up a will that designates the portions of their estate that are to pass to various heirs at their death. They can set up trust arrangements; and if they take no action governing the transfer of their estates, their properties will be distributed after their death in accordance with the laws of descent and distribution.[19]

Transfers by gift. Parents frequently give property, money, and other assistance to their children and other heirs. As long as they can afford to give away part of their property, the decision to do so often results from a simple choice between giving the property now or letting the recipient wait until their death. With this choice of alternatives, property owners frequently choose to give because (1) gifts can provide the donee with opportunities to make an earlier and stronger start in business; (2) this method of transferring property avoids some of the taxes, legal fees, and other costs normally associated with estate settlements; and (3) most donors gain considerable satisfaction from the act of giving and from seeing their children prosper.

Gifts provide a satisfactory method of property transfer as long as donors reserve sufficient property or other resources to provide for their future comfort and support. But property owners must always remember that they lose control over property the moment they give up their ownership rights and that they should retain at least partial control if the gift will in any way endanger their future security.

Some owners try to meet this situation by giving a joint tenancy interest in the

---

[19] For a more detailed discussion of these property transfer arrangements as they apply in different areas, cf. J. H. Beuscher and Louise A. Young, *Your Property—Plan Its Transfer,* University of Wisconsin Extension Circular 407, 1951; Robert Brosterman, *The Complete Estate Planning Guide* (New York: McGraw-Hill, 1964), Part II; William C. Clay, Jr., *Guide to Estate Planning,* 3rd ed. (Homewood, Ill.: Richard D. Irwin, 1980); Sidney Hess and Bertil Westlin, *Estate Planning Guide* (Chicago: Commerce Clearing House, 1977); and Marshall Harris and E. B. Hill, *Family Farm Transfer Arrangements,* University of Illinois Extension Circular No. 680, 1951.

property or by giving a deed with the reservation of a life estate to the present owner. These arrangements can prove satisfactory under certain conditions. They may complicate possible management, leasing, mortgaging, or sale arrangements, however, and definite problems can arise if the donor outlives the donee. Arrangements of this type should be used only when they fit the owner's circumstances and needs.

Sales arrangements. Property owners often find it advantageous to use sales arrangements in the disposition of their properties. If owners want to retire, they can sell their businesses to potential heirs at the going market price or at a reduced price, which involves an element of gift; or they may sell to a nonrelative, usually at the highest price they can get. By selling their properties, owners can receive money they can use to purchase annuities, invest in other enterprises, or support themselves during their declining years. Any residue left at the time of their death can be passed on as part of their estates.

Contractual arrangements. Contracts can be used to cover a wide variety of property transfer arrangements. Sons and other heirs frequently operate family businesses with verbal understandings that ownership will pass to them in the future. Oral agreements of this type are usually undesirable because of the difficulties attendant to their enforcement. When these agreements are formalized in a written contract, they can provide a highly satisfactory method of intrafamily property transfer.

Owners frequently use land contracts in the sale of businesses, farms, and residential properties. Potential heirs and even nonrelatives sometimes operate family businesses or care for a property owner with the agreement that they can buy the property at a specified price or that they will acquire it at the owner's death. Contracts can be used to transfer ownership rights to an heir and at the same time reserve certain rights, perquisites, and support for the parents.

Joint tenancies. Joint tenancy arrangements are frequently used by husbands and wives, widowed mothers and sons, or other combinations of two or more persons as a means of holding property in joint ownership. With this arrangement, the entire ownership is transferred at one's death to the surviving owner or owners. This method of transfer has obvious advantages in some cases. A husband and wife may use this approach to protect the interests of a surviving spouse. Similarly, a widowed mother may find it desirable to give her only child joint tenancy rights in her property if she feels that this arrangement will not interfere with her future management and operation plans.

When property is held in joint tenancy, an owner's rights pass automatically at his or her death to the survivor and there may be no need to probate the estate. This feature reduces the time, cost, and trouble associated with estate settlements. Yet this system of ownership transfer has certain drawbacks. Owners who enter into a joint tenancy give up the exclusive right of managerial control they could exercise as sole owners. Joint tenancy works as long as the joint tenants agree. It can be-

come a burdensome arrangement when they disagree. It provides an unsatisfactory inheritance arrangement when one has several deserving heirs, and it is hard to apply to most types of personal property. The possibility of unexpected death seldom makes it wise for a son or daughter to put all of their property under joint tenancy with a parent. In doing so, they would risk the chance of having their property go to the parent (or other joint tenant) rather than to their spouse and children.

Use of a will. Property owners who want to hold all or part of their property throughout their lifetimes can use wills to stipulate how their estates shall be divided after their death. Wills are used in the transfer of approximately one-half of the properties subject to estate settlement. Many wills, however, are deathbed documents prepared within a few days or weeks of the testator's death.

A *will* may be defined as a gift of property that takes effect on the giver's or testator's death unless revoked before that time.[20] To be valid, a will should be written in ink or be typed, though oral wills are accepted in some cases. It must be made by a testator of legal age and sound mind; and it should be signed by the testator and witnessed by two or more disinterested persons. Once a will has been prepared, the testator is free to revoke it at any time, replace it with a more recent will, or add a codicil that supplements or alters certain of its provisions.

Testators can "give, devise, and bequeath" any property they own, subject to legal claims, to any person or organization of their choice. Ordinarily, they provide that their property shall pass to their normal heirs. But they are free to establish eccentric trusts and inheritance arrangements, such as the famous Toronto baby derby of the 1920s, which started when a wealthy Canadian left the bulk of his estate to the family that would have the most children in the ten years following his death. The rights of testators in this regard are always limited, however, by the legal requirement that they recognize all of their children, not defeat the dower rights of the wife or the homestead rights of the family, not avoid the claims of rightful creditors, and not transfer properties they no longer own.

Following a testator's death, his or her will must be filed in the proper court and be admitted to probate. An executor named in the will or an administrator appointed by the court will then administer the estate until it has been settled. This settlement process ordinarily involves periods ranging from four months to a year, though it sometimes extends over periods of several years. During this settlement period, the claims of the heirs, creditors of the estate, tax officials, and others are aired and settled so that the residual estate can be divided according to the terms of the will.

Wills are often regarded as a highly desirable method for transferring property to one's heirs. They permit owners to hold and control property throughout their lifetime and give them an opportunity to apportion their estates among their heirs in what they consider to be an equitable manner. But they have the disadvantage of

---

[20] For a more detailed discussion of wills and estate settlements, cf. Kratovil and Werner *op. cit.*, chap. 18.

calling for a settlement period, which necessitates a certain amount of cost, time, and trouble. They may also delay the actual transfer of property until sometime after the period when the prospective heir could realize the greatest benefit from its receipt and use.

Even when property owners use other methods to transfer most of their property, they ordinarily find it desirable to use a will to direct the distribution of any residual portions of their estates. Wills can be used along with other transfer arrangements to handle the problem of contingent beneficiaries (as when a husband and wife holding property in joint tenancy die in the same accident), to designate a guardian for one's minor heirs, and for other similar purposes.

Settlements under the laws of descent. When property owners die intestate—without a will—their estates are settled under the laws of descent and distribution in much the same way that they would be settled under a will. The details of these laws vary from state to state. As a general rule, between a third and a half of the estate passes to the surviving spouse, and the balance is divided equally between the children. The share that would have gone to any child who has died ordinarily goes to his or her heirs, if there are any, by right of representation. When there is no surviving spouse, the entire estate goes to the direct heirs. When there are no direct heirs, estates pass to the parents of the deceased and after them to brothers, sisters, and their legal heirs.

Laws of descent usually call for equal division of estates among heirs of equal relationship. This system provides a logical average way of distributing estates. In this respect, it parallels the practices used in many wills. But it completely ignores the fact that equal division may result in inequitable sharing arrangements, with some deserving heirs receiving smaller shares than they actually deserve.

Owners who are satisfied with the distribution arrangements provided by law may find it just as well not to prepare a will. A will should be prepared, however, if one plans to favor particular heirs, wants one's spouse to enjoy complete control of an estate until after minor children are grown, or wishes to attain some other specific goal.

Living trusts. Many owners use trust arrangements as an alternative or supplement to wills. With trusts, one can turn all or part of the property one owns over to persons or institutions that then serve as trustees in administering the property in the interests of the owner, possible heirs, or others. Trust arrangements can be revocable or nonrevocable. They can be established as living trusts during one's lifetime or as testamentary trusts that become effective upon one's death.

Owners who establish living trusts for inheritance purposes can assign major portions of their property to the trusts. They can receive all of the income from their trusts while they live or share this income with others. The trusts can be dissolved following their death, or they can be continued for various periods of time with the income going to one's spouse or others. Arrangements also can be made to have them skip a generation before they are dissolved.

Living trusts are an attractive inheritance arrangement for some people because they permit the passing of property to one's heirs with considerable privacy and without need for court probating. The avoidance of probate and skip-a-generation features provide opportunities for tax savings and also some savings in legal costs. Like other inheritance arrangements, however, living trusts are not the answer to everyone's need. Owners who use this approach should make certain that it fits their circumstances and that it will operate for them as intended.[21]

Choice of an inheritance arrangement. Every family should give careful thought to its choice of an adequate and appropriate property inheritance plan. These plans are just as important to young families and operators who are still getting themselves established as they are to older people who have already acquired considerable property. No hard and fast rules can be made regarding one's choice of an inheritance arrangement. There is usually a best arrangement for every property owner, but the choice of this arrangement depends on one's goals and circumstances.

Property owners who are working out inheritance arrangements should always start by considering the amount of property they own, its value, and the manner in which their ownership rights are held. They should ask themselves what goals they want to attain in the inheritance process; how their property rights can best be distributed among their heirs; and what inheritance arrangement or combination of arrangements can best be used to attain these ends. They should seek competent advice from attorneys, bankers, or other knowledgeable people before finally deciding on an inheritance arrangement. The plan chosen should facilitate the smooth transfer of the estate to one's intended heirs while minimizing possible troubles, time delays, estate and inheritance taxes, and other estate settlement costs.

Once owners have decided on a plan, they should have an attorney draw up the proper papers, if any are needed. Then as time goes on, they should review their arrangements from time to time to make sure that they are kept up to date with their changing family and property ownership situations. As Beuscher and Young have observed, "Plans carefully worked out may save money, will give peace of mind to the family, and will save long drawn out settlement proceedings."[22]

—SELECTED READINGS

Clay, **William C., Jr.**, *Guide to Estate Planning*, 3rd ed. (Homewood, Ill.: Richard D. Irwin, 1980).

Dean, **John P.**, *Home Ownership: Is It Sound?* (New York: Harper & Row Publishers, Inc., 1945).

Fisher, **Ernest M.**, and **Robert M. Fisher**, *Urban Real Estate* (New York: Holt, Rinehart & Winston, 1954), chaps. 5, 6, and 8.

[21] Cf. Clay, *op. cit.*; and Hess and Westlin, *op. cit.*
[22] Beuscher and Young, *op. cit.*, p. 15.

Harris, Marshall, and **Elton B. Hill.,** *Family Farm Transfer Arrangements*, University of Illinois Extension Circular No. 680, 1951.

Harwood, Bruce, *Real Estate Principles*, 2nd ed. (Reston, Va.: Reston, 1980), chaps. 5, 6, 8, 14, and 17.

Kratovil, Robert, and **Raymond J. Werner,** *Real Estate Law*, 8th ed. (Englewood Cliffs, N.J.: Prentice-Hall, 1983), chaps. 6, 7, 10, and 18.

Ratcliff, Richard U., *Real Estate Analysis* (New York: McGraw-Hill, 1961), chap. 8.

Shenkel, William M., *Modern Real Estate Principles* (Dallas, Tex.: Business Publications, Inc., 1977).

14

LEASING ARRANGEMENTS

Unencumbered owner-operatorship has long been viewed as a fore most goal in American land-tenure policy. Yet a brief look at the present tenure situation in the United States shows that 34 percent of the residential units are tenant-occupied and that tenants operate approximately one of every eight farms. Furthermore, around two-thirds of those who do own houses hold their ownership rights subject to a real estate mortgage.

These statistics indicate that thousands of real estate owners and tenants have a way to go before they can benefit from unencumbered owner-operatorship. For them, property ownership can be viewed as an end product of a dynamic process. Leasing and credit arrangements provide them with opportunities for access to land resources and at the same time represent possible stepping stones on their path to possession of the full rights of property ownership.

As these comments suggest, significant land-tenure issues are associated with the leasing arrangements that bind the interests of landlords and tenants and with the credit arrangements worked out between mortgagors and mortgagees. The first of these two tenure arrangements constitute the subject matter of this chapter.

TENANCY AND LEASING ARRANGEMENTS

Every time a property is leased or rented, there is a transfer of rights from the landlord to the tenant. As the *lessor*, the landlord retains ownership rights while granting rights of use and possession to the tenant (or *lessee*) for some given period of time. In return for this delegation of rights, the tenant agrees to make periodic rental payments and accept certain responsibilities in using and possessing the landlord's property.

Widescale tenancy is often regarded as undesirable in modern society. This does not mean that leasing arrangements are undesirable per se. Far from this, they provide a valuable and serviceable function in enhancing the interests of both landlords and tenants. They provide a means by which prospective landlords can lease surplus land resources to others and receive monetary returns from properties they do not wish to occupy or operate. At the same time, they make it possible for prospective tenants to acquire use and possession rights to the properties held by others.

Tenancy encourages landlords to invest in and develop properties for the large number of businesses, office users, urban residents, and others who prefer to rent the quarters where they work and live. It also helps many owners secure a supplemental or retirement income from their properties while providing timely assistance to others who need rental properties.

Viewed in light of our tenure goals, strong objections would undoubtedly be raised if most real estate resources were owned by a small group of people and if price and social barriers prevented much movement from tenancy to ownership. But as long as these conditions do not exist, a certain amount of tenancy is needed in the tenure system to facilitate the operation of the agricultural and residential ownership ladders, to permit an orderly retreat from owner-operatorship by older owners, and to fill the continuing demand for commercial and residential rental properties in urban areas.

Most criticisms of tenancy are related to abuses or undesirable practices sometimes associated with tenancy arrangements. Problems may arise when a landlord's bargaining power makes it possible to foist a one-sided leasing arrangement on a tenant, when the landlord claims a disproportionate share of the economic returns of a property as rent, or when a tenant regards rented property as a resource to be exploited for personal benefit. These problems are most acute when the income produced by a property is too low to provide the landlord with the desired rental return and still leave sufficient income to ensure decent living standards for the tenant. Problems also arise in some instances because of the vagaries of human nature and the failure of many leasing agreements to represent an actual meeting of minds on the details of the rental arrangement.

The discussion of tenancy problems and leasing arrangements that follows begins with an overall look at the importance of tenancy in the American economy. Attention is then given to the characteristics of leases, the principal types of rental arrangements now in use, and to some economic considerations that affect the determination of rental rates.

Importance of Tenancy

Attitudes regarding the desirability of tenancy as compared with ownership vary widely from community to community and from one type of land use to another. Some residential and farming communities are composed almost entirely of homeowners; some involve a considerable intermixture of owners and tenants; and some, such as the multistoried apartment areas of our larger cities, are inhabited almost entirely by tenants.

Comparable variations occur with different types of land use. Leasing arrangements are a common phenomenon with commercial establishments, light industries, urban residences, farms, grazing land, and mineral rights. Yet they are used only on rare occasions with heavy industrial properties, forests, developed mines, and the lands used for highway and railroad transportation purposes.

A nationwide survey of landownership in the United States in 1978 indicated that 8.7 percent of the total privately owned area of 1.35 billion acres was rented to others, 4.6 percent for cash, 2.7 percent under sharing arrangements, and 0.7 percent with other types of leases.[1] Most of the rented area was operated as farmland while the largest number of leased small properties involved residential and commercial sites.

Trends in nonfarm residential tenancy. Census findings on the residential- and farm-tenure situations have been reported in the United States since the late 1880s. As Table 14-1 indicates, 52.2 percent of the nation's occupied dwelling units and

TABLE 14-1. Tenure Status of Occupied Dwelling Units, United States, 1890-1980

CENSUS YEAR	ALL OCCUPIED DWELLING UNITS (in thousands)	PERCENTAGE OF TENANT OCCUPANCY	OCCUPIED NONFARM DWELLING UNITS (in thousands)	PERCENTAGE OF TENANT OCCUPANCY
1890	12,690	52.2	7,923	63.1
1900	15,964	53.3	10,274	63.5
1910	20,256	54.1	14,132	61.6
1920	24,352	54.4	17,600	59.1
1930	29,905	52.2	23,300	54.0
1940	34,855	56.4	27,748	58.9
1950	42,826	45.0	37,105	46.6
1960	53,024	38.1	49,458	39.1
1970	63,445	37.1	60,351	38.0
1980	80,072	34.4	(1978) 74,847	35.4

Source: U.S. Bureau of the Census, *Statistical Abstract of the United States, 1971* (Washington, D.C.: Government Printing Office, 1971), p. 673; and *Statistical Abstract of the United States, 1982-83* (Washington, D.C.: Government Printing Office, 1982), p. 757.

[1] Cf. James A. Lewis, *Landownership in the United States*, 1978, U.S. Department of Agriculture, Agriculture Information Bulletin No. 435, 1980. This study also indicated that land was owned by around 34 million urban and rural owners. Some 89.9 percent of the ownerships involve sole proprietors or husband and wife ownerships. The study showed considerable concentration of ownership with 40 percent of the investment value of the owned land being held by less than 0.5 percent of the owners, while the 78 percent with the least valuable holdings held only 3 percent of the real estate values. Cf. also Thomas McDonald and George Coffman, *Fewer Larger U.S. Farms by Year 2000 and Some Consequences*, U.S. Department of Agriculture Information Bulletin No. 439, 1980.

63.1 percent of the occupied nonfarm dwellings were tenant-occupied in 1890. The proportion of tenant-occupied nonfarm units increased to 63.5 percent in 1900 and then slowly declined to 54.0 percent in 1930. Depression conditions during the 1930s brought a reversal of this downward trend and helped push the proportion of residential nonfarm tenancy up to 58.9 percent in 1940.

Between 1940 and 1950, the total number of occupied nonfarm dwellings increased from 27.7 to 37.1 million units and the proportion of residential tenancy dropped to 46.6 percent. This marked decrease in nonfarm residential tenancy can be credited primarily to three factors: (1) the effect of wartime and postwar prosperity in fostering home purchases, (2) the availability of real estate credit for the acquisition of homeownership, and (3) the impact of public rent-control programs on landlords' decisions to sell properties they might otherwise have kept as rental investments.

Nonfarm residential homeownership continued to increase after 1950 as the number of occupied nonfarm dwelling units climbed to 49.5 million in 1960 and to 74.8 million in 1978. A national housing survey in 1980 found that 58.3 million of the nation's 86.7 million year-round housing units were single-family units. Of these, 79.5 percent were owner-occupied as compared with 20.8 percent of the two- to four-family units. 14.8 percent of the multiple-family (five or more) units, and 66.3 percent of the 3.8 million mobile home units.[2]

Farm-tenure situation. The census of 1880 showed that slightly more than one-fourth of the nation's farms were tenant-operated (see Table 14–2). This proportion gradually increased during the next few decades until 1930, when 42.4 percent of the farm-operating units were reported in this tenure class. Following 1935 and more particularly following World War II, the proportion of tenancy dropped until it reached a record low of 11.3 percent in 1974.[3] Acreagewise, the proportion of farmland under lease decreased from 44.7 percent of the total in 1935 to 33.2 percent in 1950 and then rose to 34.0 percent in 1960 and 40.2 percent in 1978, before dropping to 38.8 percent in 1982.

Table 14–2 deals with numbers of farm operating units and presents a picture of decreasing tenancy since 1930. It does not follow, however, that there has been an increase in the number of full owner-operators or of the acreage they operate. As the total number of farms dropped from a high of 6.8 million in 1935 to 3.7 million in 1959 and 2.2 million in 1982, the number of full owners dropped from 3.2 million to 2.1 million and 1.3 million in 1959 and 1982, respectively, and the number of tenants from 2.9 million to 0.7 million and 0.26 million, respectively. Meanwhile, the number of part owners—operators who own some land and rent in additional acreage—increased from 688,867 in 1935 to 868,180 in 1954 and then

---

[2] *Annual Housing Survey: 1980*, U.S. Bureau of the Census Current Housing Reports, Series H-150-80.

[3] Cf. V. Webster Johnson and Raleigh Barlowe, *Land Problems and Policies* (New York: McGraw-Hill, 1954), chap. 11, for more detailed discussion of trends during the first part of this period.

TABLE 14-2. Number of Farm Operating Units and Proportion of Farmland Operated by Tenants, United States and Major Regions, 1880-1982

CENSUS YEAR	NUMBER OF FARM OPERATING UNITS	PROPORTION OF FARM UNITS OPERATED BY TENANTS			
		United States	North	South	West
1880	4,008,907	25.6	19.2	36.2	14.0
1890	4,564,641	28.4	22.1	38.5	12.1
1900	5,739,657	35.3	26.2	47.0	17.0
1910	6,366,044	37.0	26.5	49.6	14.0
1920	6,453,991	38.1	28.2	49.6	18.2
1925*	6,371,640	38.6	28.0	51.1	18.7
1930	6,295,103	42.4	30.0	55.5	21.5
1935*	6,812,350	42.1	31.8	53.5	23.8
1940	6,102,417	38.8	31.1	48.2	21.8
1945*	5,859,169	31.7	25.0	40.4	14.5
1950	5,388,437	26.9	21.1	34.1	13.4
1954*	4,782,416	24.4	20.5	30.0	12.2
1959	3,710,503	20.5	19.6	23.1	12.3
1964	3,157,857	17.1	17.0	18.5	11.5
1969	2,730,250	12.9	14.2	11.7	11.8
1974	2,314,013	11.3	12.5	10.0	10.8
1978	2,478,642	12.7	13.8	11.7	11.3
1982	2,239,300	11.6			

*Alaska and Hawaii not included.

Source: 1964 Census of Agriculture, II, p. 756; *1969 Census of Agriculture*, II, chap. 3, p. 14; *1974 Census of Agriculture*, II, pt. 3, pp. 1-10; *1978 Census of Agriculture*, I, pt. 51, p. 123; and *1982 Census of Agriculture*, I, pt. 51, p. 3.

gradually dropped to 628,224 in 1974 before rising again to 713,036 in 1978 and 655,840 in 1982. As Table 14-3 indicates, full owners have increased slightly as a percentage of all operators since 1920 but have accounted for a decreasing proportion of the total land area in farms. Their rate of decline in this respect has been exceeded only by that of the tenant group. Part owners became steadily more important as a tenure class and controlled 55.8 percent of the farmland in 1982 as compared with 18.4 percent in 1920. Managers, though few in number and no longer enumerated after 1964, also became more significant as their holdings rose from 5.7 percent of the total area in 1920 to 10.2 percent in 1964.

Most of the nation's farm tenants have operated in the cotton-producing states of the South and in the Corn and Wheat Belt portions of the Middle West and the Great Plains regions. When farm tenancy was at its peak during the 1930s, more than 60 percent of the farm units in the eight Cotton Belt states—Alabama, Arkansas, Georgia, Louisiana, Mississippi, Oklahoma, South Carolina, and Texas—were operated by tenants, mostly by sharecroppers and "third and fourth" share tenants who operated tracts within larger plantation holdings. During the same period, more than half of the farmland in Georgia, Illinois, Iowa, Kansas, Nebraska, Oklahoma, North Dakota, and South Dakota was operated under lease. Much of the

TABLE 14-3. Percentage Distribution of Operator Units and Land Area in Farms Operated by Different Tenure Groups, United States, 1920-1982

ITEM	1920	1930	1940	1950	1959	1969	1982
				percentages of total numbers			
Number of operating units							
Full owners	52.2	46.3	50.6	57.4	57.6	62.5	59.1
Part owners	8.7	10.4	10.1	15.3	22.5	24.6	29.3
Managers	1.1	0.9	0.6	0.4	0.6	n.a.	n.a.
Tenants	38.1	42.4	38.8	26.8	19.8	12.9	11.6
Area in farms							
Full owners	48.3	37.6	35.9	36.1	30.8	35.3	32.3
Part owners	18.4	24.9	28.2	36.4	44.8	51.8	55.8
Managers	5.7	6.4	6.5	9.2	9.8	n.a.	n.a.
Tenants	27.7	31.0	29.4	18.3	14.5	12.9	11.9

Source: 1969 Census of Agriculture, II, chap. 3, pp. 16, 25–27; *1978 Census of Agriculture*, I, pt. 51, p. 22; and *1982 Census of Agriculture*, I, pt. 51, p. 3.

overall change in the national farm-tenancy picture since 1935 can be credited to a marked reduction in the proportion of tenancy in these two regions.

Part owners and tenants accounted for above average proportions (49.6 and 12.6 percent, respectively, as compared with 37.8 percent for full owners) of the total volume of farm product sales in 1982. As Table 14-4 shows, considerably larger proportions of the part owner and tenant groups reported sales of $10,000 or more and of $100,000 or more than was the case with the full owner group. Above average proportions of part owners and tenants reported farming as their principal occupation while above average numbers of full owners were 55 years of age or older and reported 100 days or more of off-farm work. When farms are classified

TABLE 14-4. Comparison of Selected Characteristics of Agricultural Tenure Groups, United States, 1982

CHARACTERISTICS	ALL OPERATORS	FULL OWNERS	PART OWNERS	TENANTS
Percentage of operators reporting				
Farm product sales of $10,000 or more	51.0	36.4	74.9	65.7
Farm product sales of $100,000 or more	13.5	7.0	26.6	13.8
Farming as their principal occupation	55.1	45.6	70.7	63.8
Operator ages of 55 years or more	41.8	49.0	34.7	22.7
100 or more days of off-farm work	47.0	53.8	35.6	42.5
Type of farm organization				
Individual and family	86.9	89.5	84.7	79.2
Partnerships	10.0	7.8	11.6	17.1
Family-held corporations	2.3	1.8	3.3	2.7
Other corporations	0.3	0.3	0.2	0.5
Other arrangements	0.5	0.6	0.2	0.6

Source: 1982 Census of Agriculture, I, Pt. 51, Table 44.

according to type of organization, it appears that 86.9 percent of the farm units were operated by individuals and families in 1982, 10.0 percent by partnerships, and 2.6 percent by family or other corporations. Higher proportions of tenants and part owners were involved in partnership and corporate management arrangements than was the case with full owners.

Characteristics of Leasing Arrangements

The key feature in the landlord-tenant arrangement is the leasing agreement. This agreement involves a contract known as a *lease* by which the landlord conveys rights of use and possession in a given property to a tenant for a definite period of time in return for a specified rental payment.

A lease may be oral or written; but all leases covering periods of more than one year in some states, and more than three years in others, must be written to be legal. Leases can be limited to short periods of time such as a month or a year or may cover periods of ninety-nine years or more. Some states, such as California and Nevada, have laws limiting the duration of agricultural leases to periods of ten or fifteen years and other leases to periods of ninety-nine years. Where no time limits are prescribed by law, valid leases may be written for any length of time.[4] Their terms may be the end product of detailed negotiations; they may be dictated entirely by the landlord or in some instances by the tenant; or they may be implied by a general understanding that the landlord and tenant will accept a customary rental arrangement.

Regardless of how the landlord-tenant relationship is established, it vests the tenant with a leasehold estate, which can be enjoyed as long as the tenant fulfills his or her obligations as a tenant. These obligations ordinarily involve payment of rent in the amount and at the time agreed upon together with some responsibility for making repairs and maintaining the property against undue deterioration. The landlord in turn must protect the tenant's exclusive rights of use and possession throughout the lease period unless specific arrangements to the contrary are included in the leasing arrangement. Many leases are vague and inexplicit regarding the duties and privileges of the landlord and tenant. When legal questions arise regarding these leases, they are ordinarily interpreted according to the provisions of statutory and common law. Care should be taken in the drafting of a written lease if the landlord or tenant wishes to deviate from these usual legal interpretations.[5]

Termination of leases. Most leasehold estates terminate automatically at the end of the lease period. At this time, the lease may be renewed; the tenant may stay on with the permission of the landlord; or the tenant may move away. Leases may be

---

[4]Cf. Stanley M. McMichael and Paul T. O'Keefe, *Leases: Percentage, Short and Long Term*, 5th ed. (Englewood Cliffs, N.J.: Prentice-Hall, 1959), chap. 13.

[5]For a more detailed discussion of the legal aspect of leasing arrangements, cf. Robert Kratovil and Raymond J. Werner, *Real Estate Law*, 8th ed. (Englewood Cliffs, N.J.: Prentice-Hall, 1983), chaps. 30 and 34; Curtis J. Berger, *Land Ownership and Use: Cases, Statutes, and Other Materials* (Boston: Little, Brown, 1968), pp. 290–391; and Ernest M. Fisher and Robert M. Fisher, *Urban Real Estate* (New York: Henry Holt, 1954), chap. 7.

terminated earlier by mutual consent, by the action of the landlord or tenant, or by termination of the landlord's estate.

A landlord-tenant relationship can be ended at any time if the tenant voluntarily surrenders the leasehold and the landlord accepts this action. Many leases contain cancellation clauses, which authorize landlords to terminate leases—usually with proper notice—if they feel their properties are not being put to proper use or if they decide to sell them to others. Landlords also have an implied right to (1) terminate leases when properties are abandoned by tenants, and (2) seek legal eviction of those tenants who fail to pay rent or otherwise violate the terms of their leases. In either instance, tenants can often be held liable for any loss of rent suffered by landlords during the remainder of a lease period.

Tenants have an implied right to terminate leases whenever landlords evict them from all or part of a property and occupy and use it contrary to the tenant's wishes. They also have this right when landlords act in such a way as to make premises un-inhabitable. For example, when a landlord agrees to provide heat, electricity, and water but does not do so or fails to provide agreed-upon repairs, tenants can terminate leases on the ground that they have been subject to "constructive eviction." In addition to terminating the lease in these instances, tenants can also sue their landlords for damages.

A lease is always terminated when a tenant buys the property from the landlord. It is also terminated when the property is acquired for some public purpose by eminent domain or if other parties prove that the title belongs to them rather than to the lessor.

Security deposits. Unless other provisions are included in the lease, rent is payable at the end of the use period, for instance, at the end of the month or at the end of the year. It is common practice, however, for the lessors of residential and commercial properties to require that rent be paid in advance. When there is some chance that the tenant may abandon or damage the property, go bankrupt, or default on the rental payments, landlords may also insist that tenants make a security deposit.

Security deposits take several forms. Owners of commercial sites may require tenants to deposit money or securities of a certain value with them, with the provision that they will be returned with interest at the end of the leasehold period. Lessors of furnished houses or apartments may require a cash deposit, which will later be returned to the tenant with deductions for any damage to the owner's property.

Periodic tenancies. Although most landlord-tenant arrangements start with mutual acceptance of an oral or written lease, this relationship often continues as a month-to-month or year-to-year tenancy. Periodic tenancies of this type result when no definite termination date is specified at the time the leasing agreement is made or when a tenant is allowed to hold over at the end of a lease.

The rights and duties of landlords and tenants in these cases are governed by rules of law. These rules include the covenants described above and require certain minimum periods of notice for termination of the tenancy. A month-to-month tenant cannot be evicted without proper notice. This period of notice—usually between seven and fifteen days—is also required when a tenant wishes to move and thus end his or her liability for rent.

A year-to-year tenancy is established whenever tenants hold over or remain in possession of a landlord's property at the end of the lease year. Tenants cannot use the act of holding over to force landlords to extend their tenancies. Landlords can have them evicted if they hold over without permission. In many states, they may also hold tenants for another year's rent if they hold over voluntarily for as little as a single day. Some states have modified this rule concerning the holdover of year-to-year tenants to make them tenants at will or month-to-month tenants. Several states require periods of thirty days to six months notice to terminate year-to-year tenancies. Year-to-year tenancy is established when landlords accept rent or allow tenants to remain until the middle of the year or until they have planted their crops.

Types of Rental Arrangements

Several types of rental arrangements can be used in the leasing of landed properties. Some leases involve an elaborate array of provisions; others contain a minimum of detail. Some run for periods of less than a year; others run for periods of one or more years. Some involve all of a landlord's property; others are limited to a few acres, square feet, or rooms from a larger holding. And some involve all the use and possession rights associated with a given property; others are limited to specific interests such as an owner's mineral rights or specified rights in the use of surface land.

Many leases call for a flat cash rental; others specify definite sharing arrangements. Some give the tenant full responsibility for managing the property; others provide for varying amounts of landlord supervision or domination. Some give the tenant little incentive to improve the property, while others provide considerable security for the tenant with compensation arrangements for unexhausted improvements and with options for possible future purchase of the property.

Emphasis is given here to five important types of rental arrangements: the fixed-cash, agricultural-share, and percentage-sharing arrangements used with most leases; long-term leases; and oil and gas leases.

Cash rent. Most property leases simply call for a specified cash payment every month or every year. This arrangement is used with almost all rentals of residential properties and with most leases involving commercial and industrial properties and pasture and grazing land.

When leases call for a fixed-cash rental payment, it is ordinarily assumed that the landlord's return should be adequate to cover the costs of property ownership (taxes, insurance, upkeep, and depreciation) plus a fair return on investment. This arrangement protects the landlord's interests. But it leaves commercial, industrial, and farm tenants subject to all the risks and uncertainties associated with their activities.

Cash tenants are ordinarily entrusted with full responsibility for managing the rented properties. In addition, they are usually free to allocate their fixed and variable inputs as they wish, and to receive the full return from their marginal inputs.

Agricultural-share renting arrangements. Most farm leases in the United States involve some type of share-rental arrangement.[6] These arrangements frequently stem from customary practices and thus vary from region to region and from one type of farming area to another. Sometimes they involve a set share of all crops or all livestock products produced, sometimes shares of specific crop and livestock enterprises, and sometimes combinations of crop and livestock sharing arrangements with some enterprises and payments of cash rent for others. Regardless of the combinations used, the sharing arrangements ordinarily shift some of the risks of farming along with certain responsibilities of management to landlords.

Two of the most publicized sharing arrangements occurred with the sharecropping and "third and fourth" systems used for many years in the South. Both systems were closely tied to the plantation economy, even though they were also accepted on a customary basis on many smaller farms. In both instances, landlords ordinarily provided tenants with a house, a tract of land suitable for raising a cash crop, and frequently with sufficient credit to cover family living expenses during the crop year. With the customary sharecropping arrangement, croppers supplied the labor to work a portion of a landlord's holdings—an average of thirty-five acres per cropper family in 1954—subject to the landlord's supervision. Both parties shared the costs of variable inputs such as fertilizer, insecticides, and ginning. At harvest, the croppers received a share of the crop—usually one-half—for their services. "Third and fourth" tenants enjoyed higher economic and legal status than the croppers, supplied their own farm power and equipment, and paid one-third of the corn crop and one-fourth of the cotton and other specialty crops as rent.

Sharecropping has not been reported separately by the U.S. Bureau of the Census since 1964, but modified "third and fourth" sharing arrangements are still used on several farms and plantations in the South. Sharing arrangements in which landlords receive one-fourth, one-third, or one-half of the value of the crops harvested are also used in many Corn Belt and Great Plains areas. These leases normally allow landlords and tenants to share in the total returns in the same proportions as they supply resource inputs.[7] Comparable livestock-sharing and crop-and-livestock-sharing arrangements are used extensively in the dairying and corn-hog producing areas of the Middle West. With this arrangement, landlords and tenants usually share equally

---

[6] For other discussions of this subject, see Alvin L. Bertrand and Floyd L. Corty, eds., *Rural Land Tenure in the United States* (Baton Rouge: Louisiana State University Press, 1962), chap. 7; Virgil L. Hurlburt, *Farm Rental Practices and Problems in the Midwest*, Iowa Agricultural Experiment Station Research Bulletin 416, 1954; and Harold Hoffsommer *et. al., The Social and Economic Significance of Land Tenure in the Southwestern States* (Chapel Hill: University of North Carolina Press, 1950).

[7] Cf. Don D. Pretzer, *Crop Share or Crop Share-Cash Rental Arrangements for Your Farm*, North Central Regional Extension Publication 105 (Manhattan: Kansas State University Press, 1980).

in the livestock inventory, in the cost of producing feed crops and caring for livestock, and in the division of farm income. Landlords and tenants both have an important stake in the managerial decisions made under this type of lease. Their mutual success depends on their ability to work closely together, to work out differences amicably, and to balance or share their inputs so that each party will find it to their interest to work for the maximization of the combined return.

Percentage leases. Leases for commercial sites have usually called for stipulated cash rents in times past. But a sharing concept, known as the *percentage lease*, was introduced around 1915, attained some prominence during the 1930s and has been widely applied with the leasing of commerical sites in shopping malls and central business districts since the 1950s.[8] These leases take four principal forms. Some leases specify that tenants pay a straight percentage of their gross sales receipts as rent. Some require a straight percentage payment but also specify a minimum rent. Some accept the percentage principle but specify both a minimum and a maximum rent. And some provide that the landlord's share be based on a percentage of the net profits rather than the gross receipts.

The second arrangement with its minimum rent requirement is used with most percentage leases. The minimum is usually fixed at between 70 and 80 percent of the estimated cash rental value of the property or at a level sufficient to cover the landlord's costs for holding and maintaining the property. Landlords generally favor this provision because it gives them assurance of sufficient rental return to cover their basic owner-operator costs while leaving open the opportunity to benefit from a tenant's higher gross returns. Recapture clauses are frequently included in percentage leases to make it possible for landlords to cancel leases after reasonable notice if a tenant's business volume falls below expectations. A comparable provision may protect a tenant's interests by allowing cancellation of the lease if the business does not prove profitable.

Little use is made of the percentage-of-net-profits approach. Landlords are usually advised against this arrangement, partly because of difficulties that arise in the definition of net profits and partly because some legal advisors feel that this arrangement creates a legal partnership between landlord and tenant.

The sharing rates accepted in percentage leases vary a great deal depending on the bargaining power of landlords and tenants, the general use-capacity of the site, and the tenant's type of business. A 1 or 2 percent rate may be adequate with a chain grocery that uses a low-price markup to gain a high volume of business, while a 50 to 60 percent rate may be justified with downtown garages and parking lots.[9] Community experience often provides a guide to choices of percentage rates. But actual determination of the rates is normally left to the bargaining process. In this process, each site and each prospective use is treated as an individual case; and the

[8] "The Percentage Lease–Its Functions and Drafting Problems," *Harvard Law Review*, 61 (January 1948), 317.

[9] Cf. McMichael and O'Keefe, *op. cit.*, pp. 43–47; and National Institute of Real Estate Brokers, *Percentage Leases*, 10th ed. (Chicago: National Association of Real Estate Boards, 1961).

percentage rate decided upon frequently represents an approximation of the proportion that the landlord's desired rent is of the anticipated business receipts.

Percentage leases are often favored on the ground that they help landlords secure the true rental value of their properties and provide incentives for finding tenants who will make the best use of properties. They also have a leveling effect in reducing the tenant's costs during periods of reduced income and in gearing the landlord's return to general business conditions. In this respect, percentage leases provide a hedge against both inflation and the fixed-rent problems that develop with possible public rent controls. With the state and federal rent-control regulations used in the United States during the 1940s, the rates used in percentage leases were frozen, but rent collections increased as commercial establishments received larger volumes of business receipts.

Several disadvantages are associated with percentage leases. They make it possible for landlords to benefit from additional productivity prompted by a tenant's inputs or managerial ability rather than the site. They can call for additional bookkeeping by the tenant, landlord inspection of the tenant's books, and landlord supervision of some tenant business practices. In this last respect, landlords may require certain bookkeeping practices, insist that certain minimum sums be spent for advertising, demand that establishments be kept open during normal business hours, prevent transfer of leaseholds to other parties, or prohibit tenants from investing in competitive establishments to which they might divert business.

Long-term leases.   Leases that run for periods of fifteen years or more are ordinarily described as *long-term leases*. This type of lease is used occasionally in the operation of public utilities and in the administration and private use of certain public lands. Its most spectacular use, however, is associated with the development of commercial sites.

Many valuable sites in large cities such as those occupied by Rockefeller Center and the Waldorf-Astoria Hotel in New York City have been developed by tenants operating under ninety-nine-year leases. The owners of these sites often prefer to retain their ownership for investment or other purposes rather than sell, while the tenants prefer long-term leases to the alternative of actually buying the sites. In this respect, long-term leases frequently represent a method of real estate financing in which tenants enjoy a 100 percent loan of the value of the landlord's property.

With most long-term leases involving business sites, tenants start with an unimproved lot or with a site covered with structures that do not fit their needs. Once the leases are signed, they are expected to clear the site, erect the structures desired, and administer the properties throughout the duration of their leaseholds. They are expected to pay all real estate taxes and all the operating costs associated with their property uses.

Long-term leases must be carefully written to cover many contingencies. Some of the most important of these involve determination of the rental rate, protection of the landlord's interests, and termination of the lease. It is difficult to specify in advance the amounts of land rent that should be credited to real estate investments over long rental periods. This fact along with some disappointing experiences with past attempts to use adjustable rental rates favors the use of flat annual cash rental rates with long-term leases. These rates typically call for payments equal to 5 to 6 percent of the market value of the land at the time the lease is executed.[10]

The buildings erected by tenants ordinarily provide landlords with adequate security for future payments of rent. Security deposits of various types, however, may be used during the period that elapses before the buildings are erected, and again after a tenant's improvements become obsolete.

When no provisions to the contrary are in force, a tenant's improvements pass to the landlord at the end of the leasehold. This eventuality can have a deteriorative effect on the tenant's property-management practices during the later years of the lease. Arrangements are accordingly included in most leases for automatic renewal of the lease if the tenant so desires and for the tenant's possible purchase of the site at some designated price. If the tenant does not accept these options, leases may provide that the landlord will purchase the tenant's improvements at their appraised value. Since the interests of both the landlord and the tenant can be terminated by condemnation proceedings, provisions are also included in most leases relative to the division of any possible indemnity award.

A long-term leasing arrangement known as *ground rent* is commonly used as a real estate financing measure in parts of Maryland and eastern Pennsylvania. Considerable areas used for residential, commercial, industrial, and other purposes are leased under this arrangement for ninety-nine-year periods, with provisions for perpetual renewal. All property taxes on these sites are assessed to the tenant operators; and these operators are bound to periodic—usually semiannual—payments of specified ground rents to their landlords. If tenants default in their payments, landlords can cancel the leases and evict the tenants. These ground-rent arrangements often date back to colonial times. Legislation has been passed in both states since 1884 making it possible for tenants to purchase their leased lands at the capitalized value of the ground rent.[11] A comparable situation has existed on Oahu, Hawaii's most densely populated island, where the Bishop estate held 22 percent and 21 other owners 50 percent of the total land area in 1983. Much of this area had been developed under long-term leases for residential purposes. State legislation requiring sales of homesites to tenants was upheld by the U.S. Supreme Court in 1983.[12]

Oil and gas leases. Much of the land area of the United States is underlaid with oil and natural gas deposits. Every time drilling operations reveal or suggest the presence of commercial quantities of these products in a new area, there is usually a flurry of interest in the leasing of oil and gas rights. Owners are seldom able to develop their own properties, and they are usually inclined to lease their rights to agents of oil-producing companies. Leases are frequently taken for speculative pur-

[10] Cf. McMichael and O'Keefe, *op. cit.*, p. 100.

[11] Cf. *ibid.*, pp. 95–104; and National Institute of Real Estate Brokers, *Ground Leases* (Chicago: National Association of Real Estate Boards, 1965).

[12] Cf. *Hawaii Housing Authority* v. *Midkiff*, 104 U.S. Sup. Ct. 7 (1983).

poses. They are also used by many oil and gas companies to forestall the possible overdevelopment or overexploitation of oil or natural gas fields.

Oil and gas leases vary somewhat in details.[13] As a rule, they run for limited periods of up to ten years and for as much longer as oil or natural gas "are produced in paying quantities." Most leases provide for a fixed annual cash rental rate per acre plus a one-eighth share or royalty on any oil or natural gas that may be produced. The lessee normally secures the right to take equipment onto the lessor's surface land to drill for oil or natural gas. Fees are sometimes paid for this privilege; and provisions are often inserted in the lease concerning the location of wells, the burying of pipes, and the lessee's use of the surface land.

Mineral-rights owners must often choose between selling or keeping their rights and between leasing their rights for a relatively low rental on the strength of a "wildcat" development or waiting for a possible higher rental. Lease management problems also arise when a development company decides to drill wells on some lands to which it holds lease rights but not others. With this arrangement, some surface owners may enjoy considerable income from royalties while others receive none. Mineral-rights owners seldom have enough bargaining power to require lessees to drill wells on their properties if other wells are drilled in the area. Arrangements have been worked out in many communities, however, for the pooling of royalty interests. This arrangement makes it possible for all participating owners to share in any royalties received from oil or natural gas production on the lands of any member of the pool.

ECONOMIC CONSIDERATIONS AFFECTING RENTAL RATES

Landlords and tenants have various goals in mind at the time they strike their rental bargain. Under adverse circumstances, a landlord may be mostly interested in just finding a reliable tenant; or a prospective tenant's primary goal may be simply that of securing suitable rental quarters at an affordable price. Occasionally, one party may deliberately plan to exploit the interests of the other; a landlord may offer exceptional leasing terms to a favored tenant; or a tenant may cheerfully pay abnormally high rent with the knowledge that he or she is helping to support a needy landlord. In most instances, however, the two parties are motivated by desire to maximize their economic returns and satisfactions while participating in an arrangement that is fair to the other party.

Attainment of these goals is invariably complicated by the fact that rental bargains are made prior to the period of property use and possession. Since neither party can predict the future, both must act on faith. In this process, operators fre-

[13] Cf. L. A. Parcher, John H. Southern, and S. W. Voelker, *Mineral Rights Management by Private Landowners*, Great Plains Agricultural Council Publication 13, Oklahoma Agricultural Experiment Station, 1956. For other discussions of this subject, cf. Howard R. Williams, Richard C. Maxwell, and Charles J. Meyers, *Cases and Materials on the Law of Oil and Gas* (Brooklyn, N.Y.: Foundation Press, 1956); and the annual *Proceedings of the National Institute of Petroleum Landmen* (New York: Matthew Bender, 1960 and later years).

quently accept customary, unilateral, or vague rental arrangements without much thought concerning their ultimate consequences. Some dictate the terms they want and offer them to prospective tenants (or landlords) on a take-it-or-leave-it basis. Others analyze the overall rental proposal and bargain for the terms they desire.

With this mixed situation, it is not at all surprising that custom, the going rental rates supported by current supply and demand conditions, and hit-and-miss arrangements play major roles in the determination of most rental rates. Many landlords and tenants, however, seek a logical rationale for developing workable rental arrangements that are both equitable and fair. With this goal, some argue for straight-cash rental payments, which are equivalent to a current market rate of return on the landlord's investment. Some favor sharing arrangements that relate the returns received by landlords and tenants to the marginal value productivity of their inputs, and some endorse combinations of the cash- and share-rental approaches.

Two major sets of goals—(1) maximization of economic efficiency in resource use, and (2) attainment of distributive justice in the allocation of returns between landlords and tenants—have far-reaching effects on the determination of ideal and workable leasing systems. Economic efficiency is sometimes heralded as a principal objective to be attained with leasing arrangements.[14] This objective is secured when profits are maximized for the entire firm representing the combined interests of the landlord and tenant. Deviations from this maximum occur whenever the provisions of a rental arrangement make it more profitable for tenants to combine their resources in some manner other than that which spells maximum profit for the overall firm, and whenever rental arrangements foster a transfer of income from one leasing party to the other. Distributive justice in turn involves ethical, moral, and political values. It involves one's sense of fairness in dealings with others and goes beyond the realm of economics.

Five important economic issues affect the development of mutually acceptable and advantageous leasing arrangements. They involve (1) equitable sharing arrangements covering costs and returns; (2) comparable rental arrangements with all products; (3) opportunities for a fair return from all investment inputs; (4) flexibility, which permits adjustments for changing costs, prices, and production; and (5) recognition of social justice and welfare objectives. The first three of these issues have an economic efficiency-orientation, while the last two are concerned with social justice.

Equitable Sharing of Costs and Returns

Share and percentage leases are often advocated because they relate rental returns to the marginal value productivity of the landlord's inputs. But they can be criticized on efficiency grounds. This criticism stems from differences between the

[14]For more detailed discussions of this criterion as it relates to leasing arrangements, cf. Earl O. Heady, *Economics of Agricultural Production and Resource Use* (Englewood Cliffs, N.J.: Prentice-Hall, 1952), chap. 20; Hurlburt, *op. cit.*, and Howard W. Ottoson, "The Application of Efficiency to Farm Tenure Arrangements," *Journal of Farm Economics*, 37, (December 1955), pp. 1341–53.

income incentives of share tenants and those of owner-operators and cash tenants. Owner-operators and cash tenants ordinarily have every incentive to apply their variable inputs to the point at which $MFC = MVP$ because they receive the entire economic surplus from each of their marginal inputs. But when the share tenant "has to give his landlord half of the returns to each dose of capital and labor that he applies to the land, it will not be to his interest to apply any doses the total return to which is less than twice enough to reward him."[15]

This concept is illustrated by the two examples depicted in Figure 14-1. Figure 14-1A represents the situation that exists under an agricultural-share lease in which a landlord supplies a fixed input of land and improvements in return for a one-half share of the crop while the tenant supplies all the variable inputs. Under these conditions, the tenant pays the full marginal factor cost of each successive variable input unit but receives only half of the marginal value product.

If the operators in this example were owner-operators or cash tenants, they would find it profitable to apply S inputs (the number of inputs at which $MFC = MVP$). Share tenants would find it more profitable to stop with R inputs, because this is the point at which their marginal factor costs equal their one-half share of the marginal value product. Landlords would benefit if the tenants applied up to T inputs (the point at which the landlords would receive their highest possible return); but no rational tenant would go this far because any application of inputs beyond R would involve a transfer of income from the tenant to the landlord.

This same type of reasoning with its recognition of the problems caused by disassociation of benefits and costs applies to the percentage leases used by many commercial establishments. The example pictured in the second part of Figure 14-1 assumes the case of retail firms that use low markup policies to encourage a large volume of business. The landlords' inputs are limited to the provision of desirable commercial sites in a central business district, for which they receive a 10 percent share of their tenants' gross receipts. Here again the tenants pay the full cost of each successive variable input while their returns are limited to 90 percent of the marginal value product, a fact that makes it uneconomic for them to apply more than R inputs in their business.

Tenants in this example may recognize the opportunities owner-operators and cash tenants would have to increase total volume of business and total profits by using additional inputs—up to S input units—for advertising, delivery service, or other items. Yet as tenants operating on a percentage lease, they find it best to stop with R inputs because any additional inputs call for higher contract rent payments, the added portion of which comes from their normal returns as tenants.

[15]Alfred Marshall, *Principles of Economics*, 8th ed. (New York: Macmillan, 1938), p. 644. For an early discussion of this problem, cf. Adam Smith, *The Wealth of Nations* (1776) (New York: Modern Library edition, 1937), p. 367. For more-recent and more-elaborate discussions, cf. Rainer Schickele, "Effects of Tenure Systems on Agricultural Efficiency," *Journal of Farm Economics*, 23 (February 1941), 185–207; Earl O. Heady, "Economics of Farm Leasing Systems," *Journal of Farm Economics*, 29 (August 1947), 659–78; and D. Gale Johnson, "Resource Allocation under Share Contract," *Journal of Political Economy*, 58 (April 1950), 111–23.

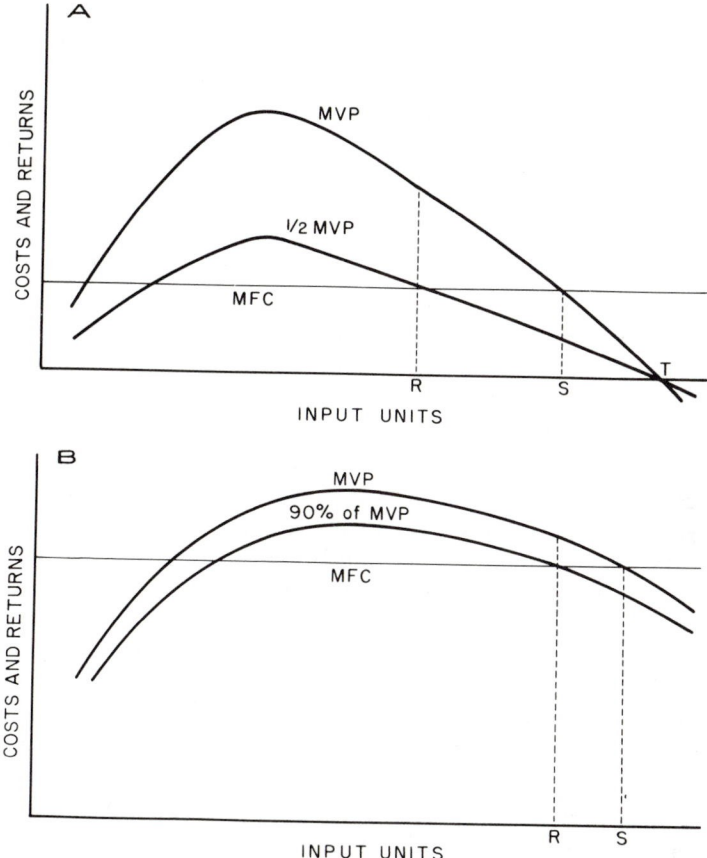

FIGURE 14-1.   Illustration of the possible effects of (A) a one-half rental arrangement and
(B) a 10 percent percentage leasing arrangement upon the tenant's willingness
to apply additional inputs in the production process.

These two examples indicate that from a theoretical view, the use of share and percentage leasing arrangements logically lead to less intensive and less efficient resource use than one might expect with owner-operatorship or cash tenancy. It is surprising, therefore, to note that little consideration has been given to this problem with percentage leases, that sharing arrangements are used by most farm tenants in the United States, and that the average share tenant tends to apply input factors and operate with approximately the same intensity as owner-operators and cash tenants.

Several reasons can be given for the failure of share tenants to act in the manner suggested above. Like other operators, they cannot predict in advance the exact combinations of inputs that will yield maximum returns. At times, they are not aware of the fact that they can maximize their short-run interests by operating differently than owner-operators. Their calculations as to the optimum number of variable inputs they should use is also complicated by the tendency of landlords to use short-term leases and restrictive leasing arrangements. Commercial landlords, for example, ordinarily specify certain managerial practices in their percentage leases, while farm landlords frequently provide portions of the variable inputs such as shares of the seed, fertilizer, pesticides, and harvesting costs. Tenants also realize that they are evaluated as tenants and as credit risks in terms of their overall performance. Percentage and share tenants who want to retain their leaseholds or who hope to move on to the operation of more productive enterprises accordingly find it to their advantage to manage their inputs in much the same way as owner-operators or cash tenants.

Another explanation of the tendency of many share and percentage tenants to operate at the same level of overall efficiency as owner-operators involves factors assumed away in the theoretical model. Figure 14-1 assumes that the labor and managerial inputs supplied by tenants are priced at going market levels and that alternative opportunities exist for the profitable utilization of any surplus variable inputs. With these assumptions, it is logical to argue that tenants will maximize their positions as producers by adding variable inputs only to the point at which their share of the marginal value product equals their share of the marginal factor costs. Under real-life conditions, tenants often find that their maximization opportunities are constrained by the terms of their leases, by the limited supplies of land resources available for their use, or by an absence of opportunities for the profitable utilization of their surplus labor and management inputs in other enterprises. Typical tenants operating under these conditions in Third World settings frequently find it to their advantage to apply inputs of labor and management somewhat beyond the point at which their share of the marginal value product is equal to the full market price of their marginal factor costs. They receive less than the full market value of the additional inputs used, but their surplus inputs have a supply price of near zero and, with no alternative opportunities for their productive use, some return is better than no return at all.[16]

Comparable Rental Arrangements with All Products

A second situation affecting the efficiency of resource use can arise when two or more rental arrangements affect a single set of operations. A simple example of this occurs when a tenant who produces two competing crops with comparable production costs and product values finds that a larger share rent will be paid with one crop than the other. With these conditions, tenants have an income incentive to shift resources to the crop that offers them the largest net return, while landlords have a comparable incentive for favoring the crop that yields the highest share rent.

This same situation applies when the tenant produces

> ... two or more crops with different unit costs, different yields and different prices with a given quantity of resources. ... Because of the opportunity to

[16]Cf. Steven N. S. Cheung, *The Theory of Share Tenancy* (Chicago: University of Chicago Press, 1969), pp. 51-55.

obtain a higher income, the operator will want to shift resources into production of that crop which gives the highest income on the factors he contributes. This will not necessarily be the one on which the lower share rental is paid because differences in unit costs may more than compensate differences in shares, and some minimum acreage of a crop like clover may be essential in the rotation to maintain the yield and income from corn. The inclination usually will be to shift more resources into production of the crop with the lower rental share.[17]

Similar situations occur when the tenant pays a share rent on some enterprises or fields and cash rent or no rent on others, or when a retailer operates two adjacent properties, one under a cash lease and one under a percentage lease. Operators under these conditions can find it advantageous to concentrate their most remunerative activities on their cash-rent or rent-free land. In so doing, they can maximize their returns at the sites where rent is a fixed cost while they carry on necessary but less productive activities at those sites where rent is treated as a variable cost.

Comparable examples ofter occur with cases of part ownership. As Ottoson has observed, farm owners who rent additional land usually tend

> . . . toward greater intensity of production on the land owned by the operator. The productivity per input of operator's resources is higher on owned land than on a tract rented under a share lease. The notion is commonly suggested that part-owners tend to spread on their owned tracts the manure resulting from feed grown on their rented tracts, thus effecting a transfer of fertility. Operators will likely give priority to the owned tracts with respect to expenditure of working capital where capital is rationed. They may also favor the owned tracts in time of crop operations during crucial periods. This tendency toward less intensity would be less true where the rented part is under cash lease, although even here the operator will have little or no incentive to make expenditures for inputs that become associated with the landlord's resources and that do not produce returns in the same year.[18]

As these examples suggest, mixed rental arrangements frequently provide tenants with an opportunity to maximize their returns at a landlord's expense. This situation can result in inefficiency in total resource use together with considerable ill will between landlord and tenant. These consequences are usually averted when the same sharing arrangement is applied to all products. When this practice is not followed, it is best to evaluate the different rental arrangements and make such adjustments as are necessary to ensure efficient use of the combined resources of the landlord and tenant.

[17]Hurlburt, *op. cit.*, p. 89.

[18]Ottoson, *loc. cit.*, pp. 1350–51. George Washington recognized this problem in a letter of February 25, 1784, in which he wrote, "From the first I laid it down as a maxim, that no person who possessed Lands adjoining, should hold any of mine as a Lease, and for this obvious reason, that the weight of their labour, and burden of their crops, whilst it was in a condition to bear them, would fall upon my Land, and the improvement upon his own, in spite of all the covenants which could be inserted to prevent it." *Writings of George Washington* (Washington: Government Printing Office, 1938), XXVII, 344.

Most landlords recognize that the opportunity tenants have to maximize personal interests is largely dependent on their power to make important production decisions. This opportunity is definitely limited when the landlord makes most of these decisions or participates jointly with the tenant in the decision-making process. Provisions are often included in farm leases specifying the acreage or scope of the principal enterprises and sometimes even the field layout and cropping practices the tenant will use. Comparable limitations are used with commercial property leases when the landlord requires a minimum cash rent in association with a percentage lease, uses a recapture clause to cancel the lease if the tenant's business falls short of the anticipated level, or acts to prevent a tenant from acquiring interests in adjacent properties.

Opportunities for a Fair Investment Return

Another important leasing problem arises when economically efficient patterns of operations call for certain types of investment inputs without providing any assurance of a fair return to the party who makes the investment. Landlords and tenants frequently find that they can use investments in certain types of improvements to increase both the productivity and the satisfactions they secure from their combined resources. With commercial business properties, these improvements may involve the shifting of wall partitions, installation of air-conditioning facilities, or erection of a new store front. With a rented house or apartment, they may call for modernizing the kitchen, tiling a bathroom, or supplying a new heating system. With farm properties, they may involve applications of lime or commercial fertilizer, the provision of an irrigation system, investments in soil-building crops, or construction of new farm service buildings.

The real problem in these cases boils down to one of who will pay for the improvements. Since improvements ordinarily add to the use-capacity and value of the landlord's property, it is often assumed that the landlord should bear the cost. Landlords may object to this reasoning on the ground that most of the benefit will accrue to the tenant. This is particularly true when no adjustments are made in the contract rent paid by the tenant. In those instances in which tenants are the principal beneficiaries, they may be willing to proceed with desired improvement programs if their lease planning horizons encompass the period when the benefits will be received. A tenant who operates under a short-term lease, however, has three important reasons for not making the desired improvements. "First, he is uncertain that he will remain on the property long enough to get full benefit from the improvements. Second, he has no assurance he would be compensated for his unused value of the improvement in case he moves. . . . Third, since improvements make the landlord's property more valuable and attractive to other tenants, an increased rent may result."[19]

As these reasons suggest, the problems of tenure instability and insecurity of

---

[19] John F. Timmons, *Improving Farm Rental Arrangements in Iowa*, Iowa Agricultural Experiment Station Research Bulletin No. 393, 1953, p. 86.

expectations may cause landlords and tenants to operate on a lower and less productive plane than their potentialities warrant. Several alternatives can be used by landlords and tenants to avoid this situation.[20] Intermediate- and long-term leases are frequently used, particularly in urban commercial districts, to protect and strengthen the income incentives tenants have for investing in property and other business improvements. Tenant improvements can be encouraged by arrangements providing that tenants be compensated for the value of any unexhausted improvements at the time they give up their leaseholds. Landlords are often willing to make property improvements if their tenants will agree to the payment of improvement rents. And joint action can be secured with programs that call for landlord-tenant sharing of improvement costs and returns, landlord provision of the materials while tenants supply the labor, or compensating arrangements under which landlords make one type of improvement while tenants make another.

A tenant's willingness to invest in property improvements without some assurance of ability to profit from the investment has its counterpart in similar situations involving landlords. Urban landlords who find themselves bound by rent controls or by long-term leases with fixed rental rates have little incentive to provide more than necessary property repairs and upkeep. Farm landlords may feel that the rental market in their communities makes no distinction in rental rates between farms with modern housing and farms with less desirable residential quarters. With this view, they may resist pressure for providing housing improvements—unless they are necessary to attract or keep desirable tenants—simply because they lack assurance of a fair return on their housing investments. Additional cash rent for housing improvements is sometimes recommended as an answer to this problem.

Adjustments for Changing Conditions

Even when a landlord and tenant accept what appears to be an ideal rental arrangement, distribution problems can arise because of unexpected changes in cost, price, and production assumptions. These changes frequently bring different distributions of the income from rental properties than were contemplated in the original leasing agreement. Problems of this order are a frequent occurrence in agriculture, where crop yields fluctuate with changing weather conditions and prices change with little notice. Problems also occur at times with urban properties. Many urban tenants during the Great Depression years found that they had agreed to fixed commercial or residential rental rates that far exceeded their reduced ability to pay. These tenants frequently found it necessary to demand rent reductions as a condition of continued business operations or residential occupancy. A reversal of this situation followed in the 1940s when many landlords found themselves bound by contract rental rates that were often definitely low relative to the rising business and personal incomes of their tenants. Some of these landlords were able to raise their rents, but many were prevented from doing so until after the end of rent controls.

Equity and welfare considerations suggest a need for leasing arrangements that

[20]Cf. *ibid.*, pp. 89–95; cf. also Pretzer, *op. cit.*

keep the returns landlords and tenants receive somewhat in line with the distribution plan envisaged in their initial rental agreements. Since rental terms are usually agreed upon prior to the tenant's occupation of the landlord's property, it is important the arrangements be sufficiently flexible to allow for changes in cost, price, and production conditions. Beyond this, they should also favor efficiency in resource use together with some protection of landlords from inferior management by the tenant.[21]

No rental system has yet been devised that has the advantage of always appearing as equitable to both parties at the time rent is paid as at the time when the rental agreement is made. Percentage leases, however, are often favored with commercial properties because of their characteristic of relating the landlord's return to the tenant's level of businss activity. Sharing arrangements are used to this same end in agriculture.

Less progress has been made in working out flexible-cash rental arrangements. Residential rents could be tied to a cost-of-living index during periods of rent control. Two similar proposals have been made for farm leases. One of these—the Iowa flexible-cash rent plan—calls for landlord-tenant agreements on a cash base rental figure that is adjusted at the end of the year for general price and production changes in the area to determine the actual rental payment. A second proposal—the Missouri multiple commodity plan—calls for a base rent associated with a mutual landlord-tenant estimate of the expected gross income of the farm. At the end of the year, this base rent would be used to compute an actual rental rate, which would bear the same percentage relationship to the base rate as the annual gross income bears to the estimated gross.[22]

Social Justice and Welfare Considerations

No discussion of the conditions needed for an ideal rental system would be complete without some mention of the overlying importance of the concepts of social justice and welfare. Most people adhere to an ethical or moral code, which causes them to regard some types of leasing arrangements and some lines of action as more right or just than others. They feel some responsibility for the welfare of their associates. They are both conscious of and responsive to the attitudes of others, and they are more inclined to do things regarded as acceptable and praiseworthy by their associates than to follow lines of action frowned on by society.

These factors play a highly significant role in limiting the extent to which landlords and tenants deliberately maximize their own interests at the expense of others. They help to make the landlord-tenant relationship a cooperative arrangement under which both parties benefit rather than a tooth-and-claw struggle for economic dominance. They explain the tendency of most landlords and tenants to

---

[21]Cf. Walter E. Chryst and John F. Timmons, *Adjusting Farm Rents to Changes in Prices, Costs, and Production*, Iowa Agricultural Experiment Station Special Report No. 9, 1955.

[22]Cf. *ibid.*, pp. 29-44; and John F. Timmons, *Landlord-Tenant Relationships in Renting Missouri Farms*, 2nd ed., Missouri Agricultural Experiment Station Bulletin No. 409, 1946.

work together, particularly when they live in the same community. By the same token, they also explain why glaring examples of inequitable and exorbitant rental arrangements on the world front usually involve instances of absentee ownership in which a landlord group has isolated itself from immediate personal contact with its tenants.

Important as the concepts of equity, social justice, and welfare are in considerations of rental arrangements, they are often neglected and under-emphasized in comparison with economic efficiency criteria. Production-oriented observers frequently equate equitability in the division of income between landlords and tenants with the marginal value productivity of the resource inputs each party contributes in production. This marginal-productivity concept of equity has some justification when landlords and tenants enjoy relatively equal bargaining power, when both possess alternative opportunities, and when both can command an adequate return from their resource inputs. It provides a superficial and meaningless measure of equity, however, when the cards are stacked in favor of either party.

History shows that tenant groups have often suffered from a lack of equal bargaining power with their landlords. When this condition has been combined with a lack of alternative employment opportunities, tenants have at times bid up the rents they were willing to pay while at the same time bidding the returns for their own inputs down to a subsistence level. By imputing the going rates of return to the tenant's variable inputs, one could use the marginal-value-productivity approach to justify the "rack rents" of nineteenth-century Ireland and the exorbitant rents of pre-1950 Egypt or India as "an equitable division of the product." Conclusions of this order run definitely counter to our prevailing concepts of equity, justice, and social welfare.

Social justice and welfare considerations play highly significant roles at times in providing the criteria for new public policies and programs that deal with leasing problems. Several countries have used agricultural and residential rent controls to secure distributive justice for tenants. Special credit programs, resettlement arrangements, and public housing programs are often used to improve the bargaining position and income status of tenant groups. Antipeonage legislation has been used in countries such as the United States to protect tenants against undesirable rental conditions. Land reform measures of various types have also been used in many nations to provide a larger measure of social justice and welfare for tenant-operators and other potential owners.

—SELECTED READINGS

Bertrand, Alvin L., and Floyd L. Corty, eds., *Rural Land Tenure in the United States* (Baton Rouge: Louisiana State University Press, 1962).

Dasso, Jerome J., and Alfred A. Ring, *Real Estate Principles and Practices*, 9th ed. (Englewood Cliffs, N.J.: Prentice-Hall, 1981), chap. 26.

Downs, James C., Jr., *Real Estate Management* 12th ed. (Chicago: Institute of Real Estate Management, 1980), chaps. 6 and 7.

Fisher, Ernest M., and Robert M. Fisher, *Urban Real Estate* (New York: Holt, Rinehart & Winston, 1954), chap. 7.

Heady, Earl O., *Economics of Agricultural Production and Resource Use* (Englewood Cliffs, N.J.: Prentice-Hall, 1952), chaps. 20 and 21.

Johnson, Bruce B., *Farmland Tenure Patterns in the United States*, Agricultural Economics Report No. 249, (Washington: U.S. Department of Agriculture, 1974).

Kratovil, Robert, and Raymond J. Werner, *Real Estate Law*, 8th ed. (Englewood Cliffs, N.J.: Prentice-Hall, 1983), chap. 37.

Timmons, John F., *Improving Farm Rental Arrangements in Iowa*, Iowa Agricultural Experiment Station Research Bulletin No. 393, 1953.

15
USE
OF REAL ESTATE CREDIT

Credit is frequently described as the "lifeblood" of the real estate business. The reasons for this characterization are easy to find. Real estate credit is used in the acquisition of a great majority of the homes, farms, and other real properties purchased in the United States. Credit arrangements have provided a shortcut to property ownership for millions of owners. It has made ownership possible for thousands of individuals who would otherwise have found their personal roads to ownership closed or filled with time-consuming detours. At the same time, it has contributed to a different type of real estate market than would exist if less credit was available.

In the discussion that follows, emphasis is first given to an overall view of the importance of real estate credit in the American economy. Attention is then focused on the chief characteristics of mortgage and land contract arrangements. These comments are followed with a brief analysis of the problems lenders and borrowers face in their administration of real estate credit.

IMPORTANCE OF REAL ESTATE CREDIT

Credit arrangements go back to the dawn of private ownership. Almost from the first recognition of exclusive private property rights in land, owners and prospective owners have found it good business to borrow and use the capital resources of others. This use of credit has persisted because it has usually provided a mutually advantageous arrangement for both lenders and borrowers. Lenders have secured payments of interest from the use of their surplus funds. Most borrowers in turn have sought credit with the expectation that they could use it to increase their productivity and maximize their returns and satisfactions.

Real estate credit has been more popular and more widely used during some periods of history than others. Throughout the Middle Ages, lending practices calling for payments of interest were frowned upon and sometimes denounced by the Church. Credit contracts were tolerated and enforced by the state, but moral considerations kept most people from borrowing money. Most of this stigma gradually disappeared with the rise of modern capitalism. With this development, more and more people sensed the significance of the role borrowed capital can play in implementing increased productivity and economic development. Governments instituted legal controls to prevent the principal abuses associated with earlier uses of credit. People were encouraged to put their savings to work by loaning them to others; emphasis was given to the provision of credit facilities; and the credit system emerged as a dominant force in the economy.

Impact of Credit on Real Estate Resources

Credit is an essential ingredient in most modern-day real estate transactions. Some buyers pay cash secured from personal savings, gifts and inheritances, or sales of other properties when they buy land. But full cash payments are more the exception than the rule and come mostly with sales that involve small considerations. Most transactions could not take place at their present scale without the use of substantial credit.

The extensive use of credit has three significant effects on the rights and opportunities associated with real estate ownership. Acceptance of credit normally involves the granting of a mortgage and a legal sharing of the rights in land by the owner and his or her mortgagee. Lenders sometimes insist that borrowers accept varying amounts of managerial control and supervision as a condition of the loan. And, most important, credit frequently makes it possible for operators to expand the scale of their operations, step up their operating efficiency, and increase their overall productivity.

Users of real estate often have sufficient inputs of land, labor, and managerial ability but lack adequate capital to operate as efficiently as they might. Credit can fill this gap and, by increasing one's productivity, make it possible for operators to pay for the credit and realize higher returns on their other factors. Inadequate capital and lack of sufficient credit are common problems in frontier and underdeveloped areas. They also pose problems for young people who are trying to establish themselves as business operators. Credit may not be as essential for established operators, but they too depend on it and find that it provides them with opportunities to operate on a larger scale than they otherwise could.

In addition to the role it plays in production plans, credit is also needed by those who wish to acquire real estate resources. Without credit, most operators would acquire property ownership later in life than they now do. Access to credit—the willingness of lenders to advance purchase funds now in exchange for an operator's promise to repay a loan over a given time period—is a necessary part of most real estate purchase arrangements. It is equally important to those operators who need additional land to enlarge the scale of their operations.

Some of the most dramatic uses of credit with real estate investments come with the dealings of large-scale land developers. These individuals often have little to offer in the land-development process except their ability as expediters and their imagination and vision concerning what can be done with given tracts of land. Yet they frequently seize the opportunities for economic leverage afforded by real estate credit to acquire and control substantial investments. At times they also use credit arrangements, options, long-term leases, and knowledge of the tax laws and of who wants what to acquire and either redevelop or resell valuable real estate holdings.[1]

Real Estate Credit Situation

The term *real estate credit* applies to loans in which real property is pledged as security for a credit grant. These loans are usually made for the purchase or improvement of property, but they may be used for other purposes as well. Real estate credit usually involves the granting of mortgages; and most discussions of real estate credit center on this type of credit. Other methods of real estate financing, however, are in use. *Land contracts* provide an important type of real estate credit for property buyers in many areas; ground rents and long-term leases are used as a real estate financing device in some localities; and a variety of real estate bond, stock, and land trust arrangements can be used to provide real estate credit.

The importance of real estate credit in the American economy can best be illustrated by a brief review of recent credit trends. At the end of 1983, property owners in the United States owed $1,826.4 billion in outstanding debt on real estate mortgages and land contracts. This total included $1,214.6 billion on one- to four-family residences, $150.9 billion on multifamily residential properties, $351.3 billion on commercial properties, and $109.6 billion on farms (see Table 15-1). The total real estate debt load for 1983 was more than ninety-eight times the level for 1945.

Examination of the real estate debt situation for nonfarm residential properties shows that the outstanding mortgage debt on these properties rose from $2.7 billion in 1896, to $4.4 billion in 1910, and to $27.6 billion in 1930.[2] This debt dropped to $22.2 billion in 1935 and then gradually increased to $24.6 billion in

---

[1] For examples of these operations, cf. Conrad N. Hilton, *Be My Guest* (Englewood Cliffs, N.J.: Prentice-Hall, 1957); and William Zeckendorf, *Autobiography of William Zeckendorf* (New York: Holt, Rinehart & Winston, 1970).

[2] Cf. L. Grebler, D. M. Blank, and L. Winnick, *Capital Formation in Residential Real Estate: Trends and Prospects* (Princeton, N.J.: Princeton University Press, 1954), Appendix N.

TABLE 15-1. **Outstanding Real Estate Debt in the United States, 1930-1983***

| YEAR | TOTAL REAL ESTATE DEBT[+] | TOTAL OUTSTANDING DEBT ON: | | | |
		Nonfarm one- to four-family residences	Multi-family residences	Commercial properties	Farm properties
1930	47.8	18.8	9.4		9.4
1935	37.3	13.1	4.4		7.4
1940	36.5	17.4	12.6		6.5
1945	35.5	18.6	12.2		4.8
1950	72.8	45.2	21.6		6.1
1955	129.9	88.2	32.6		9.0
1960	206.8	141.3	52.7		12.8
1965	325.8	212.9	91.6		21.2
1970	451.7	280.2	58.0	82.3	31.2
1975	801.5	490.8	100.6	159.3	50.9
1980	1,471.8	987.0	137.1	255.7	92.0
1983	1,826.4	1,214.6	150.9	351.3	109.6

*Data in billions of dollars.

[+]Data represent year-end totals and include mortgages, purchase-money mortgages, and land contracts but do not include corporate bonds or real estate debt on multifamily residential and commercial properties that corporate owners may owe to other nonfinancial corporations. Totals do not always add up because of rounding.

Sources: Data for 1930-1940 from *Survey of Current Business* (Washington, D.C.: Government Printing Office, 1949), October 1949, p. 11; data for 1945-1965 from *Statistical Abstract of the United States, 1968* (Washington, D.C.: Government Printing Office, 1968), p. 459; 1970-1983 data from current issues of *Federal Reserve Bulletin*, a monthly report published by the Board of Governors of the Federal Reserve System.

1945, after which it skyrocketed to totals of $228.2 billion in 1970 and $1,365.5 billion in 1983. Comparable reports on the farm mortgage debt situation show that total outstanding farm real estate debt increased from $3.2 billion in 1910 to a peak of $10.8 in 1923. This total was gradually pared down to a low of $4.8 billion at the end of 1945. Following World War II, it too rose at a rapid pace to totals of $31.2 billion in 1970 and $109.6 billion in 1983.

Census findings indicate that 28 percent of the farms and nonfarm residential properties were mortgaged in 1890. Since then the proportion of nonfarm owner-occupied single-family residential units with mortgages has risen steadily to 45.3 percent in 1940, 60.6 percent in 1970, and 64.6 percent in 1980. Farm real estate market data show that the proportion of buyers using credit in their purchases rose steadily from 58 percent in 1950, 67 percent in 1960, and 78 percent in 1970 to 91 percent in 1980 and that the average ratio of debt to purchase price rose from 57 to 78 percent between 1950 and 1980.[3]

[3]*Farm Real Estate Market Developments*, U.S. Department of Agriculture, Economic Research Service, CD-87, 1982, Tables 20 and 21.

Major Sources of Credit

Prospective borrowers can turn to several sources for real estate credit. Individual investors have always provided an important source of mortgage funds and account for almost all of the land contract credit.[4] Other important sources of mortgage credit are provided by savings and loan associations, insurance companies, commercial banks, mutual savings banks, and various public credit agencies. Table 15-2 reports the distribution of outstanding mortgage debt on residential, commercial, and farm properties at the end of 1983 by major lender groups. As this tabulation indicates, savings and loan associations were the principal source for residential properties, followed by federal agencies, individual lenders, and commercial banks. Commercial banks and insurance companies were the leading suppliers of credit for commercial properties, whereas the federal land banks and individual lenders supplied most of the real estate credit used by farmers.[5]

As one might expect, the relative significance of the various sources of credit has changed over time. Individual lenders supplied a substantial portion of the mortgage money for residential and more than half of the credit for farm mortgages during the 1920s. But losses incurred by many lenders during the 1930s, the development of new credit institutions, and the fourfold increase in mortgage volume between 1970 and 1983 contributed to a new market situation. Individuals and other nonagency lenders still accounted for over $12 billion in real

[4] Individual investors who provide sources of real estate credit are motivated by two different rationales. Many owners facilitate the sale of their properties by taking down payments offered by buyers and accepting mortgages or land contracts for the balance of the sales price. Sometimes they do this by choice. However, in periods when alternative sources of credit are in short supply, they may find it difficult to sell their properties in any other way. This situation existed during the 1970s and early 1980s and helps explain the fact that individuals held 16.7 percent of the outstanding value of mortgages on one- to four-family residences in 1982 as compared with 9.6 percent in 1970. Should these investors later find it necessary to sell their mortgages or contracts, they usually face the prospect of having to accept substantial discounts that greatly reduce the effective value of their sales prices.

A second group of investors advance funds to borrowers because they regard investments in farm and home mortgages as relatively safe investments that provide higher returns than savings accounts. The popularity of this type of investment declined during the 1970s with the higher returns and greater liquidity associated with investments in money market accounts and U.S. Treasury certificates. State usury laws, which set maximum levels on the interest rates individuals can charge on mortgages and land contracts, have also discouraged private investments in some states when investors have had higher-paying alternative investment opportunities in other states or in other investment areas.

[5] For a more detailed discussion of the facilities provided by these sources, cf. William R. Beaton, *Real Estate Finance*, 2nd ed. (Englewood Cliffs, N.J.: Prentice-Hall, 1982); Frederick E. Case and John M. Clapp, *Real Estate Financing*, 2nd ed. (New York: John Wiley, 1978); Bruce Harwood, *Real Estate Principles*, 2nd ed. (Reston, Va.: Reston, 1980), chap. 12; Henry E. Hoaglund and Leo D. Stone, *Real Estate Finance*, 6th ed. (Homewood, Ill.: Richard D. Irwin, 1977); Mary Alice Hines, *Real Estate Finance*, 4th ed. (Reston, Va.: Reston, 1983); and Maurice A. Unger and Ronald W. Melicher, *Real Estate Finance* (Cincinnati, Ohio: South Western Publishing Co., 1978).

TABLE 15-2. Outstanding Real Estate Mortgage Debt Held by Major Lenders, United States, December 31, 1983

LENDER GROUP	ALL PROPERTIES	NONFARM ONE- TO FOUR-FAMILY RESIDENCES	MULTI- FAMILY RESIDENCES	COMMERCIAL PROPERTIES	FARMS
Total outstanding debt*	$1,826.4	$1,214.6	$150.9	$351.3	$109.6
(Percentage distribution)					
Savings and loan associations	27.0	32.1	28.1	17.4	—
Mutual savings banks	7.5	7.9	11.8	6.2	.03
Commercial banks	18.0	15.0	11.9	34.1	8.5
Insurance companies	8.3	1.3	12.7	29.7	11.6
Federal land banks	2.8	0.2	—	—	43.9
Federal agencies	5.3	6.9	7.4	0.1	0.4
Mortgage pools or trusts	15.6	21.3	6.5	2.1	8.5
Individuals and others	15.5	15.3	21.6	10.4	27.1

*In billions of dollars.

Source: Federal Reserve Bulletin, Board of Governors of the Federal Reserve System, July 1984, p. A37.

estate mortgages (about one-third of the total) during the early 1940s when they held two-fifths of the farm mortgage volume and about one-tenth of that on residential properties.[6] These proportionate shares continued with slight modifications throughout the next three decades. Between 1972 and 1983 their volume of outstanding real estate mortgage loans increased fourfold from $69.0 to $284.0 billion, while their shares of the loan market rose to 16.0 percent of the residential loans and dropped to 27.1 percent of the farm loans. Much of the increase in total volume resulted from owner acceptance of land contracts and mortgages in the seller-financing of sales.

Savings and loan associations, which had accounted for one-third of the volume of mortgage credit on residential properties during the 1920s and about one-fourth of the total in 1940, moved ahead to play a dominant role as the supplier of more than one-half of the credit for residential buyers during the 1960s. Their share of the residential total dropped from 53.6 percent in 1970 to 31.7 percent in 1983, even though their loan volume rose to over $432 billion. Mutual savings banks, which operate primarily in the Northeastern states and which resemble savings and loan associations in many respects, held approximately one-sixth of the residential mortgage debt throughout most of the 1940–1970 period, after which their percentage importance declined.

Two other important private sources of financing are commercial banks and insurance companies. Commercial banks held about one-tenth of the farm mortgage debt in the early 1940s, 15.9 percent in 1950, and only 8.5 percent in 1983. Meanwhile, they held over one-fourth of the mortgage volume on residential properties during the 1940s, after which their share gradually declined to 14.6 percent in 1983. Insurance companies have shown declining interest in residential loans. Their share of that total dropped from around 20 percent of the total during the 1950s to 2.5 percent in 1983. They held 21.8 percent of the farm loans in 1950, 31.1 percent in 1960, 18.1 percent in 1970, and 11.6 percent in 1983.

Public credit programs. Notable changes in the real estate credit situation have come with the expanded use of public credit and mortgage pooling facilities. Federal credit programs started with the establishment of the federal land banks under the Federal Farm Loan Act of 1916. These quasi-public banks provided a major source of farm real estate credit during the 1930s, accounted for roughly one-fourth of the farm mortgage debt at the end of World War II and almost one-fifth of the total between 1948 and 1965, after which their share rose to 34 percent of the total in 1975 and 44.2 percent in 1982. The joint-stock land banks, also created under the Federal Farm Loan Act of 1916, played a far less important role in supplying new credit and were discontinued in 1933. The Federal Farm Mortgage Corporation—set up in 1933 to administer Land Bank Commissioner loans, which were

---

[6] The classification of "individuals and others" is made up mostly of individual investors but also includes mortgage companies, real estate investment trusts, state and local government agency and retirement funds, and noninsured pension funds.

used as second mortgages in association with land bank loans and as first mortgages on properties on which the land banks would not make loans—provided an important secondary source of funds during the 1930s and 1940s. The Farm Security Administration, now the Farmers Home Administration (FmHA), has provided an additional source of farm real estate credit since 1937. This agency was originally designed to work with farmers who could not qualify for other sources of credit but has expanded its scope to provide loans on commercial and residential properties as well as farms. It held $42.1 billion in loans in 1983 as compared with only $767 million in 1970.

Need for public action peaked during the 1930s, when both the residential and farm credit markets were caught in the paralyzing grip of the Great Depression. As the credit situation tightened, thousands of mortgagors defaulted on their payments; most conventional credit sources, which might have helped, found major portions of their assets immobilized; and owners everywhere faced the prospect of losing properties through mortgage foreclosure. In the midst of this situation, debt conciliation committees were established in numerous counties and states to facilitate voluntary debt adjustments, which would put borrowers on a sounder footing and help forestall foreclosure proceedings. The Federal Home Loan Bank Board was created in 1932 and the Farm Credit Administration in 1933 to promote better credit arrangements. Debt moratoria and farm mortgage relief laws were passed by most state legislatures and by Congress.

Federal funds were poured into the farm and residential mortgage markets through the federal land banks and the Home Owners Loan Corporation. Federal Land Bank and Land Bank Commissioner loans accounted for two-thirds of the farm loans made in 1934; and the portion of the total farm mortgage debt held in these two types of loans increased from 13 percent in January 1932 to 34 percent in January 1935. The Home Owners Loan Corporation (HOLC) used funds advanced by the Reconstruction Finance Corporation to refinance more than a million home mortgage distress cases within a three-year period. These loans were refinanced mostly on a fifteen-year basis at 5 percent interest. In its refinancing operations, this agency scaled the principal indebtedness of its borrowers down by an average of 7 percent.

The operations of the HOLC, the Federal Land Banks, and the Federal Farm Mortgage Corporation came too late to help many distressed borrowers. One and one-half million home mortgages and around one-third of a million farm mortgages found their way through the foreclosure wringer before the decade was out. But these programs did much to stabilize the real estate credit market and instill confidence in the mortgage credit system. In this sense, they prevented complete collapse of the real estate market and established important precedents for future federal action.

Following its earlier action, Congress passed a National Housing Act in 1934. This act created the Federal Housing Administration (FHA) and charged it with the responsibility of providing mortgage insurance on several types of home loans.

Comparable responsibility for guaranteeing loans to World War II veterans was later given to the Veterans Administration (VA) under the Servicemen's Readjustment Act of 1944. Provision of FHA mortgage insurance and VA loan guarantees has made it possible for federal credit agencies to play a strong leadership role in the home mortgage market. Even so, only 22.9 percent of the outstanding mortgage debt on nonfarm one- to four-family residences ($104.9 billion by FHA and $91.1 billion by VA) was covered by FHA insurance or VA guarantees at the end of 1980, compared with 41.8 percent in 1950.

An additional program for facilitating the placement of residential and other mortgages was started with the creation of the Federal National Mortgage Association (FNMA or "Fannie Mae") in 1938. This agency does not make direct loans to property owners but operates rather as a facilitator in absorbing temporary excess supplies of mortgages, which it then places with various mortgage investors. It operates as a quasi-public corporation with the power to issue securities. At the end of 1983, FNMA held $78.3 billion in unplaced mortgages plus an additional $25.1 billion held in a pool to provide backing for FNMA securities.

Two other public agencies are active participants in the secondary mortgage market. The Government National Mortgage Association (GNMA or "Ginnie Mae") was split off from FNMA in 1968 and placed in the Department of Housing and Urban Development, where it operates a tandem mortgage program and also a mortgage pool. GNMA has used its tandem program to buy mortgages in areas where HUD has wanted to stimulate housing construction at less than normal market rates of interest. It has then insured the mortgages and sold them at the market rate, with the government paying the difference between the two rates as a subsidy for housing construction. It also sells interest-bearing securities that are backed by a large pool of mortgages. This pooling arrangement covered $159.9 billion in mortgage loans (11.7 percent of the total outstanding value of residential loans) in 1983.

The Federal Home Loan Mortgage Corporation (FHLMC or "Freddie Mac") was created by Congress in 1970 to conduct secondary mortgage market operations for the Federal Home Loan Bank System. It too can buy and sell mortgages and issue certificates for investments in a mortgage pool. At the end of 1983, it held $7.6 billion in mortgages plus a mortgage pool with an outstanding value of $57.9 billion.

The secondary mortgage market programs initiated by Congress have played prominent roles in stabilizing the mortgage market and increasing its capacity for handling additional mortgage volume. Secondary market operations are also conducted by the Mortgage Guarantee Insurance Corporation, a private operator which buys and sells conventional mortgages and has mortgage pool arrangements. Like the FHA, it also insures home mortgages that meet its standards. With the development of secondary mortgage facilities and the use of mortgage pooling arrangements by GNMA, FHLMC, and FmHA, the share of the total outstanding real estate debt held by federal agencies rose from 2 percent during the 1950s and 1960s to 23.7 percent in 1983.

MORTGAGE ARRANGEMENTS

Most people contemplating the use of mortgage credit have numerous questions concerning the nature and characteristics of mortgages, the amounts of money they can borrow, their probable interest and repayment terms, and possible foreclosure procedures.

Types of Mortgages

Mortgages involve formal agreements or contracts in which owners (the mortgagors) grant certain legal claims against the titles of the properties they offer as security for the repayment of funds advanced to them by lenders (mortgagees). The *standard mortgage* of the past often called quite simply for repayment of the amount of money borrowed within a given time period with interest paid at a given rate on the outstanding unpaid principal. Some mortgages were called *straight-term mortgages* because they required periodic (monthly, quarterly, or annual) payments of interest and a single entire payment of the principal at the end of the mortgage period. Studies of the residential mortgage lending activities of banks and insurance companies during the 1920s show that approximately four-fifths of the loans involved standard mortgages and that 41 percent of the bank loans and 20 percent of the insurance company loans were straight-term loans.[7]

Other loans called for scheduled repayments of principal that would retire all or part of the outstanding debt by the end of the mortgage term. With an arrangement known as the *Springfield plan*, for example, borrowers were expected to make a series of periodic payments representing equal proportionate shares of the total principal plus current interest (see Figure 15-1A).

An important modification of this format known as the *amortized mortgage* came into general use during the 1930s. These mortgages usually provided for longer repayment periods—fifteen to twenty years at first and twenty-five to thirty years with most mortgages in the 1980s, as compared with the five- to ten-year terms used with most earlier standard mortgages. Amortized mortgages call for schedules of periodic payments of uniform size. With these schedules, minimum payments of the same dollar amount are required on each payment date, and each successive payment involves a larger sum for reduction of principal together with a gradually decreasing allowance for current interest (see Figure 15-1B). Long-term amortized mortgages are ordinarily paid in full during the contracted period of the mortgage.

Borrowers are seldom able to secure mortgage loans covering 100 percent of the value of their properties. When more funds are needed than they can secure in a single mortgage, it is possible to secure additional mortgages. When this situation

[7]Cf. C. F. Behrens, *Commercial Bank Activities in Urban Mortgage Financing* (New York: National Bureau of Economic Research, 1950), p. 50; and Raymond J. Saulnier, *Urban Mortgage Lending by Life Insurance Companies* (New York: National Bureau of Economic Research, 1950), pp 130-31.

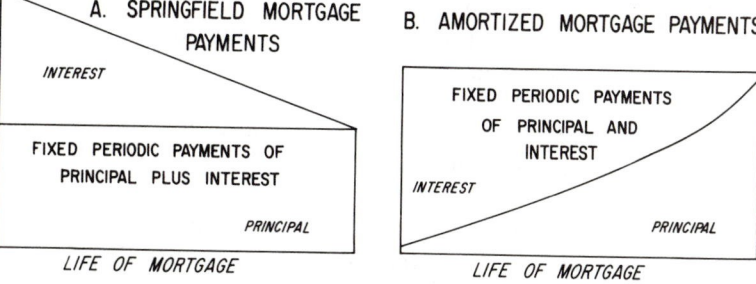

FIGURE 15-1. Comparison of the schedules for periodic mortgage repayments with the Springfield standard mortgage and the amortized mortgage plans.

exists, the first mortgage filed and recorded with the register of deeds is known as the *first mortgage*. First mortgages have top priority in their legal claims against properties offered as security in the event of foreclosure proceedings. They are better security risks than second, third, or other junior mortgages. First mortgages can be refinanced without losing their priorities; and the rights of a first mortgagee can be safeguarded by the terms of junior mortgages.

Mortgages accepted by sellers as part of the consideration in property sales are sometimes described as *purchase-money mortgages*. They may also be called *packaged mortgages* when they are used to finance household and other equipment purchases along with real property. When buyers assume existing mortgages in the acquisition of properties and arrange for extensions or expansions of the outstanding loans at different interest rates, the revised arrangements are sometimes described as *blended* or *wraparound mortgages*. Mortgages that involve two or more properties are called *blanket mortgages*, and those that provide for possible future advances of additional funds to borrowers are called *open-end mortgages*.

High interest rates and a general shortage of housing credit relative to demand during the 1970s and early 1980s prompted considerable interest in new methods of "creative financing." One approach involved *mortgage balloon* arrangements. In a typical example, a borrower would agree to make payments in accordance with a twenty-five-year amortization schedule but with the stipulation that the entire remaining unpaid portion of the loan would become due at the end of a five-year period. Buyers often accepted this arrangement with the expectation that interest rates would decline and that they would be able to refinance their mortgages before their balloon dates. Recent credit market experience seemed to justify this rationale. Many borrowers, however, became victims of their optimistic planning when their mortgages became due, interest rates remained high, supplies of refinancing credit were scarce, and their initial lenders displayed little interest in refinancing their loans.

With uncertainties about trends of interest rates, many investors are reluctant to commit their funds at set interest rates over long mortgage periods. As a substitute, several lenders have offered *adjustable-* or *variable-rate mortgages*. With this arrangement, amortized mortgages are made in the usual manner, with the provision that the rates of interest are adjusted at periodic intervals, sometimes as often as once a month, to keep them in line with some index of current market interest rates. A variation of this approach calls for amortized partial-payment mortgages with payment schedules geared to long repayment periods. The mortgages themselves are scheduled to mature in shorter periods such as five years, with guarantees that they can then be refinanced at the then current rate of interest for an additional five-year period.

Graduated-rate mortgages provide an alternative arrangement which ties the borrowers' periodic payments to their expected incomes over the life of the mortgage. Scheduled payments start out at low levels that borrowers can handle with their present income and then gradually rise as borrowers get older and hopefully will have greater ability to meet their mortgage payments. A borrower's payments during the first several months may actually be less than the amounts needed to cover current interest costs. This situation can involve a period of negative amortization, during which the outstanding unpaid principal increases as underpaid interest is added to the total debt; but the picture changes as the periodic payments increase in size. This approach is geared to ability to pay; but it typically calls for larger total outlays for payments of interest over the full life of the mortgage than would accrue with other types of mortgages. Problems can arise if the borrower's income does not increase as expected. Also, the graduated-rate feature does not lend itself to transfers of mortgages if a new buyer wants to take over an existing mortgage.

Some lenders have proposed the use of *shared equity mortgages*. This approach can involve simple agreements by borrowers and lenders to share any appreciation in property values in accordance with the proportionate equities they hold in the properties over the existence of a mortgage. Equity sharing arrangements can also involve third-party investors who enter into joint-ownership arrangements with borrowers. These investors may advance funds needed for the borrower's down payment and in return claim an equity share in the property to which they hold value appreciation rights and on which they can collect contract rents from the borrower and also claim income tax credits for property value depreciation and possible mortgage and property tax payments.

Indexed mortgages have gained acceptance in some countries that have experienced high rates of inflation. With this approach, both the interest rates and the current value of unpaid principal are adjusted at periodic intervals to reflect the effects of inflation on local cost-of-living indices. Experience in nations such as Chile has shown that assurance of periodic adjustments of this nature are essential if private investment capital is to be attracted to the real estate mortgage market.

Legal Characteristics of Mortgages

A *mortgage* may be defined as "a conveyance of land given to secure the payment of a debt."[8] Like a deed, a mortgage provides for the conveyance of certain property rights to a second party. But unlike a deed, these rights are conveyed only for the

---

[8] Robert Kratovil and Raymond J. Werner, *Real Estate Law*, 8th ed. (Englewood Cliffs, N.J.: Prentice-Hall 1983), p. 264.

purpose of providing security for the payment of a debt; and the mortgage becomes void once the debt is paid in accordance with the terms specified in the mortgage agreement.[9]

From a legal standpoint, mortgages fall into three general classes: regular mortgages, deeds of trust, and equitable mortgages. Regular mortgages call for the conveyance of a property from the borrower to the lender and for the voiding of the mortgage once the debt is paid in full. *Deeds of trust* convey the mortgage rights to a third party, who holds them in trust for the benefit of the lender or lenders until the debt is paid. Certain other arrangements such as faulty or incomplete mortgages, the giving of a deed as security for a loan, or the conditional sale of land are treated by the courts as equitable mortgages if there is written evidence that the parties intended that real estate be held as security for the payment of a debt.

Mortgages represent important legal documents and should be executed with considerable care. Every mortgage provides either specifically or by implication for the conveyance of the mortgagor's property to the lender and the voiding of this conveyance once the debt is paid. In addition, it should show the date of agreement, identify the parties to the mortgage, describe the property covered by the mortgage, indicate the amount of the debt and any arrangements for future advances of credit, specify the interest rate and the time and amount of payments, provide for the mortgagor's payment of taxes and property insurance, and itemize such other conditions as circumstances may dictate. Like a deed, the mortgage must bear the signatures of the mortgagors. In some states, it must also carry their official seal, the signatures of witnesses, or an acknowledgement made before a notary public or other official. Also like a deed, a mortgage must be delivered to the mortgagee before it becomes effective.

At the time a mortgage is signed the borrower may also be required to give the lender a promissory note, known as a *mortgage note* or *mortgage bond*. This requirement has been dispensed with in several states on the assumption that the mortgage provides sufficient security for the credit granted. Where mortgage notes are used, they make the mortgagor personally liable for the payment of the mortgage debt. This liability follows the mortgagor; and if the mortgage is ever foreclosed, the mortgagee can secure a *deficiency judgment* holding the mortgagor liable for any deficiency between the foreclosure sale price and the amount of the mortgage. Deficiency judgments were sought by numerous lenders who foreclosed mortgages during the 1930s. These judgments were often the subject of considerable bitter-

---

[9] As indicated in Chapter 12, some states hold to a *title theory* approach, which assumes that a mortgage actually vests the mortgagee with a legal title to the mortgaged property. Others hold to a *lien theory* approach, which treats a mortgage as a mere lien to secure the loan made to the mortgagor. Several states take intermediate positions between these two approaches. The legal concept of a mortgage accepted in a state has an important effect on the rights a mortgagee has to possess or take rent from mortgaged property if the mortgagor defaults on the payments. Even in the title theory states, the courts have limited the rights once exercised by mortgagees to a shadow of their pre-seventeenth-century importance. Mortgagees in these states still enjoy greater rights to the possession and rents of mortgaged properties than is the case in the lien theory states. Cf. Robert Kratovil and Raymond J. Werner, *Modern Mortgage Law and Practice*, 2nd ed., (Englewood Cliffs, N.J.: Prentice-Hall, Inc., 1981), pp. 32–33.

ness, particularly when mortgagees, as the only bidders at foreclosure sales, bid properties to themselves at prices below the amount of the unpaid mortgage debt. Some states abolished deficiency judgments as the result of this experience. Several others have limited them to the difference between the mortgage debt and the fair market values established by courts.

Potential borrowers are expected to turn their abstracts of title over to lenders for examination before loans are granted and for holding throughout the loan period. Many lenders also require credit reports, a title examination, a property survey, and title insurance. The costs of these services are normally charged to the mortgagor along with the cost of preparing and filing the mortgage and a possible charge for placing the mortgage. These closing costs may be nominal, when a lender's knowledge of the circumstances makes some charges unnecessary or when the lender absorbs part of the costs. Yet they can and sometimes do add up to substantial charges, which reduce the actual funds borrowers receive without adding anything "to the borrower's equity, the seller's receipts, or to the lender's net income."[10]

Once a mortgage is signed and delivered, it should be filed with the local register of deeds in the same manner as a deed. Early filing can prevent other possible mortgages from acquiring a prior claim status. When the mortgage is paid off, the lender normally issues a mortgage *release*, which should also be filed as legal evidence that the mortgage has been extinguished. Most states have legislation making promissory notes and mortgages nonenforceable some specified number of years (twenty or thirty years in most states) after their maturity date. When a mortgage has not been extended and no payments have been made beyond the maturity date, it can be *barred by limitations* in this manner; and its cloud against a land title can be removed even though no mortgage release has been filed.

Borrowers are free to sell or otherwise dispose of their rights in mortgaged properties. When supplies of real estate credit are readily available, borrowers who are selling typically pay off their mortgages at the time their properties are transferred to new owners while the new owners arrange their own financing. But when new credit is hard to get or mortgage interest rates have risen to higher levels, new owners often have strong economic incentives for assuming existing mortgages. Borrowers can also secure junior mortgages, and lenders can sell or assign their mortgage rights to other parties.

With high mortgage interest rates such as those experienced during the early 1980s, buyers find it to their advantage to seek sale properties with assumable long-term mortgages calling for lower than current rates of interest. This practice works to the disadvantage of the lenders and has caused many lending agencies to include alienation and due-on-sale clauses in their mortgages. Some states have tried to prevent the enforcement of these provisions, but the U.S. Supreme Court held in *Fidelity Federal Savings and Loan Association* v. *de la Cuesta* that federally chartered

[10] Ernest M. Fisher and Robert M. Fisher, *Urban Real Estate* (New York: Henry Holt, 1954), p. 386.

financial institutions can require that mortgages mature and be payable in full at the time borrowers sell or convey their property rights to others.[11]

Size of Mortgage Loans

Borrowers are frequently interested in securing loans of maximum size on their properties. Most private lenders in turn favor loan policies that are conservative enough to provide a safe margin of security on their loans and yet liberal enough to ensure they will get what they consider their fair share of the mortgage market. Private noninstitutional lenders can lend any sum they wish on mortgages. Institutional lenders, on the other hand, are usually bound by definite house rules, government regulations, or legislative authorizations, which specify the maximum loan-value ratios they can use in granting long-term real estate credit.

Throughout the 1920s it was common policy for many lending agencies to limit loans to a maximum of 40 to 50 percent of the appraised loan value of a borrower's property. A property with a current market value of $9,500, for example, might have had an appraised loan value of $8,000—a conservative estimate of its safe market value throughout the loan repayment period—and thus have qualified for a 50 percent loan of $4,000. This arrangement often provided borrowers with far less credit than they needed and invited the use of junior mortgages as supplemental sources of credit.

Since the late 1930s, higher loan value ratios have come into general use. Most lenders will now advance loans of up to 80 percent of the appraised values of properties, and 90 percent loans are sometimes available although usually at higher interest rates. The average loan-value ratios for conventional mortgages on new homes in the United States ranged from 71.7 to 76.8 percent between 1965 and 1982.

FHA pioneered the concept of high loan-value ratios on housing loans and in so doing extended real estate credit opportunities to numerous borrowers who otherwise would have lacked the necessary down payments for acquiring properties. FHA, VA, and FmHA have used 100 percent loans on various occasions to facilitate ownership acquisition. Their lending practices vary with changes in policy but usually permit higher loan-value ratios on loans up to certain total value levels than are available from conventional lending sources.

Interest Rates and Repayment Terms

Borrowers have important reasons for being concerned about the repayment terms provided by their mortgages. They enjoy a more favorable position with low interest rates and no giving of points than with the reverse situation, and they often prefer to operate with mortgages that can be paid off over longer rather than relatively short periods of time.

[11] *Fidelity Federal Savings and Loan Association* v. *de la Cuesta* 102 U.S. Sup. Ct. 3014 (1982).

Interest rates. Mortgage interest rates are determined in large measure by (1) the interaction of supply and demand conditions, (2) lender perceptions of the relative risks associated with loans, and (3) government policies and regulations. With free market conditions and the typical real estate loan, interest rates tend to be low to moderate when plentiful supplies of mortgage money are available to meet current demands. They are usually higher when the opposite circumstances apply and when lenders feel that loans involve above-average risks of nonrepayment. Government policies can affect interest rates by increasing or decreasing the supplies of money available for loans, by competing for these funds during periods of deficit financing, by prescribing minimum or maximum interest rates, and by prohibiting lender discrimination against specified classes of loans that may involve above-average risks.

During the 1940s and 1950s, real estate mortgage interest rates were low by 1980 standards. The Federal Reserve Bank discount rate was less than 2 percent during the 1940s and early 1950s, mortgage funds were available at 4.5 and 5.0 percent interest, and subsidized rates of as little as 3.5 percent were paid on federal land bank loans. Interest rates rose slightly during the late 1950s and early 1960s. By 1965 the Federal Reserve Bank discount rate had risen to 4.5 percent, but adequate credit for housing and other real estate loans was still available at 5.5 to 6.0 percent interest. The story of what happened between 1965 and 1982 is summarized in Table 15-3.

Inflationary pressures and high demands on the available supplies of investment capital led the Federal Reserve banks to raise their discount rates in 1969, 1973–1974, and again in 1979–1981. These pressures prompted an increase in real estate mortgage interest rates in 1969–1970. Interest rates declined slightly in 1971–1972 and then rose steadily to the double-digit level in 1979 and to averages in excess of 14 percent on first mortgages with conventional loans on residential properties in 1981–1982. Similar increases were experienced with mortgages written for commercial and farming properties. A shortage of real estate credit relative to demand also made it possible for lenders to increase their net returns by discounting their loans. HUD's calculations on average net yields shows that while discounting added only slightly to the cost of borrowing money during the early 1970s, it became a highly significant factor around 1980, when it pushed the effective average yield rates on conventional first mortgages on residential properties up to 16.52 percent in 1981 and 15.79 percent in 1982.

Giving mortgage points. *Interest* is the economic return paid for the use of capital borrowed from others. Lenders have known for a long time, however, that the return to capital can be increased by use of a discounting process known as the "giving of mortgage points." A *mortgage point* is equal to 1 percent of the total repayment value of the mortgage. When a borrower agrees to give two points on a mortgage for $10,000, it is understood that the mortgagor is accepting a commitment to repay $10,000 plus interest over the life of the mortgage for a current advance of $9,800.

Two functions are filled by the giving of points. Lenders receive an immediate

TABLE 15-3. Trends with Selected Interest Rates Affecting Real Estate Credit, United States, 1965-1983

YEAR	NEW YORK FEDERAL RESERVE BANK DISCOUNT RATE AT END OF YEAR	AVERAGE INTEREST RATES ON CONVENTIONAL FIRST MORTGAGES ON:		ADDED FEES PAID WITH CONVENTIONAL FIRST MORTGAGES	HUD EFFECTIVE YIELD RATE ON MORTGAGES	TYPICAL INTEREST RATES ON FARM LOANS
		New houses	Existing houses			
1965	4.50	5.74	5.87	0.54	—	—
1966	4.50	6.14	6.30	0.71	—	—
1967	4.50	6.33	6.40	0.81	—	—
1968	5.50	6.83	6.90	0.89	7.12	—
1969	6.00	7.66	7.68	0.91	7.99	7.61
1970	5.50	8.27	8.20	1.03	8.52	8.37
1971	4.50	7.59	7.54	0.87	7.75	8.12
1972	4.50	7.45	7.38	0.88	7.64	7.91
1973	7.50	7.78	7.86	1.11	8.30	8.09
1974	7.75	8.72	8.84	1.30	9.22	8.63
1975	6.00	8.75	9.01	1.54	9.10	8.93
1976	5.25	8.76	8.92	1.44	8.95	9.13
1977	6.00	8.80	8.83	1.33	8.95	9.19
1978	9.50	9.30	9.36	1.39	8.99	9.39
1979	12.00	10.48	10.66	1.66	11.15	10.70
1980	12.77	12.25	12.58	2.09	13.95	13.41
1981	13.41	14.13	14.51	2.67	16.52	16.24
1982	11.02	14.49	14.78	2.95	15.79	16.71
1983	8.50	12.11	12.29	2.40	13.43	14.02

Sources: Data on Federal Reserve Bank discount rates and average interest rates on first mortgages for conventional housing loans from U.S. Department of Commerce, *Business Statistics, 1979* and more recent issues; data on added fees paid with first mortgages with conventional housing loans and Department of Housing and Urban Development's calculation of effective average yields on conventional housing mortgages from *Federal Reserve Bulletin, 1978* and current issues; data on interest rates on typical farm real estate loans from *Agricultural Finance Databook*, Board of Governors of the Federal Reserve System, E15 (125), June 1983, p. 47.

book value return from their operations when their borrowers give points. This return along with the borrower's loan origination or mortgage placement fee is typically deducted from the funds the borrower receives rather than paid for separately.

A second function came as an outgrowth of the rise in interest rates. The FHA and VA specified maximum interest rates for the loans they serviced. These maxima were often below the current market rates, but mortgage investors were willing to take these loans, provided the borrowers would give enough points to bring their effective interest rates up to competitive market levels. The practice spread, particularly with the rise in interest rates during the 1970s. It became widely accepted even with the conventional loans made by private lending agencies because lenders saw advantages in quoting lower and seemingly more competitive interest rates with the taking of points than in quoting higher, all-inclusive rates. Table 15-3 reports average added loan fees equal to 2.95 points with the average conventional loan on residential properties in 1982.[12] The giving of points complicates the calculation of the actual costs of using borrowed capital. Under the terms of the Federal Consumer Credit Protection Act (Truth in Lending Act) of 1969, lenders are required to inform borrowers of the annual percentage rates they are actually required to pay on their loans.

Repayment provisions. Mortgages are always made for specified time periods, which may range up to forty years or more. Many typical farm and home mortgages in the United States during the 1920s called for five-year straight-term arrangements. Borrowers could pay off their mortgages at the end of these periods, but in many instances they simply extended, renewed, or refinanced them for another five-year period.

Longer-term mortgages with provisions for amortized repayment schedules were offered by federal credit agencies during the Great Depression, and with this innovation, 15- and 20-year amortized mortgage terms soon became common with conventional mortgages.[13] These terms have since been extended to permit terms ranging from 20 to 30 years. Average maturity periods of 25.1 and 28.2 years applied

[12] Points were often given during the 1970s and 1980s in "buying down" interest rates to the maximum levels permitted under FHA, VA, and state usury law regulations. With FHA and VA ceilings on the number of points borrowers could give, sellers and others also occasionally gave points to help buyers qualify for loans. To a potential buyer who lacked $3,000 in necessary down payment funds to qualify for a $60,000 mortgage and FHA's 1982 limit of 2.5 borrower points on new construction, for example, an anxious seller might have given 5 seller points to ensure the sale of a house. The seller would then have received $3,000 less in payment from the mortgage, while the buyer assumed responsibility for repaying a mortgage of $60,000 on which only $55,500 in loan funds (92.5 percent) had been advanced.

[13] The proportion of mortgages on residential properties with fully amortized repayment provisions increased from 15 to 68 percent between 1920 and the middle 1940s with bank loans and from 20 to 95 percent with insurance company loans. Cf. Behrens, *op. cit.*; and Saulnier, *op. cit.*

with conventional first mortgages on residential properties in 1970 and 1980, respectively.

Prepayment privileges. Mortgages typically allow borrowers to pay off all or part of their outstanding debt in advance of the final payment dates. Borrowers often enjoy a maximum of freedom in this regard to pay their mortgages in full "on or before" a given due date and to pay a specified sum "or more" each pay period. Lenders may require that all advance or surplus payments be made in specified amounts that fit in with their bookkeeping procedures. Most mortgages that authorize prepayment privileges also provide for specified penalties if borrowers pay the entire debt off in less than a given time period. These provisions safeguard the lender's interests by guaranteeing a minimum return for the initial trouble required for processing and servicing a loan.

Flexible-payment arrangements can be used to adapt mortgage repayment schedules to the changing fortunes of borrowers.[14] The federal land banks and a few other major lending agencies have developed reserve payment plans that allow borrowers to make advance payments, which are held in reserve for application during possible later hardship periods when borrowers may find it difficult to meet their mortgage commitments.

Foreclosure Procedures

Every mortgage is considered a good mortgage at the time it is made. Yet adverse circumstances are always possible, and some borrowers default on their mortgage obligations even in prosperous times. When defaults occur, lenders often find that protection of their interests requires that they take legal steps to extinguish the borrower's rights by foreclosing the mortgage. Most mortgages contain acceleration clauses, which provide that the entire principal becomes immediately due and payable if the borrower defaults on any one payment. Without this provision, a mortgagee could seek foreclosure action only on the successive payments as they became due.[15]

Mortgage foreclosure procedures differ from state to state. In every state, however, borrowers have an *equitable right of redemption*, which makes it possible for them to redeem their land by paying off the mortgage after the due date fixed in the mortgage. This equitable right of redemption continues until the mortgage is legally foreclosed.

Foreclosure typically starts with the mortgagee filing a foreclosure suit in the

---

[14] Cf. Aaron G. Nelson and William G. Murray, *Agricultural Finance*, 5th ed. (Ames: Iowa State University Press, 1967), pp. 170–73, 184: *Improving Land Credit Arrangements in the Midwest*, Purdue Agricultural Experiment Station Bulletin No. 551, 1950; and Virgil L. Hurlburt, "Economic Effects of Flexible Land Credit Arrangements," *Journal of Farm Economics*, 35 (February 1953), 110–21.

[15] Cf. Kratovil and Werner, *op. cit.*, p. 229.

local courts against the borrower and all persons having junior claims against the mortgaged property.[16] This suit is heard by a court; and if decided in the mortgagee's favor, a decree or judgment of foreclosure is issued. This decree verifies the borrower's right to foreclose, indicates the amount due on the mortgage, describes the property that is to be sold, and names the person who is to sell the property. Public sale of the mortgaged property—usually at public auction—is then advertised together with information regarding the time, place, and terms of sale. Mortgagees are allowed to bid at the sale and, in so doing, can bid all or part of the amount of the outstanding mortgage debt without putting up any money. This arrangement makes it possible for mortgagees to bid foreclosed properties to themselves.

Following a court review and confirmation of the sale, the highest bidder receives a deed to the mortgaged property in some states. In others, the buyer receives a certificate of sale, which ripens into ownership if the mortgagor and other claimants fail to redeem their rights by paying off the debt within a *statutory period of redemption.*[17] This statutory period ranges from two months in some states to as much as two years in others.

Foreclosure proceedings ordinarily take several months and usually involve considerable time, cost, and trouble for both parties. In some instances, they give borrowers opportunities to exploit or "milk" properties before they relinquish their possession rights. During the 1930s, many borrowers voluntarily signed their rights over to mortgagees rather than go through the foreclosure process. Many others

---

[16] As an alternative to the usual procedure used in foreclosing mortgages, some states permit foreclosure of mortgages by exercise of a power of sale without resort to court proceedings. Where this procedure is used, mortgages and trust deeds spell out the conditions of default under which the mortgagee (or trustee) or a public official, has the right to sell the mortgaged property. The sale must usually be advertised, and notice of the sale must be given to the mortgagor. Without permission granted by the mortgage, the mortgagee (or trustee) is normally barred from purchasing the property. But this permission is usually given, and mortgagees (or trustees) are thus able to make the highest bid and have deeds executed to themselves.

Connecticut and Vermont use a somewhat different procedure known as *strict foreclosure*, under which a mortgagee files a foreclosure suit to secure a court decree or judgment of foreclosure. This decree gives mortgagors a specified period of time (usually six months or less) in which they can redeem their property. If not redeemed within this period, the title vests in the mortgagee without the formality of a public sale. Two other arrangements known as *foreclosure by writ of entry* and *foreclosure by peaceful entry and possession* are used in some New England states to vest titles in the mortgagee without formal sales.

[17] Care should be taken to distinguish between the concepts of *equitable redemption* and *statutory redemption*. The borrower's equitable right of redemption—right to redeem mortgaged property by paying the mortgage debt in full with interest after the due date when it supposedly forfeited to the mortgagee—was recognized by the English courts at an early date and became a general rule after 1625. The foreclosure process was established as a means of formally extinguishing this right. The *statutory right of redemption* is a concept of more recent vintage. It was authorized in several states during the past century and provides an additional redemption period (usually a year) after the sale of the property at a foreclosure sale during which borrowers can still redeem their ownership. Cf. Kratovil and Werner, *op. cit.*, chaps. 20 and 23.

worked out arrangements with their lenders for a scaling down of debt loads and for refinancing their loans on a sound payment basis.

Debt moratoria legislation. The flood of mortgage defaults experienced during the early 1930s prompted many states to enact debt moratoria legislation. Approximately two-thirds of the states passed mortgage relief laws, which provided for postponements of foreclosure proceedings, extensions of mortgage redemption periods, or both. In Iowa, for example, the legislature authorized farm mortgage debtors to petition the courts for relief from foreclosure proceedings "when and where the default or inability of such party . . . to pay or perform is mostly due to . . . drought, flood, heat, hail, storm, or other climatic conditions or by reason of the infestation of pests . . . , or when the governor . . . by reason of a depression shall have . . . declared a state of emergency to exist.[18]

Most of these laws postponed foreclosure action for one to two years, but many were renewed and extended several times. Provisions were often included requiring debtors who took advantage of the moratorium provisions to pay their creditors a sum equal to the fair rental value of their properties. The constitutionality of these laws was upheld in the *Minnesota Moratorium Act* case in 1934.[19] In its decision in this case,

> The United States Supreme Court emphasized the restriction of the relief to the period of economic emergency, the requirement of the debtor to apply for relief, the discretion on the part of the court to grant it, and the necessity of fair compensation to the creditor in the form of a rental payment.[20]

Comparable action for the nation was accomplished by amendment of the Federal Bankruptcy Act in 1933 and passage of the Frazier-Lemke Acts of 1934 and 1935.

LAND CONTRACT ARRANGEMENTS

As mentioned earlier, mortgages are not the only sources of real estate credit. *Land contracts*, or *installment* or *purchase contracts* as they are sometimes called, provide an important source of real estate credit in many areas. With the usual examples of this arrangement, property owners help finance the sale of their properties by accepting contractual arrangements under which buyers make down payments for the properties they purchase and agree to pay the balance of the purchase price in periodic payments.[21] Once the contracts go into effect, buyers take

[18] 45 Iowa General Assembly, ch. 179 (1933).

[19] *Home Building and Loan Association* v. *Blaisdell*, 290 U.S. 398 (1934).

[20] *Improving Land Credit Arrangements in the Midwest*, p. 21.

[21] For other discussions of land contracts, cf. Marshall Harris and N. William Hines, *Installment Land Contracts in Iowa*, Agricultural Law Center Monograph No. 5 (Iowa City: University of Iowa College of Law, 1965); R. Vern Elefson and Philip M. Raup, *Financing Farm Transfers with Land Contracts*, Minnesota Agricultural Experiment Station Bulletin No. 454, 1961; and Wilfred H. Pine and Ronald K. Badger, *Buying and Selling Farms by Contract in Kansas*, Kansas State University Circular No. 390, 1963.

possession and operate the properties in much the same manner as mortgaged owners. Sellers or their assignees, however, retain legal title until all of the payments are made or until the buyers refinance their purchases by converting their contracts to mortgages.

Contractual arrangements of this type are used quite extensively in many sections of the United States in the sale of rural, suburban, and urban properties while they receive little use in other areas.

Two types of land contracts should be recognized: (1) contracts that involve strictly commercial credit arrangements, and (2) contracts between relatives or close friends. With commercial land contracts, buyers ordinarily make down payments equal to at least 10 percent of the purchase price of their properties. They usually pay higher rates of interest than would apply with a mortgage.[22] They often pay somewhat higher purchase prices for their property than would apply if they could pay cash or if they could use purchase mortgage arrangements. Sellers or their assignees also tend to take prompt legal action to void contracts and repossess properties if buyers default on their commitments.

Buyers use land contracts because they can acquire property with smaller down payments than are required by conventional mortgage lenders. But this advantage can be balanced by the higher interest rate buyers ordinarily pay and by the prospect that their rights will be cut off if they default on any payments. These conditions make it advantageous for buyers to shift from land contracts to mortgage arrangements as soon as they can build their equities up to the levels required by mortgage lenders. Many land contracts either provide or are made with the understanding that sellers will convey legal title to the buyer and take a mortgage in exchange for the contract once the buyer's equity is built up to some given level.

Some sellers hold their land contracts for investment purposes, but many sell or assign them to others. In this respect, land contracts are bought and sold in many communities in much the same way as mortgages. Depending on the nature of this market, sellers can sometimes command a premium for their contracts, but more often find that they must discount them by selling at figures below their face value. This discounting process provides justification for the higher purchase prices often associated with properties sold under land contracts.

Many of the characteristics associated with commercial land contracts do not apply with the land contracts used by relatives and close friends. Parents sometimes use contract arrangements to transfer properties to their children; and property owners occasionally use them to help a neighbor or a friend. With these conditions, down payments may range from practically nothing to high percentages of the purchase price. Interest rates are often lower than the going mortgage rate and may be purely nominal. Defaults are handled on a mutually favorable basis, and buyers usually enjoy as much security as they would have with mortgages.

---

[22] During the tight credit years of the 1970s and 1980s, many sellers accepted land contracts as the only practical alternative they could use in selling their properties. Interest rate ceilings imposed by state usury laws prevented sellers in some states from charging interest rates even as high as those being paid on current new mortgages issued by financial institutions.

Like mortgages and deeds, land contracts should be registered and care should be taken to check the validity of the seller's title at the time the contract is made. Surprisingly enough, both of these procedures are frequently ignored. The legal search of title is frequently postponed until the actual transfer of title takes place. This delay can give rise to legal problems, particularly if a seller is unable to convey a merchantable title. In like fashion, most buyers make no effort to register their contracts. Their attitudes are explained by feelings of reasonable security, a general lack of pressure for registration, and the fear that registration may complicate possible plans for selling their contracts.

ADMINISTRATION OF REAL ESTATE CREDIT

Important problems are associated with the administration and management of credit. Some of these are of primary concern to lenders; others are of immediate interest to borrowers. Considerable emphasis has been given to the standards lenders should use in their lending operations, and rules have been expounded for a borrower's wise use of credit.[23] Emphasis is given in this brief discussion to some of the leading factors lenders and borrowers should consider in their use of real estate credit.

Considerations Affecting Lending Policies

Mortgage investors are primarily interested in their prospects for receiving a safe and adequate return on their investments. They seldom have any interest in acquiring property ownership via the foreclosure route. For them, real estate credit provides a means for securing a reasonable and reliable return on their investment funds. With this basic concern, they are inclined to emphasize the three C's of credit—capacity, collateral, and character—or the three R's—returns, repayment capacity, and risk-bearing ability.[24]

Security-minded lenders carefully consider the ability and capacity of both the borrower and the mortgaged property to repay the loan. Lenders make inquiries regarding the borrower's personal character, employment status, credit history, and rating as a credit risk. With mortgages on residential properties, information is sought concerning the borrower's age, health, family responsibilities, income, and employment status and security. Comparable information is needed with loans on commercial and farm properties.

A borrower's personal capacity to pay is frequently affected by factors such as

---

[23] Stanley L. McMichael and Paul T. O'Keefe, *How to Finance Real Estate*, 3rd ed. (Englewood Cliffs, N.J.: Prentice-Hall, 1967), chaps. 18 and 19; Sherman J. Maisel, *Financing Real Estate: Principles and Practices* (New York: McGraw-Hill, 1965), chaps. 6–8; and Nelson and Murray, *op. cit.*, Part I.

[24] Cf. Nelson and Murray, *op. cit.*, pp. 505–10; and William H. Husband and Frank R. Anderson, *Real Estate*, 3rd ed. (Homewood, Ill.: Richard D. Irwin, 1960), p. 354.

advancing age, ill health, accidents, and death. Lenders accordingly find that they must look beyond the borrower to the debt-paying capacity of the mortgaged property. In this respect, they are concerned with the earning capacity and productivity of farm and business properties, with their current market value, and with their probable future values. They are concerned with the repayment capacity of residential properties, the desirability of their layouts, the neighborhoods and communities in which they are located, and their probable future rental and market values.

Since lenders are always concerned with the risk factor, it is common practice to limit loans to some portion of their estimate of the probable future market value of mortgaged properties. They often insist that borrowers follow repayment schedules that ensure full payment of loans by the end of their mortgage terms.

Most lenders are concerned with their alternative investment opportunities and with the risk factor. The amount of money that is available for real estate loans reflects both the level of savings that can be used for investment purposes and market demands for the use of these funds. It also reflects the confidence prospective investors have in real estate loans and the manner in which they rate them relative to other investment opportunities. Plentiful supplies of mortgage money are usually forthcoming when the economy is healthy, most workers are employed, and people enjoy a sense of economic security. Low incomes, high taxes, rising living costs, widespread unemployment, and deficit financing of governmental expenditures can have depressing effects on the supplies of money available for real estate mortgages. Opportunities for higher returns from investments in alternative ventures and expectations of runaway inflation can have similar effects in siphoning off funds that could be available for real estate loans.

Lenders are naturally interested in securing as much net return as they can get from their investments commensurate with what they consider sound investment planning. In choosing investments, lenders always encounter the problems of risk and uncertainty. Higher rates of return are frequently available with more risky alternatives. Investors can seldom enjoy high rates of return with maximum security. Instead, they must choose between the prospect of possible higher returns with more risk or less return with more security. Regardless of the alternative chosen, they will ordinarily insist on receipt of the prevailing market rate of return as a minimum; and, if they operate in a lender's market, they may bargain for additional advantages in the form of higher interest rates, bonuses, or some discounting of the loan.

Savings and loan associations, commercial banks, and several other lending agencies ordinarily place and service their own loans. When plentiful supplies of mortgage money are available, they often place loans without charging loan origination or placement fees. When the credit market is tight, they usually add charges for these services to the costs borrowers pay for securing loans. Mortgage placement fees are a normal cost of business when one secures a mortgage from a mortgage broker. Like real estate brokers, these operators make it their business to bring borrowers and lenders together, and they often do much of the initial paperwork for lenders. Commissions, placement fees, or finder's fees are charged for their services and are ordinarily passed on to borrowers, though they may be paid entirely or in part by lenders during periods of plentiful credit supply.

Factors Borrowers Should Consider

Real estate credit can be a great asset to borrowers if they use it wisely to acquire property or to improve their productivity and earning capacity. But it can also prove a detriment when it is poorly managed. As Thomas Nixon Carver once observed, "There is no magic about credit. It is a powerful agency for good in the hands of those who know how to use it. So is a buzz saw. They are about equally dangerous in the hands of those who do not understand them."[25]

There is no simple test of wisdom in the use of credit. From a production view, real estate credit is put to its best use when it is used (1) to provide operators with access to land resources they can use to advantage in their business operations, (2) to supply needed inputs that add to the overall efficiency and productivity of a borrower's enterprise, or (3) to provide capital for alternative uses that will add to the operator's income and overall productivity. Credit plays a second significant role in providing for consumer satisfactions and for the fulfillment of consumer goals. The use of credit to provide family housing is a common example. With some individuals, credit also has a negative value in the sense that they derive important satisfactions from its nonuse. The joint impact of these three sets of considerations on individual decisions suggests that one should use credit to the point at which the marginal utilities associated with its use in production equal the marginal utilities associated with its use for consumption purposes and the marginal utilities or satisfactions associated with its nonuse.

People must decide for themselves how much credit they should use. Conservative operators of the old school often make sparing use of credit, feel uncomfortable with debt, and place high values on freedom from debt. Most present-day operators view credit as a legitimate and often desirable means to an end. For them, real estate credit plays an essential role in helping one to acquire access to property and in providing needed operating funds, but it should be used sparingly for nonproductive and highly speculative ventures.

Long periods of relative prosperity, continuing inflation, and freedom from deflation encourage free and easy attitudes about credit. Investors would have been wise to use credit to gain maximum leverage in the acquisition of real estate almost any time during the 1945–1980 period. Many held back, however, because they lacked complete faith in the future or because they or their advisors remembered the mortgage foreclosures and defaults of earlier periods such as the early 1930s.

Figure 15-2 provides an illustration of what a borrower with a 100 percent mortgage for $10,000 running for twenty-five years at 9 percent interest could

[25] Thomas N. Carver, *How to Use Farm Credit*, U.S. Department of Agriculture Farmer's Bulletin 593, 1914, p. 1.

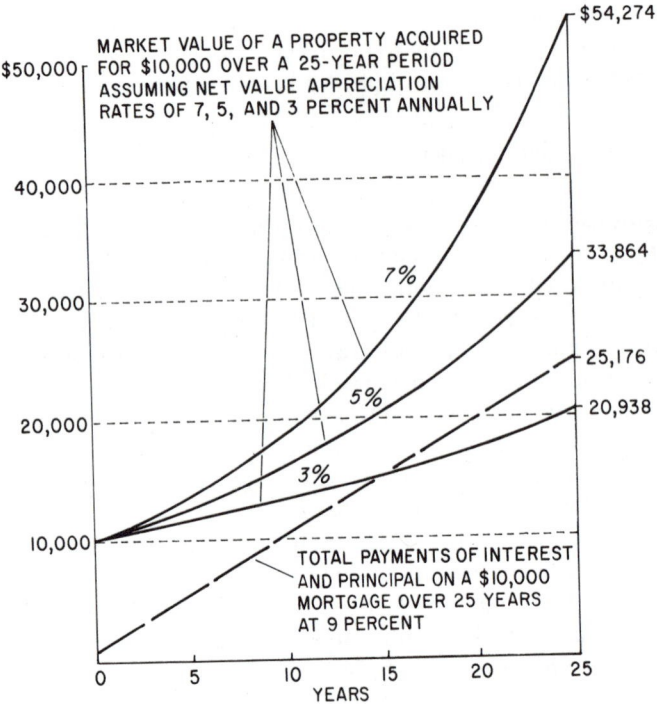

MARKET VALUE OF A PROPERTY ACQUIRED
FOR $10,000 OVER A 25-YEAR PERIOD
ASSUMING NET VALUE APPRECIATION
RATES OF 7, 5, AND 3 PERCENT ANNUALLY

TOTAL PAYMENTS OF INTEREST
AND PRINCIPAL ON A $10,000
MORTGAGE OVER 25 YEARS
AT 9 PERCENT

FIGURE 15-2.   Comparisons of a homeowner's payments of principal and interest on a $10,000 mortgage over 25 years with the value appreciation of a $10,000 property investment over the same period, assuming three different appreciation rates.

accomplish with three alternative assumptions as to future appreciation rates. Over the twenty-five years of the mortgage, the owner would pay out $25,176 for interest and repayment of principal. With the assumption of an average annual inflation rate of 5 percent and normal property depreciation of 2 percent (a net value appreciation rate of 3 percent), the property would have a market value of $20,938, or 83.8 percent of the operator's payments of principal and interest at the end of the twenty-five years. Assumption of an average annual inflation rate of 7 percent and a net value appreciation rate of 5 percent would result in a property market value of $33,864, or about 35 percent more than the owner's mortgage payments over the twenty-five years. With an average inflation rate of 9 percent and a net value appreciation rate of 7 percent, the market value would rise to $54,274, or more than double the owner's payments for interest and principal.

Hindsight comparisons indicate that farm properties appreciated at an average annual rate in excess of 8 percent and residential properties at rates of more than 5 percent in the twenty-five years from 1955 to 1980. With a duplication of this experience, one would be foolish not to use real estate credit at rates of up to 10 percent to acquire property ownership. But will this experience be duplicated during the next quarter of a century? Or might optimistic borrowers be taking unreasonable risks with the equity funds they have? One cannot assert with certainty what the future will bring. It is obvious, however, that operators can use the leverage opportunities associated with uses of real estate credit to secure and control significant property holdings and to expand the size and productivity of their business operations. These opportunities appeal to operators who are willing to take risks. They seem more of a "sure thing" with each continued year of inflation and relative prosperity. In their decisions to use credit, however, people should always remember that every time one uses real estate credit, one's property is pledged as security for the loan. One's equity in property is always held in the balance. If by good management or good luck one succeeds in increasing his or her income, the added income can be used to protect one's equity. If the reverse situation holds, one's equity can be wiped out.

Good credit management usually resolves itself into a problem of good business, farm, or home management. An operator or property owner must learn to recognize when credit is needed, when it can be used to advantage, and when it should not be used. In this process, one must always allow for risks and uncertainties. Many home, farm, and business owners feel justifiably safe in their use of credit as long as they are able to provide for their families. But they have no guarantees against accidents, ill health, or untimely death. This factor makes it desirable for most operators to use various types of term, life and health, or mortgage insurance programs to protect their families against possible mortgage default or foreclosure problems in the event of their death or incapacity.

Comparable insurance is not available to protect families against possible business reverses. Here borrowers must take definite chances. If they are willing to take risks and if their period of credit use coincides with a period of general prosperity, they can often carry a substantial mortgage debt to advantage. Comparable mortgage loans under less favorable conditions can easily lead to disaster both for the borrowers and for their families. Many borrowers find it advisable to follow a safer middle course by limiting their debt commitments at all times to the maximum loads they can reasonably carry under adverse conditions.

—SELECTED READINGS

Beaton, William R., *Real Estate Finance*, 2nd ed. (Englewood Cliffs, N.J.: Prentice-Hall, 1982).

Case, Frederick E., and **John M. Clapp,** *Real Estate Financing*, 2nd ed. (New York: John Wiley, 1978).

Fisher, Ernest M., and **Robert M. Fisher,** *Urban Real Estate* (New York: Holt, Rinehart & Winston, 1954), chap. 15.

Harwood, Bruce, *Real Estate Principles*, 2nd ed. (Reston, Va.: Reston, 1980), chaps. 9–12.

Hoaglund, Henry E., and Leo D. Stone, *Real Estate Finance*, 6th ed. (Homewood, Ill.: Richard D. Irwin, 1977).

Kratovil, Robert, and Raymond J. Werner, *Real Estate Law*, 8th ed. (Englewood Cliffs, N.J.: Prentice-Hall, 1983), chaps. 15-24.

Nelson, Aaron G., and William G. Murray, *Agricultural Finance*, 5th ed. (Ames: Iowa State University Press, 1967).

Shenkel, William M., *Modern Real Estate Principles* (Dallas, Tex.: Business Publications, Inc., 1977), chaps. 12-14.

Unger, Maurice A., and Ronald W. Melicher, *Real Estate Finance* (Cincinnati, Ohio: South Western Publishing Co., 1978).

Wiedemer, John P., *Real Estate Finance*, 4th ed. (Reston, Va.: Reston, 1983).

16

TAXATION
OF LANDED PROPERTY

Taxes can be viewed as the price we pay for civilization. They involve compulsory charges levied on persons, properties, and activities for the support of government. Some of the most important of these charges are levied on real estate resources. As taxes, they take several forms and range from small nuisance charges to much heavier levies that can press against the economic feasibility limits of given enterprises. Most land taxes have more or less neutral impacts on operator decisions; but with a continuing demand for tax revenues and an increased use of special taxing arrangements, operators often find it to their advantage to pursue investment and production strategies that limit their tax liabilities while optimizing their expected benefits and satisfactions.

Emphasis is focused in this chapter on four taxation issues that bear on operator decisions concerning the use of land. These include (1) identification of the principal taxes that affect land, (2) the question of who ultimately pays these taxes, (3) the impact taxes have on land ownership and use, and (4) the operation of the property tax.

TAXES AFFECTING LANDED PROPERTIES

Taxes impact on different things and in most instances involve levies of either uniform or graduated rates on persons, transactions, properties, incomes, or privileges. Flat-rate taxes on landowners or areal units of ownership have been used on some occasions in the past; and transaction taxes such as sales and use taxes and import duties can affect land resource management decisions. But the principal taxes that affect real estate operators in the United States involve levies on property values, incomes, and privileges.

Direct Taxes on Property Values

When one speaks of taxes on property values, most Americans think of the property tax. This tax is the most important tax on landed property. Yet it is only one of several taxes that can be levied on real estate values. Special assessments are used in many jurisdictions. Taxes on land rents and on an operator's net worth are used in some parts of the world. Estate taxes also fall in this category since they involve the value of the estates of deceased persons.

Property taxes. The general property tax, or ad valorem (according to value) property tax, as it is often called, calls for annual levies against the assessed values of all nonexempt properties located in a taxing district. These properties fall into two classes: realty or real estate and personalty or personal property. A tax on realty is usually described as a real property tax or simply as the property tax while a tax on personalty is called a personal property tax. Real property taxes apply to all classes of privately owned real estate except those that are expressly exempt by law.

All nonexempt properties are supposedly valued or assessed at some set percentage of their true market values. These assessed values are then multiplied by uniform tax millage rates for each taxing jurisdiction to determine the actual tax payments due. If these levies are not paid within a specified period of time, properties are said to be tax delinquent. Continued nonpayment ordinarily leads to forfeiture of the owner's property rights.

More attention is given to the operation of the real property tax in a later section of this chapter. At this point, though, it may be noted that this tax is "an American institution." It has precedents in the land and tax concepts the early settlers brought to the American colonies and in the annual quitrents which the settlers in some colonies paid to agents of the English crown.[1]

As it exists today, the property tax differs from these precedents in three important respects. Unlike some early taxes, it has become a tax on property rather than on persons. It has shifted from the early use of arbitrary rates to the taxation of all properties within each taxing district at uniform proportions of their assessed value.[2]

---

[1] Cf. Jens P. Jensen, *Property Taxation in the United States* (Chicago: University of Chicago Press, 1931), chap. 2.

[2] The shift to uniformity in property taxation—use of uniform tax rates and proportionate assessed values—came after 1800 at a time when the federal government's monopolization of the revenue from custom duties forced the states to shift to greater reliance on property taxation. Prior to this time, many property tax rates were quite arbitrary. In 1645, for example, Virginia taxed land at a rate of 4 pounds of tobacco (tax collected in kind) for every 100 acres. North Carolina taxed land at 2.5 shillings per 100 acres in 1715; and New Hampshire taxed land at 5 shillings per acre in 1680 and at 10 shillings per acre in 1742. Cf. *ibid.*, pp. 28–33.

And unlike the annual payments enacted under the quasi-feudal quitrent system, taxes are collected as a matter of government right and not as a symbol of superior tenure status.[3]

Since its first use in the United States, the property tax has been treated almost exclusively as a source of state and local revenues. The federal government has attempted to use it on only three occasions. And on each of these, the constitutional requirement that federal direct taxes be apportioned among the states in proportion to their population made the tax extremely difficult to administer.

States and local units of government looked to this tax for most of their revenues until the early 1900s. Since then, and particularly since the 1930s, most state governments have turned to other sources of revenue. General property taxes accounted for only 1.5 percent of the tax revenues of the states in 1982 as compared with 46.5 percent in 1913 and 71.9 percent in 1890 (see Table 16-1). Local govern-

TABLE 16-1.   Importance of Property Taxes in the United States, 1870–1982

YEAR	TOTAL PUBLIC REVENUES SECURED FROM REAL AND PERSONAL PROPERTY TAXES (millions of dollars)			PROPORTION OF TAX REVENUES FROM OWN SOURCES SECURED FROM PROPERTY TAXES		
	States and local units	States	Local units	States and local units	States	Local units
1870	$ 226	$ 55	$ 171	*	*	*
1880	314	52	262	*	*	*
1890	443	69	374	88.4	71.9	92.3
1902	706	82	624	82.1	52.6	88.6
1913	1,332	140	1,192	82.8	46.5	91.1
1922	3,321	348	2,973	82.7	36.8	96.9
1932	4,487	328	4,159	72.8	17.4	97.3
1942	4,537	264	4,273	53.2	6.8	92.4
1952	8,652	370	8,282	44.8	3.8	87.5
1962	19,054	640	18,414	45.9	3.1	87.8
1972	42,877	1,257	41,620	39.1	2.1	83.7
1977	62,527	2,260	60,067	35.5	2.2	80.5
1982	82,067	3,116	78,952	22.2	1.5	48.0

*Data not available.

Sources: U.S. Department of Commerce, *Historical Statistics of the United States, Colonial Times to 1957,* Series Y526, 590, and 658; and Bureau of the Census, Census of Governments, *Government Finances, 1962, 1972,* and *1982.*

[3]With the quasi-feudal tenure system accepted in several American colonies, the nominal owners of land held their occupancy rights subject to the seigneur rights of the crown or of a proprietary lord who was a vassal of the crown. Specified annual payments, varying from token gifts such as a red rose to cash payments, were paid each year to agents of the crown for the privilege of continued occupancy. These payments served a multiple function in recognizing the operator's inferior tenure status, in quitting or cutting off the seigneur's right of rent for the year, and in producing public revenue. The quitrent system had little application in New England but persisted in Maryland and Virginia until the time of the Revolutionary War. Quitrents were unpopular in most areas and were frequently evaded or ignored by owners.

ments have also shown interest in securing other sources of tax revenues; but their continued heavy dependence on this source is amply illustrated by their collection of $79.0 billion in real and personal property taxes in 1982.

More than half of the taxes collected by the federal, state, and local governments came from the property tax until World War I and again during the 1920s. This situation changed during the 1930s, and 1941 was the last year in which the property tax rated as the single most importance source of tax revenues. By 1980, property taxes accounted for only 7.3 percent of the nation's taxes. But they averaged $302.42 for every person in the nation and $35.47 for every $1,000 of personal income. Averages by states ranged from a low of $78.65 per capita and $11.66 per $1,000 of personal income in Alabama to highs of $900.01 and $79.03 in Alaska and $5554.91 and $62.05, respectively, in Massachusetts.

Total property tax collections in 1982 were 116 times the level for 1902. This increase reflects new construction, higher property values, and an upward trend in tax millage rates. An indication of the overall trend is provided by data reported by the U.S. Department of Agriculture on farm real estate property taxes (see Figure 16-1). These data indicate that average taxes per acre on farm real estate almost tripled between 1910 and 1921. They continued to rise until they reached a national average of 54 cents an acre in 1928, then dropped to around 37 cents an acre during the early 1930s, and remained at this level until 1944, when they started an upward climb that brought them to average levels of 69 cents an acre in 1950, $1.21 in 1960, $2.27 in 1970, $3.85 in 1980, and $4.12 in 1981.

Throughout this same time period, average taxes on farm real property reached their highest peak in proportionate value terms ($1.52 of tax on every $100 of actual market value) in 1932. Tax reductions brought a lowering of this rate between 1933 and 1935. The rate then leveled off, only to drop again during the early

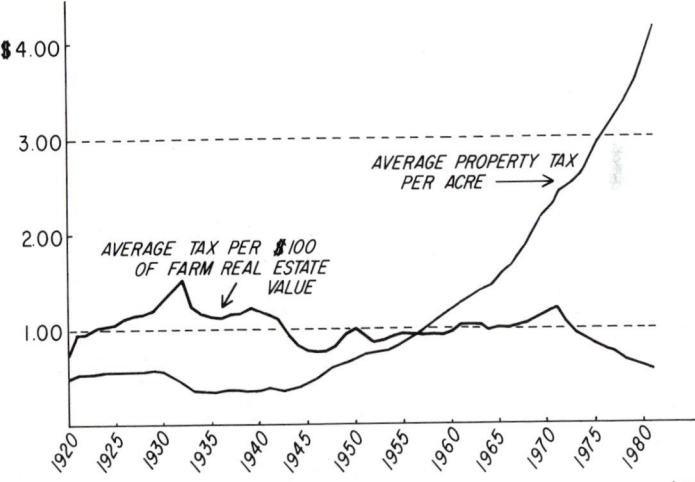

FIGURE 16-1. Trends in average property taxes per acre and average taxes per $100 of farm property value, United States, 1920–1981.

1940s when the increasing property values and relatively constant tax rates of the World War II period brought a twenty-five year low of 77 cents in taxes for every $100 of actual market value in 1946. Increasing tax rates brought this national index up to a post–World War II peak of $1.09 for every $100 of farm real estate value in 1969, after which rapidly rising land values caused it to drop to a national average of 48 cents in 1981. Averages by states in 1981 ranged from a low of 22 cents per acre and 9 cents per $100 of farm real estate value in New Mexico to highs of $33.76 per acre in Rhode Island and $2.04 per $100 of farm real estate value in New York.

Comparison of the data on taxable property values reported by the *U.S. Census of Governments* for 1957 and 1977 shows that exemptions of additional classes of personal property caused the real property portion of the locally assessed property tax base to rise from 81.1 to 86.8 percent of the national total between 1956 and 1976. Meanwhile, the composition of the real property tax base also changed to make the property tax more and more a tax on home ownership. Nonfarm residential properties provided 59.2 percent of the total real property tax base in 1976 as compared with 54.1 percent in 1956.

Numerous studies have indicated that the property tax tends to be *regressive*, that is, has a heavier impact on low-income than high-income families.[4] This regressivity has been caused by two principal factors: (1) a tendency of many assessors to overvalue low-valued properties and undervalue high-valued properties relative to properties in the middle value range, and (2) the tendency of taxpayers with low incomes to have larger portions of their capital assets in taxable properties than high-income families. Improved assessment administration is doing much to counter this type of regressivity, and "circuit-breaker" arrangements are being used in several states to reduce the burden of property taxes on low-income families.

Special assessments. Special assessments are usually treated as a type of tax, although they may also be described as a public fee. In any case, they represent "compulsory levies imposed upon owners of property for the purpose of defraying the cost of specific public improvements likely to enhance the value of assessed property."[5] Like the property tax, they constitute a claim against the property rather than the owners; and they are ordinarily added to and collected with the property tax.

Special assessments have a long history that goes back to the great fire of London of 1666. They were used in many rural areas to finance road construction and improvements prior to the 1930s. They have been used in numerous cities and subdivisions in more recent times to finance the provision of paved streets, sidewalks, sewers, water mains, and other improvements. Ad hoc units of government also use them to finance drainage, levee, irrigation, and other projects.

[4] Cf. Dick Netzer, *Economics of the Property Tax* (Washington, Brookings Institution, 1966), pp. 40–69.

[5] E. H. Hahne, "Special Assessment," *Encyclopedia of the Social Sciences* (New York: Macmillan, 1937), 14, 276.

From the standpoint of tax theory, special assessments supposedly affect only those properties that benefit from area improvements; and they never exceed the value added by the improvements. Problems can arise, however, in the allocation of costs among taxpayers. Some taxpayers can easily be overcharged or undercharged relative to others. Under certain circumstances, as when an environmentalist who is trying to maintain a wetland area is assessed for local drainage improvements, owners may also be taxed for projects that lower rather than raise the values of properties for the purpose for which they are held.

Since special assessments are supposedly allocated in proportion to benefits received, it is only proper that the allocation system fit the nature of the project. The cost of a water main may best be allocated on a house or lot unit basis, the cost of a drainage or levee project on an acreage basis, the cost of sidewalks on a foot-frontage basis, and the cost of some other projects on an assessed-value basis. With some projects such as public parks or sewer system improvements, the radius of benefit often covers a considerable area. Zones of benefit are often designated in these cases, with the costs being allocated between the properties in each zone somewhat in proportion to the benefits realized from the improvement.

Special assessments have provided a serviceable approach for financing local improvements in the past and will probably continue to function in this way in the future. A word of caution, however, is in order. Special assessments can create a considerable tax burden on property and can represent premature expenditures that are of no immediate value to property owners. Problems of this nature can assume acute proportions, as they did in many areas during the late 1920s, and can lead to wholesale tax delinquency when they are associated with depression conditions such as those of the early 1930s.

Taxation of rental values. Taxes based on the rental value of land and its improvements provide the basis for Britain's system of annual rates. These taxes have supplied much of that nation's revenue for local government operations since 1601.[6] They have also been used in various forms in Cuba, India, Iran, Pakistan, Peru, and several European nations. This approach can operate equally well as the ad valorem property tax. The only important difference is that one bases taxes on assumed rental values and the other taxes the supposed capitalized value of land rents.

No serious attempt has been made to adopt a system of land rent taxation in the United States. The concept, however, has merit and has attracted attention because of its association with Henry George's proposal of a single tax on the annual rents attributable to the site value of land. George was much concerned with the indispensable significance of land for mankind. He saw land as a primary source of wealth and as a generator of unearned surplus economic returns (land rents) that could enrich those operators who own the more productive and better located sites.

[6]Cf. J. A. Kay and M. A. King, *The British Tax System*, 2nd ed. (Oxford: Oxford University Press, 1980), chap. 10; and Alan R. Prest, "United Kingdom Land Taxation in Perspective," in Richard W. Lindholm and Arthur D. Lynn, Jr., eds., *Land Value Taxation* (Madison: University of Wisconsin Press, 1982), pp. 139–49.

He argued that this return provides a logical basis for taxation, that the economy would enjoy greater prosperity, and that we would live in a more ethical society if taxes were levied on the economic return to bare land instead of on the fruits of human labor. His reasoning led to a "simple yet sovereign remedy": Society should "abolish all taxation save that upon land values."[7]

George and his followers argued that this tax would (1) provide all the tax revenues needed by government, (2) reduce the investment cost of property ownership, (3) encourage owner-operatorship of rural and urban properties, (4) foster the improvement and more efficient use of land, (5) discourage land speculation, and (6) be a relatively painless tax, since it would affect only the "unearned" annual rent of land.

George's ideas on taxation have been accepted in varying degrees in a few countries but never on more than a token basis in his own country. His single tax could have provided a workable system of taxation had it been adopted with the first settlement of the nation; and his basic concept could still be used to advantage with a value-increment tax that would tax increases in land site rents or values after some benchmark date. Acceptance of the single tax at this time, however, would pose several problems: (1) even if applied at the 100 percent level, it would not supply all government's need for tax revenues; (2) its exclusive concern with economic returns to land would leave the equally "unearned" economic rents received by nonland factors untouched; (3) it would have no effect on former owners who have already realized their unearned land-value increments through sales of properties at their capitalized income values; (4) it would work a definite injustice on the thousands of owners who have used past savings to purchase properties at their capitalized values; (5) it would be difficult to administer in an equitable manner; and (6) elimination of the land rents now ascribed to the bare land portion of real properties would wipe out much if not all of the prospect for economic gain that now plays a dominant role in guiding the allocation of land areas among competing owners and uses.

Taxation of estate values. Taxes on net worth—on the total value of one's assets less allowances for debts—have been suggested at times as a means for taxing people in accordance with their overall abilities to pay. Annual net worth taxes are used in a few nations such as Colombia, but are not generally used in the United States. However, the federal estate tax, passed in 1976, can be viewed as a once-in-a-lifetime tax on net worth, because it places a tax levy on the full value of a deceased person's net worth.

This tax applies a graduated schedule of rates to the value of the nonexempt gifts made during one's lifetime plus the value of one's estate at death. The first $10,000 of estate and gift value is subject to an 18 percent tax. This rate rose to a maximum of 70 percent on increments of estate and gift value in excess of $5 million until 1982, but has since been lowered to a maximum of 60 percent that

---

[7]Henry George, *Progress and Poverty*, 1879 (Garden City, N.Y.: Garden City Publishing Company, 1926), pp. 403–4.

will drop to 50 percent on estate and gift values in excess of $4 million in 1988 and later years. The 1976 law also established a system of gradually increasing unified credits that provide exemptions from payment of the initial portions of the tax. By 1981 this unified credit offset the first $47,000 of tax and thus provided an effective exemption of the first $175,625 of estate value from taxation. The allowances for unified credit were raised in 1981 to permit effective exemptions of $400,000 in 1985, $500,000 in 1986, and $600,000 in 1987 and later years. An estate valued at $750,000 in 1987 would be subject to a 39 percent tax on the value in excess of the $600,000 exemption. Unlimited marital deductions are permitted for transfers of property between spouses. Gifts of up to $10,000 a year per donee are exempt from taxation, and educational and medical expenses are not counted as gifts.

High tax levies on the privilege of transferring landed property and other wealth to one's heirs can easily lead to the diminution and wiping out of large estates. Britain's death duties have worked to this end, and a similar potential exists with the federal estate tax. Most owners and their families are protected from this threat by the unified credits that provide tax exemptions for most estates. Property owners with estates that exceed the maximum unified credit exemption levels can use estate planning techniques such as irrevocable "skip-a-generation" trusts to reduce their tax liabilities and reduce the frequency with which estate taxes must be paid.[8]

Taxes on Income from Land

Taxes must always be paid out of past savings, current incomes, or borrowings against future incomes. Tithes represent the oldest direct tax on income known to mankind. Variations of this concept are now used by most nations with the taxation of specific types of income. Three taxes of this order—taxes on income, capital gains, and inheritances—have particular relevance for real estate resources in the United States.

Income taxes. Income taxes have been used by the federal government since 1913 and are now used by 44 states. Since 1941 they have provided the single most important source of tax revenues in the United States. Contract rents and other incomes from real estate are treated in the same manner as other types of income under these taxes. Real property ownership, however, offers several opportunities for the shielding of varying amounts of income from taxation. These opportunities are used to a greater extent by some operators than others and in many instances add substantially to the attraction of real estate investments.

Two of the most important of these tax advantages involve the exclusion of the imputed rental values of owned properties from calculations of net income and the privilege owners have of deducting their payments of property taxes and interest on

---

[8]For more-detailed discussions of the federal estate tax, cf. Commerce Clearing House, *1982 U.S. Master Tax Guide* (Chicago: Commerce Clearing House, 1981); Prentice-Hall *Federal Tax Handbook, 1985* (Englewood Cliffs, N.J.: Prentice-Hall, Inc., 1985); and Peter E. Lippett, *Estate Planning After the Reagan Tax Cut* (Reston, Va.: Reston, 1982), pp. 110–17.

mortgage debt from their taxable incomes.[9] Important among the other tax benefits associated with property ownership are the opportunities operators have for
equity financing, possible tax-free exchanges of some types of properties, postponement of capital gains, special treatment of casualty losses, and ability to reinvest
portions of their earnings or productivity in their properties without declaring them
for taxation. Owners of mineral deposits can use oil and mineral depletion allowances to reduce their tax liabilities. Developers and builders have at times been able
to use front-loading arrangements to charge off development costs during the initial
stages of their projects and thereby realize savings on interest-carrying charges. Investors also can charge off depreciation costs on rented properties and pay capital
gains rather than regular income taxes on incomes realized from the sale of property investments.

Capital gains taxes. Next to the property tax, capital gains tax considerations
have the greatest overall impact on operator decisions relative to real estate investments in the United States. Possible income tax savings associated with the special
treatment of capital gains provide a major incentive for many real estate investments, often encourage land speculation, and have chain reaction impacts on property values and the operation of the real estate market.

Capital gains realized from the sale or exchange of real estate and other types of
property were taxed as regular income between 1913 and 1921. Since 1921 a distinction has been made between long-term and short-term gains. Short-term gains
are treated as regular income, whereas gains or losses associated with investments in
real estate and other properties held for more than one year are treated as long-term
gains or losses. These gains and losses are subject to taxation as capital gains or
losses.

The tax rules applicable to capital gains have been changed on frequent occasions, most recently in 1976, 1978, 1981 and 1984. With the 1981 regulations, individuals with long-term gains from the sale of properties can deduct 60 percent of
their capital gains and pay their normal income taxes on the remaining 40 percent
or pay a maximum tax of 20 percent on the total amount of their capital gain.
Long-term losses can be deducted from long-term gains up to a maximum of 50
percent of the loss or $3,000, whichever is less, in any year. Unused loss allowances
can be carried over to future years.

Investors find the special tax treatment of capital gains attractive because it

[9] Cf. Chapter 10, pp. 287–90, and Chapter 13, pp. 374–76, for typical examples of calculations of these benefits. Also cf. Henry J. Aaron, *Shelter and Subsidies: Who Benefits from
Federal Housing Policies?* (Washington, D.C.: Brookings Institution, 1972), pp. 53–73; George S.
Tolley and Douglas Diamond, "Home Ownership, Rental Housing, and Tax Incentives," in *Fed-
eral Tax Policy and Urban Development*, Hearings of Subcommittee of the House Committee
on Banking, Finance and Urban Affairs, 95th Cong., 1st sess., 1977, pp. 114–95; and Frank de
Leeuw and Larry Ozanne "Housing" in Henry J. Aaron and Joseph A. Pechman, *How Taxes
Affect Economic Behavior* (Washington, D.C.: Brookings Institution, 1981), pp. 283–326. Also
cf. Paul E. Anderson, *Tax Factors in Real Estate Operations*, 6th ed., (Englewood Cliffs, N.J.:
Prentice-Hall, 1980), pp. 19–23, for a discussion of other tax advantages associated with property ownership.

assigns a superior tax-favored status to income secured from the holding and sale of property. As described in Chapter 10, they can acquire properties, use depreciation write-offs for a series of years to reduce their taxes on current income, and then pay only part of their tax savings as capital gains taxes at the time they sell their properties. This prospect of tax savings provides a major incentive for investments in rental properties and has also provided a stimulus for speculative investments.

Critics of the taxing arrangement used with capital gains argue that income from capital gains should not be accorded preferential treatment as compared with earnings from salaries and wages. In an ethical Henry Georgian sense, income secured from the mere holding of land and other investments should be taxed at higher not lower rates than incomes derived from human labor. This is particularly true when the capital gain stems from windfalls and actions of the community and others that lead to higher property values. Questions may be raised, however, about the fairness of capital gains taxes on increases in property values that stem entirely from inflation. Equitable treatment in these cases calls for indexing the property value base used in calculating the capital gains tax to a general price index, so that owners who sell properties are not taxed on increases in current dollar values that do not represent increases in real dollar purchasing power.

Numerous owners who were selling properties with the intention of acquiring newer, larger, and more productive properties were penalized by the method used in calculating capital gains taxes during the inflationary period that followed World War II. Congress recognized this situation and provided relief for residential homeowners by exempting capital gains from the sale of principal residences from taxation as long as the owners use the proceeds from their sales to build or buy replacement properties. With the modified regulations provided in 1981, owners have a rollover period extending from twenty-four months before to twenty-four months after the sale of their properties, within which they must acquire title to the replacement properties if they are to be eligible for this exemption. Similar exemptions may be enjoyed by investors in low-income rental housing who sell their units to the occupants, tenants, or nonprofit management organizations if they reinvest in or build other qualified low-cost housing. As an alternative to this exemption, owners who are fifty-five years of age or older can realize a once-in-a-lifetime exclusion of up to $125,000 of capital gain from the sale or exchange of their principal residence.[10]

Inheritance taxes. Most states have inheritance taxes that impact on the incomes heirs receive with the settlement of estates. These taxes are levied in addition to the federal estate tax and differ from them in most instances in that they tax the inheritance incomes received by individual heirs rather than the value of the entire estate. Credits from these state taxes are accepted against federal estate tax liability. The rates applied are typically lower and lack the base exemption of the federal estate tax. Inheritances are taxed separately, and the rates are usually graduated both by the amount of the inheritance and the heir's degree of relationship to the deceased.

[10]Cf. Commerce Clearing House, *op. cit.*, sections 904, 905, and 906A.

For example, cousins are taxed at higher rates than spouses or children. The inheritance tax brackets have frequently failed to keep up with inflation.

Most states treat deathbed gifts and gifts made in contemplation of death as transfers subject to inheritance taxation. Separate gift taxes are also used in approximately one-fourth of the states as a supplemental measure to tax transfers of property between persons who otherwise would escape taxation. Gifts are usually taxed at three-fourths the rate that applies with inheritances.

Taxation of Privileges Affecting Land

Although not ordinarily classified in this manner, most land taxes can be treated as taxes involving privileges—the privileges of owning or selling land, transferring it to one's heirs, and receiving incomes or an inheritance. Business franchise and activity taxes provide leading examples of taxes considered as privilege taxes. They frequently involve landed property insofar as businesses own, use, or deal with these properties. Other important examples involve property-transfer and documentary taxes and severance taxes.

Documentary and property-transfer taxes. The Federal Revenue Act of 1932 specified that all deeds, instruments, or writings under which "lands, tenements, or other realty sold shall be granted, assigned, transferred or otherwise conveyed to, or vested in, the purchaser or purchasers, or any other person or persons," shall be taxed at the rate of 55 cents "when the consideration or value of the interest or property conveyed, exclusive of the value of any lien or encumbrance remaining thereon at the time of sale, exceeds $100, and does not exceed $500" plus 55 cents "for each additional $500 or fractional part thereof." This tax was repealed in 1965. Documentary taxes of a comparable nature have been enacted by thirty-six states. Some of these apply to the recording and registration of deeds, some to mortgages, and some to both deeds and mortgages.

With these taxes, any person who grants a deed or other taxable instrument must affix and cancel the required number of revenue stamps and pay the required state tax or fee. The receipts from these taxes are small and represent a relatively insignificant source of public revenues. But they provide a valuable source of current information on property values that tax assessors and tax commissions can use in their property tax assessments, and they add to the costs and paperwork associated with property transfers.

These taxes have been supplemented in a few Northeastern states by real estate transfer taxes that call for small percentage levies on the sales prices of real properties. Revenues from this tax can be earmarked for special programs such as public acquisition of open space lands or the purchase of development rights to farm lands.

Severance taxes. Several states levy severance taxes on the privilege of harvesting, mining, extracting, or otherwise severing natural resources such as timber, metallic ores, coal, petroleum, natural gas, salt, and sulfur. These taxes normally involve either a fixed charge against each unit of product severed or a percentage charge against the gross receipts or gross market value of the raw products produced.

As a taxing device, severance taxes are sometimes combined with and sometimes used in lieu of property taxes. When used in addition to the general property tax, they are often justified on the ground that owners are exploiting a free gift of nature and they should share their profits with society. With this purpose in mind, several states levy taxes involving specified charges on every thousand board feet of lumber, every ton of coal or ore, or every barrel of oil produced within their jurisdiction.

When severance taxes are substituted for property taxes, they are usually justified as being more equitable and more favorable to conservation practices than the property tax. Realistic assessments for taxation purposes are almost impossible with hidden resources such as oil. And even when the total value of a forest or a mine can be estimated, annual taxation at its full assessed value often encourages owners to exploit or liquidate their resources as soon as possible. With commercial forest holdings, for example, high property taxes can favor "cut out and get out" policies that leave little room for sustained-yield management.

Several states have substituted severance taxes for property taxes on oil, mineral, and other underground resources. This approach is also used to a considerable extent with forest holdings.[11] Approximately a third of the states now have optional or compulsory forest-crop taxation programs that use the severance or yield tax principle. These laws call for a nominal annual land tax plus yield taxes that vary from 3 to 12.5 percent of the value of the forest crop at the time it is harvested.

SHIFTING AND INCIDENCE OF LAND TAXES

With most taxes, one can ask, Who actually pays the tax? Is the tax absorbed by the person from whom it is collected? Is it shifted forward to some eventual consumer in the form of higher prices; or might it be shifted backward in the form of lower prices for raw materials, lower rent to the landlord, or lower wages to labor? In their discussions of this topic, economicsts use the terms *shifting* and *incidence*. *Shifting* means the transfer of the tax burden from the initial taxpayer to someone else, and *incidence* refers to the persons or things that bear the actual tax burden.

Tax incidence has an important bearing on the effect taxes have on land resources. When a tax can be easily shifted, the taxpayer functions mostly as a tax collector for the government. But when a tax is only partly shifted or cannot be shifted at all, the taxpayer invariably suffers a reduction of income. With land taxes, this reduction of income can lead to lower property values. It may cause operators to

---

[11] For more detailed discussions of these taxes, cf. Harold M. Groves, *Financing Government*, 5th ed. (New York: Henry Holt, 1958), pp. 308–17; Warren A. Roberts, *State Taxation of Mineral Deposits* (Cambridge, Mass.: Harvard University Press, 1944); Ralph W. Marquis, *Forest Yield Taxes*, U.S. Department of Agriculture Circular No. 899, 1952; *Forest Taxation Symposium Proceedings* (Blacksburg, Va.: Virginia Polytechnic Institute, 1977); and Mason Gaffney, ed., *Extractive Resources and Taxation* (Madison: University of Wisconsin Press, 1967).

live in smaller or less pretentious houses than they otherwise would. It may cause them to use land more intensively or possibly exploit their available resources. It may cause them to move to lower-taxed areas; or it may even reduce their income to a point at which they find it uneconomic to continue operations or pay taxes.

Whether or not a tax will be shifted "depends on (1) the nature of the tax, (2) the economic environment in which it is levied, and (3) the taxpayers' practices in taking advantage of any possibility of shifting."[12] Some taxes are more shiftable than others. A uniform tax on each ton of iron ore produced may be shifted on and on until it is fairly evenly diffused throughout the economy. A tax on the land rent received from the use of land resources, on the other hand, may be virtually nonshiftable.

Incidence of Property Taxes

According to the "pure" theory of tax incidence, a property tax on bare land is nonshiftable, whereas a tax on land improvements can usually be shifted over time, even though it may be nonshiftable at particular moments in time. This theory assumes a simple model such as that depicted in Figure 16-2, in which bare land is viewed as a nonreproducible resource with no cost of production. The market values of the better grades of land reflect the capitalized values of their land rents, and the amount of tax levied is assumed to be directly proportional to the levels of market value and land rent. Land at the extensive margin of use pays no rent, has no capitalized value, and accordingly pays no tax. With these assumptions, the rais-

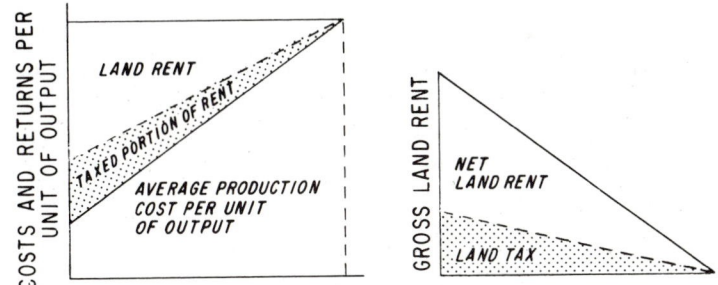

PRODUCTION SITES OF DIMINISHING USE-CAPACITY

FIGURE 16-2. Land rent models assumed with "pure" theory of land tax incidence.

[12]William J. Schultz and C. Lowell Harriss, *American Public Finance*, 8th ed. (Englewood Cliffs, N.J.: Prentice-Hall, 1965), p. 142. For other discussions of this topic, cf. Harold M. Groves and Robert L. Bish, *Financing Government*, 7th ed. (New York: Henry Holt, 1973), chap. 3; Joseph A. Pechman and Benjamin A. Okner, *Who Bears the Burden?* (Washington, D.C.: Brookings Institution, 1974), pp. 25–43; Henry J. Aaron, *Who Pays the Property Tax? A New View* (Washington, D.C.: Brookings Institution, 1975); Netzer, *op. cit.*, pp. 32–66; and Herbert A. Simons, "The Incidence of a Tax on Urban Real Estate," in Richard A. Musgrave and Carl S. Shoup, eds., *Readings in the Economics of Taxation* (Homewood, Ill.: Richard D. Irwin, 1959).

ing or lowering of a tax—the taking of a larger or smaller portion of the land rent—in no way affects the owner's operating decisions. An increase in the tax rate has no effect on the location of the extensive margin, on the total amount of land used, or on the market price of the products produced. The higher tax cannot be shifted to others.[13] It simply takes a larger share of the owner's land rent.

In contrast with this situation, it is assumed that taxes on buildings and other property improvements can be shifted over time because they have limited economic lives and must be replaced. Property improvements compete with other investment alternatives for new investment capital. When high property taxes depress current investment values, new capital is withheld from property improvements, and the existing supplies of reproducible improvements wear out faster than if new improvements are provided. With free market conditions, this situation supposedly leads to higher product prices, higher rental rates, reattraction of investment capital to these uses, and some incidental shifting of the tax burden.

This two-part theory of property tax incidence applies to some extent under real-life conditions. But its application is complicated by differences between its assumptions and the conditions of the real world. The lack of a clear-cut distinction between the concepts of land and improvements, for example, often makes it unrealistic to apply these theories to properties that represent combinations of bare land and improvements. Other problems arise with properties used primarily for consumptive purposes and when there is no expectation of replacing improvements as they wear out.

A principal weakness of the "pure" theory of land tax incidence centers on the simplicity of its assumptions. It assumes a single land use and visualizes the tax on land as a relatively uniform proportional tax on the capitalized value of land rent. A different set of conditions prevails in the real world.[14] Land areas are subject to many competing uses. Few uses are carried on to their actual extensive margins. Individual uses tend instead to be displaced at or near their margins of transference by other uses that can utilize lands of lower use-capacities to greater advantage, (see Figure 6-11). The transference and extensive margins for most uses are supramarginal for other uses and thus have considerable value for taxation purposes. This situation is further complicated by the location of properties in different tax assessment districts and by a frequent lack of uniformity in tax assessment levels. The United States has around 82,000 separate taxing districts, each of which has its own assessment levels and taxation rates. Moreover, even though assessors try to value all properties at the same proportions of their capitalized rental values, variations in assessment levels are common, and experience shows that property taxes are often regressive in the sense that they have a heavier proportional impact on lower-valued than higher-valued properties.

[13] Cf. Jensen, *op. cit.*, pp. 61–62.

[14] Martin Feldstein argues that taxes on pure rents may also be shifted in the short run as investors maneuver to balance their portfolios and over the longer run because capital stock is not limited in supply. Cf. Martin Feldstein, "The Surprising Incidence of a Tax on Pure Rent." *Journal of Political Economy*, 85 (1977), 349–60; idem, *Capital Taxation* (Cambridge, Mass.: Harvard University Press, 1983), pp. 414–26.

With these circumstances, the situation depicted in Figure 16-3 provides a more accurate representation of what happens with land taxes than does Figure 16-2. Far from having land of no taxable value, operators at the transference and extensive margins for most uses find that their sites are valuable for other uses and that they still have a tax to pay. Payment of the tax represents an additional cost of production and, insofar as production takes place to the extensive margin, product prices must rise enough to cover production costs at this margin if land is to continue its assumed use. This does not mean that prices must always rise enough at the transference margin to permit a complete shifting of the tax. But when taxes are regressive in their application or when sites at the extensive margin (use B in Figure 16-3) are subject to high local tax levies, the price of the product of this use must rise sufficiently to cover the tax cost. This price increase permits the B users to shift much of their tax and may even bring windfalls to operators in low-tax areas. The effect of the tax on the transference margin ab also necessitates an increase in the price of the product of A, which permits some shifting of the tax burden from the A use producers to the consumers of the product of the land.

How much actual shifting of taxes takes place often depends on the type of properties involved and the conditions under which they are used. Taxes on owner-occupied homes and properties used for personal or family uses are usually non-shiftable. This situation exists because individual homeowners are seldom able to shift their taxes in the form of higher wages or salaries for their services or higher prices for the products they produce and sell. An exception to this situation arises when a property owner's income is indexed to the cost of living.

With productive properties such as industrial plants, commercial establishments, farms, and forests, the extent to which one can shift property taxes depends on the universality of the tax and the degree of control one has over market prices. Under perfectly competitive conditions, product prices are set by the interplay of supply and demand and generally correspond to the productive costs encountered at the extensive margin. A property tax that affects all properties used for some given

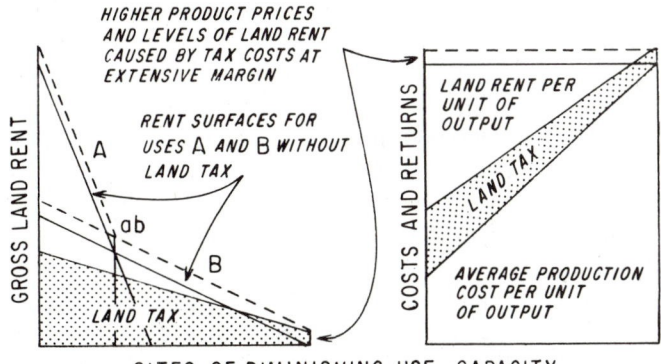

FIGURE 16-3. Illustration of impact of a land tax on production costs at the extensive margin for an assumed use of land (use B) and its probable effect on the shifting of a portion of the tax through higher product prices on higher grades of land.

purpose could thus be shifted up to the point at which it affects production costs at the extensive margin. If consumers refuse to buy the same volume of product with a tax added onto the price, producers must absorb some or all of the tax, and the long-term extensive margin for that use may shift to the left. Similarly, with higher taxes in some districts than others, operators in the higher-taxed areas must usually bear the incidence of the excess of taxes they pay, because they cannot shift these extra taxes on in the form of higher product prices and still maintain their competitive position with lower-taxed areas.

Unlike farmers and other operators who enjoy little individual control over the prices they receive for their products, some business operators are able to dictate the prices at which their products sell. In their exercise of this power, these operators can shift increases in their property taxes on to the buyers of their products, or they may even shift part of their tax load backward in the form of lower wages and lower supply prices. Even with these price-setting powers, however, these producers find it difficult to shift a tax increase if they have already set their market and supply prices at those levels that promise them a maximum net return.

Taxes on rental properties are ordinarily shifted to tenants; but landlords are frequently restrained in this shifting process by personal inertia, long-term lease commitments, rent-control regulations, and uncertainties in current rental market situations. Taxes can be shifted in the form of higher rents rather easily when the rental market is tight and when competition for properties exists among tenants. When landlords compete for tenants against other landlords who are willing to absorb all or part of a tax increase or against landlords located in taxing districts not affected by tax increases, there is some tendency to absorb the increase rather than risk the loss of desired tenants.

Tax capitalization.   Some shifting of property taxes from buyers to sellers takes place with a process known as *tax capitalization*. As noted in Chapter 6, productive properties have an economic value equal to the present discounted value of their expected future land rents. Since property taxes represent a cost factor, a tax increase will reduce the net annual land rents associated with a property and accordingly bring a reduction in its market value.

To illustrate the effects of tax capitalization, one might assume a property with an annual expected land rent of $10,000, which, when capitalized at 5 percent, gives a market value of $200,000. If the property tax in this case is increased by $500 a year and if this tax increase appears to be permanent, cannot be shifted, and is not associated with any increase in property value or income, the annual net rent will drop to $9,500. Capitalization of this figure at 5 percent gives a market value of $190,000—the drop of $10,000 being attributable to the higher property tax.

As one might expect, knowledgeable buyers frequently try to capitalize out the cost of their future property taxes at the time they acquire real estate. This is particularly true when they operate in buyers' markets and when they use an income-capitalization approach in the valuation of their properties. Insofar as the tax-capitalization process works, these operators buy themselves free of property taxes while the current owners bear the incidence of future as well as present taxes.

Several factors limit the operation of this tax-shifting process. To begin with, tax capitalization works best with cases of income-producing properties and then only to the extent that the tax is not shifted through higher product prices or lower supply prices. Furthermore, it applies mainly to cases in which buyers anticipate a definite pattern of future taxes with no compensating increases in property values. Expenditures of increased taxes for items such as better schools, police and fire protection, or civic improvements often enhance property values and counterbalance any losses that may take place with tax capitalization.

Still another complication stems from the average buyer's failure to understand the capitalization principle. Many buyers ignore the tax factor until after they have acquired their properties and received their first tax bills. These buyers compete alongside tax-conscious buyers in most markets involving real properties, and, through their bidding activities, can force better-informed buyers to take less tax capitalization than they might wish if they are to get the properties they desire.

Incidence of Other Land Taxes

Special assessments ordinarily bring improvements that benefit the properties taxed. With productive properties, these benefits often lead to higher production, lower costs, or higher business receipts and thus to a possible shifting of actual tax costs. Even when tax shifting does not take place, property owners can shift the portion of their tax costs represented by the value of their benefits to buyers at the time they sell their properties. Insofar as the tax cost exceeds the sale value of the benefit that comes with the improvement, the incidence of the special assessment lies with the original taxpayer.

Capital gains taxes, documentary taxes, and property-transfer taxes provide excellent examples of land taxes that fall almost entirely on the persons who sell property. Little tax shifting is possible because the sellers ordinarily seek the highest prices possible for their properties regardless of the tax rate. With this situation, one might argue that the incidence of these taxes falls entirely on the sellers because the sales prices received would not change even if their tax rates are doubled. Higher tax rates, however may discourage some owners from offering their properties for sale. This can reduce the supply of properties in the market and lead to new supply and demand interactions that call for higher market prices and thus permit an element of tax shifting.

Estate and inheritance taxes impact on the beneficiaries, who are deprived of portions of the inheritances they would otherwise receive. This incidence may also be shared by property owners during their lifetimes if they allow concerns about minimizing estate and inheritance taxes to affect the manner in which they manage their properties.

Severance taxes ordinarily involve either a fixed charge against each unit of natural resource produced or a fixed percentage of its market value. In either case, this tax represents an addition to the operator's production costs; and it can usually be at least partially shifted in the form of higher prices. This is particularly true when the tax affects all producers of a given resource or when it affects some producers including those at the extensive margin.

The amount of tax shifting that actually takes place varies with market conditions. Assuming perfect competition and an elastic demand schedule, any attempt to shift the tax in the form of higher prices will lead to less taking of products in the market (see Figure 2-2), a leftward shift of the extensive margin (see Figure 5-7), and the withdrawal of several marginal producers from the market. Under these conditions, one could argue that the tax is included in the costs covered at the new extensive margin. But the new equilibrium price will be somewhat below the sum of the previous price plus the tax; and the producers with the more productive properties will find that the overall price adjustment falls short of the amount necessary to permit a complete shifting of the tax.[15]

Complications arise when competing producers are subject to different tax arrangements. Under these conditions the producers in high-tax areas usually find that they must bear the incidence of their additional tax burden. Varying amounts of tax absorption also take place when producers continue their mining, forest-cutting, or other resource-severance activities during periods when gross market receipts are inadequate to cover full production costs.

EFFECTS OF TAXES ON LANDOWNERSHIP AND LAND USE

Although taxes can be considered as compulsory subtractions from income, they can also be viewed as investments in the community and social order that provide an array of public services that bring positive benefits to taxpayers and their properties. Most taxes are administered without particular regard for their impacts on economic decisions or behavior. It is usually hoped that their impacts will be neutral. Yet, although it is recognized that some taxpayers contribute more in taxes than they receive in benefits, public policy generally assumes that most taxpayers receive benefits commensurate with or in excess of their tax costs.

This situation applies with most land taxes. Tax revenues are used to provide peace, order, and security. They provide streets and highways, parks, schools, and water and waste disposal facilities along with other aspects of the community infrastructure that play a vital role in supporting our economic and political order. They finance public projects that hopefully make life more pleasant and humane, and they maintain an institutional environment within which operators can make constructive uses of their properties and live pleasant and productive lives. In addition to these favorable and routine impacts, taxes can and often do have significant effects on land ownership and land use. Five important examples occur with the

[15]For further discussion of the effects of different market conditions on this aspect of tax incidence, cf. Groves and Bish, *op. cit.*, pp. 41–60.

impacts they can have in (1) fostering more intensive uses of land, (2) promoting conservation and environmental goals, (3) advancing particular tenure goals, (4) influencing investment decisions, and (5) enhancing property values.

Fostering More Intensive Land Use

With its frontier background and its emphasis on provision of opportunities for a steadily increasing population, public policy in the United States has usually looked with favor on new resource developments. Tax policies have often shown a development bias as they have been used directly and indirectly to encourage or stimulate the more intensive use of land resources. Officials along the western frontier frequently assessed potential as well as actual farmlands at value levels reflecting their highest and best use and thereby encouraged speculators to sell, settle, or otherwise use the idle lands they were holding in anticipation of windfall profits. This same technique was used again in many forested and cutover communities to foster the clearing, sale, and agricultural development of areas held by various land and lumber companies. Similar practices that encourage the shifting of land to supposed higher and better uses are still employed around many cities when farm and forestlands are assessed at their subdivision values and residential lots are assessed at their emerging values for commercial uses.[16]

In addition to encouraging the shifting of lands to higher and better uses, tax pressures can cause operators to make more effective and intensive use of properties in their present use. Countries such as Finland, the Soviet Union, and Yugoslavia have used incentive tax levies based on levels of production expected from average producers to encourage more intensive land-use management. Similar goals are pursued when operators who are subject to higher taxes try to increase production so that they will suffer no decrease in net returns or when homeowners decide to rent out unused rooms because the rental income "will help pay the taxes."

Although property taxes can be used effectively at times to force more intensive land uses, this approach has its limitations. There is no automatic relationship between taxes and the ripening of land for particular uses. Higher taxes can favor more intensive use of lands not utilized at their optimum levels of intensity, but they have an undesirable effect when they encourage operators to go beyond this point. They can pressure lands into higher uses when demand exists for these uses and the lands in question qualify for the needed uses. When these conditions do not exist, however, taxes can have injurious effects in fostering the waste that comes with premature developments, tax delinquency, and tax forfeiture. Moreover, the development bias of tax policy can operate at cross-purposes with land-use policy

---

[16] Penalty taxes of a somewhat similar order have been used in some countries to discourage the underutilization of land. Colombia, for example, enacted a law in 1957 that called for official classification of all rural lands, specified minimum proportions of each landholding that should be in cultivation, and established an annual penalty tax starting at 2 percent of the cadastral value of the land in the first year and rising to 10 percent in the fifth year for holdings that failed to meet their cultivation quotas.

when the long-run interests of a community call for retaining lands that might otherwise be developed in open space or other socially desired uses.

Promotion of Conservation and Environmental Goals

When known reserves of timber, mineral ore, coal, or other resources are taxed year after year at their full assessed values, resource owners are sometimes impelled to follow "cut out and get out" policies. Aside from their effects in speeding up the exploitation of these resources, these policies often lead to considerable social waste. Forest owners operating under these conditions will often use more exploitive cutting practices than owners who are interested in reserving seed trees, saving younger and smaller trees for future growth and harvesting, and carrying on long-term forestry programs that preserve the economic base of local communities. In like manner, mine owners who are subject to heavy annual property taxes may feel that they should skim off the best of their resources as soon as possible. As a result, they may lack incentives to bother with marginal grades of ore; in their effort to reduce operating costs, they may use practices that make the lower-grade ore deposits less accessible for future use.

Several states have substituted severance taxes for property taxes in an attempt to promote resource conservation. Special forest-yield taxes are now used in approximately one-third of the states to encourage long-term forestry practices by exempting growing forests from annual taxation and taxing the forest crop only at the time it is harvested.

In most of these states, forest owners have the option of keeping forestland under the general property tax or entering it under the provisions of special forest crop taxation laws. The number and acreage of voluntary private entries have been somewhat disappointing in most states. The existence of this taxation alternative, however, has had a beneficial effect on all forest owners in keeping property tax levies on forestlands competitive with those available under the special forest crop taxation laws. Minnesota uses an alternative approach with its tree-growth law, which makes only the annual growth of the forests enrolled under this program subject to property taxation. Some other states have tried to encourage long-term forestry by exempting growing timber from taxation or by classifying it for special treatment under the real property tax.

Comparable measures can be used with other resources such as oil wells and mines to encourage slower rates of extraction and less social waste in the exploitation process. Some states have levied a flat tax on new oil wells, partly as a conservation measure to discourage the drilling of new wells. Special tax exemptions for investments in solar energy generation or in energy-saving facilities and in soil and water conservation expenditures may also be cited as examples of tax policies used for conservation purposes.

Tax incentive programs are used in almost every state to encourage the continued holding of farmlands in agricultural use. Most of these programs call for taxation of farmlands at their agricultural use-values.[17] But some programs, such as those in Michigan and Wisconsin, go farther to provide cooperating owners with sizable credits against their state income taxes. Comparable use-value-assessment arrangements are used with other types of open space in some states, often in combination with open-country zoning and "green acres" programs, to protect landowners from the tax pressures that might impel them to offer their lands for urban-oriented developments. Tax exemptions on air and water pollution control and treatment equipment are permitted in some areas to encourage private investments in these facilities. Some communities have also taken steps to monitor and measure the volume of wastes contributed by leading polluters to municipal waste treatment systems and to assess appropriate shares of the treatment costs to the responsible parties.

Taxing arrangements are used in each of these situations to promote environmental goals. In practice, however, tax incentives could be used to a much greater extent than they are to promote conservation and environmental goals. Tax penalties also could be used to a greater extent to discourage the use of specified substances, leaded gasoline, and practices that permit operators to secure private benefits while shifting significant costs to society. Important as their potential as a means for promoting conservation and environmental goals may be, it should be recognized that tax measures by themselves cannot bring attainment of many of the objectives expected of them. This is particularly true when private economic advantage favors contrary operating goals.

Advancement of Tenure Goals

Taxation policies can often be used to implement the advancement of particular tenure goals. Tax systems that discriminate against particular classes or groups of operators can hasten their elimination. Arrangements that are uniform and closely correlated with abilities to pay and benefits received, in contrast, can encourage capital accumulation, higher levels of living, and a wide distribution of ownership rights.

Among those people who have adhered to a communal village system of property ownership or who have regarded property rights as centering in the crown, the fixing of a tax-paying duty in particular individuals has sometimes represented the first step in the evolution of individual ownership rights. This was the situation in Czarist Russia where the tax-collection rights assigned to the early *boyars* gradually ripened into the property rights exercised by the later nobility. It was also the situation in parts of India where the British authorities designated the *zamindari* as tax

[17]Cf. Raleigh Barlowe and Theodore R. Alter, *Use-Value Assessment of Farm and Open Space Lands*, Michigan Agricultural Experiment Station Research Report 308, 1976; and Thomas F. Hady and Ann Gordon Sibold, *State Programs for the Differential Assessment of Farm and Open Space Land*, U.S. Department of Agriculture Research Service, Agricultural Economics Report No. 256, 1974.

collectors in the communal villages—a step that later led to their recognition as landlords. "Similarly, the Spanish colonial governments made systematic use of local native chieftains or tribal leaders as tax collectors, thus giving them powers which were later solidified into land-tenure rights, and ultimate ownership."[18]

Special taxing arrangements have been used on various occasions to favor particular classes of owners. Property tax exemptions make it possible for religious and charitable organizations to own properties located at some of the highest-valued sites in and near central business districts. Several states have homestead exemptions, which free owner-occupants of homes and farms from certain tax levies on the first $500 to $25,000 of their assessed property values. Similar exemptions are provided in some communities for special groups such as disabled veterans, widows, the aged, and families with low incomes. Federal income tax deductions for payments of interest and taxes plus exemption of the imputed return on the equities taxpayers have in their homes also provide an important tax subsidy for home ownership in the United States. The Congressional Budget Office reported in 1984 that removal of these subsidies would bring "large windfall losses in the values of many assets," cause a dramatic drop in the value of houses, and discourage housing sales.[19]

History records numerous examples of rulers who have used oppressive tax measures in their efforts to raise revenues. These taxes have often had an onerous effect on the great mass of the population. But they have usually favored certain individuals or groups, and in so doing have often contributed to the aggrandizement of small privileged classes, to the debasement of the average citizen, and to the rise of new tenure institutions such as the feudal system.[20]

Discriminatory tax policies are often used in the modern world as an arm of land reform. Special tax exemptions are claimed by farming cooperatives in countries such as Czechoslovakia, India, and Yugoslavia. The Soviet Union used a highly discriminatory tax during the 1930s to break the economic position of the free peasants and force them to join collective farms. Czechoslovakia and Poland have used a system of compulsory deliveries of farm products, paid in kind, and steeply graduated by size of farm to (1) encourage certain types of land use, (2) foster the breaking up of large and medium-sized farms, and (3) undermine the economic status of the more prosperous peasants. Comparable measures are used in other countries. Great Britain has used death duties to break up large landed estates. Estate taxes are used in the United States with this same result. Exemptions from land recording fees are used in Belgium to favor the consolidation of fragmented farm holdings. Graduated land taxes with provisions for higher tax rates on larger and

---

[18]Philip M. Raup, "Agricultural Taxation and Land Tenure Reform in Underdeveloped Countries," *Agricultural Taxation and Economic Development* (Cambridge, Mass.: Harvard Law School Press, 1955), p. 256.

[19]*Newsweek*, April 16, 1984, p. 69.

[20]The Danegeld tax imposed in England near the end of the tenth century is reputed to have had a crushing effect on small landholders. "The English peasant lost his freedom to his commended lord . . . who was ready to discharge the tax burden in exchange for greater service. Men were ground into soil." Cf. D. R. Denman, *Origins of Ownership* (London, U.K.: Allen & Unwin, 1958), p. 58.

higher-valued properties than on smaller holdings are used in Australia and New Zealand to discourage concentrations of land ownership.

Unbridled taxation can easily become the power to confiscate and destroy. In this sense, high taxes may lead to the forfeiture of individual ownership rights and in some cultures have even caused people to sell themselves or their children into bondage. Because of the unsavory nature of this approach, taxes are seldom used for the deliberate purpose of depriving people of their property rights. The experience of the United States, however, indicates that high property taxes can have an incidental effect in causing substantial areas of tax-delinquent land to shift from private to public ownership. This process can involve considerable social waste as happened during the late 1920s and the 1930s when millions of acres of once privately owned land tax reverted to public ownership.

Influencing Investment Decisions

Taxation arrangements can operate both directly and indirectly to encourage or discourage particular types of investment decisions. Investments in property are encouraged when tax revenues are used to provide desired services to property. Repressive taxes have the opposite effect. Tax incentives have been used to encourage land drainage in Finland and the provision of water canals in Iran. Depletion allowances have been used with the federal income tax to favor investments in mining properties. Special investment credits have encouraged corporation investments in plant modernization and new plant construction. Special income taxation arrangements also were used in the United States in the early 1980s to encourage investments in public utility stocks, the depositing of savings with institutions that made housing loans, and investments in rental housing.

Protective tariffs—the freedom of domestic producers from a tax that affects outside producers who compete in domestic markets—have favored the rise of new industries and the production of numerous products in areas where such production would not have been economically feasible under freely competitive conditions. Property tax concessions involving exemptions and preferential assessment arrangements are used in many states to induce industries to refurbish and modernize existing plant facilities and build new factories at particular sites. Comparable incentives are sometimes used by communities to attract new industries.[21] Similar approaches have a potential for facilitating the rebuilding and rejuvenation of deteriorating central city areas.

[21] Cf. Lewis H. Kimmel, *Taxes and Economic Incentives* (Washington, D.C.: Brookings Institution, 1950), chap. 7; and Paul E. Alyea, "Property-Tax Inducements to Attract Industry," in Richard W. Lindholm, *Property Taxation: USA* (Madison: University of Wisconsin Press, 1967), pp. 139–58. New York City's practice of freezing preimprovement assessment levels for twenty-five years on residential property remodeling projects provides another example of a tax concession of questionable desirability. Netzer indicates that this practice has favored the renovation and holding of old properties in preference to private redevelopment efforts that could provide new apartments. Cf. Netzer, *op. cit.*, pp. 83–85.

Taxing arrangements have also been used to discourage or prohibit certain types of investments and practices. The Soviet Union has used prohibitive taxes in its virtual elimination of the free peasantry; Chile has used penalty taxes on newly planted vineyards to discourage further expansion of the area used for this purpose; Indonesia uses a tax on tobacco production for a similar reason; the Gambians of Africa have used a "stranger farmer tax" to discourage the seasonal leasing of cropland to migrant farmers; and at one time several American states taxed purchases of oleomargarine to discourage its use in competition with butter.

The federal and state governments can use taxes for regulatory and nonfiscal purposes. But in these uses of the taxing power, they must put up a pretense of collecting revenue; and in the case of the federal government, the government cannot attempt to do by taxation that which it has no authority otherwise to do. Acting in this regard, the federal government has used tax measures to drive state bank notes out of circulation, prevent the manufacture of white phosphorus matches, discourage the sale of certain firearms to unlicensed parties, and control the importation and sale of products such as adulterated butter, marijuana, and opium. Its attempt to use taxes to control child labor and to limit agricultural production, however, were held invalid on the ground that these functions were reserved to the states under the Tenth Amendement.[22]

Enhancement of Property Values

Taxes can have important side effects that either enhance or depress property values. Enhancement of values has happened more often than not as the revenues provided by taxes have been used to provide services, such as police and fire protection; garbage collections; community planning; and good streets, parks, and school facilities, that property owners and prospective investors associate with high real estate values. As long as the benefits associated with tax payments exceed their costs and do not exceed the cost at which owners could provide satisfactory levels of services for themselves, taxing programs contribute to rising property values.

Property values can be depressed when taxes lead to higher business or household operation costs without providing services of compensating value. It is in situations of this nature that tax capitalization operates. Taxes are viewed as an additional cost, and real estate values are reduced by the capitalized amount of this cost. Taxes designed to provide new services enhance property values only when an effective demand exists for the services provided. Experience with tax programs that have provided public services for premature subdivisions clearly show that these programs cannot add to property values in the absence of consumer demands. Far from adding to real estate values, these programs can complicate the ability of owners to hold their lands and have sometimes led to tax delinquencies and forfeitures.

[22]Cf. *Bailey* v. *Drexel Furniture Co.*, 259 U.S. 20 (1922), and *United States* v. *Butler*, 297 U.S. 1 (1936).

OPERATION OF THE PROPERTY TAX

Land taxation in the United States finds its most important single application with the general property tax. In our examination of the leading issues associated with the operation of this tax, attention is centered on the steps involved in its administration, the problem of tax delinquency, an evaluation of its strengths as a tax, and proposals for its improvement. The administration of the property tax involves a series of procedures often described as the *property tax calendar*. These steps can be grouped under the following headings: (1) assessment of properties, (2) review of assessments, (3) equalization between assessment districts, (4) determination of the tax levies, (5) collection of the tax, and (6) possible appeals.

Assessment of property. As a first step in this taxation process, every taxable property must be located, listed, and appraised for taxation. This assessment step is quite simple in some districts, while it is often complex in others. Assessors who work with rural properties used for similar purposes ordinarily find it easy to locate each holding, establish who the owner is, list the owner's name and the property's legal description on the assessment roll, and then assign the property an assessed value that represents its value relative to other taxable properties. Assessments in urban areas, in contrast, are often complicated by the smaller size of individual properties and the more heterogeneous nature of their property uses. Assessors in these areas frequently find it desirable to use elaborate filing and valuation procedures in the pursuance of their duties.

In theory, all properties should be assessed at their full market value or at some specified proportion of their market value.[23] With ideal assessment conditions, all properties should also be reassessed at frequent intervals. Neither of these rules is rigidly adhered to in practice. Some properties are almost invariably assessed at higher proportions of their market values than others within the same assessment districts. And once an assessed value is accepted, it is often carried over year after year on the assessment rolls with only occasional adjustments for additions of new improvements or for losses resulting from factors such as fire. These practices naturally result in numerous inequities both within and between assessment districts.

The assessment process involves only a few days's work for a single assessor in some districts while it calls for the year-round services of a large office force in others. Regardless of when properties are visited, assessments are ordinarily made as

---

[23] Twenty states had statutory guidelines that called for assessment of properties at 100 percent of their full market value in 1968. Nineteen had statutory assessment ratios pegged to some flat percentage of full market value, six used locally determined ratios, three used property tax classification systems that called for assessing different classes of property at varying rates, and two were guided by the objectives of "just value" and "fair value." Cf. Paul V. Corusy, "Improved Property Tax Administration," in Arthur D. Lynn, Jr., ed., *The Property Tax and Its Administration* (Madison: University of Wisconsin Press, 1969), pp. 66; and D. R. Epley, "The Issue of Functional Assessment," *American Real Estate and Urban Economics Association Journal*, 2 (Spring 1974), 57–73.

of some uniform data. Differences also exist in the location of the assessment responsibility. This function is handled at the county level in more than half of the states; on a township, city, or village level in about a fourth of the states; and on a joint county and local basis in the remaining states. Most assessors have been elected to office in times past, but during recent decades, more and more units have shifted to an appointive system and the employment of assessors who have technical training in appraisal work.

Review of assessments. After all the properties in a given tax district are assessed, an assessment roll is prepared and property owners are ordinarily advised of any changes in the assessed values of their properties. At this point, property owners have the privilege of meeting with a board of review for the assessment district to protest their assessment.

This board has the power to correct assessment errors and lower those assessments that involve the inequitable or unjust treatment of particular taxpayers. In presenting protests to this board, it is not sufficient for property owners to merely assert that their taxes are too high. They must shoulder the burden of proving that their properties are assessed at higher proportions of their fair market value than are certain other properties. In this process, protesting owners often operate at a disadvantage because of their frequent lack of information concerning the assessed and market values of other properties.

Equalization between assessment districts. Although the assessment review process helps to iron out inequities within individual assessment districts, it has little effect on possible inequities between districts. These inequities exist because of the tendency of each assessment district to use its own standards in determining the proper level for its assessed values. A city, for example, may try to assess all its properties at 80 percent of market value while five neighboring units may assess their properties at 67, 50, 40, 33, and 25 percent of market value, respectively. These differences would have little significance if taxes applied to single assessment districts. But definite inequities can arise when school districts overlap two or more assessment districts or when local units bear a share of a countywide or statewide property tax.

The smoothing out of these inequities between assessment districts is called *tax equalization.* In states with local assessors, the first step in the equalization process is usually carried out by a county board of equalization. This board examines the corrected assessment rolls submitted by the local units, considers such other information on property values as it may have, and then proceeds to raise or lower the various assessments totals, so that each district has an equalized value that represents the same relative proportion of the market value of its properties.

These county-equalized values—and the county-assessed values in those states using county assessors—are in turn sent on to a state tax office, where a state equalization of county assessments is made. State equalizations are necessary in those states that levy a state property tax. They are used in computing the taxes applied in school and other taxing districts that overlap county lines. And they are also used in some states as a basis for the distribution of certain state-collected taxes and state aids.

Levying the tax. The next step in the property taxation process involves the actual levying of the tax and determination of the total tax due on each property. Acting under the powers delegated to them by the state legislatures, the governing bodies of the various units of government (1) determine the amounts of money they need for operating and other purposes during the coming year, (2) formally appropriate funds for this purpose, and (3) levy a property tax for that portion of the appropriation to be financed by property taxes. This tax levy is divided by the total assessed value of the tax district to determine the tax rate. A local unit with an assessed value of $5 million and a tax levy of $60,000, for example, would have a tax rate equal to 1.2 percent of its assessed value ($60,000 divided by $5 million = 0.012).[24] This tax rate is frequently referred to as a *millage rate*—12 mills in this example—because it represents the number of mills of tax to be collected for each dollar of assessed valuation.

Once the tax rates for all the units of government affecting a given area are known, they are totaled and a computation is made of the actual tax due on each property. A tax bill is then sent to the property owner with information concerning the various tax rates applied, the legal description and assessed value of the property taxed, the date when the tax is due, and the period within which it can be paid without penalty.

Collecting the tax. Most state tax laws specify a date when property taxes are due. From this date on, the tax constitutes a lien on the property which can be discharged only through payment of the tax or eventual tax sale and forfeiture. Property tax notices are mailed on or near the due date. Periods of several days or weeks are then allowed during which the tax can be paid without penalties. Following the final payment date, all unpaid taxes are regarded as delinquent. Discharge of the tax lien against a property then calls for payment of the full tax plus penalties plus interest.

In times past, most states had a single final payment date—usually some date after harvesttime when farmers were most able to pay. Approximately three-fourths of the states have since deviated from this pattern to permit tax payments in two or more installments. These payments are usually collected by the treasurers of the units of government that administer the assessment process. Once a tax becomes delinquent, the collection function is ordinarily shifted to the office of the county treasurer.

---

[24]This tax computation process is more complicated when a tax levy applies to properties included in two or more assessment districts. In a typical case, a county spreads its tax levy between its local assessment districts in proportion to their county- or equalized-assessed valuations. Each local district then divides its share of the county tax by its local assessed valuation to determine its county millage rate for the year. State-equalized values are substituted for local assessed values in their determination of actual tax levies in some states.

Possible appeals. Dissatisfied owners can usually appeal their assessments to a state tax agency or a court. Appeals can also be made to the courts when taxpayers feel that a tax levy is illegal. Owners in these cases ordinarily find it best to pay their taxes under protest. Then if their appeal is upheld, any overpayment made will be returned to them.

Tax Delinquency Problems

Property taxes are levied primarily for the purpose of collecting revenue; and as revenues are necessary for the conduct of local government, it is assumed that all taxes must be collected. Experience has shown, however, that many property owners fail to pay their taxes before the penalty date. Special provisions are thus necessary to cover the collection of delinquent taxes.

These provisions vary from state to state but usually allow periods of several months during which delinquent taxpayers can pay their back taxes plus penalties and interest charges. If the tax remains unpaid at the end of this period, the tax lien on the property is offered at a public tax sale. Outside buyers can then acquire tax certificates to the delinquent properties by paying a minimum price equal to the sum of the tax and its accumulated charges. With this sale, local govenments receive their tax money, and the buyer of the tax certificate acquires a legal interest in the property, which ripens into ownership if the delinquent owner fails to redeem the property within a specified redemption period. When no one bids on a delinquent tax, it is ordinarily bid off in the name of the state or county.

Tax delinquency is an infrequent problem during prosperous times. Penalties for late payment of taxes are used mostly as a prod to get people to pay their taxes on time. With depression conditions, tax delinquency often becomes a serious problem because of the inability of many owners to meet their tax obligations. This was the situation during the 1930s when the average tax delinquency rate for the nation went up to 17 percent (1932-1933) and three-fourths or more of the taxes levied by some units went uncollected in some years.[25]

[25]Tax delinquency was most serious in areas with premature and submarginal developments. Thousands of unsold lots around most major cities were sacrificed through nonpayment of taxes. Serious delinquency problems also occurred in the cutover portions of the Lake States, the Pacific Northwest, and the Ozarks; in the drained areas of the South; and in the drought-stricken portions of the Great Plains States. More than 90 percent of the rural drainage and levee district taxes were delinquent in some Arkansas counties in 1933; and 93 percent of the tax levy of the Everglades drainage district in Florida was delinquent in 1936. Several townships in Minnesota reported complete delinquency for some years. For a more-detailed discussion of this problem, cf. *Tax Delinquency and Rural Land-Use Adjustment*, National Resources Planning Board Technical Paper No. 8, 1942; Raleigh Barlowe, *Administration of Tax-Reverted Lands in the Lake States*, Michigan Agricultural Experiment Station Technical Bulletin No. 225, 1951; and A. M. Hillhouse and Carl H. Chatters, *Tax-Reverted Properties in Urban Areas* (Chicago: Public Adminsitration Service, 1942), chap. 1. For a more recent analysis of the problem as it affects urban properties, cf. Robert W. Lake, *Real Estate Tax Delinquency* (New Brunswick: Rutgers, The State University of New Jersey, Center for Urban Policy Research, 1979).

The experience of the 1930s illustrated that costly governmental services cannot be supported by an inadequate tax base. Steps were taken in many areas to curtail local government costs; state aids were extended to local units; and several local functions were shifted to the states. Tax moratoria, special tax settlement arrangements, and delays in tax sales were also used to soften the ravages of the tax-delinquency problem. Yet even with these ameliorative measures, the tax-delinquency problem remained serious until the return of better times.

Much of the confusion and indecision that surrounded the tax foreclosure policies of many states during this period gradually cleared during the late 1930s as states and counties took tax title to large areas of hopelessly delinquent land. Much of the tax-reverted land was soon restored to the tax rolls. But several states and counties broke away from their previous practice of acting mostly as land brokers in transferring forfeited properties from one private owner to another. Instead, they looked on their tax-reverted holdings as a "new public domain," which should then be adminstered according to its best use. These states and counties sold considerable portions of this new public domain for private uses but also dedicated large areas for public forests, grazing, recreation, and other uses.

Evaluation of the Property Tax

Like other taxes, the property tax has its strengths and weaknesses. These features can best be appraised when they are considered in light of the principal canons or criteria of a "good" tax. Important among these are the assumptions that a sound tax should (1) be related to the taxpayer's ability to pay, (2) reflect the benefits received from public expenditures, (3) provide a reliable and uniform yield, (4) be both easy and economical to administer, and (5) be both familiar to the taxpayer and socially expedient.[26]

As late as a century ago, most of the nation's wealth was held in tangible goods and properties that tax assessors could see and touch. But even then, many owners were "land poor" in the sense that their income potentials were not adequate to carry their tax loads and other operating expenses during rough times.

With the changes of the past century, the relationship between real estate ownership and ability to pay has become worse rather than better. Most of our larger personal incomes are now associated with salaries, fees, and dividends rather than direct income from real property. Intangibles have become more and more impor-

[26] Adam Smith argued that a sound tax should reflect ability to pay, be certain and understandable, convenient to pay, and take as little from taxpayers as necessary. [Adam Smith, *The Wealth of Nations* (1776) (New York: Random House, Modern Library Edition, 1937), pp. 777–79]. Richard and Peggy Musgrave, *Public Finance in Theory and Practice*, 3rd ed. (New York: McGraw-Hill, 1980), p. 235 assert that a good tax should be equitable with everyone paying their share, minimize interference with economic decisions, disturb the "equity of the system" as little as possible when used for nonfiscal purposes, facilitate stabilization and growth, be understandable and not subject to arbitrary administration, and have low compliance and administrative costs.

tant; and many people with high capacities to pay taxes now own very little real property. This situation has brought numerous inequities. Some property owners pay a tax commensurate with their ability to pay; some pay less than their fair share of the cost of local government; and some pay higher taxes than they should with their limited ability to pay.

Property taxes are often more closely associated with benefits received than with ability to pay. This is particularly true when a high proportion of the collected tax is used to provide protective services and community facilities that enhance property values. Inequities exist, however, between the relative benefits enjoyed by different properties. One might ask, for example, does a $20,000 vacant lot receive as much fire protection as a $20,000 building or a $20,000 automobile? Or does a new fireproofed building with an assessed value of $400,000 receive twenty times as much protection as an older, more antiquated structure with an assessed value of $20,000? Going beyond these questions, it may be noted that tax revenues are often used for nonprotective purposes such as schools, streets, and welfare services. The benefits from these expenditures accrue more to people than to properties and often have only indirect effects on property values.

Some of the leading arguments favoring the property tax center on its uniformity of yield, its ease of administration, and its long acceptance. Unlike many other taxes, the revenues collected through the property tax need not fluctuate with changes in the business cycle. They provide a generally reliable source of public revenues; and except for possible increases in tax delinquencies, collections can be as high during recessions as during periods of prosperity. This reliability and uniformity of yield is naturally desirable as far as local governments are concerned. But it can mean that individual taxpayers are hit hardest at the very time they are least able to pay.

The real property tax compares very favorably with other taxes in its ease and cost of administration. Problems naturally arise with the assessment and equalization steps. But the fixed and continuing nature of real property practically guarantees the collection of a tax once it is levied. Property taxes are seldom evaded or avoided. They are ordinarily paid on time, and the total revenues collected from this source was usually high relative to the costs of tax administration.

People seldom show enthusiasm for paying taxes. But they are more inclined to accept a tax with which they are familiar, which they think they understand, and with which they have long been accustomed than the unknown quantity of a new tax. In this respect, social expediency favors the retention of a long-accepted tax such as the property tax which, though widely criticized, has become an accepted part of our economic system.

In addition to the criticisms of the property tax listed above, the usual lump-sum or two-installment payment systems make the property tax an inconvenient tax for most people to pay. Its heavy reliance on a residential real property tax base makes it burdensome for homeownership. Its failure to respect periods of income favors those operators who enjoy a rapid turnover of capital or inventory as compared with those who suffer from crop failures or those who invest in long-term forestry ventures. High property taxes in urban communities have encouraged some migration of families and industries to lower-tax areas. And separate taxation of real properties and intangible properties such as mortgages can result in double taxation.

Every tax has its weak as well as its strong points. The property tax is probably no worse in this respect than most other taxes. Our problem in tax policy is not so much in finding a perfect tax as it is in combining taxes in such a way that they complement each other with the strong features of one tax balancing the weak features of another. Some of the weaknesses of the property tax such as its alleged regressivity are compensated when it is used alongside a graduated income tax.

Possible Modifications and Improvements

Few taxes have been subject to more criticism than the property tax. As one might expect, this criticism has been accompanied in many cases with suggestions for its modification. These suggestions fall into three classes: (1) insistence on need for better administration, (2) proposals for changes in the tax, and (3) recommendations for partial or complete replacement of the property tax with other sources of revenue.

Need for better administration. As long as the property tax is retained in its present form, every effort should be made to administer it in an efficient and equitable manner. The need for better property tax administration is nowhere more widely felt than in the case of property assessments. Many assessors do a professional job in appraising properties and keeping their assessed values up to date. But the prevailing pattern in some districts is one of haphazard assessment practices, with taxpayers sometimes reporting their own property values and with local assessors all too often limiting their function to the copying of last year's values onto this year's assessment roll.

Better administration calls for profesionalization of the assessment function. Steps are needed to either (1) train and supervise local assessors so they can carry on their assessment activities in a satisfactory manner, or (2) combine assessment districts to permit the employment of full-time salaried assessors who have both training and experience in property appraisal work. Most tax authorities favor the use of appointed rather than elected assessors, the centering of the assessment function in a county assessor, and state assessment of properties such as railroads and public utilities.[27]

Considerable progress was realized during the 1960s and 1970s in improving and updating assessment practices.[28] Improvements of a comparable nature are needed in the equalization process if taxpayers are to enjoy equitable tax treatment. The property tax collection function should be centralized in single officials who collect the taxes levied in one or more taxing districts. Wider use might also be made of optional installment or pay-as-you-go methods of making tax payments.

---

[27] Schultz and Harriss, *op. cit.*, p. 347; cf. also Daniel M. Holland, ed., *The Assessment of Land Value* (Madison: University of Wisconsin Press, 1971).

[28] Cf. *The Property Tax in a Changing Environment* (Washington, D.C.: Advisory Committee on Intergovernmental Relations, 1974).

Good property tax administration is invariably tied up with the costs of local government. These costs must always be kept in line with the tax-paying capacity of the local property tax base. When local governments fail to abide by this basic rule, even the best administration breaks down. Avoidance of the conditions that foster tax delinquency thus becomes a paramount objective of good property tax administration.

Possible modifications of the property tax. Numerous proposals have been made for modifications of the property tax. Some of the more important of these involve the use of tax rate limitations; circuit-breaker arrangements; classified property taxes; special tax exemptions; and a possible shift to taxes on net worth, site values, or unearned increments.[29]

Most states have constitutional or statutory provisions that set ceilings for the tax rates levied by their various units of government. These rate limitations take different forms. Some prescribe maximum overall rates for all property taxes or maximums for specific state or local levies. Some limitations are graduated according to the assessed valuation or population of the tax district; and some specify maximum permissible percentage increases over the tax levies of previous years. Each of these approaches is designed to prevent excessive taxation. Except for the overall limitations, they usually allow most government units considerable flexibility and freedom in their use of the taxing power.

Overall limitations have been adopted in several states for the specific purpose of keeping property taxes low. Limitations were adopted to reduce property taxes to seventeen mills in Oklahoma in 1907, ten mills in Ohio in 1911, fifteen mills in towns and cities and ten mills in rural areas in Indiana, fifteen mills in Michigan, twenty mills in New Mexico, forty mills on assessments at 50 percent of value in the state of Washington, and five to twenty mills in West Virginia in 1932 and 1933. Rapidly rising property taxes brought new demands for property tax relief during the 1970s.[30] The two most publicized limitations of this period were adopted in California in 1978 and Massachusetts in 1980. California's Proposition 13 rolled property taxes back to a maximum of 1 percent of their 1976 assessed values and limited increases to a maximum of 2 percent each year. Massachusetts' Project $2\frac{1}{2}$ set a maximum rate of $2\frac{1}{2}$ percent of the assessed values of property after July 1, 1981.[31]

These maximum rates can sometimes be boosted for specific purposes and specific time periods by popular referendum. But even with this flexibility, they are ordinarily criticized by tax and fiscal authorities because of the effect they have in

---

[29] Cf. Shultz and Harriss, *op. cit.*, pp. 365–69; Roland R. Renne, *Land Economics*, 2nd ed. (New York: Harper & Row Publishers, Inc., 1958), pp. 276–86; Netzer, *op. cit.*, chap. 8; James Heilbrun, *Real Estate Taxes and Urban Housing* (New York: Columbia University Press, 1966), chap. 6; and Mason Gaffney, "Land Rent, Taxation, and Public Policy," *Regional Science Association Papers*, 23 (1969), 141–53.

[30] Cf. Steven David Gold, *Property Tax Relief* (Lexington, Mass.: Lexington Books, 1979).

[31] Cf. George G. Kaufman and Kenneth T. Rosen, eds., *The Property Tax Revolt: The Case for Proposition 13* (Cambridge, Mass.: Ballinger, 1981); and Alvin Rabushka and Pauline Ryan, *The Tax Revolt* (Stanford, Calif.: Hoover Institution, 1982).

cramping the fiscal activities of local governments. Adoption of these limitations has often led to a rigid curtailment of government services, an increase in local governmental debt, a search for new sources of public revenue, and frequent increases in assessed valuations. On the positive side, these blanket limitations have helped to reduce property taxes and have forced many local units to look to sources other than the property tax for part of their revenues.

A second modification involves the use of circuit-breaker arrangements with property taxes to reduce the burden they pose for low- and middle-income property owners. With the typical circuit-breaker arrangement, taxpayers are allowed state income tax offsets or receive tax rebates for stated portions of the amount that their property tax payments exceed some percentage of their household incomes (for example, 60 percent up to a maximum of $1,200 of the amount that one's property taxes exceed 3.5 percent of household income in Michigan). Arrangements of this order can be used to counterbalance the frequent regressivity of property taxes. A system of graduated property taxes—taxation of high-valued properties at higher proportions of their market than low-valued properties—has been suggested as another means to attain this end but thus far has attracted little support in the United States.

Complete acceptance of tax uniformity would require use of a single standard in the assessment and taxation of all types of property. Most states have deviated from this standard by authorizing at least one of three different types of special tax treatment. Minnesota and some other states use a classified property-assessment system under which different types of real property are assessed at different proportions of their market values. Approximately half the states use classified rates in their taxation of intangibles and tangible personalty; and several use *in-lieu-of-tax-arrangements* such as severance or gross receipts taxes in their taxation of forests, minerals, and railroads. These classified property taxes have drawbacks. But they recognize that (1) some types of property have greater ability to pay taxes than others, and (2) equality in tax treatment does not necessarily mean equitable treatment.

Special tax exemptions represent another deviation from the uniformity principle in property taxation. Public properties are exempt from property taxes. This exemption causes a minor problem in most tax districts. Real problems exist, however, with those districts where most of the property is held in public ownership. The federal government has recognized its implicit obligation to help support the costs of government in those areas where it holds considerable productive property. Annual cash grants are made to the government of the District of Columbia. Twenty-eight different revenue-sharing arrangements were used with the nation's larger landholdings in 1970. These ranged from TVA's payment of 5 percent of its gross receipts from power sales to the units of government in which it operates, the U.S. Forest Service's payment of 25 percent of its receipts from sales of timber products, and the Department of Interior's payment of 37.5 percent of the proceeds from oil, gas, and mineral leases and 50 percent of the grazing fees received from the leasing of some classes of grazing lands, to the Army Corps of Engineers' payment of 75 percent of the revenues received from acquired flood control lands. No payments were made to the state or local governments for some types of land held in federal ownership, whereas payment-in-lieu-of-tax arrangements ranging up to the approximate equivalent of local property taxes were applied with others. Several states also make payments in lieu of taxes to their local units on state lands held in forests, parks, and game areas.

Comparable exemptions also apply to cemeteries and to properties used for charitable, educational, and religious purposes. In addition to these exemptions, many states have gone further to exempt personal properties and numerous specified items and products from property taxation. Homestead exemptions and veterans exemptions are used in many states to favor particular classes of citizens by exempting them from certain taxes on the first $500 to $25,000 of their assessed valuations. Exemptions are also used in some states to foster the location of manufacturing plants.

Several arguments can be advanced for granting special tax exemptions. But experience has shown that exemptions can be abused. Decisions to exempt incorporeal properties such as stocks and bonds and various classes of personal property from property tax assessment have often had political overtones and, by reducing the total tax base, have naturally had adverse effects on those taxpayers whose property consists mostly of real estate.

The most significant proposals for modifications of the property tax would shift the emphasis in property taxation from a tax on property values to taxes on (1) net worth, (2) land site values, and (3) unearned value increments. A net worth tax would be concerned with the equities owners have in their properties plus the value of other assets they own. Because of its all-inclusiveness, this tax approach could be more equitable than the property tax.[32] But it would be harder to administer, and any taxes on mortgages would probably be shifted to mortgaged owners.

Henry George's followers have championed proposals for site-value taxes that would levy taxes on building sites and bare land values while exempting the value of land improvements.[33] Variations of this proposal have been applied in some American and Canadian cities and also in Australia, Jamaica, and New Zealand.[34] This taxation approach would provide economic incentives for developing land, intensifying its use, and shifting sites to their commercial highest and best uses. Removal of the tax on improvements could have beneficial effects in encouraging owners to invest more in building and land improvements than they are willing to do under

---

[32] Cf. C. T. Sandford, J. R. M. Willis, and D. J. Ironside, *An Annual Tax on Wealth* (London, U.K.: Heinemann Educational Books, 1975).

[33] Cf. Mason Gaffney, "An Agenda for Strengthening the Property Tax," in George E. Peterson, ed., *Property Tax Reform* (Washington, D.C.: The Urban Institute, 1973), pp. 65–85; and Gaffney, "Taxation to Release Land" in Marion Clawson, ed., *Modernizing Urban Land Policy* (Baltimore, Md.: Johns Hopkins University Press, 1973), pp. 115–51.

[34] For descriptions of these programs, cf. A. M. Woodruff and L. L. Ecker-Racz, "Property Taxes and Land-Use Patterns in Australia and New Zealand," and Daniel M. Holland, "A Study of Land Taxation in Jamaica," in Arthur P. Becker, ed., *Land and Building Taxes* (Madison: University of Wisconsin Press, 1969), pp. 147–86 and 239–86; and Kenneth Taeuber, "A Century of Australian Experience with Land Value Taxation," in *1966 International Seminar on Land Taxation, Land Tenure, and Land Reform in Developing Countries* (West Hartford, Conn.: John C. Lincoln Foundation and University of Hartford Press, 1966), pp. 128–74 and 210–38.

the property tax. But site-value taxes could also have adverse environmental effects by discouraging the private holding of natural and open space areas and large land-scaped lots such as now grace many urban and suburban neighborhoods.

Proposals for land-value increment taxes had an early advocate (1848) in John Stuart Mill. This tax would impact on increases in values or rents between two spec-ified time periods—the date one acquires a property and the date one sells it. It has counterparts in the capital gains taxes levied on increases in the market value of land during one's period of ownership and Britain's land development tax of 1976, which is levied on increases in value resulting from the granting of governmental permission to develop sites for higher and better uses.[35]

Replacement of the property tax. Some advocates of property tax reform argue that it is not enough to merely patch up the present property tax. Instead of settling for modifications and administrative improvements such as those listed above, they sometimes recommend partial or complete replacement of the property tax with other sources of revenue.

Realistic appraisal of the present situation suggests little prospect for the com-plete replacement of the property tax. A strong case can be made, however, for the limitation and partial replacement of this tax in the future. In this respect it may be noted that a gradual replacement process has been going on for several decades. Most states have withdrawn almost entirely from the property tax field; and many once locally supported activities such as those that involve highways, public health, and welfare are now financed on a state basis. With this shift to new sources of revenue, the property tax accounted for only 30.7 percent of the tax revenues (not including intergovernmental aids) of the state and local governments in 1980 as compared with 88.4 percent in 1890.

Despite this decline in its relative importance as a tax, the total property tax burden has increased manyfold. The growing demand for new governmental services has brought need for more and more tax revenues in most taxing districts. This demand has resulted in tax levies in some areas that could easily provoke a wave of tax delinquency under depressed business conditions. In facing up to this problem, two points bear emphasis: (1) a conscious effort should be made to limit property taxes to the long-run paying capacity of the properties taxed, and (2) new sources of revenue should be sought to supplement the revenues now secured from the property tax for financing local governments and public schools.

—SELECTED READINGS

Groves, Harold M., and **Robert L. Bish,** *Financing Government,* 7th ed. (New York: Holt, Rinehart & Winston, 1973), chaps. 3–6.

Jeddeloh, James B., and **Cheryl G. Perkins,** *Real Estate Taxation: A Practical Guide* (Reston, Va.: Reston, 1982).

---

[35] Cf. Prest, *loc. cit.*

Lindholm, Richard W., ed., *Property Taxation: USA* (Madison: University of Wisconsin Press, 1967).

Lindholm, Richard W., and Arthur D. Lynn, Jr., eds., *Land Value Taxation* (Madison: University of Wisconsin Press, 1982).

Lynn, Arthur D., Jr., ed., *The Property Tax and Its Administration* (Madison: University of Wisconsin Press, 1969).

Musgrave, Richard A., and Peggy B. Musgrave, *Public Finance in Theory and Practice*, 3rd ed. (New York: McGraw-Hill, 1980).

Netzer, Dick, *Economics of the Property Tax* (Washington, D.C.: Brookings Institution, 1966).

Renne, Roland R., *Land Economics*, 2nd ed. (New York: Harper & Row Publishers, Inc., 1958), chaps. 13 and 14.

Shultz, Willian J., and C. Lowell Harriss, *American Public Finance*, 8th ed. (Englewood Cliffs, N.J.: Prentice-Hall, 1965), chaps. 7, 18, and 19.

17

PLANNING
FOR BETTER LAND USE

The twentieth century has blossomed forth as the age of the common man. It has unfolded as an era of wondrous technological advance and rising material standards of life. Its benefits have been extended to people of all classes and have helped to free millions of individuals from the drudgery and servile status of earlier days. It has helped average men and women to emerge as a political force with power to demand the rights of first-class citizenship.

But the twentieth century has also become a time of increasing economic interdependence. With more people, more competition for the use of resources, and the growing complexity of modern economic life, new tensions and conflicts of interest have evolved. Group action has been substituted in numerous cases for the individual laissez-faire-type action that characterized the eighteenth and nineteenth centuries. Popular opinion has made it expedient for governments to expand their functions and to coordinate, direct, and plan many activities once reserved to the private sector of the economy. Along with its other characteristics, the twentieth century has become a plan age, an era of expanded social action.

This chapter and the next chapter are primarily concerned with the role society can and should play in directing and controlling the use of land resources. Emphasis is given in this chapter to the nature of land resource planning and to the uses made of land reform and other planning measures for securing better land use. Chapter 18 examines the principal powers governments use in directing land use.

NATURE OF LAND RESOURCE PLANNING

Planning may be defined as the conscious direction of effort toward the attainment of a rationally desirable goal. Webster's dictionary defines *plan* as a "scheme of action, procedure, or arrangement."[1] With this definition, *planning* may be described simply as the devising or projection of a course of action. Ability to reason and plan ahead is an attribute of human beings. We all plan our day's work, our vacations, and the houses we build. Business operators plan when they combine resources in various ways with the intent of maximizing their returns. Corporation executives plan when they devise production and sales programs to increase their shares of a market. Families plan when they budget their incomes to permit home ownership, retirement for the parents, or college educations for the children. Governments plan every time they develop policies to achieve particular ends.

Private and public planning play a necessary role in modern society. Our present economy and high material standards of life have not just happened. They have come as the end product of the plans of millions of individuals, groups, and government bodies. Most of our planning has been and probably will continue as private planning. Yet varying amounts of overall planning and coordination are needed if people are to work peacefully together with some thought of maximizing social welfare.

Needs for private ownership and private planning are normally taken for granted. But wide differences of opinion are associated with almost every mention of public planning. These differences have had an unfortunate impact on the view some people have of the planning process. Instead of seeing planning as an essential aspect of our way of life, they sometimes regard it as an ugly symbol of possible repression. In practice,

> the word "planning" has been widely and loosely used. It has meant different things to different people. To crusaders it has been a Holy Grail leading to the sunlit hills of a better day. To conservatives it has been a red flag of regimentation heralding the dawn of collectivism and the twilight of the old order of free private enterprise and the democratic way of life. But to the humble practitioners of the art, viewing the matter with the cold eye of engineering rationality and a matter-of-fact indifference either to crusades, Red hunts, the class struggle, or the omnipotent state, it has been merely a process of coordination, a technique of adapting means to ends, a method of bridging the gap between fact-finding and policy-making.
>
> Planning is the opposite of improvising. In simple terms it is organized foresight plus corrective hindsight.[2]

In the discussion that follows, consideration is first given to the case for public

---

[1] *Webster's New Collegiate Dictionary*, 2nd ed. (Springfield, Mass.: G. and C. Merriam Co., 1953), p. 644.

[2] George B. Galloway, *Planning for America* (New York: Holt, Rinehart & Winston, 1941), p. 5.

planning, the nature of the planning process, and some examples of land resource planning. Emphasis is then focused on the uses that may be made of land reform and other similar programs to secure better land use.

The Case for Public Planning

No one can deny the need for some government planning. Like their counterparts in private business, governments must plan their functions and the uses they make of resources if they are to make order out of what might otherwise be chaos. But granting the fact that governments must plan when they provide for national defense, for the maintenance of civil order, and for balancing their expenditures against prospective revenues, why must the modern state engage in economic and social planning? Why shouldn't we return to the relatively planless conditions of the early 1800s?

Several factors have contributed to our present interest in economic and social planning. With an increase in population numbers and the development of an industrial society, people have been brought into closer contact with one another. The subsistence operator and the self-sufficient community have become things of the past. Our health, safety, and welfare as individuals often depend on the activities and decisions of people we never see, people who may live hundreds of miles away from us. This dependence on others, in combination with the problems generated by the pressure of a larger and more consumption-minded population on a limited resource base, has multiplied the prospect for individual and group conflicts of interests—conflicts that could easily upset our normal working and living arrangements were it not for our use of planned group action to designate and enforce certain minimum rules of the game.

Economic and social planning is not a recent innovation. Governments have used various types of public planning since the dawn of history. Ancient Egypt had its grain-storage and pyramid-building projects. Rome had its public works and its policy of "bread and circuses." Even at the peak of the laissez-faire period, the United States had its protective tariff, public land disposal programs, and experiments with monetary and credit policy. But while public planning is not new, much of the emphasis now given to economic and social planning can be credited to changing attitudes concerning the rightful functions of government.

Prevailing attitudes regarding the function of government in planning economic and social matters have changed considerably in the United States during the past century. Until the end of the nineteenth century, most Americans held religiously to a philosophy of rugged individualism. They believed that "that government governs best which governs least." They espoused a laissez-faire doctrine, which assumed that an "unseen guiding hand" would operate to coordinate individual actions and bring forth the optimum in social welfare.

During the late 1800s, many average people began to question and reorient their attitudes concerning the desirability of unchecked free enterprise. Abuses associated with the rise of big business and the exploitive business practices of certain operators provided convincing evidence of the failure of unbridled individualism to maximize social welfare. Instead of the basic harmony of interests assumed by the laissez-faire doctrine, the uncoordinated self-seeking of millions of individuals frequently led to frustrating conflicts of interests, to human exploitation and misery, and to something far less than optimum social welfare.

At this stage, people began to turn more and more to group action as a means of minimizing conflicts and maximizing their joint interests. Business operators joined together in corporations and trade associations; laborers organized unions; farmers formed cooperatives and political pressure groups; and large blocs of citizens looked to the government for action programs that would enhance their economic and social interests. Whereas most individuals had once looked to government primarily for protection, justice, and the preservation of civil order, the realization spread that group action could and should be used to promote the social welfare and material well-being of the great mass of the citizenry. With this change in attitudes, it was easy to justify public economic and social planning, particularly when it contributed to the well-being of large portions of the citizenry either by expanding their opportunities and liberties or by minimizing the risks and uncertainties that would otherwise afflict them.

The demand for public action that followed has brought new social controls such as the government's supervision of interstate freight rates, pure food and drug regulations, and requirements for minimum safety standards in factories and mines. It has prompted social welfare measures such as urban renewal, the provision of public credit facilities, sponsorship of public multipurpose resource developments, social security, and public use of fiscal and monetary measures to stabilize the nation's economy. The first of these two types of social action has expanded the rights and liberties of large groups of people by limiting the opportunities some groups formerly had to exploit the interests of others. The second has sometimes brought the government into active competition with certain of its citizens; but it has gained popular support by assisting people to attain economic and social objectives that could not be attained as easily, as fast, or as well through strictly private action.

Rationale for land resource planning. Some of our most important planning needs involve land resources. This is true because the welfare and well-being of everyone in our society is dependent on how we use our common land resource base. If people were completely free to use land as they chose, the end result could be as chaotic and confusing as if all members of a symphony orchestra played their own tune. Some types of group guidance are necessary to secure order. A first major step in this direction came with the recognition of private property rights. Owners, with their exclusive rights of possession, can decide how individual tracts are used. As long as most owners adhere to common goals in the productive use of their lands, there is little need for more formal planning. But when some individuals begin to pursue goals that are in direct conflict with those of other owners or that run counter to the perceived interests of members of society, a case arises for the provision of additional community guidance in land use. At this point, society can exercise its inherent vested interest in securing orderly and effective use of its resource base to discourage unwise and wasteful practices, which may prove injurious to owners, their neighbors, their communities, or society at large.

Demand for public action to direct land-use practices seldom arises as long as there are sufficient supplies of land resources at the right locations to care for the needs of prospective users. With the growing competition for land resources that comes with increasing population numbers and rising per capita demands for land products, however, the case for social action takes on new meaning. Individual operators and community groups find that the free enterprise system does not always bring the best use or allocation of land resources. They find that unguided individual action often leads to resource exploitation, social waste, and a shifting of costs to other members of society. In their search for answers, they find that land resource planning can be used like a community road map to show how to get where they want to go in land use.

Land resource plans find their justification or rationale in the objectives they seek. These objectives range from the picayune to the grandiose. Most public land resource planning in the United States has been designed to promote orderly development of the nation's land resources, minimize problems and conflicts associated with private land use, foster optimum development of the land resource base, and maximize public welfare. Our record in attaining these objectives has been good. Yet mistakes have been made. Plans have been abandoned or revised at times because they were considered overly ambitious or overly restrictive; and we have frequently erred in planning for too little rather than too much.

To plan or not to plan. Land resource planning is one of the most essential and most easily defended types of economic and social planning. Yet like other types of public planning, it is subject to varying amounts of criticism and fearful speculation. Some critics argue that public planning is less imaginative and less forward-looking than the plans of selected entrepreneurs. Others fear that public planning logically leads to more planning and that the planning road inevitably leads to a controlled economy, a police state, or George Orwell's *1984*. They question the motives of planners and ask whether we can really trust political leaders to stop with planning measures designed to enhance the public welfare. They question the willingness of the average citizens who make up the "mass mind" of our democratic society to resist the lure of public planning when it offers them "bread and circuses," job security, and other benefits in exchange for segments of their individual liberties.

In recognizing these arguments, it can be admitted that private plans can indeed be superior at times to public plans. It should also be noted that there is no inevitable conflict between the use of planning measures and personal freedom. Whether planning will mean more or less freedom for the individual always depends on the objectives and techniques of the planning process, the people who devise and administer the plan, and the individual's definition of freedom.

The term *freedom* means different things to different people. When it is defined broadly as "the absence of obstacles to the realization of desires,"[3] almost any type of public action may be regarded as a threat to the freedom of some individual. Public action can limit one's license to steal the property of others or to engage in socially undesirable practices; it can restrict personal privileges that one regards as vested rights; or it may limit one's opportunity to use private unemployment insurance or a private social security program in preference to that provided by government. By the same token, public planning can also contribute to the freedom enjoyed by large numbers of citizens by controlling forces or removing obstacles that prevent them from realizing their desires.

The concept of freedom has little meaning when treated in an abstract sense. As Erven Long has observed,

> It becomes meaningful only in specific terms which must include *a structure of opportunities* which makes the realization of the freedom possible. Any freedom enjoyed by any individual consists in a set of securities of expectations concerning the behavior of other individuals and groups. These secure expectations of an individual result from a system of rules—rights, duties, restraints—developed by processes of public action and enforceable at law—which determine what others may and may not do.
>
> . . . Any individual freedom exists in consequence of the public organization which defines this freedom and secures it for the individual against the adverse action of others. The slave became a free man, not in virtue of anything new put to him, but in consequence of a set of restraints imposed upon others. . . . By himself, an individual is anything but free—because only by joining efforts with others can he create a structure of opportunities for translating his "wishes" into realities.[4]

Viewed in this context, the concept of individual freedom is rooted in group action. Public-planning measures may be used to deprive criminals, draftees, and others of their so-called inalienable rights or to limit the vested rights of various individuals or groups. They may also be used to expand the freedom and opportunities available to large portions of the citizenry.

With the basic question of whether or not it is safe to plan, it should be noted that every society is the custodian of its own future. The planning process is neutral. A nation can plan to destroy its individual liberties; or it can use the planning process to maintain, strengthen, and expand these liberties. Democratic governments use planning techniques to enhance the public welfare and to attain socially desirable ends. Police states, in contrast, use planning as a technique of control—as a means

---

[3] Robert A. Dahl and Charles E. Lindblom, *Politics, Economics, and Welfare* (New York: Harper & Row Publishers, Inc. 1976), p. 29. This definition is also used by Barbara Wootton, *Freedom under Planning* (Chapel Hill: University of North Carolina Press, 1945), p. 4; and by Bertrand Russell "Freedom and Government," in Ruth Nanda Anshen, ed., *Freedom, Its Meaning* (New York: Harcourt Brace Jovanovich, Inc., 1940), p. 251.

[4] Erven J. Long, "Freedom and Security as Policy Objectives," *Journal of Farm Economics*, 35 (August 1953), 318-19; also Christian Bay, *The Structure of Freedom* (Stanford, Calif.: Stanford University Press, 1970), p. 15.

for repressing freedom, stifling criticism, and subjugating individuals to the will of the state. Public planning does not lead automatically to either end. If we cannot trust ourselves or our leaders, it is possible that planning will lead to certain abuses. But no nation can save itself from authoritarianism simply by refusing to plan. Far from being a threat to freedom, planning is necessary for the protection of freedom. Failure to plan can easily mean planning for failure.

Our past use of public planning in the United States has often led to more planning, particularly when the majority of the people and their representatives have felt that planning provided the best means to attain desired economic and social goals. But we have weighed each new benefit against its cost in individual liberty; and we have not hesitated to reject those types of planning we do not like. While we have placed the trust a democratic society must place in the actions of its leaders and in the majority will of its citizens, we have not forgotten that eternal vigilance is the price of liberty. The people of the United States have recognized that the willingness of a liberty-conscious people to use forward-looking, open, and well-advised planning as an instrument of public policy provides the best guarantee of continued freedom.

The Planning Process

Plans do not just happen. All planning, be it public or private, involves a process. As such, it embraces a series of important steps. These steps can be classified in different ways but normally involve (1) an assumption or determination of the objectives to be sought, (2) formulation of the plan or policy to be followed, and (3) execution of the plan.

Goals and objectives in planning. Every plan starts with some concept of a goal or objective, with some motive or purpose for the act of planning. This goal may be rooted in the individual's customary way of thinking and may be accepted without rational examination or thought; it may be the end product of the conscious thinking of an individual or a group of people; or it may be specified in the legislation that establishes a planning agency. Planning goals may be wise or foolish. They may be regarded as fixed for all time or be accepted as temporary objectives subject to future reevaluation and possible revision. Yet regardless of their nature, planning goals always play a vital role in giving direction to the planning process.

People engaging in the public-planning and policy-formation processes should always start with the question, Planning for what? They should think through their objectives and reasons for planned action. If they have time, they should formalize their goals by writing them down. They should examine them in detail and discuss them with others.

In this examination process, planners often find that they have several goals to consider, that some goals operate at cross-purposes with others, and that they must temper or compromise some objectives to attain others. The author of a housing improvement program may find that he or she must limit a goal of securing larger and more luxurious housing units if more than lip service is to be paid to another goal of reducing housing costs. The proponent of a farm-tenure improvement program may find it necessary to compromise between the objectives of securing maximum efficiency in production and a wide distribution of ownership rights among operators. Planners can often avoid considerable wasted effort by first identifying and thinking through their planning goals. By recognizing and ironing out possible conflicts between competing goals at an early stage, they can avoid later problems and gain a sense of direction that aids in the preparation of workable plans and helps them avoid embarrassing inconsistencies.

Formulation of the plan. Once the objectives are established, the next major step in the planning process concerns the actual development or formulation of the plan. The approach used at this stage varies somewhat with individual cases. Most land resource planning involves four substeps: (1) research or study concerning the nature of the problem situation, the resource base, and other factors that may be affected by the plan; (2) discovery and analysis of the various alternative solutions or approaches that can be used; (3) choice of the alternative or combination of alternatives—including the possible choice of doing nothing—to be used; and (4) the formal drafting of the plan in written or graphic form.[5]

Planning efforts frequently go astray because they are based more on wishful thinking than on a sound understanding of the facts. By accident or happenstance, some plans do succeed even though they are made in ignorance of the facts. The weight of planning experience, however, shows that a good working knowledge of the problem situation is a necessary prerequisite to successful planning. This is particularly true with land resource planning. Area planners always find it desirable to assemble as much information as possible concerning their problems and the nature and potentialities of the resource bases with which they work. They often devote major portions of their time to objective research and to the accumulation and analysis of factual data.

It is sometimes impossible for planners to assemble and analyze all the information that should be considered in their decisions. Policy decisions are frequently needed on short notice, and the processes of government and business cannot always wait for the final word in scholastic research. Planners and policymakers often find that they must make spot decisions without a full knowledge of the facts. Insofar as it is practicable, however, they should always base their decisions on findings that represent a reasonably accurate picture of the situation at hand.

This responsibility creates problems in planning administration. Most successful planning directors recognize that every individual has blind spots. Regardless of how well balanced their education has been, people are the products of the "roundaboutness" of their training. Their thinking and the points they emphasize reflect personal interests and backgrounds. In recognition of this situation, planning officials frequently utilize the services and advice of people with quite different backgrounds in the hope that their combined judgment will provide a balanced and reasonably realistic analysis of the actual problems at hand.

Going beyond the problem of securing a balanced view of the facts, planners

[5] Cf. Galloway, *op. cit.*, p. 6.

have a definite responsibility to seek out, analyze, and consider the alternative approaches that may be used for attainment of their objectives. In this process they must avoid panaceas. They should consider the legality and constitutionality of each proposal. They should reject those formulas that appear economically, politically, or socially unacceptable or unsound.[6] Where possible, they should seek workable solutions that can be dramatized for popular acceptance; and they should always ask themselves, Is this proposal workable from an administrative standpoint?

The planning process reaches its climax when planners choose between alternatives and commit themselves to given courses of action or a given set of recommendations. Up to this point, the planning process consists of preparation for planning, not planning itself. Real planning calls for more than the collection of factual data and analyses of situations and trends. It calls for formulation of courses of action, for choices between alternatives, for definite recommendations concerning what should or should not be done. These decisions involve the substance of planning and provide the real test of a planner. Plans are sometimes ambitious or visionary, sometimes inconsequential or faulty. But good or bad, they stand as monuments to the planner's ability or lack of ability to prepare workable plans.

A key problem in the planning process centers on the question of who makes the planning decision. Planning officials and consultants are often called on to formulate and recommend plans for action. Emphasis is placed in these instances on specific planning recommendations, and little or nothing may be said of the alternatives that were considered but rejected. Plans of this nature are sometimes accepted and carried out. Unfortunately, they often bask in the limelight for a brief day of glory, after which they are shelved, filed away, and forgotten. An alternative to this approach is used by those planners who regard the planning decision as a legislative function. These planners spell out the alternative solutions to planning problems together with their advantages and disadvantages and leave the actual choice of the plan to legislative bodies. This approach frequently leads to action because the planning decision is made by the officials who are responsible for accepting the plan and putting it into effect.

Successful land-use planning calls for recognition of the political facts of life. Planners must usually operate in terms of the present, the next decade, and occasionally the next quarter of a century. They need to gear their plans to the possible. In this process, they must be willing to adapt their goals and plans to political realities. They must recognize the power groups in the community and adjust their roles and techniques to the sociopolitical system in which they operate.

Effectuation of planning. Many people feel that the planning process stops as soon as planners decide on a given operational plan, draw a land-use map, prepare a planning report, or issue a policy directive. Actually, there is little point in planning if no action follows. Planners have a natural interest in the effectuation or carrying out of their plans. It is at this stage that a plan can prove itself, show need for reformulation or revision, or possibly fail.

The successful administration of any public plan calls first for a sound and rea-

[6] Cf. *ibid.* pp. 44–46.

sonable proposal. Beyond this, it calls for capable and understanding administration, the cooperation of officials in other agencies and other levels of government, popular interest in the plan and its results, and continued political and public financial support.[7]

Administrative problems frequently arise with the recruitment and training of capable administrative personnel. Care must be taken to emphasize results, to avoid bogging down in "red tape," to keep policies or programs working in their intended way, and to prevent individuals from scuttling programs or using them as stepping-stones to promotion. Coordination and integration are needed to keep individual officials or agencies from proceeding with policies that work at cross-purposes and to avoid jurisdictional conflicts over who functions where. Emphasis must also be given to the maintenance of good public relations both with citizens of the community and with other public officials.

Some Examples of Land Resource Planning

Land resource planning takes many forms. Individual planning occurs when a farmer plans a field layout or a subdivider draws a plat map. Private group planning takes place when a neighborhood association requires architectural approval of all new houses built in an area, or when a service club develops a summer camp. Public planning occurs every time we build a highway or enact a zoning ordinance. With public as well as private planning, plans may be far-sighted and successful. They may also be faulty and result in failures. Significant examples of public planning occur with city and metropolitan planning; federal, state, and regional planning; and rural land-use planning.

City and metropolitan planning. Archaeological findings show that the origins of city planning go back some 5,000 years. These findings suggest that directional measures have been used to provide for orderly and functional urban developments since people first started living together. Until recent decades, however, city planning has been limited for the most part to street layouts, the reservation of some areas for parks and public uses, and the provision of public water supplies and systems of defense fortification. Except for constraints forced by defense considerations, most of the world's cities have followed laissez-faire patterns of growth that illustrate the often myopic concerns of individual developers with their own special interests.

Formal plans dealing mostly with street layout were prepared for several American cities at fairly early dates. William Penn devised a plan for Philadelphia in 1682; and plans were made for Savannah in 1733, Washington, D.C., in 1791, Buffalo in 1801, and Manhattan Island in 1811. Additional pioneering in city development came during the middle 1800s with the appointment of boards in several cities to deal with water, sanitation, park, street, and transportation problems. Even

[7]Cf. Galloway, *op. cit.*, pp. 34–35; also Roland R. Renne, *Land Economics*, 2nd ed. (New York: Harper & Row Publishers, Inc., 1958), pp. 540–41.

with these beginnings, city planning, as we know it, did not get its start until the end of the nineteenth century.

Modern city planning began with the birth of the "city beautiful" movement at the Chicago World's Fair in 1893. Visitors at the fair were greatly impressed with the beauty of the grounds and the classic splendor of the architecture. Thousands went away imbued with a desire to beautify their own drab cities. This desire blossomed into civil improvement plans in several cities. The central thrust of this early planning effort was on municipal aesthetics and improvement of the surface appearances of public buildings, parks, and major streets. Little emphasis was given to "the more deep-seated social problems which rapid urbanization entailed."[8]

A second milestone year came in 1909 with the holding of the first national conference on city planning and with the publication of the *Plan of Chicago*. This city plan had a broader orientation than the earlier city plans and signaled a gradual shift of emphasis from the "city beautiful" to the "city practical." This shift was implemented by New York City's adoption of the nation's first comprehensive city zoning ordinance in 1916.

City planning came of age during the 1920s as cities found that they could use land-use zoning, building codes, and subdivision controls as planning tools to guide their future development. Numerous cities hired planning consultants and organized city planning commissions, and progress was made in preparing "comprehensive" city plans. Yet at the end of the decade, most city plans were still concerned with only six major items: civic appearance, parks and recreation, streets, transit problems, transportation facilities, and zoning.[9] Only passing notice was given to the economic and social basis of the city or to important problems such as housing.

Curtailment of operating budgets brought the city planning movement to a near standstill in the early 1930s. This situation gradually changed after the launching of the New Deal in 1933. Attention was soon focused on the role city planners could play in caring for the problems of the "lower third" of the population. Staff assistance—often financed with public relief roll funds—was provided for planning surveys and research. Federal, state, and local programs were set in motion to help cities clear away their slums, provide public housing, and undertake large-scale redevelopment programs.

With these developments, the scope of city planning was again broadened. Planners began to think more in terms of the overall city and of the changes and redevelopment work needed to eliminate blight and slums, to improve living conditions, and to raise standards of urban life. Master plans were adopted in more and more cities to guide the future pattern of land use and to give direction to programs for urban renewal and property conservation.

Emphasis was still given to the problems of parks, public buildings, civic centers, and municipal aesthetics. But more attention was given to the economic base of the

[8] Robert A. Walker, *The Planning Function in Urban Government*, 2nd ed. (Chicago: University of Chicago Press, 1950), p. 12.

[9] Cf. Theodora K. Hubbard and Henry V. Hubbard, *Our Cities Today and Tomorrow* (Cambridge, Mass.: Harvard University Press, 1929), p. 109.

city, to the impact of this base on planning, and to the use of planning measures to maintain and strengthen this base.[10] Increased emphasis was given to the problems of industry and commerce, to the provision of expressways and more adequate downtown parking facilities, to the location and building of schools and shopping centers, to securing more adequate water and sewerage facilities, and to numerous other factors that bear on the economic and social life of the city.

By 1971 virtually every city of more than 10,000 people in the United States had a planning agency or consultants.[11] Complications were arising with the increasing demands of disadvantaged groups, with the continuing flight of thousands of substantial citizens to the outlying suburbs, with growing public awareness of the presence of ghettos and slums near the urban core, and with the outward sprawl and scatteration of urban-oriented developments around cities. Instead of finding their problems neatly circumscribed by city boundaries, city planners found that they were dealing with segments of the larger problems of metropolitan regions.

Most central cities are at least partly surrounded by "dormitory" communities that resist annexation but have a definite interest in the employment, shopping, and other services provided by the central city. Both levels usually recognize the joint nature of their land-use problems and the futility of trying to solve these problems on an individual community basis. This fact together with the offer of federal financial aids has led many urban communities to work together in establishing government councils and metropolitan area planning commissions that can plan for their mutual land-use interests. Some regions have extended the metropolitan approach to set up special metropolitan districts that administer park or water and sewerage programs. Others such as Toronto have gone further to set up metropolitan area governments with powers to deal with a wide range of metropolitan governmental problems.

Federal, state, and regional planning. Some of the most important developments in land-use planning during the 1960s and 1970s took place at the state and federal levels. From their beginnings, these governments have engaged in activities that have had both direct and indirect effects on land use. Throughout the 1800s, they were concerned mostly with the sale and settlement of the public domain and the provision of internal improvements. New types of resource planning became necessary after 1900 with the reservation and acquisition of public forests and parks. Some attention was given during the 1930s, and later during the 1960s and 1970s, to implement federal, state, and regional comprehensive planning. But the principal resource planning efforts of this century have involved piecemeal

[10] Cf. Richard U. Ratcliff, "A Land Economist Looks at City Planning," *Journal of Land and Public Utility Economics*, 20 (May 1944), 106–8; and Ratcliff, *Urban Land Economics* (New York: McGraw-Hill, 1950), pp. 408–13.

[11] A 1971 survey of 967 cities with 10,000 or more people showed that all but 3 had a planning commission, department, or hired consultants. Cf. *The Municipal Yearbook, 1972* (Chicago: International City Managers' Association, 1972), pp. 66–79.

programs that deal, for the most part, with specific areas, problems, or types of resources.

Leading examples of state and federal land resource planning programs are found in the plans developed and used in administering state forests and parks and the holdings of the U.S. Forest Service, Bureau of Land Management, National Park Service, Bureau of Reclamation, and Bureau of Sports Fisheries and Wildlife. Some of the principal problems associated with resource planning on the federal lands were highlighted in *One Third of the Nation's Land*, the 1970 report of the Public Land Law Review Commission. Resource planning was emphasized with passage of the Forest and Rangeland Renewable Resources Planning Act (RPA) of 1974. RPA requires the U.S. Forest Service to conduct and report periodic studies in which it assesses the nation's expected needs for these resources. The Soil and Water Resources Conservation Act (RCA) of 1977 requires similar action by the U.S. Department of Agriculture in appraising trends and presenting alternative strategies for developing, managing, and conserving the nation's soil, water, and related resources.

Regional land-use planning is being conducted on a multicounty basis in most states and also on a multistate regional basis. An excellent example of the latter is provided by the Tennessee Valley Authority. This organization was established by Congress in 1933 as a federal corporation to serve the seven-state area included in the drainage basin of the Tennessee River. Among its objectives, TVA was designed to control floods and improve navigation on the Tennessee River and its tributaries, contribute to the national defense, develop and produce new types of commercial fertilizer, produce and distribute hydroelectric power, promote desired research, facilitate resource development, and improve the economic welfare of the people in the area. In fulfilling its purpose, TVA has built numerous dams along the Tennessee River and its tributaries to control floods, improve navigation, and produce public power. It has stimulated new economic development in the area, encouraged better farming practices, promoted soil conservation and reforestation measures, provided improved recreational opportunities, helped to stamp out malaria and other diseases, and encouraged community planning. It has also produced fertilizers and nitrates for explosives and contributed greatly to the development and use of nuclear power.

Hawaii set an example for state action by adopting a statewide land-use planning and zoning program in 1961. Several states have since created state planning agencies. Individual states also have enacted programs that deal with specific land-use problems. Maryland adopted the nation's first use-value assessment legislation for agricultural lands in 1956. Other innovative programs include California's coastal zone development program (1972), Delaware's coastal zone act (1971), Florida's environmental land and water management act (1972), Iowa's conservancy district law (1971), Maine's industrial site location law (1970), Massachusetts' coastal wetlands legislation (1963), Michigan's land sales act (1972), Minnesota's Twin City

Metropolitan Area growth management program (1967), New York's agricultural districts and Adirondack Park reserve programs (1971), Oregon's growth control program (1973), and Wisconsin's shoreland protection act (1966) and farmland protection program (1977).[12]

Interest in national resource planning blossomed during the 1930s, died out during World War II, and gradually revived again during the 1950s and 1960s.[13] National commissions were established to study and make policy recommendations on housing, outdoor recreation, material resources, water resources, and public land legislation problems. Congress provided "701" local and regional planning funds under the Housing Act of 1954, "208" local planning and operations funds under the Water Pollution Control Act amendments of 1972, and county planning and program management funds under the Coastal Zone Management Act of 1972.

Legislation was enacted by Congress during the 1960s and 1970s on such land-use-related topics as air and water quality, coastal zone management, environmental protection, forestry, housing, parks and recreation, transportation facilities, and the protection of natural and wilderness areas. Proposals for a national land-use policy and planning assistance program, which would have given the federal government an integrative role in financing and facilitating state and local land-use planning, were also considered by Congress but were not passed.

Rural land-use planning.   Much of the emphasis in rural land-use planning since 1970 has centered on the need for protecting and restricting the development of "critical" and "essential" land-use areas. These areas typically involve lands that environmentalists feel should be reserved for future agricultural, forestry, recreational, wetland, wilderness, or other open-space uses. Rural landowners are sometimes annoyed, sometimes pleased, with the interest nonresidents show in their land-use practices. Many now look to local land-use planning, however, as an opportunity they can use to shape the future of their communities.

Active land-use planning programs are now sponsored on a local governmental

---

[12] For discussions of these and other state programs, cf. Fred Bosselman and David Callies, *The Quiet Revolution in Land Use Control* (Washington, D.C.: Government Printing Office, 1971); Robert G. Healy and John S. Rosenburg, *Land Use and the States*, 2nd ed. (Baltimore, Md.: Johns Hopkins University Press, 1979); and W. Wendell Fletcher and Charles E. Little, *The American Cropland Crisis* (Bethesda, Md.: The American Land Forum, 1982), chaps. 3 and 4.

[13] The National Resource Board was established in 1934 to operate on an interagency basis and to take over the functions of the National Planning Board, which had been set up in the Public Works Administration in 1933. This agency was reorganized as the National Resources Board in 1935, and again as the National Resources Planning Board in the Executive Office of the President in 1939. The board sponsored considerable national and regional research on land, water, urban, and other problems before its demise in 1943. Cf. John D. Millett, *The Process and Organization of Government Planning* (New York: Columbia University Press, 1947), pp. 137–52; also Charles E. Merriam, "The National Resources Planning Board: A Chapter in American Planning Experience," *American Political Science Review*, 38 (December 1944), 1075–88.

basis in numerous rural counties and communities. Professional planners are used in most counties with considerable tax bases or when the planning function is handled on a multicounty regional basis. Volunteer citizen participation in the planning process is a more common practice in communities that lack the resources needed for hiring professional planning assistance. The "grassroots" planning efforts in these areas have an important precedent in the nationwide rural land-use planning program that was sponsored by the U.S. Department of Agriculture and the Land Grant Colleges between 1937 and 1941. This early experience involved the efforts of around 200,000 citizens in 10,000 rural communities in two-thirds of the nation's counties and showed that concerned citizen groups can accomplish a great deal in evaluating their community assets, identifying local problems and their causes, appraising possible remedies, and recommending and sometimes acting to secure desired public programs.[14] Similar volunteer local planning efforts continued in several states after 1941 and foreshadowed the later designation of official planning agencies and offices that now operate in rural areas in much the same way as in more urban environments.

PROGRAMS FOR BETTER LAND USE

Some of the most pertinent issues in public planning for better land use involve the questions of planning for whom, for what, and how. The first of these issues recognizes the fact that plans must be carried out where the land is and that their success calls for the cooperation, support, and involvement of the people whose interests are most directly affected by the plans. Beyond this, attention must be focused on the objectives of the proposed plans and on identification of the approaches that can best be used to secure their attainment. Emphasis is given in the following discussion to the following four topics: (1) identification of policy goals, (2) recognition of the range of policy measures governments can use to attain their goals, (3) examination of the uses made of land reform programs, and (4) America's experience with programs for securing better land use.

[14] Cf. Bushrod W. Allin, "County Planning Project—A Cooperative Approach to Agricultural Planning," *Journal of Farm Economics*, 22 (February 1940), 292–301; and Howard R. Tolley, *The Farmer Citizen at War* (New York: Macmillan, 1943), chap. 5. Three successive stages were envisaged in this planning process. During the initial *preparatory* stage, emphasis was given to general discussions of goals and problems, organization of committees and laying the groundwork for further efforts. Committees operating at the second *intensive planning* stage were expected to survey and classify their land resource base, prepare land-use maps, and identify and evaluate possible solutions to pertinent local problems. At the third or *unified program* stage, they were expected to take an overall view of their problems and recommend possible solutions or programs of action. Some committees naturally accomplished more than others; but planning efforts were carried on in most instances by interested local citizens, with some technical research and planning assistance supplied by state and federal agencies.

Goals in Land Policy

Public and private policy goals concerning the possession and use of land resources ordinarily reflect the thinking and objectives of the decision makers. These goals at times have stressed the aggrandizement of a single head of state or a ruling class. During recent centuries, they have frequently emphasized nationalistic ends and more recently the welfare interests of large numbers of the citizenry. With this democratization process, most nations have gradually reoriented their thinking about who should control land and how it should be used.

Several nations now accept attainment of widespread opportunities for higher and more satisfying levels of living for all their citizens as the primary master goal of public policy. Within the framework provided by this master goal, five general goals in land policy may be identified:

1. Widespread distribution of ownership, operatorship, occupancy, and use rights among citizens who wish to exercise these rights
2. Landholdings of appropriate size and productive potential to permit a maximizing of production opportunities
3. Orderly and equitable operating arrangements that encourage efficiency in use of land resources
4. Arrangements that offer economic opportunities, security, and stability to land operators
5. Arrangements that lead to the development and conservation of land resources.

Other formulations of policy goals are possible. Schickele, for example, speaks of "maximization of social product over time" and "optimization of income distribution among people" as the twin master goals of economic policy.[15] Johnson uses a less abstract approach in highlighting the following eight goals in land policy: (1) military security, (2) political stability, (3) maximum national production, (4) maximum income, (5) economic security and stability, (6) individual freedom, (7) conservation of human resources, and (8) conservation of natural resources.[16]

Significant as these groups of goals may appear, the list is not all-inclusive, and individual goals can conflict with one another. Goals must sometimes be reformulated when attention is given to specific problems, and priorities must also be assigned to individual goals at times to indicate the direction in which programs should move when two or more goals are in conflict.

Successful program administration calls for periodic reexamination of the pri-

---

[15] Rainer Schickele, "Objectives in Land Policy," in John F. Timmons and William G. Murray, eds., *Land Problems and Policies* (Ames: Iowa State College Press, 1950), pp. 6–10.

[16] Cf. V. Webster Johnson and Raleigh Barlowe, *Land Problems and Policies* (New York: McGraw-Hill 1954), pp. 8–13. For other formulations of goals, cf. Joseph Ackerman and Marshall Harris eds., *Family Farm Policy* (Chicago: University of Chicago Press, 1947), pp. 9–11; John F. Timmons, "Land Tenure Policy Goals," *Journal of Land and Public Utility Economics*, 19 (May 1943), 165–79; and Conrad Hammar, "The Land Tenure Ideal," *Journal of Land and Public Utility Economics*, 19 (February 1943), 69–84.

orities assigned to individual goals. Without broad vision in the choice of priorities, careful program planning, and willingness to act, efforts to attain particular goals can result in bitter disappointments. "Land to the tillers" and "land for the landless," for example, have been popular rallying cries for land reform in many countries. But sole emphasis on these objectives can lead to inefficient resource use, the creation of small holdings of uneconomic size or holdings with too little equipment or capital to function as effective production units, and thus to wasteful and undesirable uses of both human and land resources.

Measures for Improving Land Use

Most of the land policies and programs found throughout the world today are the products of a long, evolving process in which various measures have been devised, tested, and used to meet changing needs and conditions. The gradual development of the concept of fee simple ownership that followed the breakdown of the feudal system in England and the more recent acceptance of social controls over private property rights in the countries that accept the English common law provide classic examples of evolutionary land reform.

Here and there, particularly in areas where governments have resisted demands for change, new land policies and programs have sometimes been adopted in the aftermath of revolutionary action. Examples include the new land policies that followed the French Revolution of 1789, the Mexican Revolution of 1910, the Bolshevik Revolution of 1917, and the Egyptian Revolt of 1952. In these cases as with the examples of more evolutionary reforms, precedents existed either in the earlier history of the country or in other countries for most of the policies adopted.

Governments can choose from a wide range of possible measures in the development of new land policies and programs. Their principal alternatives call for selections from the following list of policy measures for improving land-tenure and land-use conditions:

1. Legal recognition of the scope of public and private ownership rights including arrangements for sharing rights in land and for transferring rights among parties by sale, gift, inheritance, leasing, and mortgaging
2. Provision for a system of land surveys and land title registration to define the areal scope of ownership rights, minimize conflicts over boundaries, and protect property owners and operators in the possession and retention of their rights
3. Policies for the transfer of specified public lands to private ownership and for the private settlement of these lands
4. Programs that encourage public or private development or redevelopment of land resources through clearing, drainage, reclamation, terracing, and area renewal efforts
5. Provision of public credit facilities where needed to help finance property acquisition and provide needed operating capital
6. Provision of educational and technical assistance to land operators to in-

struct them in improved management techniques and under some circumstances to require their acceptance of managerial guidance

7. Legal clarification of landlord-tenant arrangements with provisions for standardizing leasing terms, providing compensation for unexhausted improvements, and enhancing the right of tenants to security and stability

8. Encouragement and provision of assistance in the organization of cooperatives to deal with and provide self-help in handling production, buying, and marketing problems

9. Creation of special government programs, boards or commissions, and districts to deal with particular problems such as the need for local area planning, the provision of water supplies and waste disposal facilities, the regulation of grazing areas, and the operation of irrigation systems

10. Taxation of landed property, its annual rents or total values, and/or the incomes derived from them to provide revenues for government operations and at times to influence the uses made of lands

11. Enactment of government regulations (e.g., zoning ordinances, building codes, provisions limiting the minimum and maximum sizes of landholdings, rent controls and land price ceilings, and controls over eligibility to buy or own lands) to direct the uses made of land resources

12. Provide for the exercise, when needed, of the public right to take private lands with or without compensation for desired public purposes

13. Public management and operation of specified areas of educational, forest, military, park, and other lands in the public interest

14. Use of public grants and subsidies as inducements to get private land operators and owners and other governmental units to carry out desired land-use practices

The scope of this list of policy options shows that a wide assortment of land-use policies can be used to modify and improve existing land-tenure and land-use situations. Some alternatives involve mild reforms; others may be characterized as radical in that they may require the overturning of existing institutional arrangements. Both classes of measures can be enacted through either evolutionary or revolutionary means.

Far-reaching reform programs involving mixtures of mild and more radical reforms are often needed, particularly in developing countries, to bring about desired adjustments. Revolutionary approaches sometimes offer the quickest way to bring about change. This approach, however, often brings major disruptions in national economies and opens political wounds that take years to heal. The weight of history generally favors use of a more evolutionary and gradualistic approach. Far-sighted leaders can recognize and promote needed adjustments in land-use policies. More emphasis can be given to evaluating objectives, to testing policies, and to learning from past mistakes. As long as nations really follow through on commitments to develop and carry out improved policies, gradualism can often produce more permanent accomplishments, more stability, and less disruption than more revolutionary approaches.

Land Reform

Considerable attention has been centered in recent decades on the use of land-reform measures as a means for alleviating undesired economic and social conditions in rural areas and at the same time triggering economic development. The demands for these reforms come from many sources. Some come from groups who feel outraged by the inequities found in society and who demand reforms to reduce the disparity of opportunities and privileges enjoyed by landlords as compared with tenants and landless workers. Some see reforms quite rationally as a necessary first step in providing better living conditions for large masses of the citizenry. Others argue that reforms are needed to implement industrialization and to better incorporate rural people into the stream of national life; still others see the espousal of land reform measures as a road to political power and a means for punishing political enemies.

This combination of interests has served to make land reform a burning issue in many parts of the world. As an action program, it has been new to many areas. Yet the problems that give rise to land reform are not new. They have existed for centuries in many cases. Only the techniques and will for action are new. History is filled with accounts of peasant revolts and other attempts of rural people to throw off or at least lighten the yoke of slavery, serfdom, or peonage. Most of these strivings for agrarian reforms have failed, but some have succeeded and have brought added advantages or privileges to peasant populations. The first successful large-scale land reform movement of modern times followed the French Revolution and led to a general freeing of land-tenure conditions in western Europe. Land reforms spread to central and eastern Europe following the popular uprisings of 1848, the Russian peasant uprisings of the period before 1861 and again in 1905–1906, and the armistice of 1918. Land reform became an issue of worldwide significance following World War II when the reform movement spread from the more developed nations to the developing regions of Africa, Asia, and Latin America.

Meaning of land reform. Differences of opinion exist concerning the meaning of the term *land reform*. In a broad technical sense, any program that leads to a change, presumably for the better, in the manner in which rural or urban land resources are held and used may be described as land reform.[17] Common practice, however, has limited usage of this term to programs designed to bring improvements in agricultural economic institutions. This is the sense in which the term *land reform* is used here and in which it was used by the U.N. Department of Economic and Social Affairs in 1951.[18]

Doreen Warriner, a noted authority on land reform, rejects this definition. In her view, "land reform means the redistribution of property or rights in the land for

[17]Cf. Raleigh Barlowe, "Land Reform and Economic Development," *Journal of Farm Economics*, 35 (May 1953), 173–87.

[18]U.N. Department of Economic and Social Affairs, *Land Reform: Defects in Agrarian Structures as Obstacles to Economic Development* (New York: United Nations, 1951), p. 89.

the benefit of small farmers and agricultural labourers."[19] Still other observers limit their definitions to reform programs they personally approve or oppose. As a means of clarifying this confusion over the meaning of land reform, Johnson and Kristjanson have recommended the use of three terms: land distribution, land reform, and agrarian reform. *Land distribution* would involve programs designed to either break up or combine existing landholdings. *Land reform* would involve "the rearrangement of ownership rights and other institutions associated with land in the interests of the many rather than the few." *Agrarian reform* would "include overall improvements of rural life and improved relationships of rural people to the land."[20]

Significance of land reform. Two goals are frequently cited for land reform. One of these centers on improvement of the lot of the average land operator or worker, whereas the second emphasizes the possible contributions of reforms to national economic development.[21] Both of these spring from the fact that many countries are trying to implement a rapid "transition from a subsistence agrarian economy of status to a market economy in which varying proportions of the gainfully employed are engaged in agriculture."[22] Many of them seek to establish agricultural economies similar to those of western Europe. Unlike Great Britain, however, where the shift from an earlier agrarian feudal economy to the present situation was accomplished through several centuries of gradual change, they hope to realize similar progress within the lifetimes of present land operators.

The most urgent need for land reform is found in the less-developed areas of the world where the great majority of the people are dependent on agriculture and where outmoded tenure systems support the selfish interests of small classes of owners rather than the general welfare of the workers and the nation. Most of the rural population in these areas live in the traditional way of their ancestors. Typically, they look to land for their livelihood; they are no more than moderately productive; their incomes are low; and they contribute little to the market economy. The rights and opportunities they enjoy usually reflect their "relationship to the ownership and use of land and their place in the kinship or family group."[23]

Frequently the rural people in these areas are exploited by large landowners or merchants who are able to use their positions to live in comparative luxury while

---

[19] Doreen Warriner, *Land Reform in Principle and Practice* (Oxford, U.K.: Clarendon Press, 1969), p. xiv. A comparable definition is accepted by Edmundo Flores, "The Economics of Land Reform," *International Labour Review*, 92 (July 1965), 30.

[20] Cf. V. Webster Johnson and Baldur H. Kristjanson, "Programming for Land Reform in the Developing Agricultural Countries of Latin America," *Land Economics*, 40 (November 1964), 355.

[21] Erven J. Long, "The Economic Basis of Land Reform in Underdeveloped Economies," *Land Economics*, 37 (May 1961), 113–24.

[22] Kenneth H. Parsons, "Land Reform and Agricultural Development," in Kenneth H. Parsons, Raymond J. Penn, and Philip M. Raup, eds. *Land Tenure: Papers of the First Conference on World Land Tenure Problems* (Madison: University of Wisconsin Press, 1956), pp. 4–5.

[23] *Ibid.*, p. 8.

substantial numbers of small operators, tenants, and landless workers live on a near subsistence basis. Possession of ownership rights determines the individual's economic, political, and social status. Changes in tenure institutions can open doors to new opportunities. As Raymond J. Penn has observed,

> . . . the ownership of land carries with it ownership to government—the right to tax, the right to judge, the power to enact and enforce police regulations. It dominates every crucial decision about investments in social capital—education, transportation, hospitals, power projects.
>
> To the campesino, ownership of land is more than a source of wealth. It is the source of prestige and political power and social justice. It gives him the right to build his own house in which to raise his family. It gives him, too, the right to tax himself to build a school. It lets him share in the bundle of rights which have so long been a prerogative of the large landholder and denied to the landless.[24]

Land reform is significant in that it provides a necessary bridge that large segments of the population in some countries can use in shifting to a position where they can utilize their available resources to greater advantage, contribute more to national production, and share more in the bounties of modern life. Its potential as a tool for enhancing the worth and well-being of individuals as well as for facilitating national economic development has received more recognition in some countries than others. The recognition process has been facilitated since World War II by developments such as the creation of the United Nations, the phasing out of colonialism, the extension of economic technical assistance and medical aids to developing areas, and the promotion of economic development programs, which have helped raise the aspirations of average citizens. Almost from the beginning, it has been obvious that

> . . . traditional agrarian structures and especially land-tenure institutions throughout a large part of the nonindustrialized world are a major obstacle to the kind of development postulated by the United Nations. The increased incomes and other benefits of technological progress do not reach the majority of the rural people. No mass markets are generated in the countryside to support dynamic industrial growth. Food production lags, further slowing industrialization. The vicious circle of poverty is not broken in rural areas. The peasantry cannot organize and participate actively in national affairs. Somehow these traditional social structures have to be changed before real rural development can proceed.[25]

[24] Raymond J. Penn, "Understanding the Pressures for Land Reform" in *Economic Developments in South America*, Hearings before the Subcommittee on InterAmerican Economic Relationships, 87th Cong., 2nd sess., May 10, 1962, p. 15.

[25] Solon Barraclough, "Why Land Reform?" *Ceres*, 2 (November–December 1969), 23.

Land reform programs. A broad range of land reform programs has been adopted in different countries.[26] Some of these involve single reforms; others embrace combinations of reforms. Some provide for mild adjustments; others are more sweeping in their effects.

A variety of mild reforms has been used both in the more-developed and in the less-developed countries to secure particular tenure policy objectives. Early examples are provided by the Ordinances of 1785 and 1787, under which the United States prescribed a policy for surveying its public lands, offering them for sale to settlers at a nominal price, providing government for newly settled areas, and opening the way for these areas to eventually be admitted as states to the Union. Denmark made itself a nation of small landowners in the 1800s by granting liberal long-term, low-interest loans for the tenant purchase of farms and by encouraging the organization of farmer cooperatives. Great Britain used another type of mild reform in 1870 when it enacted a Landlord and Tenant Act, which greatly enhanced the bargaining position and security of farm tenants and facilitated their purchase of the lands they operated. Holland and Italy have carried on large-scale land drainage and settlement programs that have provided homesites for many operators. Other prominent examples of mild reforms include the provision of agricultural extension programs, the development of reclamation projects, and the establishment of voluntary land consolidation commissions.

Nations have frequently gone further with their programs to use regulatory measures to attain particular land-tenure and -use goals. Western Germany and Holland have stringent land-use regulations that determine the uses owners can make of their lands. Germany, Switzerland, and Great Britain had policies during World War II that authorized local boards to evict ineffective managers of farmlands and replace them with operators who could contribute more to national food production. Israel has encouraged the organization of cooperative farming settlements (*kibbutzim*) in which land is held and worked by families in common. These settlements operate alongside other cooperative villages (*moshavim*) and villages of private farmers.

France, Holland, and Italy have established agricultural rent controls and in some cases have specified legal maximum rental rates for certain sharing arrangements. Agricultural rents were limited to a maximum of 37.5 percent of the crop in Nationalist China in 1949. Egypt adopted similar regulations in 1952 that limited rents, if paid in cash, to not more than seven times the real estate tax and share rents to 50 percent of the crop. Similar laws requiring sizable reductions in farm rents have been enacted in Japan and Pakistan.

---

[26] Numerous accounts have been published that describe the land-reform experiences of different countries. Several of these appear in various issues of *Land Economics*. The experiences of several nations are covered in Parsons, Penn, and Raup, *op. cit.*, in various reports published by the Food and Agriculture Organization of the United Nations; and in the publications of the Land Tenure Research Center at the University of Wisconsin. Cf. also Russell King, *Land Reform: A World Survey* (Boulder, Colo.: Westview Press, 1977); and Howard Handelmn, ed., *The Politics of Agrarian Change in Asia and Latin America* (Bloomington: Indiana University Press, 1981).

Parcellation of farm holdings has been discouraged in Czechoslovakia, Denmark, Israel, and Switzerland by laws that specify minimum farm areas below which further subdivision cannot take place. Programs for the mandatory consolidation and rationalization of existing holdings have been undertaken in countries such as France and Western Germany. Some of these projects involve extensive programs for rural renewal and contain provisions for the relocation of roads, canals, buildings, and utilities as well as the creation of smaller numbers of operating tracts.

Limitations on areas and types of farm ownership provide another example of land reform. Czechoslovakia limited private land holdings following World War II to a maximum of 50 hectares; owner-cultivators have a 7.5-acre limit in Japan; a 33-acre limit applies in Bangladesh; a 15-hectare limit in Poland; and a 100-feddan (103-acre) limit in Egypt. Holland and the Philippines have regulations that give tenants the first right to buy farmland if it is offered for sale. France, Germany, and Sweden prohibit sales of farmland to foreigners, nonfarmers, speculators, and investment "hedgers."

Some of the most far-reaching land reforms have called for the expropriation of certain landholdings with their redistribution in smaller-sized tracts to actual farmers. Land expropriation and redistribution programs were carried out in most of the nations of Eastern Europe following World War I. A large program of this type was started in Mexico during the interwar period, and large-scale programs have been carried out in Japan, India, Pakistan, Iran, Italy, Algeria, Egypt, Kenya, Bolivia, Chile, Cuba, and Peru since World War II. Some landowners have been divested of all their holdings; others have been allowed to retain specified maximum areas. Compensation has been paid for the expropriated lands in most countries but several variations in compensation arrangements have been used. Some countries, such as Finland, have paid their expropriated owners in bonds based on current or recent land values. Egypt based its compensation on pre–World War II values. The expropriated landlords in Mexico received long-term bonds for 110 percent of the assessed value of their land. Italy's program calls for compensation based on the tax-assessed value of the land two years prior to expropriation. The payments made to expropriated owners in India and Pakistan were graduated according to the income-producing value of the expropriated areas. In the United Provinces, for example, large *zamindars* (landowners) received a nominal compensation equal to as little as twice the value of their annual net rents. Zamindars with small holdings, however, were authorized to receive payments up to the equivalent of twenty times the annual value of their net rents. A notable exception has existed in the Iron Curtain countries, where the properties of former landowners were often confiscated without compensation.

Land expropriation programs have usually been followed by redistribution of the acquired farmlands among tenants, landless workers, and small peasant proprietors. Forest and other nonfarm holdings have often been retained in public ownership, and estate holdings with extensive building improvements have sometimes been retained as a unit and used for educational, research, and cooperative purposes. Individual allotments of the redistributed lands have usually been small, and recipients have normally been required to make some payment for their lands.

Purchase payments have ranged from nominal charges equal to the value of one or two years' crops to the current market value of the land. Long-term credit and liberal repayment schedules have been provided to accommodate the new owners.

Another variant of land expropriation has been applied on a large scale in the Soviet Union, Communist China, and some other countries. Collectivization rather than private ownership programs are used in the administration of agricultural properties in these countries. Two types of programs are in use in the Soviet Union. Most of the farmland is operated in state-controlled collective farms (*kolkhoz*), while some areas are administered in large state farms (*sovkhoz*). Residents of the collective farms work together in farming the properties held by the collectives together with possible additional areas leased from the state. Through their joint efforts, they benefit from higher-quality management and the use of more farm equipment and capital than presumably would be available if they operated as individuals. Workers have individual houses and garden plots and share in the income produced by the collectives in proportion to their labor inputs.

Consequences of land reform.[27] Most nations have enacted land reform measures of one type or another in recent decades. As one might expect, these measures have been popular with most rural constituencies. Indeed, reform programs of a palliative nature have often been pushed by public officials who have wished to capitalize on the political fervor for land reforms without greatly disturbing existing conditions.

Overall, the effects of most land reforms may be regarded as favorable. They have often contributed to individual and family well-being; added to their dignity as individuals; and fostered actions such as the cooperative use of machinery or sinking of wells, the provision of operating credit, or the lowering of rents, which have added to their productivity as workers and to their welfare as individuals. Opportunities provided by land reforms have given millions of people reasons to hope for a better future. In many countries such as Egypt, Japan, Mexico, Nationalist China, and Pakistan, land reform programs are creating a more productive agriculture plus economic development.

Evidence from various countries indicates that past land reforms have helped to create viable rural institutions and have paved the way for the establishment of a

[27] For other discussions of this topic, cf. Warriner, *op. cit.*; Philip M. Raup, "The Contribution of Land Reform to Agricultural Development," in Herman M. Southworth and Bruce F. Johnson, eds., *Agriculture and Economic Development* (Ithaca, N.Y.: Cornell University Press, 1967), pp. 267–314; John W. Mellor, *The Economics of Agricultural Development* (Ithaca, N.Y.: Cornell University Press, 1966), chap. 14; V. Webster Johnson, *Man and Land* (Bangkok, Thailand: 1970); Johnson and Kristjanson, *loc. cit.*; Solon Barraclough, "Alternative Land Tenure Systems Resulting from Agrarian Reform in Latin America," *Land Economics*, 46 (August 1970), 217–28; Peter Dorner, ed., *Land Reform in Latin America* (Madison: University of Wisconsin Land Tenure Center, 1971); Erich H. Jacoby, *Man and Land* (London, U.K.: Andre Deutsch, 1971); Sein Lin, *Land Reform Implementation: A Comparative Approach* (Hartford, Conn.: John C. Lincoln Institute, 1974); Louis J. Walinsky, ed., *Agrarian Reform: The Selected Papers of Wolf Ladejinsky* (Oxford: Oxford University Press, 1977); and Ronald J. Herring, *Land to the Tiller* (New Haven, Conn.: Yale University Press, 1983).

commercial agricultural economy. Reforms have brought better housing and farm production conditions in many areas. They have prompted the cultivation of larger areas and brought more intensive use of properties whose former owners had limited interests in farming. They have encouraged operators to be more productive, to apply innovations in their work, and to adopt conservation practices. By adding to rural production and causing landlords to shift their investments from land to industrial, commercial, housing, and other urban enterprises, they have also helped to generate economic development.

Yet the case for land reform is not entirely one-sided. Some so-called reforms have involved palliative measures that have done little for rural people. The breaking up of large holdings has sometimes destroyed economies of scale, created voids when services provided by landlords have not been forthcoming from other sources, and sometimes resulted in an initial decline in marketable surpluses as operators have elected to care for their own consumption needs and have had little incentive to produce the marketable surpluses formerly needed to pay their rents. Doreen Warriner indicates that reforms are most apt to lead to increased production when they are rapidly implemented, when provisions are made for operating units of economic size, and when the reforms are desired by the rural people. Production is apt to decline when periods of uncertainty bring delays in the application of reforms, when units of uneconomic size are created, when no provisions are made for replacement of functions formerly performed by landlords, or when peasants lack incentives for increasing production.[28] Half-hearted reforms have led to continued uncertainty and unrest and to periods of insurgency in which conflicts between rival interest groups have often victimized local populations and disrupted their attempts to use land in a productive manner.

Land reforms have a considerable potential for bringing about desirable objectives, but there is no guarantee of these results. They can improve the lot of one generation of land operators only to be swallowed up in the problems created by rising population numbers if they are not associated with industrialization and urbanization movements that draw people off the land. They can add to the self-esteem of farm operators, but this may be a hollow attainment if operators are left to "lift themselves by their bootstraps" without additional efforts to involve them in the workings of modern society. They can provide a basis for a more democratic society, but this process must be implemented with educational programs and measures to create democratic institutions. One of their greatest potentials lies in the role they can play as a catalyst for economic development, but this process involves the interplay of many factors, many of which are just as necessary for success as land reform.

U.S. Programs for Securing Better Land Use

It is frequently asserted that the United States has no need for land reform. This is a valid assumption if one is primarily concerned with popular demands for the redistribution of agricultural and other land holdings. But it does not mean that the

[28] Warriner, *op. cit.*, p. 54.

nation is forever immune to this problem.[29] Trends toward greater concentrations of business and landownership in the hands of a relatively small number of owners can easily foreshadow future land reform problems.[30]

Although many people do not think of them as reforms, a considerable number of programs have already been initiated that have had significant effects on how land is owned and used. Examples of rural land reforms include the ejection of the Tories from landownership during the Revolutionary War, the acquisition of land titles from the Indians, the disposal of the public domain, the emancipation of the slaves, the establishment of national forests and parks, the initiation of public land reclamation projects, and the provision of public credit facilities for the agricultural sector. During the early 1980s, the sale of public forestlands and grazing lands to private owners was proposed as another type of land reform.

Nonagricultural reform programs have been used on numerous occasions in recent decades to promote the attainment of urban housing and land-use goals. Examples include federal insurance of housing loans; public support of the residential mortgage market; encouragement of public housing; the provision of rent control and rent subsidy programs; the outlawing of racial discrimination in housing; increased emphasis on area planning; and federal support for highway construction, urban renewal, and open-space acquisition programs.

Progress has been realized with most of these programs. Yet much remains to be done. Strong programs are needed for both urban and rural lands to promote and ensure orderly and efficient land use, safeguard the huge investments society has in the infrastructure of its cities, revitalize central cities and forestall the development of slums, improve the quality of both existing and new housing, protect the quality of the environment, preserve open-space and greenbelt areas, acquire areas needed for public recreation, and protect the nation's agricultural and forest production capabilities. Coupled with the attainment of these objectives is a need for a broader understanding on the part of landowners of the rights and responsibilities associated with property ownership.

Time alone can tell how far or in what directions the nation will move in planning its land-use programs. With increasing urbanization, rising population numbers, and

---

[29] Some writers have called for land reforms in the United States. Cf. Charles C. Geisler and Frank J. Popper, eds. *Land Reform, American Style* (Totowa, N.J.: Rowman & Allanheld, 1984); Robert C. Fellmuth, *Politics of Land* (New York: Grossman Publishers, 1973); and Richard Stack, "Agrarian Reform, U.S.A." *Hunger Notes*, 4, no. 10 (March 1979).

[30] A 1978 study found that half of the nation's privately owned land was held by 2.7 percent of the owners and 72.7 percent by 10 percent of the owners. Cf. James A. Lewis, *Landownership in the United States*, U.S. Department of Agriculture Information Bulletin No. 435, 1980. Data presented to the House Committee on the Budget indicate that 1 percent of the population owned 22.2 percent of the value of land and buildings held by households in 1975. Of the value of properties owned by persons, this top 1 percent also held a fourth of the wealth, a third of the bonds, half of the corporate stock, and almost all of the trust assets and tax-exempt municipal bonds. Cf. *Data on Distribution of Wealth in the United States*, House Committee on the Budget, 95th Cong., 1st sess., 1977), pp. 9, 57–59. *Fortune* magazine's top 500 industrial firms of 1975 had more than ten times the combined sales and profits and almost eight times as many employees as the second 500 firms. Cf. *Fortune*, 93 (May and June, 1976).

the passing of time, it could follow the examples of several western European nations in exercising stronger controls over private land use. With these developments, privately owned lands may well be treated as a type of public utility, as resources that individuals must use in accordance with public regulations. Regulations could be adopted concerning the maximum and minimum sizes of various types of individual holdings, their sale prices, the determination of possible buyers, and possibly the uses to which certain properties must be put. Steps could be taken to open beaches, forests, and other privately held properties for limited public use and to limit the rights of private owners to modify or destroy publicly enjoyed scenic or amenity features of their properties without public consent.

Whatever course is followed, it must be remembered that the planning process is neutral. Plans can take us in any of many directions. Whether they turn out to be good or bad, right or wrong, successes or failures, depends in large measure on the understandings of the goals and issues and the quality of the planning efforts and skills that go into them. Ability to plan is a human attribute. It is the tool we can use to make desired things happen. But it also calls for acceptance of discipline and restraints in guiding efforts that would otherwise be free to go their own way to work together for a common goal. Planning for better land use offers a means people can use to make their properties, communities, nations, and the world better places in which to live and work.

—SELECTED READINGS

Beatty, Marvin T., *et al.*, eds., *Planning the Uses and Management of Land* (Madison, Wis.: American Society of Agronomy, 1978).

Chapin, F. Stuart, Jr., and Edward J. Kaiser, *Urban Land Use Planning*, 3rd ed. (Urbana: University of Illinois Press, 1979), chap. 3.

Clawson, Marion, and Peter Hall, *Planning and Urban Growth: An Anglo-American Comparison* (Baltimore, Md.: Johns Hopkins University Press, 1973).

Dorner, Peter, ed., *Land Reform in Latin America* (Madison: University of Wisconsin Land Tenure Center, 1971).

Galloway, George B., *Planning for America* (New York: Holt, Rinehart & Winston, 1941).

Healy, Robert G., and John S. Rosenburg, *Land Use and the States*, 2nd ed. (Baltimore, Md.: Johns Hopkins University Press, 1979).

Hite, James C., *Room & Situation: The Political Economy of Land-Use Policy* (Chicago: Nelson-Hall, 1979), chap. 8.

Johnson, V. Webster, and Raleigh Barlowe, *Land Problems and Policies* (New York: McGraw-Hill, 1954), chaps. 1, 14, and 15.

Ottoson, Howard W., ed., *Land Use Policy and Problems in the United States* (Lincoln: University of Nebraska Press, 1963), chaps. 11–17.

Parsons, **Kenneth H.**, **Raymond J. Penn**, and **Philip M. Raup**, eds., *Land Tenure: Papers of the First Conference on World Land Tenure Problems* (Madison: University of Wisconsin Press, 1956).

Reitze, **Arnold W., Jr.**, *Environmental Planning: Law of Land and Resources* (Washington, D.C.: North American International, 1974).

Warriner, **Doreen**, *Land Reform in Principle and Practice* (Oxford, U.K.: Clarendon Press, 1969).

18

PUBLIC DIRECTION
OF LAND USE

When individuals, groups, and governments plan for better land resource use, it is normally assumed that they are able to follow through with their plans. This follow-through process logically starts with the individuals or agencies that do the planning, particularly when a plan concerns management of the resources that they control. Most public land-use planning, however, involves activities and resources over which planners have little direct control. Effective follow-through in these cases calls for public exercise of directional measures that can influence, guide, or control the decisions operators make concerning their use of land resources.

Land-use directional measures involve private-private, private-public, public-private, and public-public relationships. Private-private and private-public actions occur when individuals or groups decide how their properties or any properties they turn over to governments shall be used. The simplest arrangement occurs when private operators decide how they and others will use their own properties throughout their tenure as owners. Ordinarily, they can use their properties as they wish. But this freedom may be limited by easements, deed restrictions, or conditions of title imposed by previous owners. Acting on their own, they too can use deed restrictions to direct the future use of any lands they transfer to others. They can write conditions into inheritance arrangements or donate properties to public or private organizations with the stipulation that they be used for specific purposes.

Individuals frequently work together on a formal or informal basis to secure particular land-use objectives. They can join with others in associations to create and maintain country clubs or private hunting and fishing domains, establish and enforce neighborhood architectural standards, or control transfers of association properties to new owners. Acting informally, but collectively, they can use public opinion, the power of good example, and social pressures to influence the land-use decisions and practices of other people and of various levels of government.

Formal and informal approaches are also used by governments in their public-private and public-public efforts to direct land-use practices. Informal techniques are emphasized when public officials use their powers of persuasion and negotiation or opportunities to influence public opinion to secure desired ends. More formal approaches are exercised when they use the government's taxation, spending, proprietary, eminent domain, or police powers to induce owners to accept desired practices or to discourage and prohibit undesired activities. Emphasis is given in the discussion that follows to the uses made of these five formal powers.

TAXATION, SPENDING, AND PROPRIETARY POWERS

Some of the most fundamental functions of government are directly related to the public powers they hold to lay and collect taxes, spend public funds, and own and manage real properties. These powers are exercised on a day-to-day basis and as such have important routine impacts on land resources and how they are used. They also have a significant potential for directing possible future uses of land.

The Power to Tax

Taxes have been used in the United States in the past primarily for the single purpose of securing needed public revenues. On some occasions, they have influenced operator decisions concerning land use, but it has only been during recent decades that much consideration has been given to their deliberate use as a tool for favoring or discouraging particular land-use practices. Evaluation of their potential as a tool for securing land-use objectives calls for consideration of the scope of the taxing power and of the uses that can be made of it in directing land use.

Scope of the taxing power. No modern government could operate without the power to collect taxes. As the Supreme Court of the United States has observed, "The power to tax is the one great power upon which the whole national fabric is based. It is as necessary to the existence and prosperity of the nation as the air he breathes is to the natural man. It is not only the power to destroy but also the power to keep alive."[1]

While governments can use taxes for regulatory and nonfiscal as well as revenue collection purposes, the taxing power has its limits. Most governments are constrained by constitutional principles that specify the circumstances under which this power can or cannot be exercised. In the absence of these limits, governments are also limited in an ultimate sense by the fact that they cannot tax property beyond the point of confiscation and taxpayer compliance.

Throughout the United States, the taxing power is vested in the legislative branch

[1] *Nichol* v. *Ames*, 173 U.S. 509, 515 (1899).

of government and cannot be delegated to other agencies.[2] Except for constitutional restrictions, every legislature is free to tax the persons or property subject to its jurisdiction at rates of its own choosing. It can group or classify particular persons, properties, privileges, or incomes for taxation purposes. But these classifications "must be reasonabe, not arbitrary, and must rest upon some ground of difference having a fair and substantial relation to the object of the legislation, so that all persons similarly circumstanced shall be treated alike."[3] The courts have also held that all taxes must be levied for public purposes and that they must be levied in an equitable and reasonable manner.

In addition to these judicial limitations, the taxing powers of the federal, state, and local governments are subject to constitutional provisions. The federal and state governments are sovereign within their own spheres. Neither level has a right to infringe on or interfere with the other's legitimate functions, and neither can tax the agencies or instrumentalities of the other. The federal government has only those taxing powers delegated to it by the U.S. Constitution. State and local governments in turn have only those taxing powers not prohibited by the U.S. Constitution and not limited by state constitutional provisions.

Congress has the power under the Constitution "to lay and collect taxes, duties, imposts, and excises." This broad grant of power is subject to five specific limitations. (1) No tax or duty can be laid on exports from any state. (2) Except for income taxes, which are provided for by the Sixteenth Amendment, all direct taxes must be apportioned among the states according to population numbers. (3) All direct taxes must be applied uniformly throughout the nation. (4) Discriminatory taxation cannot be used to deprive any person of "life, liberty, or property without due process of law." And (5) from the standpoint of purpose, taxes can be collected only "to pay the debts and provide for the common defense and general welfare of the United States."

The provision that taxes can be collected for "the general welfare" both expands and limits the scope of the federal taxing power. In practice, legal questions are seldom raised about those taxes whose revenues go into the general fund, because the courts ordinarily refuse to question the character of expenditures made from this fund. But when a tax is used for regulatory purposes or when its receipts are earmarked for a particular activity or function, the courts may consider the nature of the tax and its objectives and determine whether it actually provides for the general welfare. Should a court decide in the negative, the tax would be declared unconstitutional.

Except for a requirement that no state levy import, export, or tonnage taxes

---

[2] For more detailed discussions of the scope of the taxing power in the United States, cf. William J. Schultz and C. Lowell Harriss, *American Public Finance*, 8th ed. (Englewood Cliffs, N.J.: Prentice-Hall, 1965), chap. 7; Harold M. Groves, *Financing Government*, 5th ed. (New York: Holt, Rinehart & Winston, 1958), chap. 19; Alpheus T. Mason and William M. Beaney, *American Constitutional Law*, 7th ed. (Englewood Cliffs, N.J.: Prentice-Hall, 1983), pp. 274–6; and Ralph A. Rossum and G. Alan Tarr, *American Constitutional Law*, (New York: St. Martins Press, 1983), pp. 254–5.

[3] *F. S. Royster Guano Co.* v. *Virginia*, 253 U.S. 412 (1920).

without the approval of Congress, the Constitution makes no direct references to the taxing powers of the state and local governments. But indirectly, these units are limited by the constitutional requirements that they (1) pass no laws "impairing the obligation of contracts," (2) treat federal treaties as "part of the Supreme law of the land," (3) take no action to discriminate against the citizens of other states, (4) grant "equal protection of the laws" to all persons, (5) deprive no persons of "life, liberty, or property without due process of law," and (6) adhere to the judicial doctrine of noninterference with interstate commerce.

Most state constitutions contain provisions that limit the taxing powers of state and local governments. One of the most widely accepted provisions is that of "uniformity" or "equality" in taxation. This requirement usually applies primarily to property taxes, although it can be applicable to other taxes as well. As usually applied, it requires that all properties, or all properties of the same class, within any taxing district be taxed at the same millage rate and according to the same assessment-value ratio. State constitutional provisions are also used at times to provide that (1) all taxes be for public purposes, (2) property be assessed at its "fair market value," (3) tax levies not exceed specified maximum millage levels, and (4) property not be subject to double taxation.

Use to direct land use. As indicated in Chapter 16, taxation measures can be used to encourage intensive land utilization, attain conservation and environmental goals, promote ownership as a tenure goal, favor particular types of investments, or enhance property values. Taxes can be used as a club, as a penalty device, to discourage or prohibit actions or situations that run counter to these goals. Exemptions and preferential taxation arrangements can also be used as special inducements to get operators to adopt or accept desired practices.

Public policy frowns on the use of taxes as a control or prohibition device. Except for the taxes that have been placed on a few items such as opium, marijuana, unsafe matches, and state banknotes, high tax levies have been used on only rare occasions in the United States to actually prevent particular activities. The "sin taxes" levied on alcoholic beverages and tobacco may have some effect in discouraging the use of these products, but they are viewed primarily as revenue-producing measures. In a similar manner, the impact property taxes based on highest and best use assessed valuations have in causing land areas to shift to new uses is ordinarily treated as an indirect effect of uniform property taxation, not as a direct means for influencing land-use decisions.

Tax measures can have coercive impacts, and they can be used to prevent specified activities. But when they are used for these purposes, they can easily run afoul of the uniformity rules required by most state constitutions. If exemptions from these rules were authorized, tax penalties could be used, as they are in South Korea and Taiwan, to discourage the location of new and large industrial plant facilities in specified areas. They could discriminate against certain classes of operators such as nonresidents, citizens of other states or nations, owners with more than given acreages or values of investment, or nonparticipants in particular programs. They could also be used to prevent specified property uses.

Exemptions and preferential taxation arrangements have provided a more popular policy alternative. With this approach, exemptions from taxes that otherwise would be levied or taxes with lower levy alternatives are used to provide monetary incentives that hopefully will induce operators to carry on certain types of practices. Examples include the exemption of specified types of industrial, commercial, urban redevelopment, or housing developments from full taxation for varying time periods; exemptions of woodlots and qualifying open-space natural areas from taxation; the enrollment of qualifying forestlands under special forest-yield tax programs or farmlands under use-value assessment and other farmland protection programs; and partial exemption of specified types of investment incomes from income and capital gains taxation. Exemptions can also favor certain classes of citizens such as homeowners, senior citizens, veterans, or households with low incomes.

Tax incentive arrangements will probably be used to a greater extent in the future to direct land use. This approach has the strong support of the citizen groups that benefit from the incentives and is usually accepted by the general public. Before more use is made of incentives, however, searching questions should be raised concerning their effectiveness in securing the ends expected of them. Many incentive programs originally designed to bring commercial and industrial developments to certain cities and states have been largely neutralized by similar incentives provided by other cities and states.[4] A primary result in many cases has been a transfer of sizable segments of tax burden from commercial and industrial owners to homeowners without real increases in local employment opportunities. Fifty years of experience with forest-yield taxes suggest that it will take more than special tax incentives to induce many owners of wooded lands to accept long-term forest management programs. Similarly with the special tax programs used to promote farmland protection, one can ask how much farmland has actually been saved by these programs for continued agricultural use.[5]

The Spending Power

Along with the power to levy taxes and collect revenues, governments have an inherent right to spend money. This power does not involve a stick in the bundle of property rights or a direct control over the use of property. But the experience of recent decades demonstrates that the "carrot" approach associated with exer-

---

[4] Cf. Paul E. Alyea, "Property-Tax Inducements to Attract Industry," in Richard W. Lindholm, ed., *Property Taxation: USA* (Madison: University of Wisconsin Press, 1967), pp. 139–58; W. J. Stober and L. H. Falk, "Property Tax Exemptions: An Inefficient Subsidy to Industry," *National Tax Journal*, 20 (1967), 386–94; D. A. Hellman, G. H. Wassall, and L. H. Falk, *State Financial Incentives to Industry* (Lexington, Mass.: Lexington Books, 1976); and Michael Wolkoff, "Tax Abatement as an Incentive to Industrial Location," in Harvey E. Brazer and Deborah S. Laren, eds., *Michigan's Fiscal and Economic Structure* (Ann Arbor: University of Michigan Press, 1982), pp. 281–305.

[5] Cf. Philip D. Gardner and Donald N. Frazier, "The Michigan Farmland Preservation Program: An Evaluation," *Journal of Soil and Water Conservation*, 36 (1981), 344–47.

cise of the public power of the purse can be used effectively to influence the uses made of land resources.

The Supreme Court has held that Congress can spend public funds for almost any purpose as long as its spending is for the public welfare.[6] Powers of a comparable nature are held by the state legislatures subject to possible limitations imposed by their constitutions. Local governments in turn enjoy all the spending privileges granted to them by their charters and enabling acts. Acting with these powers, governments use their spending authority to provide a wide range of services extending from fire, police, and military protection to the financing of education, public works, and land resource development programs.

During recent years, the United States has used its spending power in five important ways to direct land-use practices. It has used public funds to (1) acquire lands for various public purposes, (2) carry on public resource developments, (3) provide public credit facilities, (4) subsidize desired private practices, and (5) finance part of the cost of various state and local projects.

The federal government has used its spending power on numerous occasions to acquire (and reacquire) title to land. The Louisiana Territory, Florida, southern Arizona, Alaska, and the Virgin Islands were acquired directly by purchase. Payments of various types have been used to extinguish the claims of Indian tribes, foreign nationals, and foreign governments to considerable portions of the public domain. The federal spending power has been used along with eminent domain to acquire land for public buildings, roads, and military sites. Some 11.3 million acres were bought under the submarginal land purchase program of the late 1930s. Large areas have been purchased for national forest, wildlife, and other purposes. Altogether, slightly more than 25.8 million acres of the 755 million acres held by the federal government in 1970 was "acquired" land.[7]

Federal funds have been used to finance several types of land resource developments. They have been used to construct public buildings and highways, provide for river and harbor improvements, and dig the Panama Canal. They have provided the basic financing for TVA and for numerous reclamation, power, flood control, and multipurpose projects. They have also been used for public research, reforestation, range improvement and recreation facilities.

Thousands of farm and homeowners have benefited from the public credit facilities provided through the federal spending power. Congress provided the initial capital used to set up the federal land bank system under the Farm Loan Act of 1916. It established the Reconstruction Finance Corporation in 1932 (disbanded in 1954) to make loans to corporations and businesses. A Federal Home Loan Bank Board was created in 1932 and the Farm Credit Administration in 1933 to promote better credit arrangements. Provisions have been made for agricultural production credit, loans to farm cooperatives, loans and rehabilitation grants to disad-

---

[6] Cf. *United States* v. *Butler*, 297 U.S. 1 (1936); and *Helvering* v. *Davis*, 301 U.S. 619 (1937).

[7] Cf. Public Land Law Review Commission, *One Third of the Nation's Land* (Washington, D.C.: Government Printing Office, 1970), pp. 327-34.

vantaged farmers, loans to small business operators, and loans to victims of floods and other disasters. The FHA mortgage insurance and VA loan guarantee programs have also been set up to help prospective owners acquire residential properties with liberal credit terms and nominal down payments.

Government subsidies are used from time to time to promote particular practices.[8] Subsidies were paid to sugar producers under the McKinley Tariff Act of 1890. Public funds have been used to promote low-rent public housing. Consumer subsidies are used in the federal food stamp and school lunch programs. Government-owned plants, cost-plus contracts, and other subsidy arrangements were used during World War II to stimulate industrial promotion. Federal commodity purchase and loan programs have been used to buoy up and support farm market prices, and subsidies have been used since 1936 to provide farmers with benefit payments for accepting conservation practices.

Subsidies are often incorporated in cost-sharing arrangements that offer federal support for specified types of programs that meet federal standards. Assistance of this type has been provided for the construction of state highway projects since 1916. With the Federal Aid Highway Act of 1956, the federal government pays up to half the cost of highways located entirely within states and up to 90 percent of the cost of those highways and expressways classified as parts of interstate systems (up to 95 percent in states with large acreages of unappropriated and unreserved public lands or untaxed Indian lands). Provisions are made in the Watershed Protection and Flood Prevention Act of 1954 for federal assistance to local communities to facilitate the development and operation of small watershed programs.

Federal assistance for slum clearance and public housing projects has been available to cities since 1937. The National Housing Acts of 1949, 1954, and 1956 authorized federal grants to finance two-thirds of the cost of approved local urban renewal projects. Later Housing Acts authorized the Department of Housing and Urban Development to pay up to 40 percent of the cost of local open-space land-acquisition programs and 80 percent of the cost of model city projects. Federal agencies have also made grants to state agencies for acquiring recreation lands; paying substantial portions of the cost of state, regional, metropolitan, and local planning efforts; facilitating regional economic development programs; and providing waste water treatment and toxic waste disposal facilities.

Use of the Proprietary Power

The federal government's proprietary or public ownership power is clearly recognized in the Constitution. This power, which is also exercised by the several states, involves the right of governments to acquire, develop, manage, and dispose of properties and also their right to carry on production and marketing programs in

---

[8] A related practice known as *sanctions* has also been used in recent years to secure state compliance with federally designated objectives. With this approach, federal grants for specific programs such as highway construction and maintenance may be withheld if a state fails to accept other regulations such as billboard controls or the fifty-five mile per hour speed limit.

competition with private operators. Neither level of government has moved very far in extending the scope of this power. Their experience in recent decades, however, indicates that the power provides a potent tool for securing desired land-use objectives.

Public landownership in the United States. Approximately 40 percent of the surface land area of the United States is held in public ownership—more than 33 percent by the federal government; about 6 percent by the states; and 1 percent by counties, municipalities, school districts, and other local units of government.[9] Almost 40 percent of this total area is found in Alaska. The remaining areas are used primarily for forestry and grazing purposes and are concentrated mostly in the eleven western states, where they account for approximately 60 percent of the land area. The amount of federal ownership by states in 1970 ranged from 0.3 percent of the area of Connecticut and 0.6 percent of Iowa to 66.5 percent of Utah, 86.4 percent of Nevada, and 95.3 percent of Alaska.

Approximately 30 percent of the forestland and nearly half of the grazing land in the forty-eight contiguous states are held in some form of federal, state, or county ownership. Public ownership affects 3 million acres used for experimental purposes or operated in connection with penal or public welfare institutions. It also accounts for between 30 and 40 percent of the area of the average city; for substantial areas used for transportation, military, park, wildlife refuge, and other service areas; and for a sizable residual area of barren and wasteland.

Large portions of the land area of the United States have been held in public ownership since the birth of the nation. Yet it was not until the 1890s that much attention was given to the development of long-range public land management programs. Until that time, it was assumed that practically all the public domain would soon pass into private ownership—that the plow would follow the ax across the wooded frontier and that farmers would soon supplant cattlemen on the western range.

Two factors caused a change in this philosophy. A gradual awakening of public interest to the need for forestry and other land resource conservation measures led Presidents Harrison, Cleveland, McKinley, and Theodore Roosevelt to reserve 192 million acres of forest, mineral, park, and other lands from public disposal. And the bitter failure of thousands of high-hearted settlers in some cutover and range areas finally convinced the general public that these areas were not suited for agricultural development under current cost and price conditions.

With this change in outlook, the state and federal governments gradually reoriented many of their public land policies. Instead of holding land on a temporary caretaker basis, emphasis was given to the development of long-term management programs. New agencies such as the Forest Service, Grazing Service, Fish and Wildlife Service, and the various State Conservation Departments were established to administer the public lands. These agencies were authorized to buy and exchange lands—and in the case of the states, to use tax-reverted lands—in blocking

[9] An additional 2 percent is held in trust by the federal government for various Indian tribes.

up their holdings. This acquisition program brought the creation of a "new public domain," which accounts for approximately 10 percent of the lands held in federal ownership in the forty-eight contiguous states and for well over half of the public lands administered by many states.

Most of the large tracts of open-space land held in federal and state ownership can be regarded as residual areas that remained in or reverted to public ownership because private owners did not choose to acquire or hold them when they had the opportunity. Much of this "land that nobody wanted" a half century ago has now had time to grow valuable stands of timber and become attractive for recreation and other uses. With management and changes in the demand situation, many of these tracts have become "land that everyone wants." They provide recreation and environmental values for large numbers of people and also supply increasingly important sources of public revenues from the leasing and sale of grazing, mining, and timber-cutting rights. These lands are coveted in many instances by individuals and groups who argue for new public land disposal programs. But as is the case with streets and highways, parks, and public service areas, they should be viewed as valuable public resources and not be disposed of without careful evaluations of the effects of such practices on the public interest.

Public ownership and the direction of land use. Since 1900 the proprietary power has been used in many ways to secure particular ends in land resource use. The federal government has used this power to reserve large areas of forest, mineral, grazing, and other public domain lands that might otherwise have been homesteaded or sold to private operators. Together with the states and local units, it has reserved and acquired substantial areas for park and recreation purposes. Along with these other units, it has established public forests and undertaken reforestation projects, partly to get land areas back to their highest and best use, partly to set an example for private operators, and partly because it was feared that private operators could not or would not do the job needed for the nation's future welfare.

The federal and state governments have joined with private operators in organizing grazing districts and drawing up regulations for the orderly and efficient use of their combined range holdings.[10] They have undertaken reclamation and range improvements to shift large areas to higher and better uses. The federal government has used its spending and proprietary powers to shift submarginal farmlands to more appropriate uses and to promote other desired land-use adjustments.[11] It has used its spending and proprietary powers to develop single-purpose and multipurpose flood control, power, and reclamation projects—projects that have often

---

[10] Cf. C. W. Loomer and V. Webster Johnson, *Group Tenure in Administration of Public Lands*, U.S. Department of Agriculture Circular No. 829, 1949; and Layton S. Thompson, *Montana Cooperative State Grazing Districts in Action*, Montana Agricultural Experiment Station Bulletin No. 481, 1951.

[11] Cf. Lloyd Glover, *Experience with Federal Land Purchases as a Means of Land Use Adjustment*, South Dakota Agricultural Experiment Station, Agricultural Economics Pamphlet No. 65, 1955.

come closer to representing the optimum scale of development than those that probably would have been constructed by private enterprise.

The proprietary power has been used in many urban areas to provide public water, transportation, power, and housing facilities. It is used along with the spending and eminent domain powers in the clearance of slum and blighted areas for urban redevelopment and renewal programs. Most of the lands acquired with these programs are sold again for private redevelopment. But in their sale, restrictive covenants are ordinarily used to ensure future compliance with the redevelopment plan for the area.

Nationalization or threats of nationalization of certain types of business and property ownership have sometimes been used by some nations to direct resource use. Except for examples such as the nationalization of air navigation rights, little use of this approach has been used in the United States. Our states also have avoided the Scandinavian and Germanic policy of encouraging cities to own large areas both within and outside their municipal boundaries.[12] These approaches offer opportunities for broader uses of the proprietary power in the direction of land use, but run counter to currently accepted political views concerning the desirability of private ownership.

Acquisition and transfer of development rights. Land-use objectives can often be attained through the acquisition and transfer of less than the full fee simple rights of ownership. With this approach, private landowners can continue to possess, manage, and use their land for a wide range of activities while governments or others can acquire and hold specific property rights that bear on public policy objectives. The National Park Service used this approach during the 1930s when it acquired conservation or scenic easements that restricted the rights of owners to erect billboards or use their lands for nonnatural scenic uses on properties located along the Blue Ridge Parkway and the Natchez Trace.[13] Wisconsin acquired similar easements along the Great River Road during the 1950s, and the Fish and Wildlife Service has acquired easements that protect potholes and the wetland character of properties around national wildlife refuges.

---

[12] Many cities in the Scandinavian and Germanic countries of northern Europe have held large tracts in municipal ownership almost since feudal times. Many cities in Finland hold title to their sites with restrictions against the alienation of full titles to others. Stockholm acquired some 20,000 acres (five times its original area) between 1904 and 1937, mostly for housing developments; and the next five largest Swedish cities owned between 47 and 80 percent of their corporate areas. Oslo, Norway; Copenhagen, Denmark; The Hague, Netherlands; and Vienna, Austria, owned large areas; and Berlin owned 75,000 acres (around a third of the total area) within its city limits and another 75,000 acres outside. The areas held outside the cities were ordinarily used for park, agricultural, or forestry purposes pending their possible future need for urban development. Cf. Harold S. Buttenheim, "Urban Land Policies," in *Urban Planning and Land Policies*, Supplementary Report of the Urbanism Committee to the National Resources Committee, Vol 2 (Washington, D.C.: Government Printing Office, 1939), pp. 228-29, 312-20; and Roy J. Burroughs, "Should Urban Land Be Publicly Owned?" *Land Economics*, 42 (February 1966), 11-20.

[13] Cf. William H. Whyte, *The Last Landscape* (Garden City, N.Y.: Doubleday, 1968), pp. 83-84.

Proposals for public acquisition and holding of development rights have attracted considerable notice since the early 1970s. The New Jersey Blueprint Commission recommended in 1973 that that state purchase the development rights to 1 million acres of farmland as a means of protecting its remaining agricultural base. A comparable program was initiated the next year in Suffolk County on the eastern end of Long Island in New York. Similar programs have since been started in other New York communities and in Maryland, New Jersey, Connecticut, Massachusetts, New Hampshire, and Washington.[14] Private trusts have also been established in Virginia and some other states to acquire and hold farmland development rights and thus ensure the protection of farmlands from urban development.

A more complicated arrangement involving transfers of development rights (TDRs) has also been used in some instances to protect farmlands but has attracted more notice as a possible means for compensating owners, at least in part, for *wipeouts* (losses of economic value that might otherwise have been received or retained) associated with their compliance with zoning regulations.[15] New York City and Washington, D.C., have used this approach with their protective zoning of historic landmarks by granting assignable floor area ratios to owners that may be transferred to other developable sites within these cities. TDR arrangements may also be used with the development of planned communities to permit all participating owners to share the windfall benefits that can come with their mutual acceptance of plans that call for the development of some sites within the total area for economically more valuable uses than others. With this arrangement, every owner receives TDR shares in proportion to the portion of the total land area they own. A double market is envisaged in which prospective developers must buy both the sites they need and sufficient TDRs to cover their operation plans before they can proceed with land developments.

EMINENT DOMAIN POWER

Eminent domain provides a fourth important power governments use in their direction of land resource use. This concept, which literally means "highest authority or dominion," involves "the power of the sovereign to take property for

---

[14] Cf. J. Dixon Esseks and Robert E. Coughlin, "Land Use Controls: Purchase of Interests in Land," in *The Protection of Farmland: A Reference Guidebook for State and Local Government* (Washington, D.C.: National Agricultural Lands Study, Government Printing Office, 1981), pp. 148–73.

[15] Cf. Robert E. Coughlin, "Land Use Controls: Working with the Private Market to Retire Development Rights in Agricultural Areas," *The Protection of Farmland: A Reference Guidebook for State and Local Governments*, pp. 174–86; Jerome G. Rose, ed., *Transfer of Development Rights* (New Brunswick: Rutgers, The State University of New Jersey, Center for Urban Policy Research, 1975); David E. Ervin *et al., Land Use Control* (Cambridge, Mass.: Ballinger, 1977), chap. 7; and Donald Hagman and Dean Misczynski, *Windfalls for Wipeouts: Land Value Capture and Compensation* (Washington, D.C: American Planning Association, 1978), chap. 23.

public use without the owner's consent."[16] It is rooted in the ultimate authority of the state over property and is commonly accepted and exercised as an inherent right of government. As a community control over property, it can be exercised alone or in combination with other powers of government. Its exercise is necessary for the orderly acquisition of sites needed for highways, streets, utilities, and other public improvements. Without it, individual property owners could block the will of the majority simply by refusing to sell land needed for desired public developments.

Scope of Eminent Domain

The U.S. concept of eminent domain has its roots in the natural law movement. The term *eminent domain* was coined by Hugo Grotius, a Dutch philosopher-statesman, in his *De Jure Belli et Pacis* (1625) to describe the power of the state over all private property within its bounds.[17] As used by Grotius and later writers, this concept had two important facets. It envisaged the state as the only taker of property; and it assumed the age-old doctrine of immunity of the sovereign from liability for the taking of property.

"Eminent domain" was first mentioned by the U.S. courts in 1831.[18] Until then, there were few examples of the public taking of private property; what cases there were dealt mostly with roads and mill pond flowage areas; and little effort was made to develop a legal philosophy to justify the public taking of land. Following its recognition by the courts, the power of eminent domain was soon accepted as an established feature of American law subject to three modifications: (1) the federal and state governments can delegate this power to other units of government and to public and private corporations; (2) the power must always be used for a public purpose; and (3) just compensation must be paid for all properties taken.

Like many legal concepts, eminent domain has an elastic scope. Almost from the start, state legislatures regarded eminent domain as a power they could delegate to private corporations and groups as well as to state agencies and local governments. The courts did not object to these delegations and for several years treated the meaning of "public purpose" as a legislative question. As a result, the concept of public purpose was stretched to include many types of land resource developments. Eminent domain powers were delegated to railroads to help them acquire rights of way for their tracks, to mill-dam owners to acquire flowage rights, to mine owners to acquire land for tramways and other surface facilities, and to drainage districts for ditches and drains.

---

[16] Julius L. Sackman and Russell D. Van Brunt, *Nichols' The Law of Eminent Domain*, rev. ed. (New York: Matthew Bender, 1964), I, 4.

[17] Cf. *ibid.* pp. 6–7; also Arthur Lenhoff, "Development of the Concept of Eminent Domain," *Columbia Law Review*, 42 (April 1942), 596–638. The term *eminent domain* is not used in English law. But it is matched with a somewhat comparable concept known as *compulsory purchase*.

[18] Cf. *Beckman v. Saratoga and Schenectady Railroad*, 3 Paige (N.Y.) 45, 73 (1831).

With the rising tide of American industry during the 1840s and 1850s, it soon became evident that the eminent domain power could be abused. Fears were expressed concerning the increasing political power wielded by some corporations and the possibility that they might use grants of eminent domain to seize the property of competitors and others who stood in their way. At this point, the courts gradually assumed the prerogative of deciding what is "public use." In the process of distinguishing between public and private uses, many courts narrowed the meaning of "public use," particularly in cases involving private users, to mean use by the public.[19]

This judicial limitation of the concept of "public use" resulted in complications, uncertainties, and some legal inconsistencies. Some courts continued to hold for broad interpretations of public use, while others applied more limited definitions. Most American courts since the 1930s have agreed with Justice Holmes that "use by the public" is an inadequate test of public purpose.[20] The U.S. Supreme Court took the position in 1954 that, "Once the object is within the authority of Congress, the right to realize it through exercise of eminent domain is clear."[21]

With this liberalized concept of "public use," the courts have recognized that "the promotion of beauty or national sentiment may warrant the exercise of the power of eminent domain where it would not warrant the exercise of police power."[22] They have upheld use of eminent domain to acquire sites for airports, take blighted and open-space areas for urban redevelopment projects, secure absentee owner-held farmlands in Puerto Rico for redistribution in smaller-sized tracts to resident operators, and acquire urban residential lands in Hawaii for resale to individual homeowners. As one legal journal puts it,

> It is now established that a private corporation or individual may be granted the power of eminent domain as long as the general objective of the use of this power contributes to the public welfare and advantage. A logging railroad to serve a single lumber company, an irrigation system primarily for one landowner's benefit, and an aerial bucket line for a private mine have been held sufficient public uses to warrant the use of eminent domain for their construction. Approval of condemnation of property for slum clearance and rehousing typifies the liberal view, the legitimacy of the purpose as a whole rather than the intended use of the particular property being the criterion almost universally adopted by the courts.[23]

Payment of just compensation has been accepted as a recognized principle in

---

[19]Cf. Philip Nichols Jr., "The Meaning of Public Use in the Law of Eminent Domain," *Boston University Law Review*, 20 (November 1940), 615–41.

[20]*Mt. Vernon Woodbury Cotton Duck Co.* v. *Alabama Power*, 240 U.S. 30 (1916).

[21]Cf. *Berman* v. *Parker*, 348 U.S. 26 (1954).

[22]Ernest Freund, "Eminent Domain," *Encyclopedia of the Social Sciences* (New York: Macmillan, 1931), V. 494.

[23]"Public Land Ownership," *Yale Law Journal*, 52 (1943). Reprinted by permission of the Yale Law Journal Company and Fred B. Rothman & Company from *The Yale Law Journal*, vol. 52, p. 634.

U.S. eminent domain law for more than two centuries.[24] Provisions of this order were included in several state constitutions between 1776 and 1800 and were inserted in the Bill of Rights as part of the Fifth Amendment in 1791. The right to compensation for property taken stems from the natural- and higher-law concepts of "reasonableness," "natural equity," and "dictates of natural justice," and as such was accepted by courts in virtually all of the states before this principle was formally required by the Fourteenth Amendment.[25]

Use of Eminent Domain

From a public point of view, eminent domain may be regarded simply as a necessary power that must be used from time to time to facilitate the acquisition of particular sites for desired public purposes. But the exercise of this power can involve problems. Owners often find that they are expected to give up land resources they may need to maintain their economic scales of operation or resources intimately associated with their hopes, aspirations, and outlook on life. Administrative procedures must be followed in the property condemnation process. And emphasis should always be given to protection of the interests of those owners whose property is taken as well as the interests of the taker.

Five important problems arise in the use of eminent domain: (1) the initial decision to take property for public use, (2) a proper delegation of authority for this purpose, (3) compliance with prescribed condemnation procedure, (4) determination of just compensation, and (5) the question of how much property can legally be taken.

The question of whether eminent domain should or should not be used in particular instances calls for a weighing of public versus private interests. This weighing process starts with the initial justification and authorization of a proposed development, at which time public hearings are usually held on the decision to take. Careful consideration should be given at this point to the relative merits and costs of alternative development plans and sites and to any objections or counterproposals operators may have concerning the proposed projects. Once the decision to take property for a project has been made, public considerations ordinarily over-

[24] There have been instances in which federal takings of property have not been associated with payments of compensation. No payments were made, for example, with the emancipation of the slaves, prohibitions of the manufacture and selling of alcoholic beverages, nationalization of air navigation rights and of the radio spectrum, or the 1933 regulations limiting private holdings of gold bullion.

[25] Cf. Lenhoff, *loc. cit.*, pp. 599–600; J. A. C. Grant, "The 'Higher Law' Background of the Law of Eminent Domain," *Wisconsin Law Review*, 6 (1930), 67–85; and Emily Dodge, "Eminent Domain and Related Problems," in Jacob H. Beuscher and Robert R. Wright, *Cases and Materials on Land Use* (St. Paul, Minn.: West Publishing, 1969), pp. 709–22. Grotius recognized that sovereigns had a right to take property but argued that the taking involved a moral obligation to pay compensation for the property taken.

ride private interests, and questions may be raised only concerning the precise location of the lands to be taken and the values assigned to them. At this point, steps must be taken to acquire easements or property rights to the lands needed for a highway, power lines, canal or other proposed developments.

Except for military uses and some highway developments, the needed rights must be acquired before construction begins. This means that the needed land must be surveyed; an appraisal must be made of its value; owners must be contacted; and offers must be made for purchase of the needed property rights.[26] If an owner refuses to accept the terms offered, condemnation proceedings must be started to force the sale. But before an agency can exercise eminent domain, it must have a proper delegation of eminent domain authority from Congress or the state legislature. It must follow the procedures and operate within the limitations of its grant of power, and its use of eminent domain must be for a public use.

There is no simple standardized eminent domain procedure. Every legislature is free to outline its own method, and it can vary its requirements in its delegations of eminent domain power to different agencies and groups. Some states have acted to standardize their procedures. But as late as 1943, 320 different procedures affecting the acquisition of land for highways were in use in fifty-five jurisdictions.[27] Even with these variations, condemnation procedures always call for court action. Property owners must be informed that condemnation proceedings have been started against them; they are entitled to a legal hearing; and the determination of compensation is ordinarily made either by a jury or a group of commissioners who hear legal evidence and who may visit the property being condemned.[28]

Determination of just compensation. The question of what constitutes just compensation can pose a controversial problem in property valuation. Owners often complain that they are forced to sacrifice properties at less than their true value. Taking agencies in turn feel that they are required to pay more than fair market value. Most of this problem stems from different standards of measurement. Owners who do not wish to sell are naturally inclined to think in terms of "value to the owner," whereas takers base their calculations on the going free market prices paid for comparable properties.

In their determination of what constitutes just compensation, most courts reject

[26] Some delegations of eminent domain authority require that all properties needed for given projects be acquired by condemnation proceedings without the submission of prior offers to purchase to affected owners. Some others require the agencies to acquire specified minimum proportions of the property rights needed for their projects by direct purchase before they can exercise eminent domain to acquire any remaining properties.

[27] Cf. David R. Levin *Public Land Acquisition for Highway Purposes* (Washington, D.C.: Government Printing Office, 1943).

[28] Cf. Glenn Lawrence, *Condemnation: Your Rights When Government Acquires Your Property* (Dobbs Ferry, N.Y.: Oceana Publications, 1967) for a popular discussion of the law and procedures of eminent domain.

the concepts of "value to the taker" and "value to the owner." Just compensation is usually defined as "fair market value" or "the price a willing buyer would pay a willing seller."[29] Payment is ordinarily refused for consequential damages such as expense of moving, personal inconvenience, interruption of business operations, or loss of goodwill. Yet while payment is normally denied for consequential damages, compensation may be allowed for any severance damages associated with a diminution of the market value of the holding left with the owner.

Condemnation awards ordinarily range between the lowest estimate of value provided by a qualified witness and the price claimed by the owner. Some courts and commissions are more liberal in their awards than others, and some make allowances in their calculations of fair market value for personal inconveniences while others do not. Legislative action has been taken in some states to authorize compensation of property owners for some types of consequential damages. Michigan, for example, modified its highway eminent domain procedures in 1965 to permit nominal payments to owners for moving expenses.

Differences also exist in the land-acquisition policies of different agencies, with some agencies giving more consideration to owner problems than others. The Tennessee Valley Authority, for example, provides a good example of a public agency that operates on the premise that "the land owner should be at least as well off after taking as he was before."[30] In its determination of the purchase prices offered for land, TVA has considered not only the market value of the land but also what it would cost to reestablish the owner at some other location with no impairment of economic position.[31] This procedure has applied only as long as the owner was willing to accept what TVA considered a fair purchase price. When an owner refused to sell, TVA pressed for condemnation at the court-accepted price of what a willing buyer would pay a willing seller. Thus

TVA may offer $10,000 for a farm. If the landowner rejects this offer, the case is heard by a commission. TVA may then present evidence to show that the land is worth only $8,000 or the figure that selected witnesses will testify to be its fair market value. . . . Among the reasons advanced for this shift to a willing buyer–willing seller concept under condemnation are (1) a belief that the courts generally favor the landowner as against the Government, (2) a desire to discourage litigation, and (3) a feeling that it is under no obli-

---

[29] Cf. Lewis Orgel, *Valuation under the Law of Eminent Domain*, 2nd ed. (Charlottesville, Va.: Michie Company, 1953), I, 56; and Dodge, *loc. cit.*, pp. 531–38. In contrast with the emphasis placed on payment of just compensation for the properties taken, some Latin American nations use valorization taxes assessed against expected increases in the value of affected properties, including those that give up land, to help finance street-widening, and other public improvement projects. Cf. William G. Rhoads and Richard M. Bird. "The Valorization Tax in Colombia," in Arthur P. Becker, ed., *Land and Building Taxes* (Madison: University of Wisconsin Press, 1969), pp. 201–37.

[30] Kris Kristjanson, *TVA Land Acquisition Experience Applied to Dams in the Missouri Basin*, South Dakota Agricultural Experiment Station Bulletin 432, 1953, p. 9.

[31] Cf. Charles J. McCarthy, "Land Acquisition Policies and Proceedings in TVA," *Ohio State Law Journal*, 10 (Winter 1949), 56.

gation in litigation to offer anything other than just compensation as defined by the courts.[32]

Allowance of compensation for severance damages can pose unique problems in condemnation valuations. In theory, whenever the act of taking property from owners greatly reduces the productive value of their remnant tracts, they should receive sufficient compensation in excess of the market value of the property to "leave them whole." Differences of opinion can arise, however, as to the extent of severance damages. Some of these problems can be illustrated by an example, such as that depicted in Figure 18-1, in which a right of way is condemned across properties *A, B, C,* and *D* for the construction of a limited-access highway.

In the calculation of just compensation, the owners of properties *A* and *B* can claim severance damages only if they can show that their marginal losses of land area have a disproportionate effect in reducing the economies of scale at which they can operate. A stronger case for severance damages can be made with property *C* because the owner is left with a small severed tract that is isolated from the principal tract. Acquisition of the small severed tract for later use or disposal by the acquiring agency can often entail lower costs to the public than procedures that call for simple acquisition of the right of way plus payment of severance damages. A still stronger case for severance damages arises with tract *D*. Here, the owner's holding may lose its economic viability as it is split into two tracts of uneconomic size. Combination of the two severed tracts with those of neighboring properties may provide the most practicable solution for all concerned. If construction of the highway results in dead-ending the road that formerly ran between properties *B* and *E* and *C* and *D*, it may also add to the operating costs of property *E*. The fact that this operator and others along the road now find it less convenient to travel to a nearby shopping center, however, gives them no claim to severance damages because none of their property is taken by condemnation.

Excess condemnation. Questions are frequently raised concerning the amounts of land that can be taken by eminent domain. Is this power limited to the taking of only the land necessary for a proposed development, or might it be extended to cover additional areas? The requirement of "public use" has often been interpreted

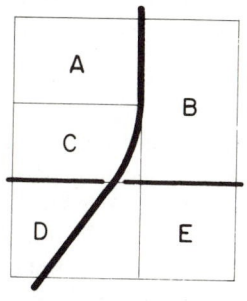

FIGURE 18-1. Illustration of impact of condemnation of a right-of-way for a limited-access highway on five properties.

[32] Kristjanson, *op. cit.,* p. 10.

as limiting all takings to the land areas needed for specific projects. But good administrative management often favors the condemnation of additional areas. Several states have constitutional provisions that expressly authorize excess condemnation; and the courts in many others now accept excess condemnation when its use appears reasonable.

Excess condemnation can be justified as (1) a means of protecting the values associated with resource developments, (2) the best way to handle remnant tracts, and (3) a means of cost recoupment. With highways and certain other developments, it is often desirable to acquire buffer tracts, which can be used for project protection or for aesthetic or safety purposes. Buffer strips can be used to separate playground areas from busy streets and to preserve the natural scenery along highways and around recreation areas. Tracts along highways can sometimes be landscaped or developed as roadside parks. In some instances they may be leased or sold back to operators with restrictions against billboards, the cutting of trees, or practices that destroy the rustic beauty of the countryside. Restrictions may also be used to control roadside developments and prevent uses that contribute to the gradual conversion of through highways into hazardous streets.

When only the minimum areas needed for specific projects are taken, owners may find that they are left with small, odd-shaped remnant tracts that have little value by themselves. Some states authorize taking agencies to acquire these extra tracts and either sell them to adjacent property owners or regroup them with other tracts to form units of greater economic value. This type of excess condemnation can often pay for itself by bringing in sales revenues and preventing claims for severance damages that may equal or exceed the taking price of the remnant holdings.

Excess condemnation may also be used to recoup all or part of the costs of new resource developments. With this approach, a public agency could acquire more land than it needs for a street, highway, or reservoir and then resell the extra land as building sites after the project is completed. This approach is used in the regrouping and sale of remnant tracts, but has received limited use, partly because of the fear that it may be rejected by the courts as a use of eminent domain for private purposes, partly because of the traditional notion that profits arising from eminent domain should accrue only to private owners, and also because of objections against public agencies engaging in real estate operations. Some people feel that land value-increment taxes should be used in cases of this nature to help recoup the costs of public project developments.

Use to direct land use.   Eminent domain represents a significant and necessary public interest in land. This power has been used on frequent occasions to acquire property rights needed for various public and public interest-associated projects. The availability of this power has also brought considerable cooperation on the part of owners in making rights available for various projects. But important and necessary as it is for the attainment of many public development objectives, exercise of this power has not provided a highly effective means for directing land use, for the simple reason that its application ordinarily affects only small numbers of properties.

More use would probably be made of eminent domain if the police power could not be used to require operator compliance with various land-use rules and regulations. Some critics of police-power procedures have suggested substitution of a zoning by eminent domain (ZED) approach.[33] With this option, economic values would be assigned to the possible wipeouts associated with zoning and other regulations; owners would be compensated for any wipeouts they experience; eminent domain would be used to acquire "rights" associated with wipeout values when operators object to their voluntary sale; and special assessments would be levied on properties that benefit from the regulations to finance the eminent domain costs.

POLICE POWER

Of the several powers state and federal governments hold over landed property, the *police power* is often regarded as most important. This power involves the basic right of governments to make regulations and centers in their inherent right to legislate for the advancement, preservation, and protection of public health, safety, morals, convenience, and welfare. As one court has observed, the police power

> . . . is of vast and undefined extent, expanding and enlarging in the multiplicity of its activities as exigencies demanding its service arise in the development of our complex civilization. It is a function of government solely within the domain of the legislature to declare when this power shall be brought into operation, for the protection or advancement of the public welfare.[34]

In some respects, the police power is a singular American institution. No mention of it appears in the U.S. Constitution. The concept has evolved instead as a product of judicial interpretation. The term *police power* was first used by the Supreme Court in 1827.[35] Its recognition as a potent source of legal authority came two decades later when Chief Justice Taney defined it broadly as "the power of government inherent in every sovereignty . . . to govern men and things within the limits of its dominion."[36] Four years later, the Massachusetts Supreme Court defined the *police power* as "the power vested in the legislature . . . to make, ordain and establish all manner of wholesome and reasonable laws . . . for the good and welfare of the commonwealth, and the subjects of the same."[37] Mason and Beaney indicate that the police-power doctrine developed as a "juristic expression of popular sovereignty," and as a legal means governments can use to control actions

[33] Cf. Hagman and Misczynski, *op. cit.*, chap. 22; and Ervin *et al.*, *op. cit.*, chap. 6.

[34] *Motlow* v. *State*, 125 Tenn. 547, 589 (1911).

[35] *Brown* v. *Maryland*, 12 Wheaton 419 (1827). Chief Justice Marshall used the term in passing to indicate the residual powers—in addition to the eminent domain and taxation powers —held by the states.

[36] *License Cases*, 5 Howard 504, 583 (1847).

[37] *Commonwealth* v. *Alger*, 7 Cushing (Mass.) 53, 85 (1851).

that endanger or threaten the community welfare.[38] Since then the courts have gradually expanded the scope of what can be accomplished with the police power to permit an ever-widening latitude of social action in areas involving the public interest and welfare.

Since its judicial beginnings, police power has been treated as a residual power of the states, which the federal government can also exercise in its administration of federal territories and as a delegated power incident to its commerce, postal, taxation, and war powers. It has provided the legal basis for a variety of public measures that limit or regulate vested property rights. But while the courts have accepted an expanding concept of police power, they have held rigorously to the view that every exercise of this power must be reasonable, must enhance the public welfare, must not be arbitrary or discriminatory, and must not deprive persons of their rights without "due process of law" or "the equal protection of the laws." The "due process" and "equal protection" restrictions against state action were adopted as part of the Fourteenth Amendment to the U.S. Constitution following the Civil War in 1868. These provisions were designed ostensibly for the protection of individual civil rights. In practice, they have often served as a check against the police power of the states.

The police power can be employed in many different ways to influence and direct the use of land resources. Emphasis is given here to (1) zoning ordinances, (2) subdivision regulations, (3) rent controls, and (4) air and water quality standards as four leading examples of its application in the United States. Other examples of its application for land-use directional purposes include its use with building and construction codes; fire protection and sanitation ordinances; and regulations affecting the provision of necessary landscaping and parking spaces, strip mining operations, destruction of diseased animals, maintenance of area quarantines, and transportation and storage of dangerous substances.

Zoning Ordinances

Zoning is a foremost example of the use of the police power to direct land use. Far from being a complicated concept, *zoning* means "the division of land into districts having different regulations."[39] It involves the designation of land-use districts within which specific regulations apply concerning the uses that may be made of land; the height, size, and use of buildings; and maximum densities of population. Zoning is a tool for carrying out a land-use plan, not a substitute for planning; and its worth and effectiveness always depend on the character of the planning on which it is based.

Beginnings of zoning. Land district regulations have been used sporadically in urban areas since ancient times. Fire districts permitting the construction of wooden buildings in some areas but not others were established at an early date in many

---

[38] Cf. Mason and Beaney, *op. cit.*, p. 298.

[39] Edward M. Bassett, *Zoning* (New York: Russell Sage Foundation, 1940), p. 9.

cities. Some of our colonial villages banned the manufacture and storage of gun-powder from their built-up residential and business sections for fire protection and public safety reasons. Massachusetts enacted a law in 1692 that authorized certain towns "to assign places in each town, where least offensive, for slaughterhouses, stillhouses, and houses for trying tallow and currying leather."[40]

Zoning regulations were adopted in several California cities during the late 1800s, often as a means of restricting the location of Chinese laundries.[41] Other cities followed with regulations that controlled practices regarded as nuisances or near-nuisances. Police-power measures were adopted to protect residential neighborhoods from the undesired environmental impacts associated with the operation of such businesses as slaughterhouses, brick kilns, dairies, livery stables, stone crushers, and carpet-beating establishments.[42]

The principal emphasis with these early examples was on public safety and the control of nuisances. Little attention was given to the possible use of zoning as an instrument of planning for orderly community growth. A first step in this direction came in 1904–1905 when the Massachusetts legislature established two types of building-height districts in Boston—one with an 80-foot height limit and the second with a 125-foot height limit. Los Angeles followed in 1909–1911 with a series of zoning ordinances that identified seven industrial districts and designated the remaining area of the city as a residential district from which certain uses were excluded. The "building-height" and "land-use" regulations adopted in these two cities represent our first real use of zoning for direction of land-use purposes.

Acceptance of comprehensive zoning. New York City adopted the nation's first comprehensive zoning ordinance in 1916. This ordinance had its beginnings with the appointment of a Commission on Heights of Buildings in 1910 to investigate the impact of skyscrapers on community health and safety. A commission report recommending regulations to control the height, bulk, and use of buildings in different districts of the city was issued in 1913. Legislative approval was secured the next year for a change in the city charter to permit district regulations; and a special commission was appointed shortly thereafter to prepare a districting resolution and map.

New York City at this time provided a glaring, but typical, example of the need for land-use regulations. As Bassett indicates,

> A building could legally rise to any height whatever, assume any form, be put to any use, and cover 100 percent of the lot from the ground to the sky. Ten-

---

[40] Erling D. Solberg, *Rural Zoning in the United States*, U.S. Department of Agriculture Information Bulletin No. 59, 1952, p. 2.

[41] Laundries and washhouses were regarded as a public nuisance in these instances because of their frequent dumpings of washwater in the streets, the fire hazard associated with their often flimsy construction and their storage of inflammable materials, and the moral hazards associated with congregations of large numbers of people in these establishments.

[42] Robert A. Walker, *The Planning Function in Urban Government* (Chicago: University of Chicago Press, 1941), p. 56.

ement houses were the only structures that might not cover the whole lot. High office buildings not only covered their entire lots and had the same floor space in their top stories and their first stories, but cornices projected into the street from eight to fourteen feet. Buildings of this sort in the southern part of Manhattan made dark canyons of narrow streets, but what was perhaps even more harmful they produced chaotic building conditions. The first sky-scraper to be erected in a block would cover the entire lot up to the roof and open its windows on neighboring lots. A high building so erected prevented other similar buildings . . . in its immediate vicinity . . . [and thus] obtained a virtual monopoly of the light and air.

. . . Improper uses caused injury to homogeneous areas and were especially productive of premature depreciation of settled localities. One-family, de-tached home districts, possessing trees and lawns, were invaded by apartment houses occupying nearly their entire lots. These in turn were damaged by the building of stores, garages, and factories. . . .

Invasion of apartment houses by stores on their ground floors lessened the desirability of neighboring apartment houses because of the increase of noise, vehicles, fire hazard, litter, and street congestion. Business streets lined with retail stores were invaded by factories, garages, and junk shops. Localities devoted to light industry, perhaps employing women and children, were in-vaded by heavy industries producing noise, smoke, and fumes. No land owner in any part of the city could erect a building of any sort with assurance that in ten or twenty years the building would not be obsolete by reason of an unnecessary and undesirable change in the character of the neighborhood. Sometimes these changes left a blighted district behind.[43]

With this situation, it was obvious that the public welfare called for some sys-tem of regulations that would control the height, area, and use of buildings and also stabilize the land-use patterns found throughout the city. Land- and building-use regulations were suggested as appropriate means for achieving these ends. These districting—later called zoning—regulations provided for height and area restrictions on new building construction and for reserving some areas for single-family dwell-ings; some for multifamily housing; and others for commercial, industrial, and other uses.

Questions were raised at this point concerning the constitutionality of the pro-posed regulations. Comprehensive zoning was a new thing; many lawyers were un-certain about the legality of its proposed use; and some argued that zoning called for a taking of property for public use that could best be accomplished through exercise of the eminent domain power.[44] This alternative was rejected because of (1) the high cost and "clumsy and ineffective" nature of the eminent domain ap-proach; (2) the conviction that effective zoning could not be accomplished in this way; and (3) a feeling that there was more than an even chance that the courts would uphold comprehensive zoning as a legal use of police power.

[43] Bassett, *op. cit.*, pp. 22–25.
[44] Cf. *ibid.*, pp. 26–27.

A full decade passed following the adoption of the New York City ordinance before the legality of comprehensive zoning was finally established in the courts. During this decade, provisions were made for zoning in forty-three states and the District of Columbia; zoning ordinances were enacted in some 420 municipalities; and several important decisions, mostly upholding but some rejecting the constitutionality of zoning, were rendered by state courts.[45]

The constitutionality issue was settled by the Supreme Court in *Village of Euclid* v. *Ambler Realty Company* in 1926.[46] In his majority opinion, Justice Sutherland observed that

> the line which separates . . . the legitimate from the illegitimate assumption of [the police] power is not capable of precise delimitation. It varies with circumstances and conditions. A regulatory zoning ordinance, which would be clearly valid as applied to the great cities, might be clearly invalid as applied to rural communities. . . . Thus the question whether the power exists to forbid the erection of a building of a particular kind or for a particular use, like the question whether a particular thing is a nuisance, is to be determined . . . by considering it in connection with the circumstances and the locality. . . . A nuisance may be merely a right thing in the wrong place,—like a pig in the parlor instead of the barnyard. If the validity of the legislative classification for zoning purposes be fairly debatable, the legislative judgment must be allowed to control. . . .
>
> There is no serious difference of opinion in respect of the validity of laws and regulations fixing the height of buildings within reasonble limits, the character of materials and methods of construction, and the adjoining area which must be left open, in order to minimize the danger of fire or collapse, the evils of overcrowding, and the like, and excluding from residential sections offensive trades, industries and structures likely to create nuisances.

Justice Sutherland indicated that the police power could be used to exclude industries "which are neither offensive nor dangerous" from designated areas. The

---

[45] Cf. Walker, *op. cit.* pp. 48–77, for a popular account of the more important court cases associated with the legal development of zoning law.

[46] 272 U.S. 365 (1926). The *Euclid* case concerned the constitutionality of a zoning ordinance in Euclid, a suburb of Cleveland, Ohio. At the time this case was appealed to the Supreme Court, many friends of zoning felt that the facts at issue were such as to make an adverse decision almost inevitable. Several members of the National Conference on City Planning, however, argued that the case would provide a leading precedent and that every effort should be made to secure a favorable decision. Arrangements accordingly were made for Alfred Bettman to submit a brief *amicus curiae* on behalf of the conference. Owing to an unfortunate oversight, Mr. Bettman was not advised of the time of the court hearing and thus was not heard by the Court. He reported this situation to Chief Justice Taft and was granted the unusual privilege of submitting his brief at a rehearing of the case on October 12, 1926. It is understood that a majority of the Court had decided against the validity of zoning after the first hearing. After the second hearing, however, the Court held six to three for the constitutionality of comprehensive zoning. Cf. *ibid.*, pp. 77–78; and Alfred Bettman, *City and Regional Planning Papers* (Cambridge, Mass.: Harvard University Press, 1946), pp. xv, 51–57, and 157–93.

real question, as he saw it, involved "the validity of what is really the crux of the more recent zoning legislation, namely, the creation and maintenance of residential districts, from which business and trade of every sort, including hotels and apartment houses, are excluded." After reviewing the impact of these establishments on fire, traffic, and noise conditions and their parasitic effect in destroying "the residential character of the neighborhood and its desirability as a place of detached residences," he concluded that "apartment houses, which in a different environment would not be entirely objectionable but highly desirable, come very near being nuisances."

Much of the dicta in the decision is phrased in terms of the law of nuisances. But the Court indicated that zoning can be used for a broader purpose than merely the control of nuisances. With its endorsement of this view, the Supreme Court opened the door for the widescale use of zoning as a tool for the effectuation of land-use planning.

The zoning process. Following the adoption of the New York zoning ordinance in 1916, several cities tried to zone under their home-rule powers. These efforts were rejected by several courts on the ground that they involved the exercise of a power vested in the state legislatures that could not be legally used without legislative authorization. Appropriate enabling legislation was thus recognized as a necessary first step in the zoning process. Enabling legislation was soon passed in most states; and by 1975, every state had zoning ordinances. Approximately three-fourths of the 3,000 counties in the United States along with townships and other minor civil divisions in a third of the states had received authority to zone.[47]

The details of the zoning process vary somewhat from state to state. As a rule, they follow the pattern recommended by the Advisory Committee on Zoning, which Secretary of Commerce Hoover appointed in 1921 to draft a standard zoning enabling act. Assuming that a city, county, or other local unit has authority to zone, the zoning process begins with the decision to appoint a zoning board or commission. This action is usually initiated by the city council, county board, or other local governing board. In some states it may also be initiated by petitions signed by given proportions of the number of eligible voters.

Once appointed, the zoning commission has the responsibility of drafting a zoning ordinance and preparing a map that shows the boundaries of the districts or zones within which different regulations apply. Most ordinances now contain several sections which provide (1) a statement of purposes that usually correspond closely with those listed in the enabling act; (2) general provisions including definitions of terms, a clause prohibiting the construction or alteration of buildings or the use of land or buildings except in conformity with the provisions of the ordinance, and a recognition of the right of property owners to continue nonconforming uses; (3) a statement that identifies the classifications of districts—residential, business, industrial, or agricultural—and that either describes the boundaries of each district or

[47] *Rural Zoning in the United States: An Analysis of Enabling Legislation*, U.S. Department of Agriculture Economic Research Service, Miscellaneous Publication No. 1232, 1972.

establishes them as set forth on the zoning map; (4) a detailed description of the regulations that apply in each district; (5) provisions for administration and enforcement including requirements for building and occupancy permits, arrangements for possible appeals, and a declaration that violations are misdemeanors punishable by fine or imprisonment; and (6) provisions for possible changes and amendments of the ordinance.[48]

Zoning commissions are usually appointed as independent bodies, though their functions are sometimes delegated to planning agencies. In either case, the process of drafting a zoning ordinance calls for a type of land resource planning and for decisions concerning the land uses that should be permitted in different districts. This situation makes it desirable for zoning commissions to note any planning done in their communities and to supplement these plans with such additional planning as may be needed to provide a reasonably sound basis for their zoning recommendations.

Once the zoning commission has tentatively formulated its recommendations, it must hold public hearings at which affected property owners can voice their opinions. The commission then considers these opinions in the preparation of the proposed ordinance, map, and final report, which it submits to the local governing body.[49] This body is usually required to advertise and hold a second set of public hearings. Following these hearings, the local governing boards are free to debate and either accept or reject the proposed ordinances. Some enabling acts require citizen approval of ordinances by voters at a general or special election; and some provide for possible referendums on board approval of ordinances.

Zoning commissions often disband following the submission of their reports. Provisions may be made, however, for their continuation as bodies charged with the responsibility of considering possible future changes and amendments. This arrangement provides a means for keeping zoning ordinances up to date and in tune with changing needs and conditions. But while reviews and adjustments are needed from time to time, frequent changes can be a sign of poor planning or weak and vacillating administration. A different function is provided by the board of appeals, which is appointed following the adoption of the zoning ordinance to hear and decide on appeals made from the actions and orders of the administrative officials who enforce the zoning ordinance. This board can reverse the decisions of the enforcing officers and authorize variances and minor modifications of the rules to fit individual needs and convenience.

Courts are inclined to question the legality of any zoning ordinance not adopted in strict conformance with the procedures specified in the state enabling act. Since

[48] Cf. Frank E. Horack, Jr., and Val Nolan, Jr., *Land Use Controls* (St. Paul, Minn.: West Publishing, 1955), pp. 46–48; and Herbert H. Smith, *The Citizen's Guide to Zoning* (West Trenton, N.J.: Chandler-Davis Publishing Co., 1965).

[49] Some states have enabling-act provisions requiring that the zoning commission submit its proposals to a county or state agency for review and approval before they are officially submitted to the local governing board. This review process provides a check against faulty ordinances and can prevent conflicts that can develop between governmental units if adjacent areas are zoned for noncompatible uses.

this process is time-consuming, communities sometimes find that they cannot zone fast enough to prevent imminent objectionable uses. *Interim zoning*, which permits the adoption of temporary ordinances, has been authorized in several states as a means of dealing with this problem. This type of zoning can suffer from the imperfections of hastily adopted stopgap ordinances, but it can also help local governments forestall undesired land-use developments during the weeks or months it may take to draft and adopt a standard zoning ordinance.

Zoning regulations typically prohibit uses not accepted by the ordinance, but they cannot prevent the continued operation of *nonconforming uses* that are carried on at the time ordinances are adopted. The police power can be utilized to stop these uses if they constitute public nuisances; and some communities have taken steps to place time limits on their continued life or to purchase or condemn them. As a general rule, however, nonconforming uses can continue as long as they are not expanded or abandoned and as long as the nonconforming structures are not altered, repaired, or reconstructed.[50] They are normally regarded as established uses, which must be accepted for the time being but which will eventually be discontinued. Administrative action is needed with every ordinance to list these nonconforming uses and to take such future action as may be appropriate to ensure their eventual discontinuance.

Zoning regulations in urban areas.   Zoning ordinances are now used in all but a few of the nation's larger cities and in most smaller cities as well. As was the case in earlier years, zoning is still used to prevent "growing congestion, impairment of access to air, light, and sunshine, and invasion by improper uses."[51] It is also used in a more positive sense both within and outside city limits to promote orderly resource development, stabilize and preserve desired land-use patterns, and help carry out master plans for future developments.

Four principal types of zoning regulations are used by cities: (1) regulations concerning the uses permitted in different districts, (2) provisions controlling the size of lots and the proportions of their surface areas that can be covered with buildings, (3) height and bulk restrictions for buildings and other structures, and (4) population density requirements. Most municipal zoning ordinances recognize three major classes of use districts—residential, business, and industrial—together with numerous subclasses. Additional classifications such as agricultural, recreation, or unrestricted may also be utilized when a city includes areas it desires to retain for these uses or to leave unrestricted until its future needs are more clearly defined.

Residential districts are ordinarily divided into several subclasses such as single-family, two-family, and multiple-family districts. Of these, the single-family districts are usually the most exclusive with all other uses being excluded. Provisions are usually made, however, for possible location of schools, churches, humanitarian insti-

---

[50] Cf. "Amortization: A Means of Eliminating the Nonconforming Use in Ohio," *Case Western Reserve Law Review*, 19 (1968), 1042–48; Horack and Nolan, *op. cit.*, pp. 151–63; and Beuscher and Wright, *op. cit.*, pp. 439–59.

[51] Bassett, *op. cit.*, p. 45.

tutions, parks, playgrounds, golf courses, and even farms in these districts. A two-family district often permits all the uses of a single-family district plus two-family residences, while the multiple-family districts allow three- or more-family residential units.

Provisions are made for commercial zones that accommodate downtown central business districts, local and regional shopping centers, and neighborhood stores or shops. Industrial districts may also designate areas for heavy industries, light industries, and the so-called clean industries. Apartment houses and new residential construction may be permitted in or be excluded from these areas. Special zones can also be provided for such uses as farming, open space, recreation, cemeteries, flood plains, or mobile homes.

Zoning ordinances in the past have often employed a cumulative approach that prohibited all except one specified use in the most exclusive zone (single-family housing) and then relaxed the prohibited uses one by one as they moved to each succeeding zone. With this approach, all types of housing were permitted in commercial zones and virtually all in industrial districts. This approach has not been conducive to good land-use planning, as it has permitted residential and other uses to move in and preclude the planned uses of many areas. Most zoning authorities now favor the designation of exclusive zones. This approach permits the planning and zoning of industrial parks, commercial areas, shopping centers, and multiple-family housing areas where only these types of developments will be permitted.

Building-height regulations vary from city to city. A maximum height of $2\frac{1}{2}$ stories applies in many single- and two-family residential districts. Higher structures are permitted in the multiple-family, business, and industrial districts of most large cities. Building-bulk and setback regulations are used in combination with height regulations in many central business districts to prevent individual skyscrapers from infringing on the rights of other property owners for reasonable access to air and light. Building-height regulations are also applied in most areas adjacent to airports.

Area regulations are frequently used to specify maximum densities of population. These regulations may limit the number of families housed per acre, require a minimum number of square feet of lot area or building space per family, or require minimum lot sizes or frontages. Comparable area regulations may be used to prevent owners from building on more than some given proportion of their lot area; to require minimum building setback lines from the front, sides, and rear of lots; to establish minimum sizes for inner or rear court areas; or to prescribe minimum off-street parking requirements.

Urban leaders have found that zoning is a flexible tool that can be applied in different ways to attain desired results. Some cities have substituted performance standards for land-use-oriented regulations. These standards recognize that commercial and industrial uses of property can be compatible with adjacent residential uses if they can meet minimum performance standards concerning noise levels, smoke and odors, fire hazards, traffic, parking, landscaping, and generation of wastes.

Step or graded zoning is used by some communities to facilitate managed future growth at rates at which programmed municipal improvements will be supplied. Cluster zoning has been proposed as a means of grouping housing to minimize water and sewerage costs and optimize the reservation of open space and natural areas. Arrangements have been made for planned unit developments (PUDs) in which a variety of residential, commercial, and service uses can be provided for residents. Transfers of development rights and trade-off arrangements have also been used to permit the relaxation of zoning standards in some districts if developers will agree to protect historical sites or provide desired facilities.

Rural zoning. Zoning has somewhat less appeal and less application in rural areas than in cities, mainly because the land-use pressures that give rise to demands for zoning are less acute in these areas. But as cities have increased in size, their problems have often spilled over into rural areas, and demands have emerged for the enactment of zoning ordinances. Several states have authorized rural area zoning.[52] Hawaii has gone farther to authorize a statewide system of land classification and districting, which amounts to state zoning.

Two distinctly different types of rural zoning are now in use. Numerous counties and townships use a suburban type of comprehensive zoning that involves regulations very similar to those used in cities. Wisconsin pioneered this type of zoning with an enabling act (1923), which authorizes counties to use comprehensive zoning measures. With many populous counties and townships, this type of zoning is hardly distinguishable from the zoning found in cities, except that provisions may be made for agricultural, forestry, grazing, and recreation districts as well as for residential, business, and industrial districts.

A second type of zoning known as open-country zoning was also pioneered in Wisconsin when that state amended its county zoning enabling act in 1929 to authorize the use of zoning to "regulate, restrict, and determine the areas in which agriculture, forestry, and recreation may be conducted." This type of county and township zoning emphasizes land-use regulations but makes little, if any, use of building-height and area restrictions. The first open-country zoning ordinance of this type was adopted in Oneida County in northern Wisconsin in 1933. This ordinance and the ordinances which followed in numerous other counties in the northern Lake States region recognized three principal types of use districts—forestry, recreation, and unrestricted (or agricultural) districts.

Many rural landowners still view zoning with considerable suspicion. It enjoys considerable support in most areas where it has been applied, however, because local people see it as a means they can use to shape the future of their communities. It can be used around cities to bring orderly suburban developments and prevent the rise of rural slums. And it can be used in more strictly rural areas to protect the continued use of farmlands for agricultural uses, prevent human residence in flood plain areas, control roadside environments, and enhance community values.

[52] Cf. Erling D. Solberg, *The Why and How of Rural Zoning*, U.S. Department of Agriculture Information Bulletin No. 196, rev. 1967; and William J. Block, *Rural Zoning: People, Property, and Public Policy*, U.S. Department of Agriculture Federal Extension Service, ESC-563, 1967, for general discussions of the case for rural zoning.

Some problems with zoning. Zoning can be used in many ways to promote socially desired goals. But like other public powers, it can be and sometimes is abused. The courts are inclined to accept a broad concept of zoning as long as it involves reasonable regulations that protect or promote the public health, safety, morals, comfort, convenience, or welfare.[53] But they have taken a strong stand against improper uses of zoning. They have objected to spot zoning and to the use of zoning for racial- and economic-class-segregation purposes. They have usually agreed that zoning cannot be used retroactively to prohibit already existing uses. And they have held that, though aesthetic factors such as scenic beauty or architectural uniformity should be considered, zoning cannot be justified on aesthetic grounds alone.[54]

Several zoning ordinances have been voided by the courts as arbitrary and unreasonable for (1) excluding uses not regarded as nuisances, (2) providing use regulations that create monopoly rights for a few property owners, (2) restricting land areas to uses for which they are not suited, (4) creating small island districts that are more restricted than the properties around them, (5) excluding uses that are incidental to permitted uses and that do not conflict with the purposes of the zoning plan, and (6) permitting unfair or discriminatory administrative practices.[55]

By and large most of our experience with zoning can be characterized as successful and desirable. On the discord side, however, many communities have learned to their regret that the decision to zone can be postponed too long if they hesitate to act until undesired situations arise. Some have zoned too little or too much land for particular purposes. Several have abused the concept of zoning by using it to freeze existing situations without really trying to plan for the best future use of their land resources. Many have seen the stability suggested by their ordinances whittled away as local officials have failed to take a firm stand against demands for variances and exceptions. And some have allowed special-interest groups to use or change zoning ordinances for selfish purposes.[56]

Communities often face serious problems in keeping their zoning ordinances up to date and in tune with changing conditions. Zoning ordinances tend to be relatively

[53] Cf. Beuscher and Wright, *op. cit.*, pp. 326–32; Smith, *op. cit.*; Horack and Nolan, *op. cit.*; Robert M. Anderson, *American Law of Zoning*, 4 vols. (Rochester, N.Y.: Lawyers Cooperative Publishing Company, 1968); Curtis J. Berger, *Land Ownership and Use: Cases, Statutes, and Other Materials* (Boston: Little, Brown, 1968); Daniel R. Mandelker, *The Zoning Dilemma* (Indianapolis, Ind.: Bobbs-Merrill, 1971); and Clan Crawford, Jr., *Strategy and Tactics in Municipal Zoning* (Englewood Cliffs, N.J.: Prentice-Hall, 1969).

[54] Cf. Leighton L. Leighty, "Aesthetics as a Legal Basis for Environmental Control," *Wayne Law Review*, 17 (1971), 1347–95; also Anderson, *op. cit.*, I, 520–30; and Beuscher and Wright, *op. cit.*, pp. 541–63.

[55] Cf. Horack and Nolan, *op. cit.*, pp. 166–67.

[56] Cf. Marion Clawson, *Modernizing Urban Land Policy* (Baltimore, Md.: Johns Hopkins University Press, 1973), pp. 221–39. Cf. also Richard F. Babcock, *The Zoning Game* (Madison: University of Wisconsin Press, 1966; John A. Gardiner and Theodore R. Lyman, *Decisions for Sale: Corruption and Reform in Land-Use and Building Regulation* (New York: Praeger Publishers 1978); and Hugh O. Nourse, "A Cynic's View of Zoning Reform," *American Real Estate and Urban Economics Association Journal*, 6 (1978), 327–34.

fixed and inflexible. They lack built-in mechanisms for adjusting to changing conditions and the needs of a dynamic society. Attempts have been made to itemize all permitted or excluded uses, but these lists have often failed to allow for the unexpected. In facing up to this problem, most communities find that they must reexamine and reappraise their zoning ordinances from time to time if they are to keep them up to date.

A review of the nation's experience with zoning shows that far too many communities have been prone to accept zoning as a negative or defensive measure designed primarily to prevent undesired developments. More emphasis should definitely be given to its use as part of a larger, more positive, more dynamic plan for attaining better use of land resources. This more positive approach calls for the integration of zoning with other measures for the social direction of land use. It recognizes that there is no magic in planning. We have some good planning and some planning that is best characterized as insufficient or incompetent. It is not enough that zoning be merely tied to a land-use plan. For optimum results, it must be tied to a comprehensive, realistic area plan that is geared to a thorough and continuing study of the community's resources, goals, and potentialities.

Land Subdivision Regulations

Land subdivision controls provide a means governments can use to establish and enforce minimum standards in the development of new urban and suburban areas. These controls were first authorized in Wisconsin in 1849 but received limited use until the 1920s.[57] Since then cities have become increasingly aware of the fact that the layout of residential, business, and industrial areas can have indelible impacts on their future character and development and that adherence to platting regulations can do much to prevent future traffic congestion, high public maintenance costs, the emergence of slums, inadequate open space, and similar land-use problems.

Subdivision controls can play a vital role in promoting and protecting the interests of the community, the buyer, and the subdivider. They can enhance the value, desirability, and long-time worth of the areas developed.[58] Everyone benefits when lots in a new subdivision are sufficiently attractive to favor a high standard of development; and everyone can lose when they are too narrow or too deep for effective use, water-supply or sewerage-disposal conditions create health problems, building sites are subject to periodic flooding, streets are too narrow to handle

---

[57] Cf. Marygold S. Melli "Subdivision Control in Wisconsin," *Wisconsin Law Review*, Vol. 1953, 389–457.

[58] For other discussions of subdivision controls and their value, cf. Beuscher and Wright, *op. cit.*, pp. 258–322; *Suggested Land Subdivision Regulations*, rev. ed. (Washington, D.C.: Housing and Home Finance Agency, 1960); Anderson, *op. cit.*, Vol. 3, chap. 19; *Control of Land Subdivision* (Albany: New York Office of Planning Coordination, 1968); and Roger A. Cunningham, "Land Use Control—The State and Local Programs," *Iowa Law Review*, 50 (Winter 1965), 367–457.

fire-fighting equipment or to permit reasonable parking, or street designs contribute to traffic hazards or general inconvenience.

Communities suffer from uncontrolled subdividing when they find themselves compelled to provide public services for blighted or partially developed subdivisions that offer little promise of ever producing sufficient tax revenue to pay their way. Subdivision controls can help protect lot buyers and home builders from the declining property values and broken hopes that often come when neighborhoods fail to develop as expected. They can help subdividers by discouraging investments in poorly planned developments that could leave them "holding the bag" and by encouraging improved subdivision designs that enhance their prospects of profit.

Nature of subdivision controls. Subdivision controls vary somewhat depending on the type of legislative authorization, the powers granted to local governing boards, and local administrative policies. Two principal types of statutes are in use. Several states have mandatory requirements affecting new subdivisions, while others authorize cities and sometimes other units to adopt subdivision regulations. Some states with mandatory requirements have also acted to authorize cities to map future improvements and supervise subdivision developments within (and sometimes around) their corporate limits.

Most of the mandatory state laws permit exemptions. Several apply only to subdivisions in which more than given numbers of lots are offered for sale in given time periods; some apply only to lots of less than a specified size; and most exclude tracts that are sold for agricultural use. Some enabling acts are selective in that they apply only to cities of a certain size. Some also authorize extraterritorial powers that permit cities to regulate subdivision developments for distances ranging from one-half to ten miles from their municipal limits.

Most subdivision regulations require that new subdivisions be surveyed by licensed surveyors, that plat maps be prepared to show the location and boundaries of every lot and the location of streets and other areas dedicated to public use, that the subdivision plat be approved by certain officials, and that it be officially registered. Plats must usually be approved by the local planning commission or the governing body of the local unit. Some states also require approval by county road commissioners or the state highway commission (when they must administer the streets dedicated to public use), the state or county board of health, a county plat board, and certain state officials.

Subdivision controls can be used to prevent developments in areas regarded as unhealthful because of flood hazards, improper drainage, or other conditions. They may require that subdividers conform to an overall street plan, provide right-angle intersections with through streets, or avoid jogging street connections. Minimum construction standards, maximum grades, and definite widths are usually specified for streets and street rights of way. Lots of minimum size and depth are often required. Regulations or administrative guidance can also be used to discourage plats involving a single row of long, narrow, "rifle-range" lots along an existing road, blocks of insufficient width to permit two rows of lots, dead-end streets that are more than 500 feet in length, or long blocks with infrequent cross-street connections to parallel streets.

City subdivision ordinances usually require subdividers to provide a complete water supply system or water main connections to the city system. Sanitary and storm sewers must be provided; and subdividers may be required to pave the streets, provide sidewalks and street lamps, and plant trees along new streets. Subdividers are sometimes required to dedicate or reserve minimum areas in addition to streets for public uses. When a city maps the streets, parks, and other public improvements planned in an undeveloped area within its jurisdiction, it can usually require future subdividers to adhere to its plan. Cities do not secure ownership of the areas mapped for public use until they are acquired by dedication, purchase, or eminent domain. But the courts have held that they need not compensate owners for buildings and other structures built on areas that have previously been mapped for public use.

Subdivision regulations have been accepted as a valid use of the police power. In the cases that have come before them, the courts

> . . . have upheld as reasonable requirements that streets be much wider than the adjoining portions of the same streets, that park areas be dedicated, that land be dedicated to widen abutting streets, that improvements such as grading streets and installing utilities be made, that a bond be filed to cover such improvements, and that fees be paid to cover the cost of examining and checking the plat.

> From these cases it appears that . . . the limits on what conditions can be required for approval of a plat are very broad. But they must be imposed for the good of the community as a whole and not for the sole benefit of the surrounding property owners.[59]

Administration of subdivision controls.   Like most land-use regulations, the real test of the usefulness of subdivision controls comes in their administration. Some communities have made far more effective use of these controls than others. Among the states with mandatory regulations, some officials have treated their approval function as an automatic formality. Others have pressed for high standards that protect or promote the public health, safety, and welfare.

Subdivision controls are used most effectively when they are integrated into a larger planning program. In this way, they can be used to help carry out the provisions of an area master plan. They cannot be used to force the reservation of open spaces, but they can be used to support patterns of orderly development. Planning agencies and local officials can often help this cause along by counseling with local subdividers and assisting them in upgrading their subdivision plans.

Since the process of surveying, registering, and developing a full-fledged subdivision always involves time and cost, it is quite natural that many landowners should try to circumvent the usual requirements. This avoidance process is encouraged by

[59] Melli, *loc. cit.*, pp. 389–99.

the exemptions provided in some laws and by the tendency of many property owners to sell nonplatted building lots. These practices can be discouraged by (1) revising state platting laws to require registration of all plats involving two or more properties, (2) prohibitions against the use of unregistered plat descriptions in the legal transfer of properties, (3) more rigid requirements affecting metes and bounds transfers, and (4) use of the state's licensing of real estate brokers to discourage this type of subdividing.[60]

Local units have two potent powers they can use to discourage unauthorized subdivision developments. They can refuse to grant building and occupancy permits for structures built on lots that fail to measure up to their subdivision standards. And they can refuse to provide unauthorized lots with public services such as access to city water and sewers, street maintenance, and sometimes access to public streets or roads.

Rent Controls

Rent controls represent an important use of the police power to regulate private practices affecting the leasing of real estate. As used in the past, these controls have taken two major forms. Agricultural rent controls have been used in several parts of the world to protect tenants from possible landlord exploitation. They have also been applied in Egypt, Japan, Nationalist China, and Pakistan as features of far-reaching land reform programs. Residential rent controls in turn have been widely used in Europe, the United States, and various other nations to limit the contract rents tenants must pay for the use of residential properties.

The United States has had considerable experience with residential rent controls but virtually no experience with agricultural rent controls. Several states adopted residential rent controls during World War I. Federal residential rent controls were put into effect in 1919 after the armistice, discontinued in the early 1920s, reinstituted during World War II, and then phased out between 1949 and 1954. Controls have been continued or reinstituted since then on a local option basis in more than 200 cities including Baltimore, Boston, Los Angeles, New York, and Washington, D.C.

Case for rent controls. Public rent controls involve a substitution of legal and political determinations of fair or warranted levels of contract rent for the economic levels that supposedly should emerge from the interaction of supply and demand factors in the market. They usually assume that landlords possess unwarranted bargaining power in their dealings with tenants and that public constraints are needed to ensure a measure of equity and distributive justice for tenants.

Strong cases can be made for the application of residential rent controls in wartime or other emergency periods when the movement of large populations to certain sites or the destruction of much of the housing at those sites bring heavy demands

---

[60] Cf. *Local Planning Administration*, 2nd ed. (Chicago: International City Managers' Association, 1948), pp. 253–55; and Horack and Nolan, *op. cit.*, p. 214.

for the use of limited supplies of housing. Similar demands may develop over longer time periods when land ownership is monopolized by a small class of owners or when tenant demands for agricultural or residential properties exceed the supplies of the resources that are or can be made readily available for tenant use.

A strong case for residential rent controls existed during World War II when the mobilization of 12 million people into the armed forces, the attempts of wives and families to follow their men from one military post to another, and the sudden creation of thousands of jobs in defense industries caused approximately one-half of the nation's families to move to new locations. This population movement, in combination with a rise in the marriage and birth rates, higher family incomes, and an undoubling of families from the doubled-up housing arrangements of the 1930s, brought tremendous new demands for housing in most defense and metropolitan areas. The problem of meeting these demands was complicated both by the time needed for new residential construction and by wartime restrictions on private building activities.

Severe housing shortages provided major opportunities for landlord profiteering. Some landlords regarded rent gouging as unethical and unscrupulous. But many were willing to press their advantages, and their activities frequently had a contagious effect on the attitudes of others. For them, the sudden increase in demand relative to the supply of housing provided opportunities for realizing an economic rent in addition to their normal contract rents. Landlord exercise of this opportunity prompted demands for rent controls. Rental rates were frozen at existing (or sometimes earlier) levels, arrangements were made for establishing fair rental ceilings for new properties, and procedures were devised for reviewing complaints of abuses and hardship cases in which landlords were caught with unreasonably low rent ceilings.

Relatively few objections were raised regarding rent controls during the wartime emergency period. Many landlords were willing to accept restrictions in the public interest as long as other segments of the national economy were subject to wage and price controls. Their sentiment changed, however, once the war was over and wage and price controls were lifted. At this point, landlords often argued that they had been singled out for repressive treatment. Tenants meanwhile argued for continued controls, often on the ground that insufficient new housing had been provided to permit equitable bargaining between landlords and tenants.

Effects of residential rent controls. Residential rent controls often appeal to tenants because they freeze contract rental rates at less than market levels. With rising incomes, they may also make it possible for tenants to afford larger and more luxurious quarters than they otherwise could. In this sense rent controls may permit and even encourage economic overconsumption of residential space.

These advantages for tenant groups are always conditioned by their ability and luck in finding suitable housing. During the most critical phase of the World War II housing shortage, the supply of residential units was inadequate to meet the demand and many workers had to leave their families in rural areas, accept housing at inconvenient locations, live in furnished rooms, or double up with other families. Their ability to secure better housing by bidding up rents was strictly limited. Landlords sometimes accepted bonus payments, but black market practices of this type were illegal. When landlords did have vacancies, they were able to choose from long lists of applicants. Property maintenance considerations often caused them to favor single persons and childless couples, and families with children were commonly discriminated against.

Freezing contract rental rates at levels that were low relative to rising property values also gave landlords strong economic incentives for selling their properties. Thousands of landlords accepted this alternative during and after World War II. Tenants in single-family dwellings, in duplexes, and in apartment houses found that they had the unpleasant choice of either obtaining other rental quarters or buying their units at what they considered inflated prices.

Tenants as a group can definitely benefit from rent controls over the short run and sometimes find it to their advantage to marshal their political power to perpetuate controls. Serious problems can arise, however, when controls are continued over extended time periods. Continuation of controls can depress property values and cause a shifting of the portion of the property tax load that could otherwise be collected from the controlled properties to other property owners. Continuation of controls with only nominal upward adjustments for inflation and rising operating costs can also deprive landlords of their normal economic incentives for renovating, refurbishing, and improving their properties. Without these incentives, outlays for maintenance are often reduced, and some landlords may reach the point at which they simply abandon their properties.[61] Rent controls and the threat of rent controls also have far-reaching effects in discouraging new residential investments. Potential investors find it less risky to make their investments in communities that do not have rent controls. When they do build in controlled communities, they tend to favor investments in high-rent complexes. Little new housing is provided for low- or medium-income tenants without special subsidies.

Air and Water Quality Standards

State and federal regulations affecting air and water quality provide another example of police-power measures that have special implications for land use. Until the present century, little attention was given to the impacts individual and group activities have on the quality of air and water resources. The states exercised most of the responsibility for dealing with pollution problems, and they took little action

[61]For other discussions of the effects of rent controls, cf. Monica R. Lett, *Rent Control: Concepts, Realities, and Mechanisms* (New Brunswick: Rutgers, The State University of New Jersey, Center for Urban Policy Research, 1976); Ira S. Lowry, Joseph S. De Salvo, and Barbara M. Woodfill, *The Demand for Shelter* (New York: Rand Corporation, 1971); and Frederick A. Hayek *et al., Verdict on Rent Control: Essays on the Economic Consequences of Political Action to Restrict Rents in Five Countries* (London, U.K.: Institute of Economic Affairs, 1972).

because the relationship between air and water pollution and public health was not well understood. Significant changes in this approach came in the 1960s and 1970s when emerging concerns about the deteriorating quality of the environment caused Congress to launch new programs that employ the joint efforts of the states and the federal government in working for improved air and water quality.

Water quality programs. Four phases of activity have occurred with the nation's water quality programs. Throughout the first phase, the states were left alone to develop water pollution control programs within their jurisdictions. The only federal legislation during this period was the Refuse Act of 1898, which limited the dumping of wastes into navigable waters. A second phase started with the enactment of the Federal Water Pollution Control Act of 1948. This act and its amendments in 1956 and 1961 provided for technical assistance to agencies working on water pollution issues, for grants to the states, and for construction grants for municipal waste treatment facilities. Pollution abatement was treated primarily as a public health problem; and emphasis was given to state enforcement of pollution abatement orders, though provisions were made for providing some federal backup enforcement authority.

Phase three started with the passage of the Water Quality Act of 1965. This act provided for the establishment of ambient water quality standards for interstate and intrastate waters. The states were called upon to develop water quality standards for interstate waters which after review could be accepted by the secretary of the Department of the Interior as federal standards. Primary responsibility for enforcing the standards was left with the states. But if states failed to act, the secretary was empowered to establish and enforce water quality standards for them. Standards were established for all states under this act.

A fourth phase of the program was mandated by the Federal Water Pollution Control Act Amendments of 1972, which shifted emphasis from the provision of ambient water quality standards to prescription of discharge requirements. Municipal waste treatment plants were required to provide secondary treatment of wastes by 1977 and to adopt the "best practicable waste treatment technology" by 1983. The program aims to eliminate the discharge of wastes into navigable waters by 1985. It provides that implementation "be carried out largely by the states, but most of their actions are subject to extensive Federal guidelines and backup enforcement authority."[62] Strong efforts involving state agency determinations (and possible court orders with penalties of fines and imprisonment) have been pushed under this program to minimize water pollution from waste treatment plants, industries, feedlots, and other "point" sources of pollution. Comparable programs are also being considered for dealing with pollutants from storm runoff and other "nonpoint" sources of pollution.

Air quality programs. Until recent years, little emphasis has been given by the states to the control of air pollution. Federal legislation was passed in 1955 to finance research and provide for technical assistance to state and local governments

[62] *Environmental Quality—1973*, Fourth Annual Report of the Council on Environmental Quality (Washington, D.C.: Council on Environmental Quality, 1973), p. 174.

on air pollution problems. With the Clean Air Act of 1963, provisions were made for federal grants to air pollution agencies and for federal enforcement authority to deal with interstate air pollution problems. The Clean Air Act of 1965 went further to permit federal regulation of air pollution from new motor vehicles.

A major step in the development of a national air quality protection program came with the passage of the Air Quality Act of 1967. This act authorized the designation of air quality regions; the promulgation of air quality criteria; and state establishment of standards which, once approved, would become the basis for air quality enforcement programs. The Clean Air Act Amendments of 1970 stated a national goal of reducing automobile pollutants by 90 percent in five years and called for submission, review, and approval of state plans for implementing ambient air quality standards. Primary and secondary national ambient air quality standards were established by the Environmental Protection Agency in 1971 for particulates, sulfur oxides, hydrocarbons, carbon monoxide, oxides of nitrogen, and photochemical oxidants. As with the water quality standards, the major responsibility for enforcing the air quality standards rests with the states, but EPA provides extensive guidelines for the state programs together with federal backup enforcement authority.

Effects on land use. The adoption and enforcement of air and water quality standards have had far-reaching effects on the control of air and water pollution. Environmental enhancement goals are being attained, and major negative externalities have been reduced. The programs have also had important side effects on the uses made of land resources. Environmental enhancement has made some sites better places to live, work, or play. But enforcement of the standards has also meant that certain types of developments and activities will no longer be tolerated. Internalization of externalities has led to higher production costs with many industrial activities; regulation-associated changes in costs have affected the land rents operators attribute to certain sites; and changes in land rent situations have brought shifts in the economic highest and best uses of numerous sites.

Air and water quality controls provide an indirect approach to land-use regulation. Air and water quality standards can place definite constraints on several types of development. Residential developments may be prohibited in areas with poor soil permeability unless water and sewerage facilities are provided, and new industrial developments may even be prohibited at rural sites having only minimal air pollution. Regulations affecting "point" and "nonpoint" sources of pollution may force operators to install costly waste treatment facilities or cease business operations. EPA reviews of the potential for air pollution generation with projected industrial, shopping center, highway, or other land uses may also result in denials of development permits. Each of these controls may be justified, but the fact that the controls are mandated by bureaucratic state and federal agencies rather than by locally controlled and administered ordinances creates acceptance problems.

Legislated air and water quality goals can also involve problems that spring from failure to recognize the threefold framework within which successful policies must be developed. Clean air and water are politically popular and institutionally acceptable goals. But one can ask if a legislatively mandated goal that water run clear by a given date is physically or biologically possible. Technological and economic feasibility questions may also be raised concerning whether we have the technical knowhow to do what our projected standards require and whether the expected benefits will exceed their costs. An objective of removing 80 percent of an undesired substance from air or water can often be accomplished at reasonable cost. Raising this goal to the 95 percent level requires significantly larger cost inputs for each additional increment of substance removed, while policies that require the removal of 99 percent or all of the undesired substances can easily entail incremental costs that far exceed the value of the added benefits realized by society.

FUTURE DIRECTIONS IN LAND-USE POLICY

The decade from the middle 1960s to the middle 1970s was one of expanding awareness of environmental issues. Dozens of laws were passed and new programs started at the federal, state, and local levels that dealt directly with environmental and land-use issues. Much of this action involved acceptance of public directional measures to secure particular land-use and environmental goals.[63] Some of these policies and programs commanded considerable popular support; but some also met with criticisms and strong organized opposition.. By the end of the decade, "land use" had become a symbol of the unwanted for many people. For many advocates and opponents of land-use programs, it was as Frank Popper has indicated, "a neutral term that covers a multitude of highly charged and dangerous matters."[64]

A feeling of the national pulse in the early 1980s indicates a wide divergence of opinions on the goals the nation, states, and local governments should seek with future land-use policies. Some groups see land as a scarce and fragile resource, the use of which is closely intertwined with community interests that must be protected and advanced even if this requires public intervention. Others at the opposite pole view land resources as a commodity individual owners should be able to possess and use as they wish. For them, land ownership is an opportunity for wealth, for operator free-agency, and for freedom from community and outside constraints. Between these two extremes are thousands of citizens who are sometimes willing to be sold on the need for additional programs, sometimes have little discernible interest in them, and sometimes are philosophically opposed to actions that may expand the role played by governments.

As attempts are made to sort out the factors that influence decisions concerning

[63] For descriptions and analyses of these programs, cf. Fred Bosselman and David Callies, *The Quiet Revolution in Land Use Control* (Washington, D.C.: Council on Environmental Quality, 1971); Robert G. Healy and John S. Rosenburg, *Land Use and the States*, 2nd ed. (Baltimore, Md.: Johns Hopkins University Press, 1979); Elaine Moss, ed., *Land Controls in the United States* (New York: The Dial Press/James Wade, 1977); Frank J. Popper, *The Politics of Land-Use Reform* (Madison: University of Wisconsin Press, 1981); and *Environmental Quality—1971* through *1980*, annual reports of the Council on Environmental Quality (Washington, D.C.: Council on Environmental Quality, 1971 through 1980).

[64] Popper, *op. cit.*, p. 8.

the initiation of new programs and the modification or abandonment of present programs, two basic facts should be kept in mind. First, much as some people may wish otherwise, the land-use problem is not about to go away. The factors that have made land use a political issue are becoming more rather than less significant. With increasing population numbers, separation of more and more people from the intimate day-to-day contacts with nature and the soil that were once accepted as normal features of life, more competition between potential uses and users of land resources, and the resultant greater potential for conflicts of interests, the future will probably bring more rather than fewer demands for community action to direct the uses made of land.

Second, it is also apparent that the trend toward expanded public exercise of the interests governments hold in private uses of land can bring significant modifications in the rights we hold in landed property. In this respect, the bundle of rights shown in Figure 18-2A represents an idealized view of the private property rights fee simple owners have held in the past. Many people feel that we are moving toward the situation depicted in Figure 18-2B. The private owner's bundle contains as many separable rights as earlier but seems somehow smaller because of the greater emphasis now placed on public interests in property. Some of the lesser sticks from the private bundle such as control over navigation through one's air space, the "right" to discharge pollutants into air or water, and the "right" to burn down one's house have also moved to the public sector.

Few problems would arise in the future direction of land-use practices if we had wide acceptance of common goals and if everyone endorsed and followed a stewardship philosophy in their use of land resources. Education programs that emphasize the significance of our relationships with land and the society-based nature of the legal rights we hold in landed property can provide important first steps for movement in this direction. Lacking these ideal conditions, we must deal as best we can with the conflicting concerns and interests of different operators and citizen groups. These conditions call for frank recognition of the problems associated with the taking of property rights often associated with applications of the police power and the devising of strategies for attaining public goals at a minimum of social cost.

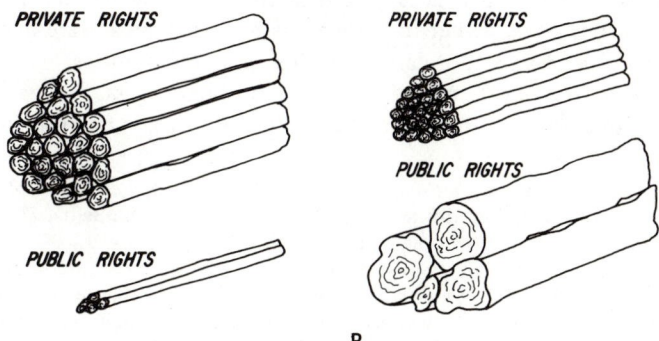

A   **B**

FIGURE 18-2.   Two views of the distribution of private and public rights in land resources.

The Taking Issue

Much of the opposition to public programs for directing land-use centers in concerns regarding uses of the police power with zoning and other regulatory measures that deprive owners of rights they have, or think they have, without payments of compensation. The taking of property in this manner raises ethical questions concerning the fairness of programs that protect and promote the interests of the public and some owners while limiting the rights of others. Economic issues arise because programs can easily bring economic wipeouts of existing property values as well as losses of opportunities for private economic gain. Legal questions also arise concerning how far governments might proceed before they must invoke the eminent domain power and pay compensation for rights taken.

No hard and fast rules have been laid down that define a precise boundary between the police and eminent domain powers. The courts have retained the prerogative of judging each case on its merits as to whether applications of the police power are justified and legally valid. Zoning ordinances have typically been upheld as legitimate exercises of the police power, because they leave owners in full physical possession of their properties even though certain use rights may be restricted in the interests of public health, safety, morals, and welfare. In this respect, a regulation prohibiting an urban landowner from using a vacant lot as a rifle range would be accepted as a protection of public health and safety. An ordinance limiting use of the same lot to a neighborhood playground, in contrast, would be viewed as a taking requiring compensation.

Some analysts argue that "ownership of private property historically has never implied complete and unencumbered control over assets" and that many ideas owners have concerning the extent and sanctity of their property rights are based more on myth than fact.[65] With this view, it is possible that courts may accept broader applications of the police power in the future.[66] Many observers, however, feel that serious consideration must be given to the rights owners think they have in land resources and that possible compensation arrangements must be devised to protect operator equities when owners with properties located within the "red areas" on zoning maps must accept restrictions on the economic uses they can make of properties while owners of properties in "green areas" are encouraged to develop their properties for more profitable higher and better uses.

Several alternatives have been suggested as appropriate means for dealing with this problem. One option calls simply for the elimination of zoning ordinances.[67] With

---

[65] James C. Hite, *Room & Situation: The Political Economy of Land-Use Policy* (Chicago: Nelson-Hall, 1979), p. 29; and Fred Bosselman, David Callies, and John Banta, *The Taking Issue* (Washington, D.C.: Council on Environmental Quality, 1973), pp. 318–19.

[66] A possible step in this direction is suggested by the case of *Just v. Marinette County*, 56 Wisconsin 2nd 7, 201 N.W. 2d 761 (1972), in which the Wisconsin Supreme Court upheld the county's power to restrict owners of lake frontage to natural uses of their frontage property.

[67] Cf. Bernard H. Siegan, *Land Use Without Zoning* (Lexington, Mass.: Lexington Books, 1972), chap. 15.

this approach, more use presumably would be made of the spending and eminent domain powers in acquiring needed public rights. The compensation issue also could be sidestepped by designing programs for the public sale of zoning rights.[68] Similar results could be attained by requiring payments for development permission.[69] Compensation arrangements might also be devised to pay owners for their loss of rights with funds appropriated from governmental general funds, from special tax assessments on benefiting owners, or from taxes on capital and real estate windfalls.[70] Transfers of development rights (TDRs) could also be used to compensate disadvantaged owners. These could involve grants of TDRs for utilization at other sites or for pooled ownership arrangements that permit all owners within given areas to benefit from the windfalls associated with the assignment of development priorities to some sites but not others.[71]

Strategies for Future Policy Developments

Assuming that adequate demand exists for the adoption of new or stronger policies or for the launching of new programs, careful consideration should be given to devising the new policies and programs and to developing strategies for their successful adoption and effectuation. The first step logically calls for an evaluation of what has already been accomplished and of the strengths and weaknesses of earlier programs. Emphasis should be given to identification of the specific goals to be attained and the choice of directional measures to be used. It should be noted that various policy tools can often be used to attain the same objective and that some approaches ordinarily offer advantages over others. Community education, tax incentive, and public subsidy programs, for example, usually provide more positive approaches than police-power measures. Combinations of techniques involving different powers may also be used to advantage. Efforts must also be expended on generating public and legislative support for the proposed programs.

No single strategy is suitable for all proposed programs. The following guidelines, however, represent essential building blocks that can be used to advantage in devising successful land-use policies and programs.

1. Educational efforts to inform citizens and members of legislatures of the nature of current problems, alternative solutions, their strengths and weaknesses, and the case for suggested lines of action
2. Public recognition, acceptance, and endorsement of the specific goals to be attained

[68] Cf. Robert H. Nelson, *Zoning and Property Rights*, (Cambridge: Massachusetts Institute of Technology Press, 1977), chap. 8; and Madelyn Glickfeld, "Sale of Development Permission: Zoning on the Auction Block," in Hagman and Misczynski, *op. cit.*, pp. 376–421.

[69] Cf. Frederik Jacobsen and Craig McHenry, "Exactions on Development Permission," in Hagman and Misczynski, *op. cit.*, pp. 342–66.

[70] Cf. Hagman and Misczynski, *op. cit.*, pp.. 256–335, 437–88, and 517–31; and Ervin *et al.*, *op. cit.*, chap. 6.

[71] Cf. p. 519 in this chapter.

3. Intergovernmental cooperation in working for attainment of these goals and suspension or reexamination of all public programs that involve developments, grants, or other actions that work at cross-purposes with the stated goals

4. Use of the spending power, tax incentives, and other positive measures where possible to attain the goals

5. Applications of police-power measures such as zoning and land-use ordinances as positive tools of public policy to provide objectives and standards with resort to the use of their control features only when such action is necessary to protect the overriding interests of the community

6. Careful planning to ensure that policies and programs can accomplish their objectives, that these objectives not be lost in legislative compromises or in a bureaucratic shuffle, and that adequate operating funds be provided to permit their smooth operation.[72]

Whatever the course of future land policy, it must be remembered that we live in a world where the manner in which land resources are developed, managed, used, and conserved has important implications for the future welfare of nations, communities, and their citizens. Our society has found it desirable to vest certain of its members with private rights of property ownership. Possession of these rights represents more than a privilege; it involves responsibilities to use land in the interests of the community and society as well as those of the individual owner. Individual creativity and initiative should by all means be encouraged. But in our use of land resources, public and private goals must mesh and pursue a common course over time. Communities and their landowning citizens must share a common goal if we are to enjoy a full measure of the benefits, productivity, and satisfactions that can come with the orderly, effective, and efficient use of our land resource base.

—SELECTED READINGS

Beuscher, Jacob H., and Robert P. Wright, *Cases and Materials on Land Use* (St. Paul, Minn.: West Publishing, 1969).

Bosselman, Fred, David Callies, and John Banta, *The Taking Issue* (Washington, D.C.: Government Printing Office, 1973).

[72]For examples that illustrate the importance of this guideline, see Popper, *op. cit.*, chaps. 6 and 7. Popper describes how several state land use programs adopted during the early 1970s were watered down in the adoption process through the skillful tactics of intrenched, well-financed opposition groups. Exemptions were frequently provided for important groups of uses and users; high threshold levels were provided; grandfather clauses were used; and representation of opposition groups provided for on administrative boards. The success of programs was further hampered by underfunding, and their administration was often turned over to representatives of opposition groups.

Ervin, David E., *et al.*, *Land Use Control* (Cambridge, Mass.: Ballinger, 1977).

Hagman, Donald, and Dean Misczynski, *Windfalls for Wipeouts: Land Value Capture and Compensation* (Washington, D.C.: American Planning Association, 1978).

Hite, James C., *Room & Situation: The Political Economy of Land-Use Policy* (Chicago: Nelson-Hall, 1979).

Popper, Frank J., *The Politics of Land-Use Reform* (Madison: University of Wisconsin Press, 1981).

The Protection of Farmland: A Reference Guidebook for State and Local Government (Washington, D.C.: National Agricultural Lands Study, Government Printing Office, 1981).

Reitze, Arnold W., Jr., *Environmental Planning: Law of Land & Resources* (Washington, North American International, 1974).

Smith, Herbert H., *The Citizen's Guide to Zoning* (West Trenton, N.J.: Chandler-Davis Publishing, 1965).

INDEX